SEIFERT AND THRELFALL:
A TEXTBOOK OF TOPOLOGY

and

SEIFERT:
TOPOLOGY OF 3-DIMENSIONAL
FIBERED SPACES

This is a volume in
PURE AND APPLIED MATHEMATICS

A Series of Monographs and Textbooks

Editors: SAMUEL EILENBERG AND HYMAN BASS

A list of recent titles in this series appears at the end of this volume.

SEIFERT AND THRELFALL: A TEXTBOOK OF TOPOLOGY

H. SEIFERT and W. THRELFALL

Translated by **Michael A. Goldman**

and

SEIFERT: TOPOLOGY OF 3-DIMENSIONAL FIBERED SPACES

H. SEIFERT

Translated by **Wolfgang Heil**

Edited by **Joan S. Birman** and **Julian Eisner**

 1980

ACADEMIC PRESS

A Subsidiary of Harcourt Brace Jovanovich, Publishers

NEW YORK   LONDON   TORONTO   SYDNEY   SAN FRANCISCO

ACADEMIC PRESS, INC.
111 Fifth Avenue, New York, New York 10003

United Kingdom Edition published by
ACADEMIC PRESS, INC. (LONDON) LTD.
24/28 Oval Road, London NW1   7DX

Mit Genehmigung des Verlager B. G. Teubner, Stuttgart,
veranstaltete, allein autorisierte englische Übersetzung,
der deutschen Originalausgabe.

Library of Congress Cataloging in Publication Data

Seifert, Herbert, 1897–
    Seifert and Threlfall: A textbook of topology.
    Seifert: Topology of 3–dimensional fibered spaces.

    (Pure and applied mathematics, a series of mono–
graphs and textbooks ;)
    Translation of Lehrbuch der Topologie.
    Bibliography: p.
    Includes index.
    1. Topology. I. Threlfall, William, 1888–
joint author. II. Birman, Joan S., 1927–
III. Eisner, Julian. IV. Title. V. Title: Text–
book of topology. VI. Series.
QA3.P8 [QA611] 510s [514] 79–28163
ISBN 0–12–634850–2

PRINTED IN THE UNITED STATES OF AMERICA

80 81 82 83   9 8 7 6 5 4 3 2 1

CONTENTS

SEIFERT AND THRELFALL: A TEXTBOOK OF TOPOLOGY

CHAPTER ONE
ILLUSTRATIVE MATERIAL

CHAPTER TWO
SIMPLICIAL COMPLEXES

v

CHAPTER THREE

HOMOLOGY GROUPS

CHAPTER FOUR

SIMPLICIAL APPROXIMATIONS

CHAPTER FIVE

LOCAL PROPERTIES

CHAPTER SIX

SURFACE TOPOLOGY

CHAPTER SEVEN

THE FUNDAMENTAL GROUP

CHAPTER EIGHT
COVERING COMPLEXES

CHAPTER NINE
3-DIMENSIONAL MANIFOLDS

CHAPTER TEN
n-DIMENSIONAL MANIFOLDS

PREFACE TO ENGLISH EDITION

The first German edition of Seifert and Threlfall's "Lehrbuch der Topologie" was published in 1934. The book very quickly became the leading introductory textbook for students of geometric-algebraic topology (as distinguished from point set or "general" topology), a position which it held for possibly 30 to 35 years, during which time it was translated into Russian, Chinese, and Spanish. An English language edition is, then, long overdue. The translation presented here is due to Michael A. Goldman.

In spite of the fact that with the passage of time our understanding of the subject matter has changed enormously (this is particularly true with regard to homology theory) the book continues to be of interest for its geometric insight and leisurely, careful presentation with its many beautiful examples which convey so well to the student the flavor of the subject. In fact, a quick perusal of the more successful modern textbooks aimed at advanced undergraduates or beginning graduate students reveals, inevitably, large blocks of material which appear to have been inspired by if not directly modeled on sections of this book. For example: the introductory pages on the problems of topology, the classification of surfaces, the discussion of incidence matrices and of methods for bringing them to normal form, the chapter on 3-dimensional manifolds (in particular the discussion of lens spaces), the section on intersection theory, and especially the notes at the end of the text have withstood the test of time and are as useful and readable today as they no doubt were in 1934.

This volume contains, in addition to Seifert and Threlfall's book, a translation into English, by Wolfgang Heil, of Seifert's foundational research paper "The topology of 3-dimensional fibered spaces" ["Topologie dreidimensionales gefaserter Raum," *Acta Mathematica* **60**, 147–288 (1933)]. The manuscript treats a simple and beautiful question: what kinds of 3-dimensional manifolds can be made up as unions of disjoint circles, put together nicely? Using practically no technical machinery, Seifert takes hold of the question, lays down a set of ground rules, and with exceptional clarity of thought and particular attention to detail weaves a theory which culminates in the complete classification of his "fibered spaces" up to fiber-preserving homeomorphism.

The manuscript is a model of mathematical exposition, and on these grounds alone it is worthy of translation, preservation, and study by new generations of mathematicians. But the primary reason for publishing this translation is, in our minds, because the ideas in the paper have proved to be at the origins of so many different flourishing areas of current research. We give a brief review.

The question that Seifert asks and answers is not an idle question. The first paragraph of the paper shows that Seifert understood well the crucial role that the manifolds which have since come to be known as Seifert fibered spaces were likely to play in the classification problem for compact 3-manifolds. In fact it could be said that Seifert anticipated the theory of geometric structures on 3-manifolds, as it is evolving at this writing.

Although Seifert only classified his fibered spaces up to fiber-preserving homeomorphism, the classification has turned out (with some exceptions) to be topological (Orlik, Vogt, and Zieschang [10]). This classification was generalized by Waldhausen [11] to a related classification of "graph manifolds" which are Seifert fiber spaces (now with nonempty torus boundaries) pasted together along their boundaries, much as one pastes together 3-manifolds along 2-spheres in the boundary to obtain connected sums. Conversely, it has been proved that one may *decompose* a 3-manifold into simpler pieces by splitting along essential 2-spheres and tori. This procedure is summarized in the following theorem (Kneser [7], Milnor [8], Johannson [6], Jaco and Shalen [4]):

> Let M^3 be closed, orientable. Then M^3 has a unique decomposition as a connected sum of prime 3-manifolds $M_1 \# M_2 \# \cdots \# M_r$. Moreover, for each prime summand M_i, there is a finite collection of disjoint essential tori in M_i such that M_i split along these tori consists of pieces that are either Seifert manifolds or prime M_j that contain no essential tori. Moreover, a minimal such collection is unique up to isotopy of M_i.

Now Thurston has announced the remarkable result that if M is prime, Haken, and contains no essential tori, then M admits a unique complete hyperbolic metric. If the word "Haken" could be removed from Thurston's theorem, we would have a complete classification theorem for 3-manifolds. At this writing, a beautiful classification seems very close, and Seifert's contributions appear to have been seminal.

Next, we remark on connections with the theory of fiber bundles. Seifert has given, in effect, a classification of $O(2)$ bundles over surfaces. The question of the existence of a cross section for such a bundle emerges in

Seifert's paper and is illustrated with beautiful examples. All of this is quite remarkable, because in 1933 the concept of a fiber bundle was not known, and it is only with hindsight that we recognize it as such.

In another direction, we see in Seifert's work the origin of the theory of Lie groups acting on manifolds, generalizing the action of the circle group on 3-manifolds. Seifert classified the standard free actions of S^1 on S^3. Later it was shown in Jacoby [5] that in fact the only free S^1 actions on S^3 are the standard ones, so that Seifert's result gives a complete classification. Toral actions with only finite isotropy groups turn out to be a beautiful generalization of Seifert's ideas (Conner and Raymond [2]). Seifert showed that the complicated structure of a Seifert fiber space unwinds in an appropriate covering space, and similar behavior occurs for $S^1 \times S^1$ actions. (Raymond described this to us as a "global uniformizing process.")

Singularity theory is yet another area that can be traced, in its early origins, to Seifert's paper. We mention in this regard first the work of Hirzebruch and Von Randow, who, in resolving singularities for surfaces, were led to consider graph manifolds. The connection between these manifolds and singularities was further developed by Neumann, Orlik and Wagreich, and thence generalized further to the study of isolated singularities of algebraic surfaces admitting more general actions by Arnold, Dolgachev, and others.

Yet another application to 3-manifolds appears in the recent work of Morgan [9], which shows how the existence of a nonsingular Morse–Smale flow on a 3-manifold is related to the singular fibrations of Seifert.

Finally, we note recent work of Bonahon [1] deriving from ideas of Montesinos, which shows that the decomposition of a 3-manifold into Seifert manifolds and prime M^3 which contain no essential tori has an analog in knot and link theory, with a "union" of "rational tangles" (Conway [3]) replacing "Seifert fiber space."

Seifert's paper is, in brief, a mathematical classic. We recommend it to graduate students and research mathematicians for its beauty, originality, clarity, and (as it seemed to us when we first encountered it) freshness of ideas, as well as for insight into the historical foundations of mathematics.

JOAN S. BIRMAN

REFERENCES

1. F. Bonahon, "Involutions et fibres de Seifert dans les varietes de dimension 3." These de 3eme cycle. Orsay, 1979.
2. P. E. Conner and F. Raymond, Actions of compact Lie groups on aspherical manifolds. *In* "Topology of Manifolds" (J. C. Cantrell and C. H. Edwards, Jr., eds.), pp. 227–264. Markham, Chicago, Illinois, 1970.
3. J. H. Conway, An enumeration of knots and links and some of their algebraic properties, *In* "Computational Problems in Abstract Algebra" (J. Leech, ed.), pp. 329–358. Pergamon, Oxford, 1970.

4. W. Jaco and P. B. Shalen, "Seifert fibered spaces in 3-manifolds," Amer. Math. Soc. Memoirs No. 220. Amer. Math. Soc., Providence, Rhode Island, 1979.

5. R. Jacoby, One-parameter transformation groups of the three-sphere, *Proc. Amer. Math. Soc.* **7** (1956), 131–142.

6. K. Johannson, Homotopy equivalence of knot spaces and 3-manifolds. Preprint, 1976.

7. H. Kneser, Geschlossene Flachen in dreidimensimalen Mannifaltigkeiten. *Jahresber. Deutsch. Math.-Verein.* **38** (1929), 248–260.

8. J. Milnor, A unique factorization theorem for 3-manifolds. *Amer. J. Math.* **84** (1962), 1–7.

9. J. Morgan, Non-singular Morse–Smale flows on 3-dimensional manifolds. *Topology* **18** (1978), 41–53.

10. P. Orlick, E. Vogt, and H. Zieschang, Zur topologie gefaserter dreidimensionales Mannigfaltigkeiten. *Topology* **6** (1967), 49–64.

11. F. Waldhausen, Eine Klasse von 3-dimensionalen Mannigfaltigkeiten, I, II. *Invent. Math.* **3** (1967), 303–333; **4** (1967), 87–117.

ACKNOWLEDGMENTS

The publication of an English language edition of Seifert and Threlfall's "Textbook of Topology," 45 years after the first German edition appeared, is the result of the effort of many individuals. Michael Goldman's persistent enthusiasm for the task, his expertise, and his serious and sustained hard work was the principal reason that this project came to a successful conclusion. In the early phases, Dr. Jerome Powell (now Fr. Jerome Powell) helped in the editing, and we thank him. When the English language edition was essentially ready to go to press, we learned that simultaneously and independently of our work, Professor John Stillwell of Monash University, Australia, had completed his own very fine translation of the book. Professor Stillwell graciously offered his help, sending us a copy of his own version of the text. It was impossible for us to incorporate his translation, which duplicated our own, however we did consult it from time to time when we were unclear about the best way to express an idea, and we thank him for that and for his generosity. Professor Frank Raymond also contributed useful ideas. Lastly, we thank the staff of Academic Press for the smooth and orderly way in which all details have proceeded.

JOAN S. BIRMAN
JULIAN EISNER

PREFACE TO GERMAN EDITION

The original stimulus to the writing of the present textbook was a series of lectures which one of us (Threlfall) presented at the Technische Hochschule, Dresden. But only a part of this course was incorporated into the text. The main content has been developed subsequently in a close daily exchange of thoughts between the two authors.

In this book we attempt to give an introduction and overall survey of the presently flourishing discipline of topology; however, we do not want to reach our goal merely by stating and proving propositions in their full generality. Rather, we wish to describe those concepts which have proven to be of value, and those methods which have been successful and appear promising to us, by means of worked-out examples.

Specialized knowledge is not assumed in advance. References to the literature are given in footnotes whenever less-well-known propositions are given without proof, so that one may find the proof in the form required.

We restrict our treatment to combinatorial or algebraic topology but make wide use of these methods, avoiding set theoretical difficulties wherever possible. Accordingly, the concepts of the simplicial complex and of the manifold, as introduced by L. E. J. Brouwer are central to our treatment. For the reader who is not intimately acquainted with group theory and its presently used notation, we have appended the group theoretical propositions that are used in a concluding chapter. If required, it might be read between the second and third chapters. As far as possible, the chapters have been made independently readable. A detailed alphabetical subject index is given. The comments presented at the conclusion of the book give an indication of further literature on the subject and should guide the way to a deeper study of those topics which have only been given a first treatment in this book.*

---

* Numbers in square brackets refer to the alphabetical references in the Bibliography; small superscripted numbers refer to the comments following Chapter XII.

Several topics could not be treated at all, because of restricted space. It was especially painful to us to have to omit the Alexander duality theorem as well as the Alexandroff theory of closed sets and projection spectra. We hope to close these gaps in a further volume, in the event that other textbooks of topology do not do so in the interim.

For the appearance of this book we are indebted, first of all, to Professor E. Trefftz who not only sacrificed his leisure time to its development but also provided encouragement by involving himself in our problems with understanding and by contributing practical advice. In the same spirit we thank the Arbeitsgemeinschaft der Dresdener Mathematiker for their cooperation, and Professor C. Weber in particular. We were assisted by our outside colleagues L. Bieberback and K. Reidemeister who read the proofs, and F. Hausdorff, H. Kneser (cf. §58 in particular), and B. L. van der Waerden whose Prague lecture [1] and verbal stimulation were of influence upon the arrangement of the book. In addition we thank Candidates Math. Mr. W. Hantzche and Mr. H. Wendt of Dresden for numerous particular improvements in the proof.

Dresden, January 1934   H. SEIFERT
                                                                       W. THRELFALL

SEIFERT AND THRELFALL:
A TEXTBOOK OF TOPOLOGY

CHAPTER ONE

ILLUSTRATIVE MATERIAL

1. The Principal Problem of Topology

The subject of topology deals with those properties of geometric figures which are unchanged by topological mappings, that is, by mappings which are bijective (i.e., one-to-one correspondences) and bicontinuous (i.e., continuous, with continuous inverses). A geometric figure is understood to be a point set in 3-dimensional space (or in a higher dimensional space). A continuous mapping is a mapping which is realized by continuous functions in a Cartesian coordinate system for this space. The mapping functions need be defined only for the points of the figure, and do not have to be defined over the whole of the space. Those properties which remain unchanged under topological mappings are called the topological properties of the figure. Two figures which can be mapped topologically onto each other are said to be homeomorphic.

For example, the surface of a hemisphere and a circular disk are homeomorphic, because one can map the hemisphere topologically onto the disk (shaded in Fig. 1) by means of an orthogonal projection. More generally, any two surfaces which can be deformed one to the other, by means of bending and distortion, are homeomorphic. As examples: the surface of a sphere, the surface of a cube, and an ellipsoidal surface are homeomorphic; an annulus and a cylindrical surface of finite length are homeomorphic.

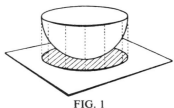

FIG. 1

It is an easy matter to discover arbitrarily many examples of homeomorphic figures, including among them figures which do not appear to be homeomorphic on first glance. Such is the case for the Euclidean plane and for a spherical surface from which one point has been removed. Each can be mapped topologically onto the other, by means of a stereographic projection. They are each, moreover, homeomorphic to an open disk (§6, second and third examples).

In general, it is more difficult to prove that two point sets are *not* homeomorphic than to prove that two points sets are homeomorphic. It is clear that a point and a line interval are not homeomorphic, because as point sets they cannot even be put into one-to-one correspondence. It is also easy to see that a line interval and a disk cannot be mapped topologically one onto the other: For if A, B, C are three arbitrary points of the disk we can transform A continuously to B without passing through C; this property of the disk is preserved under a topological mapping. However, it does not hold for the line interval because, when one continuously transforms one endpoint of the interval to the other endpoint, it must necessarily pass through the midpoint of the interval. This simple reasoning fails, on the other hand, when we compare a disk to a solid sphere. For if we characterize the disk by the property that it contains closed curves* which separate it, then we must prove that there is no closed curve which separates the solid sphere. This is not obvious! Why cannot a closed curve lie in a solid sphere in such a way that it passes through all points of some surface which separates the solid sphere? *In fact, this cannot happen*; however, it no longer suffices to use ideas as simple as those used for the line segment and the disk to prove that the disk and the ball, and analogous pairs of figures in higher dimensions, are non-homeomorphic (see §33 for the proof).

The concept of homeomorphism plays the same role in topology that the concept of congruence plays in elementary geometry. Just as two figures are not essentially different in elementary geometry when they are congruent, two figures are not essentially different in topology when they are homeomorphic. In contrast, however, while two congruent figures can always be moved rigidly one onto the other by means of a congruence transformation of the whole space, in the case of homeomorphic figures there need not exist a topological mapping of the whole space which carries one onto the other. For example, it is not possible to transform a circle topologically to a knotted curve, in particular the trefoil knot (Fig. 2), by means of a topological mapping of the whole of 3-space (and certainly not by means of a deformation). On the other hand, the two curves are homeomorphic, since one can map the points of the knot bijectively and bicontinuously onto the points of the circle. The relation of homeomorphism thus holds for the points of the knot but not for the embedding space.

---

* *Editor's note*: "Closed curve" means "simple closed curve" in this discussion.

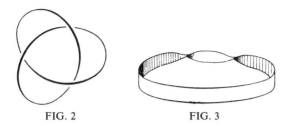

FIG. 2   FIG. 3

Likewise, an untwisted closed band and a closed band which is twisted by an integer multiple of 2π radians (Fig. 3) can be mapped topologically one onto the other, but not by means of a topological mapping of the whole of 3-space. One can see this by cutting the band into two congruent rectangular strips, and then identifying corresponding points of the strips with one another.

Knots and circles, twisted and untwisted bands, are topologically equivalent figures and differ only in the manner in which they are embedded in 3-dimensional space. The difference between a knot and a circle, or a twisted and an untwisted band, will vanish if one considers 3-dimensional space to be a subspace of 4-dimensional space and allows deformations in the latter space. In that case circle and knot, as well as circle and ellipse, can be deformed to one another without self-intersection.[1]

From now on, then, we will make a distinction between the intrinsic topological properties of a figure, that is, those properties which are preserved under all topological mappings of the figure, and the remaining properties, which depend upon the figure's placement in space, and which are preserved only under topological mappings of the whole space.

We shall illustrate this difference by an additional example. When a circle is rotated in space about a line in its plane which does not intersect the circle, it will sweep out a ring-shaped surface, or *torus*. A generating circle will pass through a given point O of the torus. We will call this circle a *meridian a* of the torus. In addition, the point also lies on the circle which was swept out by the point during the rotation; we will call this latter circle a *longitude b* of the torus (Fig. 4). Meridian and longitude circles can be distinctively characterized by observing that a meridian can be contracted to a point by shrinking through the interior of the solid ring which is bounded by the torus, while a longitude cannot. The figure consisting of a torus and a meridian cannot, then, be mapped to the figure consisting of a torus and a longitude by means of a topological mapping of the whole of space onto itself. This distinction between a meridian and a longitude is not, however, an intrinsic property of the torus. For the torus can be mapped topologically onto itself in a way such that meridian and longitude circles are interchanged, though not by means of a deformation in 3-space. To do this, let us consider the torus to be an elastic skin, and let us cut it along a and b. We can then bend and

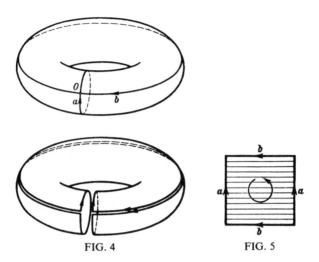

FIG. 4   FIG. 5

distort the cut surface to form a plane square (Fig. 5). If we now rotate the square by 180° about a diagonal, we obtain a topological mapping of the square onto itself such that a and b are interchanged. A like mapping of the torus onto itself will correspond to this mapping of the square. Another typical example, illustrating the distinction between intrinsic topological properties and topological properties of the embedding, is that of orientability and two-sidedness of a surface; this example will be discussed in the next section.

There is, then, a distinction among topological properties, which is analogous to that occurring in differential geometry, where it is well known that the intrinsic metric properties which belong to a surface, independent of its position in space, are determined by its first fundamental form, in contrast to the metric properties of the figure consisting of surface and space together, which are determined by the second fundamental form.

The principal problem of topology is to decide whether two given figures are homeomorphic and, when possible, to enumerate all classes of nonhomeomorphic figures. Although extensive theories exist which treat arbitrary subsets of Euclidean space,* we will not deal with the concept of a figure in that generality. To do so would entangle us in set theoretic difficulties. The concept of a complex, as introduced by L. E. J. Brouwer, and further narrowed down during the course of our investigations to that of a manifold, will be sufficiently restrictive so that it bypasses the set theoretic difficulties but will also be broad enough to include almost all figures of interest. The topology which we treat here is not, then, set theoretic topology but is a topology of complexes and manifolds.

---

* References to the literature are given in Tietze–Vietoris [1, I].

The property which distinguishes a complex from an arbitrary point set of a space is its triangulability: a complex is a point set consisting of finitely many or countably infinitely many not-necessarily straight-line intervals, triangles, tetrahedra, or corresponding higher dimensional building stones, assembled together as a structure. This assembly does not necessarily have to occur in an embedding space. If we so desire, we can fully dispense with the embedding space. As a consequence of the triangulability property, most so-called pathological point sets will be excluded from our considerations. A close connection with objects of geometric interest is then achieved, even to the extent that the topology of complexes has been called an India-rubber topology. Examples of complexes are: all Riemann surfaces; Euclidean space of arbitrary dimension; open subsets and algebraic curves and surfaces lying within that space; the projective plane and projective 3-space; all Euclidean and non-Euclidean space forms,[38] regions of discontinuity of metric groups of motions* and, finally, position and phase spaces of mechanical systems.

In order to approach the solution of the principal problem of topology, we seek topological invariants and suitable properties of complexes which can serve to indicate differences between them. The most important of those properties are the homology groups and the fundamental group belonging to a complex. They stand at the center of our investigations.

For the time being, however, we will try to see how far we can take the principal problem without using these aids. Without further preparation, we move to a subproblem. We ask: What nonhomeomorphic closed surfaces exist?

2. Closed Surfaces

As in the preceding section, let us cut apart the torus along meridian and longitude circles, to give a plane square. This square, taken together with the specification that opposite edges are to be regarded as equivalent (that is, are not to be regarded as different), is called the Poincaré fundamental polygon of the torus. This polygon will completely determine the torus, with regard to intrinsic topological properties. On the other hand, metric and embedding properties of the torus, such as surface area and position in space, are not determined. Surface topology does not concern itself with such properties. From the standpoint of surface topology, all tori obtained from the fundamental polygon by bending and joining of corresponding edges are equivalent. As an example, a rotationally symmetric torus is not regarded, topologically, as a different surface than a knotted tube! The procedure illustrated here, of cutting a surface into one or more polygons, can be directly generalized to provide the definition of a closed surface.

---

* *Editor's note*: See Threlfall and Seifert [1] for a discussion of this concept.

We consider a closed surface to be a structure which can be assembled from finitely many polygons by joining the polygon edges pairwise. In this way, the closed surface can be lifted out of the space surrounding it and can be given an independent existence as a "2-dimensional manifold." This is a concept which will later be given an exact definition.

The cutting of the torus into a square is the first step of a procedure which leads to the representation of an infinite set of closed surfaces by polygons, each polygon having a pairwise association of edges. We demonstrate this procedure as follows. Let us cut an approximately circular hole out of a torus. Let the boundary l of the hole pass through the point O. After making a deformation, we obtain a perforated torus or *handle* (Fig. 6). We form a pentagon by cutting open the handle along curves a and b and spreading it flat. We can also form the pentagon by cutting the hole from a torus which has previously been cut into a square (Fig. 7), and then cutting the hole boundary l at point O. The pentagon has one free edge, the hole boundary l, while the other edges are associated pairwise with one another (Fig. 8).

Let us now take two handles which have been cut into pentagons and join them along their hole boundaries (Fig. 9). Upon erasure of the common hole boundary l we obtain an octagon having pairwise association of edges (Fig. 10). We generate the *double torus* (also called a "pretzel surface") by bringing the corners together at one point and rejoining corresponding edges.[2] In the figure, we express the pairwise association of edges by marking the paired edges with the same letter and placing arrows on them, so that the arrows fall in coincidence tip on tip when the edges are joined. We can describe the association of edges, and thereby describe the closed surface, by means of a single formula. It is obtained by running around the edges in a given sense and furnishing an exponent to each edge, which is either $+1$ (usually not written explicitly) or -1, the sign depending upon whether the edge in question is traversed in the sense of the arrow or in the sense opposite to the arrow. For the case of the double torus, using the orientation of edges shown in the figure and using an appropriate sense of traversal, the expression reads:

$$a_1 b_1 a_1^{-1} b_1^{-1} a_2 b_2 a_2^{-1} b_2^{-1}.$$

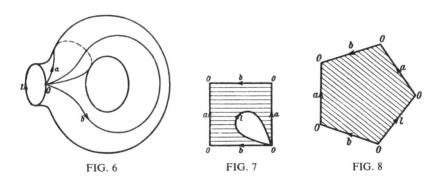

FIG. 6   FIG. 7   FIG. 8

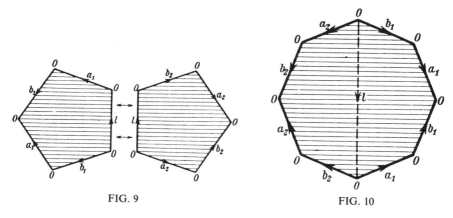

FIG. 9   FIG. 10

If we cut a circular hole out of the double torus, in such a way that the boundary of the hole passes through point O, we can attach an additional handle to the double torus, along the hole boundary. Upon erasing the hole boundary we obtain a 12-gon. Since one can regard the double torus as being a sphere with two handles attached (after making an inessential deformation), we see that this 12-gon will, after assembly, be a sphere with three handles attached. *In general, one will obtain a 4h-gon after cutting apart a sphere with h handles attached, or after cutting apart a torus with h − 1 handles attached.* The pairwise association of edges over the 4h-gon, which is the fundamental polygon of the sphere with h handles, is determined by the sequence of edges

$$a_1 b_1 a_1^{-1} b_1^{-1} \cdots a_h b_h a_h^{-1} b_h^{-1}$$

on the boundary circle. Each expression $a_i b_i a_i^{-1} b_i^{-1}$ corresponds to a handle, which can be obtained as a pentagon when one cuts four edges from the polygon by making a diagonal cut l.

The sphere surface, that is *the 2-sphere*, can also be defined by means of a bijective and bicontinuous identification of the edges of a fundamental polygon. We obtain the polygon by cutting the 2-sphere along an arc a having endpoints O and P. This gives a 2-gon having the boundary circle

$$aa^{-1}$$

which, in contrast to the other fundamental systems, contains two distinct vertices, which are not to be identified (Fig. 11). To generate the 2-sphere the edges of the 2-gon are folded together like the wings of a coin purse, about hinge joints at O and P, whereupon the circular disk closes to form a spherically shaped sack.

The sphere with h handles attached provided us with only half of the topologically distinct closed surfaces. Just as these surfaces are derived from the handle, the remaining surfaces are derived from the Möbius band.

The Möbius band is a closed band which has been twisted by π radians (Fig. 12). It is swept out by a line segment c having initial point O and

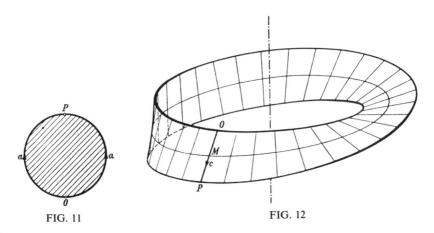

FIG. 11   FIG. 12

endpoint P, when the midpoint M rotates about an axis which does not intersect the segment and the segment moves in such a way that the axis and the segment always remain coplanar and, in addition, the segment rotates in that plane by an angle of π radians about its midpoint during a full rotation of this plane about the axis.

Upon cutting the Möbius band along the segment c we obtain a rectangle having the boundary circle

$$cr'cr''^{-1}$$

and having two distinct (nonequivalent) vertices O and P (Fig. 13). The free edges r' and r'' form the boundary of the Möbius band; this boundary is a single closed curve, topologically a circle. The remaining two edges c and c, which are to be joined together, do not appear with different exponents on the boundary circle, as was the case for the fundamental polygons encountered previously. On the contrary, they have the same exponent with respect to a traversal of the boundary circle. One accordingly says that they are inversely identified (or identified in the second way), while the previous case would be called direct identification (or identification in the first way).

The handle and the Möbius band are each bounded by a single topological circle, that is, by a curve homeomorphic to a circle. But they differ in the fact

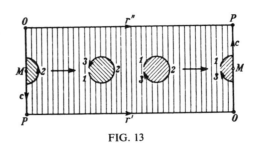

FIG. 13

that the handle is two-sided and the Möbius band is one-sided in Euclidean 3-space. This means the following. If a fly were to creep along the Möbius band without crossing the boundary it could at some time—for example, by running along the middle line of the band—arrive at a position antipodal to its initial position. Thus, unlike the handle, the Möbius band does not have two sides, separated by its boundary, which could be painted black and white, respectively, so that the colors do not meet, except at the boundary. On the contrary, it has only a single side.

One can justifiably raise the objection that this one-sidedness has not yet been established to be an intrinsic property of the surface and, therefore, no valid reason exists for the impossibility of topologically mapping the band twisted by π radians onto the untwisted band. If, however, instead of allowing the fly to creep along the surface, one pushes a small directed circle having three numbered points 123 within the surface, where one can no longer allow any thickness, then one can bring the circle on the Möbius band into inverse cover with itself so that the points 123 of the surface fall, respectively, upon the points 321 (Fig. 13). If this is possible for a surface, then we call the surface nonorientable and in the other case we call the surface orientable, because one can then extend the orientation defined by a small directed circle in the neighborhood of a point to every point on the surface in a unique way. Orientability is an intrinsic property of a surface. Two-sidedness, on the other hand, can only be defined when one embeds the surface in a 3-dimensional space; it depends upon the type of embedding and must not be confused with orientability. It will be shown later (§76) that orientable surfaces cannot, in fact, be embedded in Euclidean 3-space in such a way that they lie one-sided in this space; however, they *can* be embedded as one-sided surfaces in *other* 3-dimensional manifolds.

One may regard the torus, which is a closed surface, as having been obtained from a handle (see Fig. 6) by capping the circular handle boundary with a disk or, equivalently, by capping that boundary with a perforated sphere. One can close a Möbius band in the same way, by attaching a circular disk to the boundary of the band. The latter operation cannot, of course, be accomplished in 3-dimensional space unless self-intersections are allowed (§64). One can show, however, that this stitching together can be accomplished without self-intersection in 4-dimensional space; the closed Möbius band is thus a surface lying without self-intersection in 4-dimensional space.[3] The possibility of embedding the surface in a space is not at all essential, however, for the study of the intrinsic topological properties of a surface. We will, in fact, be able to fully discuss the intrinsic topological properties of the closed Möbius band after we have obtained its representation by topologically mapping the boundary of a circular disk onto the boundary of a Möbius band and by regarding corresponding points as being identical to one another. In the neighborhood of each point the closed Möbius band will behave like a piece of plane surface. The question of whether the closed band can be inserted as a whole into 3-space, or whether the entire boundary of the disk can be jointed to the entire boundary of the Möbius band in 3-space, is a separate question.

The closed Möbius band is, along with the 2-sphere, a fundamentally important closed surface in mathematics. It is called the projective plane. In projective geometry one is accustomed to seeing it introduced not as a closed Möbius band, but rather by closure of the Euclidean plane by adding an improper line. In this interpretation, the points of the projective plane (described by a projective coordinate system) are in one-to-one correspondence with equivalence classes of 3-tuples of real numbers $x_1:x_2:x_3$, where only the class $0:0:0$ is to be excluded.* If one interprets x_1, x_2, x_3 as homogeneous Cartesian coordinates,** so that $x = x_1/x_3$ and $y = x_2/x_3$ are ordinary Cartesian coordinates on the Euclidean plane, then the Euclidean plane together with the "improper" (infinitely distant) line $x_3 = 0$ gives the projective plane. Alternatively, if one interprets x_1, x_2, x_3 as Cartesian coordinates of 3-dimensional space, then the points of the projective plane can be placed in one-to-one correspondence with the manifold whose "points" are lines which pass through the origin.

We will now demonstrate the topological equivalence of the manifold of these lines and the manifold of points of a closed Möbius band. Let us describe a sphere of unit radius about the center of the pencil of lines. Each line of the pencil will intersect the sphere in two, diametrically opposite, points. The points of the projective plane can be mapped, in this way, bijectively and bicontinuously onto pairs of diametrically opposite points of the unit sphere. If, in the set of all points of the sphere, we regard those points which lie diametrically opposite to one another as being the same point, then we obtain the projective plane. In describing the projective plane, we can restrict ourselves to points of the lower hemisphere as representatives of projective points. If we cross the boundary of the lower hemisphere, that is, the equatorial circle, then we jump over to the diametrically opposite point of this circle. If we now project the lower hemisphere normally upon the tangent plane E at the south pole S (Fig. 14), we will then have mapped the projective plane onto the surface of the unit circular disk which has been closed by identifying diametrically opposite points of its boundary circle. The points of the boundary circle will thereby correspond, in pairs, to points of the improper (infinitely distant) projective line which closes the Euclidean plane to give the projective plane.

---

* *Editor's note*: The interested reader will find a more detailed discussion of this point of view in Section 14. The equivalence relation intended here is

$$(x_1, x_2, x_3) \sim (x_1', x_2', x_3') \qquad \text{if} \quad x_i' = \lambda x_i,$$

$i = 1, 2, 3$, for some nonzero real number λ. The symbol $x_1:x_2:x_3$ is used to describe the equivalence class.

** See Kowalewski [1, Section 14, p. 32]. For general purposes one can find the basic facts of projective geometry described in a form suitable for our use in, for example Klein [2, Chapter 1], Bieberbach [1], Hilbert and Cohn-Vossen [1], and Weyl [4].

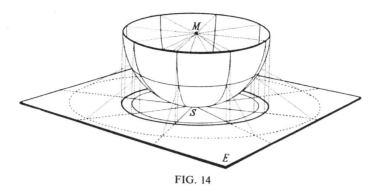

FIG. 14

We can recognize from this mapping, first of all, that the projective plane corresponds to the closed Möbius band. We need only to cut a central band out of the disk, along parallel line segments r' and r'' equally distant from the center point of the disk (Fig. 15). Upon identification of diametrically opposite points of the boundary circle, this band becomes the Möbius band. The two remaining disk segments can be joined along the two b edges to form a circular disk having boundary $r'r''$, which closes the Möbius band (Fig. 16).

Since the projective plane is represented topologically by a closed Möbius band, the Möbius band can then be regarded as a punctured projective plane. This gives rise to a new representation of the Möbius band. Topologically, it is obviously unimportant where the circular hole is punched into the projective plane, with regard to the bordered surface which arises. We can then locate the hole at the middle of the circular disk which closes to the projective plane when diametrically opposite points on its boundary are identified. This gives the representation of the Möbius band shown in Fig. 17. Diametrically opposite points of the outer boundary circle are to be identified. The inner circle is the boundary of the Möbius band. The original form of the Möbius band (Fig. 13) is produced by cutting the circular annulus along the dotted segments and joining it together along the outer semicircles.

We see, second of all, that the mapping onto the circular disk provides us with the *fundamental polygon of the projective plane*. We regard the unit disk as a polygon, that is, a 2-gon which is bounded by the edges a and a. As before,

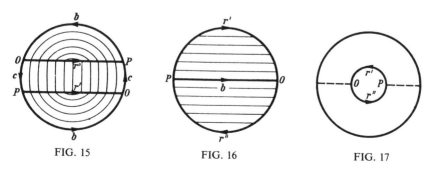

FIG. 15   FIG. 16   FIG. 17

the identification of diametrically opposite points on the boundary circle is given by means of the formula

$$aa.$$

The initial point of the edge a falls in coincidence with its endpoint, at the point O of the projective plane (Fig. 18), and a is the image of a projective line through which one can cut the projective plane to give the fundamental polygon. With this, the projective plane has been included in the series of closed surfaces which can be represented by a fundamental polygon. The fact that the projective plane is a closed surface in Euclidean 3-space is not nearly so well known as the fact that a torus is a closed surface in Euclidean 3-space. This is so because the projective plane cannot be inserted into Euclidean 3-space without having self-intersections. Neither can any other nonorientable closed surface (for proof see §64).

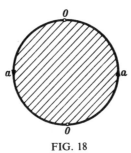

FIG. 18

The Möbius band, the punctured projective plane, and the annulus with identification of diametrically opposite points along a boundary circle are topologically equivalent surfaces. It follows from this equivalence that one can generate one and the same new surface, from an arbitrary closed surface, when one cuts a hole in the latter surface, and sews in either a Möbius band or a punctured projective plane along the hole boundary, or closes the boundary by the process of identifying diametrically opposite points. The Möbius band sewn into a circular hole is sometimes called a cross-cap.*

All other closed surfaces are now obtained by sewing an arbitrary number of Möbius bands onto a multiply punctured 2-sphere. We use the same procedure as for the handles. We cut a hole, having boundary l and passing through the point O, in the projective plane (Fig. 18). When we cut the projective plane at point O the fundamental polygon of the projective plane is transformed to a triangle having boundary circle aal. Let us join two such triangles, having boundary circles a_1a_1l and $a_2a_2l^{-1}$, along their hole boundaries and erase the seam line l; this generates a quadrilateral with boundary circle

$$a_1a_1a_2a_2$$

*The justification for this name, as well as an explicit pictorial description, can be found in Hilbert and Cohn–Vossen [1, p. 279].

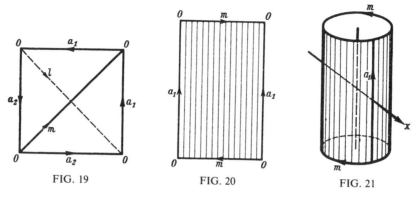

FIG. 19   FIG. 20   FIG. 21

(Fig. 19). This quadrilateral is the fundamental polygon of a sphere with two attached Möbius bands.

The surface just described is known by the names of "one-sided tube" and "nonorientable ring surface."* If we cut its fundamental polygon along the diagonal m opposite to the seam line l and join the two resulting triangles along their edges a_2, then we get a new quadrilateral, having the boundary circle $a_1 m a_1^{-1} m$ (Fig. 20), which can obviously be sewn together to give the same surface. Upon joining the two edges a_1, the quadrilateral can be regarded as a cylinder (Fig. 21). Its two boundary circles m are to be identified pointwise, but in such a way that they cannot be superimposed by making a displacement parallel to the surface lines, for this would generate a torus. The identification is to be made so that those points coincide which come together by tipping a circle over about a line x normal to the cylinder axis and passing through the cylinder midpoint.

This identification by superposition of boundary circles can be carried out in Euclidean 3-space only if we allow the cylinder to intersect itself (Fig. 22).

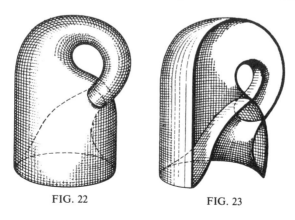

FIG. 22   FIG. 23

* *Editor's note*: This surface is commonly referred to as a *Klein bottle* in the English language literature.

If we slice the one-sided tube which results along its length, it will split into two Möbius bands, one of which is illustrated in Fig. 23.

When we attach k Möbius bands to the 2-sphere we obtain a closed surface, which can be given the fundamental polygon having boundary circle

$$a_1 a_1 a_2 a_2 \cdots a_k a_k.$$

We are not sure, yet, whether the closed surfaces which we have generated exhaust all of the closed surfaces, nor have we proven that no two of them can be mapped topologically one onto the other. The first doubt can be removed easily; we will do so in §38. To prove topological distinctness, on the other hand, requires the concept of the homology groups and proof of their topological invariance (§39).

The surfaces just described do not exhaust all surfaces. We have obtained only closed surfaces; these are characterized by the two properties that they can be covered by finitely many polygons and that they have no boundary. Surfaces which require infinitely many polygons for a polygonal covering are said to be *infinite*.[4] The Euclidean plane and the hyperboloid of one sheet are examples of infinite surfaces.

3. Isotopy, Homotopy, Homology

The methods which we will use to prove nonhomeomorphism of manifolds, for example, closed surfaces, rest, roughly speaking, upon a classification of the ways in which lower dimensional objects can be mapped continuously into these manifolds. We illustrate this with the simplest example, curves on surfaces.

We first consider curves which are free of double points and have a definite sense of traversal, that is, topological images of oriented circles. One obvious classification of all such closed curves is to consider two curves a and b to be equivalent if each can be continuously deformed to the other on the surface. We consider first of all, isotopic deformations. These are deformations such that the curve a remains free of double points in all of its intermediate positions during its transformation to b. In that case a and b are said to be isotopic. For example, any two meridian circles having the same orientation on a torus are isotopic. Likewise, the plane curves I and II of Fig. 24 are isotopic but I is not isotopic with III. Isotopic deformations are difficult to deal with mathematically and they will play a subordinate role in comparison to homotopic deformations.

In a homotopic deformation of a to b it is not required that the curve remain free of double points at all intermediate positions. Rather, a can intersect itself arbitrarily during its transformation to b. If a and b, which we no longer assume are free of double points, can be transformed into one another by a homotopic deformation they are said to be homotopic, or more precisely, freely homotopic. Isotopic curves are of course homotopic. All four

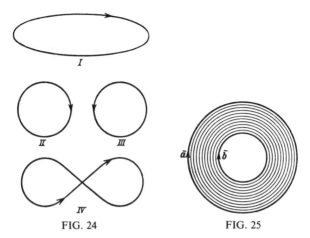

FIG. 24   FIG. 25

curves of Fig. 24 are homotopic to one another. Each curve is also null homotopic, that is, can be shrunk to a point. A meridian circle of the torus and a second meridian circle traversed in the opposite sense are not homotopic. Likewise, a meridian and a longitude of the torus are not homotopic and no meridian or longitude is null homotopic.

We can also explain the homotopy of two curves without making reference to a deformation. We do this as follows: two curves a and b are homotopic on the surface $\mathfrak{F}$ if an annulus (Fig. 25) can be mapped continuously (but not necessarily topologically) into the surface $\mathfrak{F}$ so that its two oriented boundary circles $\bar{a}$ and $\bar{b}$ are mapped to a and b. If this is possible then there will exist a homotopic deformation of a to b corresponding to a concentric transformation of $\bar{a}$ to $\bar{b}$ and conversely: when one transforms a homotopically to b then a "singular annulus," that is the continuous image of an annulus, will be swept out.

A generalization now appears obvious, which will lead us to the coarsest and certainly most important classification of closed curves: the homology classes. We need only to replace the annulus, which is a twice punctured 2-sphere, by a twice punctured orientable surface of arbitrary genus h (Fig. 26 shows the case $h = 1$) and map it continuously (but not necessarily bijectively) into $\mathfrak{F}$. If this mapping can be accomplished so that the hole boundaries $\bar{a}$ and $\bar{b}$ (oriented as in the figure) transform to two given curves a and b, then a and b are said to be homologous to one another. For example, the curves a and b on the surface $\mathfrak{F}$ shown in Fig. 27 (sphere with three handles) are homologous to one another because they form the boundary of each of the two twice punctured tori into which $\mathfrak{F}$ is separated by a and b. As we will see later the homology classes can be regarded as elements of an Abelian group, the 1-dimensional homology group. This group is a topological invariant of the surface $\mathfrak{F}$. With its aid we will be able to prove the distinctness of the surfaces classified in the previous section. By way of

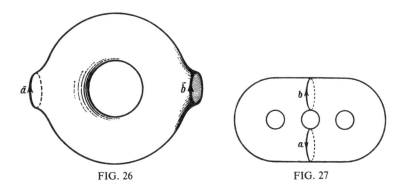

FIG. 26   FIG. 27

example: there exists only one homology class on the sphere, two exist on the projective plane, while the homology groups of the remaining closed surfaces are infinite. Of course, these concepts, which are discussed here only superficially, require more precise definition and the theorems require proof. This will be the main objective of a later chapter.

4. Higher Dimensional Manifolds

There is a complete theory answering the question of how to classify the topologically distinct surfaces; however, the corresponding problem in 3 or more dimensions is unsolved. In fact, one cannot even completely classify the topologically distinct 3-dimensional spaces. The higher dimensional case is, moreover, not just of abstract interest. Problems in the theory of differential equations and in the theory of two independent complex variables lead back to this question. Admittedly, the subject matter of such a theory is difficult to grasp intuitively. All of the orientable *closed surfaces* appear in complex variable theory, as Riemann surfaces, but only two closed 3-dimensional spaces have played a prominent role in mathematics, outside of topology. These are *projective* 3-*space* and the 3-*sphere*.

Projective 3-*space* arises from the requirement that one completes Euclidean 3-space by adding new points, so that projective transformations become one-to-one correspondences in the new point set, that is, are bijective. As is well known, one accomplishes this by closing Euclidean 3-space by adding an improper plane, the image plane of the vanishing plane of a projective mapping of Euclidean 3-space. In contrast to Euclidean 3-space, projective 3-space is a closed space; it can be covered with finitely many tetrahedra (§14).

Euclidean 3-space closes to form the 3-*sphere*, when we require that conformal mappings, which are circle preserving mappings of Euclidean 3-space, (Klein [1, Section 50]; Blaschke [1, Section 40]), be one-to-one correspondences. Among these mappings is the mapping by reciprocal radii. Under this mapping, the center point of the unit ball (the ball of inversion) has no image in Euclidean 3-space. We now wish to close Euclidean 3-space in order to generate the 3-sphere by adding the image point of this one point (and not the points of a whole improper plane as we did to obtain projective 3-space). The closing used here is the 3-dimensional analog of the closing of the Euclidean plane to generate the 2-sphere. We will see in §14 that the 3-sphere is the natural generalization of the 2-sphere to a space of 3 dimensions.[5]

The other higher dimensional manifolds which occur in mathematics are often not point manifolds, and their elements are objects of other types. We are already familiar with the case of a manifold which does not consist of points but is, nonetheless, 2-dimensional: the set of all nonoriented lines passing through a point of 3-dimensional Euclidean space. This manifold can be mapped onto the projective plane so that neighboring lines map to neighboring points. Another example is given by the set of all of the positions of a mechanical system. A particularly simple case, for example, is the plane double pendulum. This consists of two rigid rods l_1 and l_2 bound together by a hinge joint B. One of the rods is hinged at its free end to a fixed support point A (Fig. 28). Otherwise, the double pendulum is freely movable in the plane. The totality of different positions which it can assume may be described by means of the two angles which the rods subtend with respect to the vertical direction. Each position is described by the values of the two parameters φ and ψ, each of which is determined up to a multiple of 2π. The totality of positions can then be described by means of a square in the (φ, ψ)-plane, having side length 2π, whose opposite edges are to be identified. The torus can be mapped continuously onto the same point set. We can then relate all of the positions of the double pendulum to the points of the torus in such a way that points lying close to one another on the torus correspond to positions of the double pendulum which are close to one another. A periodic motion of the double pendulum which returns it to its initial position will map to a closed curve on the torus.[6]

Instead of fastening the two rods with hinges we could use spherical joints. The totality of positions of the resulting spherical double pendulum can be mapped to pairs of points on two spheres, so that neighboring positions correspond to neighboring points on each of the spheres. Each position is

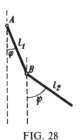

FIG. 28

specified by four parameters, for example, the geographic latitude and longitude on each of the spheres. The "position space" of the spherical double pendulum is, therefore, a 4-dimensional manifold.

The set of all oriented lines of projective 3-space forms a 4-dimensional manifold which is topologically equivalent to the manifold consisting of pairs of points on two spheres, as we now show. We consider real projective 3-space to be embedded in the complex projective space whose points are the equivalence classes of ratios of four complex numbers $x_1:x_2:x_3:x_4$, where only the quadruple $0:0:0:0$ is to be excluded. We first claim that the set of oriented real lines embedded in this space can be mapped in one-to-one correspondence and bicontinuously onto the points of the null sphere, that is, the set of points (x_1, x_2, x_3, x_4) which satisfy

$$x_1^2 + x_2^2 + x_3^2 + x_4^2 = 0,$$

where each x_i $(i = 1, \ldots, 4)$ is a complex number. Each real line cuts the null sphere in two distinct complex points P and $\bar{P}$, which are conjugate. Conversely, each pair of conjugate complex points of the null sphere determines a real line g which connects the points. If P_1, P_2, and P_3 are three real points on g, then when we orient g we specify a particular cyclic ordering of the three points, for example, $P_1 P_2 P_3$. Since the cross ratios

$$(PP_1P_2P_3) = \lambda \qquad \text{and} \qquad (\bar{P}\bar{P}_1\bar{P}_2\bar{P}_3) = \bar{\lambda}$$

are conjugate nonreal complex numbers, then just one of them, let us say λ, has a positive imaginary part. This property of the cross ratio λ remains unchanged on cyclic permutation of $P_1 P_2 P_3$. This can be seen by inspection, for we have

$$(PP_2P_3P_1) = \frac{1}{1 - \lambda}$$

and

$$(PP_3P_1P_2) = \frac{\lambda - 1}{\lambda}.$$

The oriented line g is therefore uniquely associated with the point P. The line g with its orientation reversed is then directed toward the point $\bar{P}$ since

$$(PP_1P_3P_2) = \frac{1}{\lambda}$$

has a negative imaginary part. This establishes a one-to-one correspondence between the oriented lines and the points of the null sphere. The null sphere is covered by two families of complex lines.* Let r be a line of the "right" family and let l be a line of the "left" family. Through each point P of the null sphere there will pass exactly one line of the left family which intersects r at a particular point P_r; likewise, exactly one line of the right family will pass through P and intersect l at a particular point P_l. Thus each point P on the null sphere will be in one-to-one correspondence with a pair of points, one lying on r and the other lying on l. Our claim that the oriented real lines of projective 3-space can be put in one-to-one correspondence with point pairs on two spheres will be demonstrated when we have shown that all of the points of r constitute a (real) 2-sphere and all of the points of l constitute a 2-sphere; that is, all of the points, real and complex, on a (complex) projective line constitute a real sphere surface. One can convince oneself of this by introducing a projective coordinate system $\mu_1 : \mu_2$ on the projective line; $\mu_1 : \mu_2$ will run through all possible complex numbers, including ∞. But these numbers will just correspond to the 2-sphere.

---

* Klein [1, Section 45], Study [1], or Bieberbach [2].

Like the position space of the spherical double pendulum, the manifold of all real oriented lines of projective 3-space is related to the manifold of pairs of points on two spheres in such a way that neighboring lines correspond to neighboring point pairs.

Not only the positions of mechanical systems but also their states of motion give rise to multidimensional manifolds, the so-called *phase spaces*.[7] By the state of motion of a point mass we mean its position in space and the magnitude and direction of its velocity. Let us assume, for example, that a single mass point is constrained to move on the surface of a solid sphere in the presence of a gravitational field. Let us also assume that its energy, which is the sum of its kinetic and potential energies, is constant and independent of time and is sufficiently large that the particle can run through the highest point of the sphere on its given path. Every possible state of motion is then described by means of the position of the point on the sphere—this requires two parameters—and by the direction of its velocity—this requires a third parameter, for the magnitude of the velocity is already determined from the initially given total energy and the particle position. The states of motion are therefore in one-to-one correspondence with the directed line elements of the sphere surface. The phase space is a 3-dimensional manifold. We will see (§14, Problem 2) that this manifold can be mapped onto projective 3-space. The states of motion thereby correspond to the points of (real) projective 3-space, in the sense that states of motion which differ only slightly will be mapped to neighboring points.

Earlier we departed from the naive concept of a figure as a point set in Euclidean 3-space when we closed the Möbius band to give the projective plane, and again when we described surfaces by sewing polygons together. While embedding in 3-space is justifiable as a natural intuitive aid to the description of 2-dimensional manifolds, freedom from a surrounding space becomes a necessity when one describes higher dimensional manifolds. To represent them by means of points sets and imbed them in a higher dimensional Euclidean space would be an artificial and impractical procedure.* Up to now, our concept of topological mappings and homeomorphism has gone together with embedding in a space, because we defined the continuity of a mapping of two figures as the continuity of the corresponding coordinate mapping functions. But, in contrast, we have also stated that the manifold of double pendulum positions coincides with the point set which forms the torus. Thus we need to define the concept of continuity in a way which is independent of coordinate mapping functions.

What is it, then, that position and phase spaces have in common with point sets of Euclidean spaces, making them capable of supporting continuous

---

* *Editor's note*: It is, however, true that every manifold can be embedded in Euclidean space of sufficiently high dimension.

mappings? In these spaces one knows which elements (positions and states of motion) give rise to the immediate neighborhood of an element! For each element there exist subsets (which admittedly can be chosen in a multiplicity of ways) that form the neighborhoods of that element. A mapping which is a one-to-one correspondence will now be continuous if it transforms neighborhoods to neighborhoods!* It is known that the point sets of the 2-sphere and the torus can be put in one-to-one correspondence, because they have the same cardinality. The fact that the two surfaces are non-homeomorphic means that the bijective mapping cannot be chosen so that neighboring points always transform to neighboring points. In order, then, that a mathematical entity be a topological object, and thereby keep a spacelike character, one must determine its neighborhoods. In general, one will not be in doubt about which subsets form the neighborhoods of an element. For example, in Euclidean 3-space, a neighborhood of a point must always contain all points of a solid sphere, that is, a ball, described about the point. We will get a larger or a smaller neighborhood according to whether we choose the ball to be larger or smaller. In the 3-sphere, a point set will be a neighborhood of the newly introduced improper point only if it contains all points outside of a sufficiently large ball. The neighborhood of a projective line must contain all lines which lie interior to a sufficiently "slender" one-sheeted hyperboloid containing the line in question as its axis, and so forth.

By tracing continuity and topological mapping back to the concept of a neighborhood we shall have liberated the figures in question from their embedding space to such an extent that the embedding space can be set aside, and itself be regarded as a figure equivalent to the other spacelike objects of topology, such as closed surfaces. If, as we will want to do, we retain the idea of space as a continuous point set, in order to introduce a geometry, then the concept of a neighborhood will be tied most deeply to the intrinsic nature of space itself. We will become acquainted with significant concepts and theorems which make no mention of distance, rectilinearity, or even of the dimension of a space. But the concept of the points which lie in a neighborhood of a point must be retained if we are still to speak of a point set as a space. If no other properties of the point set are specified, then we will call it a neighborhood space (§5).**

The concept of neighborhoods serves, first of all, to define mathematically the most general concept of space; that is, to construct it, to make it independent of vague intuition, and to trace it back to the fundamental

---

Editor's note: This heuristic statement is not quite correct. Actually, a bijective mapping is continuous if its *inverse* transforms neighborhoods to neighborhoods. It then follows that a one-to-one correspondence is bicontinuous, and hence a homeomorphism if, under the correspondence, neighborhoods correspond to neighborhoods.

**Editor's note*: The concept of a *neighborhood space*, introduced here, will ultimately be related to the more familiar concept of a *topological space*.

concepts of set theory. But more than this, the concept of neighborhoods provides a mathematical tool of the greatest utility and breadth; wherever we can determine neighborhoods for a set of mathematical objects, so that certain axioms are satisfied, then we will be able to speak of this set as a space in the widest sense and we will be able to apply concepts and theorems derived for arbitrary neighborhood spaces to any particular object in question.

When we treat the concepts and theorems of this introductory chapter rigorously, using the methods of synthetic geometry, and we subordinate them to a theory of neighborhood spaces, we shall have to begin anew and remove ourselves from all immediate intuition. Only occasionally will examples allow us to recognize the relationship between our general investigations and the geometric problems of this chapter. Individual geometric problems will not appear in the foreground again until Chapter VI (on surface topology). In spite of this, the theme of the next chapter is not exclusively general neighborhood spaces. Rather, we shall deal with special neighborhood spaces, the complexes already mentioned. The area between our very broad concept of neighborhood spaces and the restrictive concept of complexes is occupied by set theoretic topology, not treated by us, which includes the theory of point sets (see Hansdorff [1, 2]; Alexandroff [12]) and dimension theory (see Tietze and Vietoris [1, V]; Nöbeling [1]) in general topological spaces.

CHAPTER TWO

SIMPLICIAL COMPLEXES

5. Neighborhood Spaces

A finite or infinite nonempty set of mathematical objects, which will be called points, is called a *neighborhood space* if, to each point, certain subsets are assigned as neighborhoods of that point; these neighborhoods must satisfy both of the following axioms:

AXIOM A. *Each point P of the neighborhood space has at least one neighborhood; each neighborhood of P contains P.*

If $\mathfrak{M}$ is the neighborhood space we denote a neighborhood of P by

$$\mathfrak{U}(P \mid \mathfrak{M}).$$

AXIOM B. *Given a neighborhood $\mathfrak{U}(P \mid \mathfrak{M})$, each subset of $\mathfrak{M}$ containing this neighborhood is also a neighborhood of P.*

Examples of neighborhood spaces are (1) the set $\mathfrak{M}$ of all integers such that, for each integer, any subset of $\mathfrak{M}$ containing that integer is declared to be a neighborhood of that integer; (2) the same set when one declares that each subset which contains an integer itself and the integers immediately preceding it and following it is a neighborhood of that integer; (3) the same set when the only neighborhood of each point is declared to be the entire set.

These examples are presented only to indicate the broad generality of the concept of a neighborhood space.[8] Otherwise they play no role in the investigations to follow.

On the other hand, the example of n-dimensional Euclidean space or *Euclidean n-space* $\mathfrak{R}^n$ is important. A point in $\mathfrak{R}^n$ is an n-tuple of real numbers $x_1, x_2, \ldots, x_n$. $\mathfrak{R}^n$ consists of the totality of these numerical n-tuples.[9] We call $x_1, x_2, \ldots, x_n$ the coordinates of the point $(x_1, x_2, \ldots, x_n)$. We will agree that a subset of $\mathfrak{R}^n$ will be a neighborhood of a point $(\bar{x}_1, \bar{x}_2, \ldots, \bar{x}_n)$ if it contains all interior points of some cube surrounding $(\bar{x}_1, \bar{x}_2, \ldots, \bar{x}_n)$, i.e., if it contains all points $(x_1, x_2, \ldots, x_n)$ which satisfy the inequalities

$$|x_i - \bar{x}_i| < \eta \qquad (i = 1, 2, \ldots, n). \tag{1}$$

Thus, a given subset of $\mathfrak{R}^n$ is a neighborhood of the point $(\bar{x}_1, \bar{x}_2, \ldots, \bar{x}_n)$ if and only if there exists an η such that the η-cube about $(\bar{x}_1, \bar{x}_2, \ldots, \bar{x}_n)$ belongs to that subset. In particular, the whole of $\mathfrak{R}^n$ is a neighborhood of each of its points. With this specification of neighborhoods Euclidean space becomes a neighborhood space.

The distance between two points $(x_1, x_2, \ldots, x_n)$ and $(\bar{x}_1, \bar{x}_2, \ldots, \bar{x}_n)$ will be understood to be the nonnegative number

$$\sqrt{\sum_{i=1}^{n} (x_i - \bar{x}_i)^2} \tag{2}$$

A definition of neighborhoods which is equivalent to the one given previously is the following: Each subset of $\mathfrak{R}^n$ which contains all points of a ball neighborhood about $(\bar{x}_1, \bar{x}_2, \ldots, \bar{x}_n)$ is a neighborhood of $(\bar{x}_1, \bar{x}_2, \ldots, \bar{x}_n)$; the points of the ball neighborhood are those points whose distance from the given point $(\bar{x}_1, \bar{x}_2, \ldots, \bar{x}_n)$ is smaller than an appropriately chosen $\varepsilon > 0$; they therefore satisfy the inequality

$$\sum_{i=1}^{n} (x_i - \bar{x}_i)^2 < \varepsilon^2. \tag{3}$$

In particular, the points of this ball neighborhood, taken by themselves, form a neighborhood of $(\bar{x}_1, \bar{x}_2, \ldots, \bar{x}_n)$. We sometimes call such a neighborhood a (ball) ε-neighborhood of the point $(\bar{x}_1, \bar{x}_2, \ldots, \bar{x}_n)$; we will have a larger or a smaller ε-neighborhood depending upon the choice of ε. In ordinary language, a neighborhood is thought to be a set of points which does not contain any points which are too distantly situated. In contrast, here, we call a point set a neighborhood whenever we can find all points of a sufficiently small ball neighborhood in it.

Given two sets $\mathfrak{A}$ and $\mathfrak{B}$ we define their *set theoretic union* $\mathfrak{A} + \mathfrak{B}$ to be the set of those points which belong either to set $\mathfrak{A}$ or to set $\mathfrak{B}$ or to both sets. We define the *set theoretic intersection* of sets $\mathfrak{A}$ and $\mathfrak{B}$ to be the set of those points which belong to both $\mathfrak{A}$ and to $\mathfrak{B}$. If the intersection of two sets is empty then we say that the sets are *disjoint*.

If $\mathfrak{N}$ is a nonempty subset of a neighborhood space $\mathfrak{M}$ we will define, once and for all, the neighborhood $\mathfrak{U}(Q \mid \mathfrak{N})$ of a point Q of $\mathfrak{N}$ to be the intersection of any neighborhood $\mathfrak{U}(Q \mid \mathfrak{M})$ with $\mathfrak{N}$. In this way $\mathfrak{N}$ becomes a neighborhood space. Axiom A is obviously satisfied. Axiom B is satisfied since any subset $\mathfrak{B}$ of $\mathfrak{N}$ which contains the neighborhood $\mathfrak{U}(Q \mid \mathfrak{N})$ is the intersection of the neighborhood $\mathfrak{U}(Q \mid \mathfrak{M}) + \mathfrak{B}$ of Q in $\mathfrak{M}$ with the subset $\mathfrak{N}$, and is therefore a neighborhood of Q in $\mathfrak{N}$. *Each non-empty subset of a neighborhood space is, again, a neighborhood space. Since Euclidean n-space has already been made into a neighborhood space, with this prescription all subsets of Euclidean n-space become neighborhood spaces and the neighborhoods in them are completely determined. This is true in particular for all curves and surfaces in Euclidean 3-space.*

As an example, let $\mathfrak{M}$ be the real number line $\mathfrak{R}^1$ and let $\mathfrak{N}$ be the point set $0 \leqq x < 1$. Here a subset of $\mathfrak{N}$ is a neighborhood of the point $x = 0$, $\mathfrak{U}(0 \mid \mathfrak{N})$, whenever there exists an ε such that all points in $0 \leqq x < \varepsilon$ belong to this subset.

Now let $\mathfrak{N}$ be an arbitrary, possibly empty, subset of a neighborhood space $\mathfrak{M}$ and let P be a point of $\mathfrak{M}$. P is called

an *accumulation point of* $\mathfrak{N}$ if each neighborhood $\mathfrak{U}(P \mid \mathfrak{M})$ of P contains infinitely many points of $\mathfrak{N}$,

a *boundary point of* $\mathfrak{N}$ if each neighborhood $\mathfrak{U}(P \mid \mathfrak{M})$ contains points which lie in $\mathfrak{N}$ and also points which do not lie in $\mathfrak{N}$,

an *interior point of* $\mathfrak{N}$ if there exists a neighborhood $\mathfrak{U}(P \mid \mathfrak{M})$ which is contained in $\mathfrak{N}$; in this case P necessarily belongs to $\mathfrak{N}$.

The set of all boundary points of $\mathfrak{N}$ forms the *boundary* of $\mathfrak{N}$. Each point of $\mathfrak{N}$ is either a boundary point or an interior point.

In the example above, for the subset $\mathfrak{N}$: $0 \leqq x < 1$ on the real line $\mathfrak{M}$, all points $0 \leqq x \leqq 1$ are accumulation points of $\mathfrak{N}$; $x = 0$ and $x = 1$ are boundary points; the points $0 < x < 1$ are interior points.

The empty subset has neither accumulation points, boundary points, nor interior points.

If $\mathfrak{N}$ coincides with $\mathfrak{M}$, then all points are interior points.

The subset $\mathfrak{N}$ of $\mathfrak{M}$ is said to be *open* relative to $\mathfrak{M}$ if no boundary point of $\mathfrak{N}$ belongs to $\mathfrak{N}$, and *closed* relative to $\mathfrak{M}$ if all boundary points of $\mathfrak{N}$ belong to $\mathfrak{N}$.

It makes no sense to say that a particular neighborhood space $\mathfrak{N}$ is open or closed or to say that a point of $\mathfrak{N}$ is a boundary point or an interior point if one is not at the same time given the enveloping neighborhood space $\mathfrak{M}$, since these concepts are defined only with reference to the latter space. A subset $\mathfrak{N}$ may, for example, not be open relative to $\mathfrak{M}$ while it is open relative to a subset of $\mathfrak{M}$. In particular, the subset $\mathfrak{N}$ is open relative to itself.

The interval $0 \leqq x \leqq 1$ is a closed interval of the real number line; $0 < x < 1$ is an open interval; $0 \leqq x < 1$ is neither an open nor a closed interval. The empty subset, just like the subset $\mathfrak{N} = \mathfrak{M}$, is both open and closed at the same time.

It is clear that the set theoretic union of arbitrarily many open subsets is itself open; the intersection of arbitrarily many closed sets is closed.

An open subset is also characterized by the fact that it contains a neighborhood of each of its points. A closed subset is characterized by the fact that it is the complement of an open subset.*

---

* One defines the complement of the subset $\mathfrak{N}$ of a set $\mathfrak{M}$ as the totality of all points of $\mathfrak{M}$ which do not belong to $\mathfrak{N}$. It is denoted by $\mathfrak{M} - \mathfrak{N}$.

One defines the *closed hull* of a subset $\mathfrak{N}$ of $\mathfrak{M}$ to be the intersection of all closed subsets of $\mathfrak{M}$ which contain $\mathfrak{N}$. The closed hull of $\mathfrak{N}$ is therefore the smallest closed subset of $\mathfrak{M}$ which contains $\mathfrak{N}$.

It is important to distinguish the concept of a *limit point of a sequence* from that of a *boundary point of a subset*. An infinite sequence (in which one and the same point may occur repeatedly) *converges to a point*, which is a limit point of the sequence, whenever almost all points of the sequence (that is, all but finitely many points) lie in an arbitrary neighborhood of that point. With our basic concept of a neighborhood space it can happen that a sequence is convergent and still converges to different points,* an inconvenience which point sequences in Euclidean spaces and complexes do not exhibit.

6. Mappings

If, to each point P of a neighborhood space $\mathfrak{A}$ one assigns exactly one point P' of a neighborhood space $\mathfrak{B}$, then one has a *mapping T* of $\mathfrak{A}$ into $\mathfrak{B}$. The point $P' = T(P)$ is called the *image point* of P. The set $\mathfrak{A}'$ of all image points of $\mathfrak{A}$ is called the *image set* of $\mathfrak{A}$. The set $\mathfrak{A}'$ is a subset of $\mathfrak{B}$ and may coincide with the whole of $\mathfrak{B}$. The mapping of $\mathfrak{A}$ into $\mathfrak{B}$ is said to be one-to-one if distinct points of $\mathfrak{A}$ always have distinct images in $\mathfrak{B}$. In this case there exists an *inverse mapping* of $\mathfrak{A}'$ onto $\mathfrak{A}$, which associates the original point P to each point P' of $\mathfrak{A}'$. This mapping is denoted by T^{-1}. One says that a mapping takes the set $\mathfrak{A}$ *into* the set $\mathfrak{B}$ when the image points comprise a subset of $\mathfrak{B}$ (possibly coinciding with $\mathfrak{B}$) and one speaks of a mapping of $\mathfrak{A}$ *onto* $\mathfrak{B}$ whenever each point of $\mathfrak{B}$ is an image point, that is, $\mathfrak{A}' = \mathfrak{B}$.

If $\mathfrak{A}$ is transformed by a mapping T into a subset of $\mathfrak{B}$ and $\mathfrak{B}$ is transformed by an additional mapping U into a subset of a neighborhood space $\mathfrak{C}$, then $\mathfrak{A}$ is mapped into $\mathfrak{C}$ at the same time. This mapping is called the *product of the mappings T* and *U* and is denoted by UT. One must pay attention to the order of the factors, which is determined so that the image of a point P is given by $U(T(P)) = UT(P)$.

A mapping of $\mathfrak{A}$ into $\mathfrak{B}$ is said to be *continuous at a point P* of $\mathfrak{A}$ if, for each neighborhood $\mathfrak{U}(P' \mid \mathfrak{B})$ of the image point P' there exists a neighborhood $\mathfrak{U}(P \mid \mathfrak{A})$ whose image is contained in $\mathfrak{U}(P' \mid \mathfrak{B})$. The mapping is said to be continuous if it is continuous at each point.

This definition is in agreement with the classical definition of continuity. A function $y = f(x)$ maps the x-axis, a real number line, into the y-axis, another real number line; this mapping is said to be continuous at the point $P = \bar{x}$ if, for a given $\varepsilon > 0$, there exists a $\delta > 0$ such that $|f(x) - f(\bar{x})| < \varepsilon$ for all x in the interval $|x - \bar{x}| < \delta$. That is, to a given ε-neighborhood $\mathfrak{U}(P' \mid \mathfrak{B})$ of the image point $y = f(x)$ there corresponds a δ-neighborhood $\mathfrak{U}(P \mid \mathfrak{A})$ of the original point $\bar{x}$ which maps entirely into the ε-neighborhood (Fig. 29).

---

*This occurs, for example, when the whole set is defined as the only neighborhood of each point. Each infinite sequence will then converge to every point.

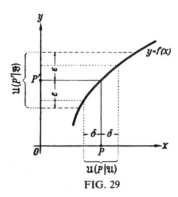

FIG. 29

More generally, each mapping T of a subset $\mathfrak{A}$ of Euclidean m-space $\mathfrak{R}^m$ having coordinates $x_1, x_2, \ldots, x_m$ onto a subset $\mathfrak{A}'$ of $\mathfrak{R}^n$ can be carried out by means of n functions

$$y_k = f_k(x_1, x_2, \ldots, x_m) \qquad (k = 1, 2, \ldots, n). \qquad (1)$$

We find

THEOREM I. *The mapping T is continuous if and only if the mapping functions* (1) *are continuous.*

As is well known in analysis, a function of several variables $y = f(x_1, x_2, \ldots, x_m)$ is said to be continuous at a point $\bar{x}_1, \bar{x}_2, \ldots, \bar{x}_m$ if for each given $\varepsilon > 0$ there exists a $\delta > 0$ such that

$$|f(x_1, x_2, \ldots, x_m) - f(\bar{x}_1, \bar{x}_2, \ldots, \bar{x}_m)| < \varepsilon$$

for all points of the domain of definition of f which satisfy the inequalities

$$|x_i - \bar{x}_i| < \delta.$$

Proof. Call the δ-cube neighborhood $\mathfrak{W}_\delta(P \mid \mathfrak{A})$ of a point P of $\mathfrak{A}$ the set of all points of $\mathfrak{A}$ which lie inside the δ-cube about P. In each neighborhood $\mathfrak{U}(P \mid \mathfrak{A})$ there lies a δ-cube neighborhood when δ is chosen to be sufficiently small. For $\mathfrak{U}(P \mid \mathfrak{A})$ is the intersection of $\mathfrak{A}$ with a neighborhood $\mathfrak{U}(P \mid \mathfrak{R}^m)$ and the latter neighborhood contains a δ-cube.

(a) Let the mapping T be continuous. Let $\mathfrak{W}_\varepsilon(P' \mid \mathfrak{A}')$ be a given ε-cube neighborhood of the image point P' of P. Because of the continuity of T there exists a neighborhood $\mathfrak{U}(P \mid \mathfrak{A})$ whose image lies in $\mathfrak{W}_\varepsilon(P' \mid \mathfrak{A}')$. In $\mathfrak{U}(P \mid \mathfrak{A})$ there lies a δ-cube neighborhood which then will also map into the given ε-cube neighborhood. This means that the mapping functions (1) are continuous.

(b) Let the mapping functions (1) be continuous.

For a given neighborhood $\mathfrak{U}(P' \mid \mathfrak{A}')$ there exists an ε-cube neighborhood

$\mathfrak{W}_\varepsilon(P' \mid \mathfrak{A}')$ contained within it. Thus there exists a δ-cube neighborhood $\mathfrak{W}_\delta(P \mid \mathfrak{A})$ whose image belongs to $\mathfrak{W}_\varepsilon(P' \mid \mathfrak{A}')$ and hence also to $\mathfrak{U}(P' \mid \mathfrak{A}')$, so that T is continuous.

THEOREM II. *If a mapping of the neighborhood space $\mathfrak{A}$ into the neighborhood space $\mathfrak{B}$ is continuous, then the mapping of $\mathfrak{A}$ onto the image $\mathfrak{A}'$ of $\mathfrak{A}$ is also continuous. Conversely, if the mapping of $\mathfrak{A}$ onto $\mathfrak{A}'$ is continuous, then the mapping of $\mathfrak{A}$ into $\mathfrak{B}$ is continuous.*

Proof. An arbitrary neighborhood $\mathfrak{U}(P' \mid \mathfrak{A}')$ is the intersection of a neighborhood $\mathfrak{U}(P' \mid \mathfrak{B})$ with $\mathfrak{A}'$. When $\mathfrak{A}$ is mapped continuously into $\mathfrak{B}$ there exists a neighborhood $\mathfrak{U}(P \mid \mathfrak{A})$ whose image lies entirely in $\mathfrak{U}(P' \mid \mathfrak{B})$ and therefore lies within the intersection of $\mathfrak{U}(P' \mid \mathfrak{B})$ and $\mathfrak{A}'$, that is, in $\mathfrak{U}(P' \mid \mathfrak{A}')$. Since $\mathfrak{U}(P' \mid \mathfrak{A}')$ was chosen arbitrarily this implies that the mapping of $\mathfrak{A}$ onto $\mathfrak{A}'$ is continuous.

Conversely, if the mapping of $\mathfrak{A}$ onto $\mathfrak{A}'$ is continuous, then for an arbitrary neighborhood $\mathfrak{U}(P' \mid \mathfrak{B})$ we form the neighborhood $\mathfrak{U}(P' \mid \mathfrak{A}')$, which is the intersection of $\mathfrak{U}(P' \mid \mathfrak{B})$ and $\mathfrak{A}'$. Since the mapping of $\mathfrak{A}$ onto $\mathfrak{A}'$ is continuous there exists a neighborhood $\mathfrak{U}(P \mid \mathfrak{A})$ whose image belongs entirely to $\mathfrak{U}(P' \mid \mathfrak{B})$. In other words: *the definition of continuity is the same, whether one chooses the neighborhoods of P' in $\mathfrak{A}'$ or in $\mathfrak{B}$.*

THEOREM III. *If $\mathfrak{A}$ is mapped continuously onto $\mathfrak{A}'$ and if a subset $\mathfrak{N}$ of $\mathfrak{A}$ is thereby transformed to the subset $\mathfrak{N}'$ of $\mathfrak{A}'$, then the mapping of $\mathfrak{N}$ onto $\mathfrak{N}'$ is also continuous.*

Proof. Corresponding to an arbitrary neighborhood $\mathfrak{U}(P' \mid \mathfrak{A}')$ of an image point P' there exists a neighborhood $\mathfrak{U}(P \mid \mathfrak{A})$, of the point P of $\mathfrak{N}$ which maps into $\mathfrak{U}(P' \mid \mathfrak{A}')$ because of the continuity of the mapping of $\mathfrak{A}$ into $\mathfrak{A}'$. The intersection of $\mathfrak{U}(P \mid \mathfrak{A})$ with $\mathfrak{N}$ is a neighborhood of P in $\mathfrak{N}$ which then also maps into $\mathfrak{U}(P' \mid \mathfrak{A}')$. Thus the mapping of $\mathfrak{N}$ into $\mathfrak{A}'$ is continuous and therefore, by the previous theorem, the mapping from $\mathfrak{N}$ onto $\mathfrak{N}'$ is continuous.

THEOREM IV. *If $\mathfrak{A}$ is mapped continuously onto $\mathfrak{A}'$ and $\mathfrak{A}'$ is mapped continuously onto $\mathfrak{A}''$, then $\mathfrak{A}$ is also mapped continuously onto $\mathfrak{A}''$.*

This is obvious.

A mapping T of $\mathfrak{A}$ onto $\mathfrak{A}'$ is said to be *topological** when T is one-to-one and both T and its inverse mapping T^{-1} are continuous. If $\mathfrak{U}(P \mid \mathfrak{A})$ is a neighborhood of a point P, there exists a neighborhood $\mathfrak{U}(P' \mid \mathfrak{A}')$ which maps into $\mathfrak{U}(P \mid \mathfrak{A})$ due to the continuity of T^{-1}. This implies that the image of $\mathfrak{U}(P \mid \mathfrak{A})$ under the mapping T is a point set containing $\mathfrak{U}(P' \mid \mathfrak{A}')$ and is again a neighborhood of P'. Thus we have:

---

* *Editor's note:* Cf. the first footnote in Section 1.

THEOREM V. *A topological mapping and its reciprocal mapping each transform neighborhoods to neighborhoods. Conversely, each one-to-one mapping having this property is topological.*

We speak of a topological mapping of $\mathfrak{A}$ *into* $\mathfrak{B}$ if $\mathfrak{A}$ maps topologically onto a subset $\mathfrak{A}'$ of $\mathfrak{B}$, which may also coincide with $\mathfrak{B}$.

Two neighborhood spaces are said to be *homeomorphic* to one another if they can be mapped topologically onto one another.

Theorems III and IV are still valid when stated for topological mappings instead of continuous mappings. It follows from this that two neighborhood spaces which are each homeomorphic to a third neighborhood space are also homeomorphic to one another. (Homeomorphism is a transitive property.)

We are interested only in those properties of neighborhood spaces and their subsets which are preserved by topological mappings. As a counterexample, the property that a subset of a Euclidean space is a straight line is not of this nature because the straightness is generally lost after mapping topologically to another Euclidean space.

In contrast, the property that a subset of a neighborhood space $\mathfrak{M}$ be a neighborhood of a point, say P, is a topologically invariant property. For under a topological mapping of $\mathfrak{M}$ onto $\mathfrak{M}'$, a neighborhood of point P of $\mathfrak{M}$ will transform to a neighborhood of the image point P' in $\mathfrak{M}'$. The same is true for the concepts: boundary point, interior point, and accumulation point of a subset; limit point of a sequence; and open subset and closed subset. For example, the property that a point R of $\mathfrak{M}$ is the boundary point of a subset $\mathfrak{N}$ is not destroyed by a topological mapping of $\mathfrak{M}$ onto $\mathfrak{M}'$. That is, if R transforms to R' and $\mathfrak{N}$ to $\mathfrak{N}'$ by this mapping, then R' is also a boundary point of $\mathfrak{N}'$. To see why this is true observe that the neighborhoods of R in $\mathfrak{M}$ are in one-to-one correspondence with those of R' in $\mathfrak{M}'$. Thus, in each neighborhood of R' there exist both points which belong to $\mathfrak{N}'$ and points which do not belong to $\mathfrak{N}'$ since the corresponding statement is true for R in $\mathfrak{M}$. One proves the topological invariance of the other concepts mentioned above in similar fashion. The property that a mapping of $\mathfrak{A}$ into $\mathfrak{B}$ is topological or continuous is itself a topological invariant; that is, if one replaces $\mathfrak{A}$ and $\mathfrak{B}$ by homeomorphic neighborhood spaces $\mathfrak{A}'$ and $\mathfrak{B}'$, then the corresponding mapping is, respectively, always topological or continuous.

EXAMPLE 1. *The topological mapping of the interval $0 \leqq x \leqq 1$ of a real number line onto the interval $0 \leqq y \leqq 1$ of another real number line.* From Theorem I the mapping is achieved by means of a continuous function $y = f(x)$. The function $f(x)$ must assume every value between 0 and 1 exactly once. It is known from the subject of analysis that such a function must be either monotonically increasing or monotonically decreasing. It follows that $f(x)$ takes either its largest or its smallest value at $x = 0$, so that boundary points of the one interval transform to those of the other interval.

EXAMPLE 2. *Stereographic projection of the punctured 2-sphere onto the Euclidean plane.* Considered as a subset of Euclidean 3-space, the unit 2-sphere

$$x_1^2 + x_2^2 + x_3^2 = 1$$

is a neighborhood space and remains so when one punctures it, that is, removes one point, for example, the north pole $(0, 0, 1)$. *The punctured 2-sphere is homeomorphic to the Euclidean plane.* A topological mapping is produced by stereographic projection of the punctured sphere from the north pole onto the equatorial plane, i.e., the Euclidean plane $x_3 = 0$. The mapping is one-to-one, since the only point of the sphere which cannot be mapped in this manner, the north pole, is removed. To show continuity of the mapping from the sphere to the plane we must find, for any neighborhood of a point P' of the plane, a neighborhood of the original sphere point P that maps entirely into the former neighborhood. An arbitrary neighborhood of P' will always contain an ε'-neighborhood, that is, a circular disk with P' as center point. A spherical cap about the original point P, formed by the intersection of the sphere with a spatial ε-neighborhood of P, is also a neighborhood of P. As is known from the properties of stereographic projection, this latter neighborhood maps into a circular disk which contains P' as an interior point but not necessarily as center point. If one chooses the cap to be sufficiently small, then one can insure that its image falls in the previously mentioned ε'-neighborhood of P'. One can show in a like manner that the mapping from the plane to the sphere is continuous. Thus it is topological.

Another proof follows from Theorem I if one constructs the mapping functions and proves their continuity. In order to distinguish image point coordinates from coordinates of the original point we will introduce a ξ_1, ξ_2-coordinate system in the equatorial plane, coinciding with the (x_1, x_2)-system. For a sphere of radius r, where $r = 1$ in our case, the mapping formulas become (where we can just as well write them out simultaneously for the case of n dimensions):*

$$\xi_i = \frac{r}{r - x_n} x_i \qquad (i = 1, 2, \ldots, n - 1),$$

$$x_i = \frac{2r^2}{\xi_1^2 + \xi_2^2 + \cdots + \xi_{n-1}^2 + r^2} \xi_i,$$

$$x_n = \frac{\xi_1^2 + \xi_2^2 + \cdots + \xi_{n-1}^2 - r^2}{\xi_1^2 + \xi_2^2 + \cdots + \xi_{n-1}^2 + r^2} r.$$

These are in fact well defined and continuous for all points of both the punctured sphere and the equatorial plane.

EXAMPLE 3. *The Euclidean plane is homeomorphic to the interior of a disk.* To produce the topological mapping one projects the points of the lower unit hemisphere, excluding the equatorial boundary circle, first radially outward from the center (Fig. 14) onto the plane and a second time perpendicularly onto the Euclidean plane tangent to the south pole. Both mappings are topological. Since two topological mappings carried out in succession yield a topological mapping, it follows that the inverse of the second mapping composed with the first mapping yields a topological mapping of the interior points of the unit disk onto the Euclidean plane.

Problems

1. Show that the following are homeomorphic: a cylindrical surface of finite height h, excluding its two boundary circles; a hyperboloid of one sheet in Euclidean 3-space; an annulus, excluding its two boundary circles, and a twice punctured sphere.

2. The boundary circle of a disk has been mapped topologically onto itself. Show that this mapping can be extended to a topological mapping of the entire disk onto itself.

3. Let $\mathfrak{M}$ be a subset of a Euclidean space and let P be a point not belonging to $\mathfrak{M}$. Prove that if P is a boundary point of $\mathfrak{M}$, then P is also a point of accumulation, and conversely.

---

* *Editor's note:* The $(n - 1)$-sphere is regarded here as the set of points $(x_1, x_2, \ldots, x_n) \in R^n$ which satisfy the equality $x_1^2 + x_2^2 + \cdots + x_n^2 = r^2$. Euclidean $(n - 1)$-space is regarded as the set of points $(\xi_1, \xi_2, \ldots, \xi_{n-1}, 0) \in R^n$.

7. Point Sets in Euclidean Spaces

The point sets with which we will deal later are of a special type. They are homeomorphic to subsets of a Euclidean space. Their structure is far more specialized than the structure of a general neighborhood space. This is expressed in the fact that many theorems valid for subsets of Euclidean spaces cannot be derived from Axioms A and B of §5. In this section we shall be concerned with neighborhood spaces $\mathfrak{M}$ which are themselves subsets of Euclidean spaces.

Based upon the definition of neighborhoods in Euclidean spaces (§5) and the definition of neighborhoods in subsets (§5), a subset $\mathfrak{N}$ of $\mathfrak{M}$ is a neighborhood $\mathfrak{U}(P \mid \mathfrak{M})$ of a point P in $\mathfrak{M}$ if and only if all points of $\mathfrak{M}$ whose distance from P is less than a certain $\varepsilon > 0$ belong to $\mathfrak{N}$. The set of all points of $\mathfrak{M}$ which are less distant than ε from P, that is, the intersection of the ε-ball described about P with $\mathfrak{M}$, is called an ε-neighborhood $\mathfrak{U}_\varepsilon(P \mid \mathfrak{M})$. $\mathfrak{U}_\varepsilon(P \mid \mathfrak{M})$ is obviously an open subset of $\mathfrak{M}$ relative to $\mathfrak{M}$, because the sign $<$ and not $\leq$ occurs in the inequality (3) of §5.

An important property not possessed by every neighborhood space but possessed by every subset $\mathfrak{M}$ of a Euclidean space is the property that two distinct points P and Q have distinct neighborhoods.* Such neighborhoods are cut from $\mathfrak{M}$, for example, by sufficiently small ε-balls about P and Q.

It follows from this that a convergent sequence, that is, a sequence almost all of whose points lie in each neighborhood of P, has the point P as the only limit point. This is true because, due to the existence of disjoint neighborhoods, almost all points cannot lie simultaneously in each neighborhood of P and also in each neighborhood of a different point Q.

Among the theorems of elementary analysis concerning subsets of Euclidean spaces, we shall need the accumulation point theorem, the maximum value theorem, and the uniform continuity theorem.

THEOREM I (ACCUMULATION POINT THEOREM). *Each bounded infinite subset of a Euclidean space has at least one point of accumulation.***

A point set is said to be *infinite* if it contains infinitely many points. A subset of a Euclidean space is said to be *bounded* if the absolute values of all coordinates are smaller than some finite bound.

THEOREM II. *The continuous image of a bounded closed subset of a Euclidean space in another Euclidean space is again bounded and closed.*

Proof. Assume that the continuous image $\mathfrak{M}'$ of the closed bounded set $\mathfrak{M}$ is not closed and bounded. Then there exists a point sequence $P_1', P_2', \ldots,$

---

* *Editor's note*: In current usage a space which has this property is said to be a *Hausdorff space*.

** The proof is found in Knopp [I, Part I, p. 22]. The proof given there for 2 dimensions can be extended immediately to n dimensions.

consisting only of points of $\mathfrak{M}'$, which either converges to a limit point $\overline{R}$ not in $\mathfrak{M}'$ or does not converge at all to a point of accumulation. If one chooses a preimage point P_i in $\mathfrak{M}$ for each P_i', then by assumption the set $P_1, P_2, \ldots$ possesses a point of accumulation R belonging to $\mathfrak{M}$. In each neighborhood of its image point R' there will then lie infinitely many points of the sequence $P_1', P_2', \ldots$, which contradicts the assumed nature of this point sequence.

An immediate corollary is

THEOREM III (MAXIMUM VALUE THEOREM). *A continuous function which is defined in a bounded closed subset of a Euclidean space takes a maximum and a minimum value.*

This is true because the subset can be mapped by means of the continuous function into the real number line.

One can formulate the *uniform continuity theorem* as follows:

THEOREM IV. *If a bounded closed point set $\mathfrak{A}$ of a Euclidean space is mapped continuously into a neighborhood space $\mathfrak{B}$ and if a particular neighborhood $\mathfrak{U}^*(Q \mid \mathfrak{B})$ is arbitrarily assigned to each point Q of $\mathfrak{B}$, then there will exist a $\delta > 0$ such that the image of the δ-neighborhood of each point P of $\mathfrak{A}$ will be a subset of $\mathfrak{U}^*(Q \mid \mathfrak{B})$, where Q is the image of P.*

Proof. Assume that the theorem is false. Then for each δ of the sequence $1/2, 1/3, 1/4, \ldots, 1/i, \ldots$ there will exist a point P_i whose $(1/i)$-neighborhood will not be mapped into a neighborhood $\mathfrak{U}^*(Q \mid \mathfrak{B})$. The sequence $P_1, P_2, \ldots$ has an accumulation point P in $\mathfrak{A}$, because $\mathfrak{A}$ is closed and bounded. If $\mathfrak{U}^*(P' \mid \mathfrak{B})$ is the previously chosen neighborhood of the image point P' of P, then from the continuity of the mapping there will exist a neighborhood $\mathfrak{U}(P \mid \mathfrak{A})$ whose image is covered by $\mathfrak{U}^*(P' \mid \mathfrak{B})$. Since infinitely many points of the sequence lie in $\mathfrak{U}(P \mid \mathfrak{A})$, there will exist an i which is sufficiently large so that the $(1/i)$-neighborhood of P_i will lie entirely in $\mathfrak{U}(P \mid \mathfrak{A})$, so that its image is completely covered by $\mathfrak{U}^*(P' \mid \mathfrak{B})$. This gives a contradiction.

If the sets $\mathfrak{A}$ and $\mathfrak{B}$ coincide and the continuous mapping is the identity, one gets the corollary

THEOREM V. *If an arbitrary neighborhood $\mathfrak{U}^*(Q \mid \mathfrak{A})$ is assigned to each point Q of a bounded closed Euclidean space, then there exists a $\delta > 0$ such that the δ-neighborhood of each point P of $\mathfrak{A}$ will be covered by a neighborhood $\mathfrak{U}^*(Q \mid \mathfrak{A})$ of an appropriately chosen point Q.*

The usual formulation of the uniform continuity theorem is

THEOREM VI. *When a bounded closed subset $\mathfrak{A}$ of a Euclidean space is mapped continuously into another Euclidean space $\mathfrak{B}$, then for each $\varepsilon > 0$ there exists a $\delta > 0$ such that any two points of $\mathfrak{A}$ whose distances from one another is*

$< \delta$ *are transformed to two points of* $\mathfrak{B}$ *which are at a distance* $< \varepsilon$ *from one another.*

Proof. In Theorem IV choose the $(\varepsilon/2)$-neighborhood of Q as $\mathfrak{U}^*(Q \mid \mathfrak{B})$ for all points Q of $\mathfrak{B}$. There will then exist a $\delta > 0$ such that the δ-neighborhood of an arbitrary point of $\mathfrak{A}$ will map into an $(\varepsilon/2)$-neighborhood of an appropriately chosen point Q of $\mathfrak{B}$. If P_1 and P_2 are points of $\mathfrak{A}$ having a distance $< \delta$ from each other, then the δ-neighborhood of P_1 will include P_2. Since the image of the δ-neighborhood in $\mathfrak{B}$ is covered by an $(\varepsilon/2)$-neighborhood, the image points P_1' and P_2' in $\mathfrak{B}$ will have a distance from each other $< \varepsilon$.

For later use we present some additional propositions and definitions related to point sets in Euclidean spaces.

THEOREM VII. *Let* $\mathfrak{M}$ *be a subset of a Euclidean space or be homeomorphic to such a subset. Let* P *be a point of* $\mathfrak{M}$, *let* Ω *be a neighborhood of* P *relative to* $\mathfrak{M}$, *and let* Ω_1 *be a neighborhood of* P *relative to* Ω. *Then* Ω_1 *is a neighborhood of* P *relative to* $\mathfrak{M}$.

Proof. Ω contains an ε-neighborhood $\mathfrak{U}_\varepsilon(P \mid \mathfrak{M})$ as a neighborhood of P in $\mathfrak{M}$. Likewise, Ω_1 contains a δ-neighborhood $\mathfrak{U}_\delta(P \mid \Omega)$. If η is the smaller of the two radii ε and δ, then $\mathfrak{U}_\eta(P \mid \mathfrak{M})$ will belong to Ω and $\mathfrak{U}_\eta(P \mid \Omega)$ will belong to Ω_1. In that case $\mathfrak{U}_\eta(P \mid \mathfrak{M})$ will also belong to Ω_1. The neighborhood Ω_1 thus contains all points of $\mathfrak{M}$ which are at a distance $< \eta$ from P and is therefore a neighborhood of P relative to $\mathfrak{M}$, as required.

Now let $\mathfrak{N}$ be a subset of $\mathfrak{M}$. The closed hull $\mathfrak{N}'$ of $\mathfrak{N}$ relative to $\mathfrak{M}$ was defined in §5 to be the intersection of all closed subsets of $\mathfrak{M}$ which contain $\mathfrak{N}$. Thus all boundary points of $\mathfrak{N}$ relative to $\mathfrak{M}$ belong to $\mathfrak{N}'$. It is in fact true that *the closed hull* $\mathfrak{N}'$ *is formed exactly by* $\mathfrak{N}$ *and its boundary,* $\overline{\mathfrak{N}}$. To demonstrate this we must show that the set union $\mathfrak{N} + \overline{\mathfrak{N}}$ is a closed subset of $\mathfrak{M}$. If R is a boundary point of $\mathfrak{N} + \overline{\mathfrak{N}}$ relative to $\mathfrak{M}$, then an ε-neighborhood $\mathfrak{U}_\varepsilon(R \mid \mathfrak{M})$ will contain a point of $\mathfrak{N} + \mathfrak{N}$ which is either a point of $\mathfrak{N}$ or a boundary point $\overline{R}$ of $\mathfrak{N}$. In the latter case $\overline{R}$ will lie at a distance $< \varepsilon$ from R. Since there exist points of $\mathfrak{N}$ which are arbitrarily close to $\overline{R}$ (from the definition of boundary points), it then follows that there will exist points of $\mathfrak{N}$ which are at a distance $< \varepsilon$ from R. That is, R is a boundary point of $\mathfrak{N}$ and therefore belongs to $\overline{\mathfrak{N}}$. One can prove in like manner that the boundary of $\mathfrak{N}$ relative to $\mathfrak{M}$ is a closed subset of $\mathfrak{M}$.

Let $\mathfrak{M}_1$ and $\mathfrak{M}_2$ be two disjoint bounded closed sets in $\mathfrak{R}^n$ and let $d(P_1P_2)$ be the distance of a point P_1 of $\mathfrak{M}_1$ to a point P_2 of $\mathfrak{M}_2$. The greatest lower bound δ of all the distances $d(P_1P_2)$ is called *the distance between the sets* $\mathfrak{M}_1$ *and* $\mathfrak{M}_2$. The greatest lower bound is adopted because $\mathfrak{M}_1$ and $\mathfrak{M}_2$ are closed and bounded. It is > 0 because they are disjoint.

The *diameter* of a closed bounded set in $\mathfrak{R}^n$ will be defined to be the least upper bound of the distances of any two points of the set.

8. Identification Spaces

It frequently occurs in geometry that one must split up the points of a neighborhood space into classes of equivalent points and must then introduce these classes as "points" of a new neighborhood space. For example, if one projects the points of Euclidean 3-space onto a plane, this process can be regarded as a division of 3-space into classes: all points having the same image under the projection, that is, all points of a projection line, form one class; one no longer regards the points as being different but instead one identifies them with the projection point. Another example: We are given a group of translations of the Euclidean plane; the group is generated by a displacement $x' = x + a$ in the x-direction and a displacement $y' = y + b$ in the y-direction. We decide to regard points as equivalent and put them into one class if they can be transformed to one another by means of translations belonging to the group. That is, points are equivalent if their x- and y-coordinates differ by integer multiples of a or b, respectively. One may select an appropriately chosen representative from each class in such a way that these representatives comprise the interior points of a rectangle to which a corner point and two sides are added (Fig. 30), that is, all points whose coordinates satisfy the inequalities $0 \leqq x < a$ and $0 \leqq y < b$. When one now "identifies" equivalent points of the Euclidean plane the torus is generated, since the side $x = a$ is equivalent to the side $x = 0$ and the side $y = b$ is equivalent to the side $y = 0$.

Finally, we want to use the procedure of identification to construct surfaces by joining the sides of polygons. For example let us consider the point set consisting of two triangles in the Euclidean plane (Fig. 31). We will declare that points are equivalent if they correspond to one another in a particular topological mapping of a side of one triangle onto a side of the other triangle. Let us identify equivalent points. The two triangles will then join together to form a quadrilateral.

We shall now define the process of identification in terms of the fundamental concepts of set theory: sets, subsets, and mappings.

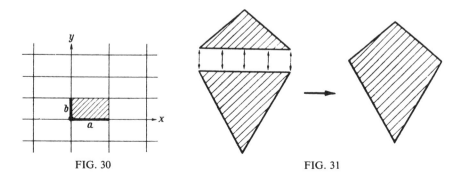

FIG. 30   FIG. 31

Let the points of a neighborhood space $\mathfrak{M}$ be divided into classes (subsets) so that each point belongs to exactly one class. Let us call the points of a class *equivalent points*. We shall not exclude the possibility, for example, that one class consists of a single point, while another class consists of infinitely many points. We form a new set $\mathfrak{M}'$ whose "points" are the classes of equivalent points of $\mathfrak{M}$. This gives a natural mapping from $\mathfrak{M}$ onto $\mathfrak{M}'$; one assigns as image of a point P of $\mathfrak{M}$ its class P' of $\mathfrak{M}'$. The set $\mathfrak{M}'$ will be made into a neighborhood space by means of the following prescription: If $P_1, P_2, \ldots$ are all of the points of $\mathfrak{M}$ which map into one and the same point P' of $\mathfrak{M}'$, choose a neighborhood $\mathfrak{U}(P_i \mid \mathfrak{M})$ at each point P_i and form the set theoretic union of these neighborhoods $\mathfrak{U}(P_1 \mid \mathfrak{M}) + \mathfrak{U}(P_2 \mid \mathfrak{M}) + \cdots$. The image of this union set is declared to be a neighborhood $\mathfrak{U}(P' \mid \mathfrak{M}')$ of P' in $\mathfrak{M}'$. The neighborhoods in $\mathfrak{M}'$ defined in this manner obviously satisfy Axiom A. They also satisfy Axiom B. For if $\mathfrak{U}'$ is a subset of $\mathfrak{M}'$ containing $\mathfrak{U}(P' \mid \mathfrak{M}')$, then the preimage set $\mathfrak{U}$ giving rise to $\mathfrak{U}'$, that is, the set of all points which map into points of $\mathfrak{U}'$, is a neighborhood $\overline{\mathfrak{U}}(P_i \mid \mathfrak{M})$ of the arbitrary preimage point P_i of P'; for $\overline{\mathfrak{U}}(P_i \mid \mathfrak{M})$ contains the neighborhood $\mathfrak{U}(P_i \mid \mathfrak{M})$. $\mathfrak{U}'$ is the image of the union set $\overline{\mathfrak{U}}(P_1 \mid \mathfrak{M}) + \overline{\mathfrak{U}}(P_2 \mid \mathfrak{M}) + \cdots = \mathfrak{U}$ and therefore by definition is a neighborhood of P'.

We say that $\mathfrak{M}'$, as well as any neighborhood space homeomorphic to $\mathfrak{M}'$, arises by means of identification of equivalent points.

If one replaces $\mathfrak{M}$ by a homeomorphic neighborhood space **M** and declares points in **M** to be equivalent if they correspond to equivalent points in $\mathfrak{M}$, then by identifying equivalent points in **M** one obtains a neighborhood space which is obviously homeomorphic to $\mathfrak{M}'$.

For each neighborhood $\mathfrak{U}(P' \mid \mathfrak{M}')$ there exists a neighborhood of the point P_i, $\mathfrak{U}(P_i \mid \mathfrak{M})$, whose image belongs entirely to $\mathfrak{U}(P' \mid \mathfrak{M}')$. Thus we have

THEOREM I. *The mapping of $\mathfrak{M}$ onto $\mathfrak{M}'$ is continuous.*

If, on the other hand, a neighborhood space $\mathfrak{M}$ has been mapped onto an arbitrary neighborhood space $\mathfrak{M}'$, the question may be asked: when can $\mathfrak{M}'$ be considered to arise from $\mathfrak{M}$ by means of identifying points which map to the same point of $\mathfrak{M}'$. To answer this we have

THEOREM II. *When a neighborhood space $\mathfrak{M}$ is mapped continuously onto a neighborhood space $\mathfrak{M}'$, the image space $\mathfrak{M}'$ will arise by means of identification of points of $\mathfrak{M}$ having the same image point in $\mathfrak{M}'$ only if, for every point P' of $\mathfrak{M}'$ and arbitrarily chosen neighborhood $\mathfrak{U}(P' \mid \mathfrak{M}')$, the neighborhood $\mathfrak{U}(P' \mid \mathfrak{M}')$ is the image of a set theoretic union $\mathfrak{U}(P_1 \mid \mathfrak{M}) + \mathfrak{U}(P_2 \mid \mathfrak{M}) + \cdots$ of neighborhoods of the points $P_1, P_2, \ldots$ which map to P'.*

Proof. It is to be shown that each neighborhood $\mathfrak{U}(P' \mid \mathfrak{M}')$ is the image of a set union of the form $\mathfrak{U}(P_1 \mid \mathfrak{M}) + \mathfrak{U}(P_2 \mid \mathfrak{M}) + \cdots$. Because we assume continuity there exists for each of the points P_i a neighborhood $\overline{\mathfrak{U}}(P_i \mid \mathfrak{M})$ which maps into $\mathfrak{U}(P' \mid \mathfrak{M}')$. The set $\mathfrak{U}$ of all points of $\mathfrak{M}$ that map into

$\mathfrak{U}(P' \mid \mathfrak{M}')$ contains $\overline{\mathfrak{U}}(P_i \mid \mathfrak{M})$ as a subset and is therefore itself a neighborhood $\mathfrak{U}(P_i \mid \mathfrak{M})$ of P_i, from Axiom B. Thus $\mathfrak{U}(P' \mid \mathfrak{M}')$ is the image of the set union $\mathfrak{U}(P_1 \mid \mathfrak{M}) + \mathfrak{U}(P_2 \mid \mathfrak{M}) + \cdots$.

We shall now prove formally the almost self-evident fact that one can proceed stepwise with the identification of points and will arrive at the same result regardless of the order of the steps:

If $\mathfrak{M}'$ arises from $\mathfrak{M}$ by means of identification of equivalent points and $\mathfrak{M}''$ arises from $\mathfrak{M}'$ by means of identification of equivalent points, this results in a mapping of $\mathfrak{M}$ onto $\mathfrak{M}''$; $\mathfrak{M}''$ arises from $\mathfrak{M}$ by identifying all points of $\mathfrak{M}$ which map to the same point of $\mathfrak{M}''$.

Proof. A point P'' of $\mathfrak{M}''$ is the image of the points $P_1', P_2', \ldots$ of $\mathfrak{M}'$, P_i' is itself the image of the points $P_{i1}, P_{i2}, \ldots$ of $\mathfrak{M}$. The mapping of $\mathfrak{M}$ onto $\mathfrak{M}''$ is continuous, and thus from Theorem II it suffices to show that $\sum_i \sum_j \mathfrak{U}(P_{ij} \mid \mathfrak{M})$ maps to a neighborhood of P''. The neighborhood $\sum_j \mathfrak{U}(P_{ij} \mid \mathfrak{M})$ transforms to a neighborhood of $\mathfrak{U}(P_i' \mid \mathfrak{M}')$ since $\mathfrak{M}'$ arises by means of identification of the points $P_{i1}, P_{i2}, \ldots$. Furthermore, $\sum_i \mathfrak{U}(P_i' \mid \mathfrak{M}')$ transforms to a neighborhood $\mathfrak{U}(P'' \mid \mathfrak{M}'')$ since $\mathfrak{M}''$ arises from $\mathfrak{M}'$ by means of identification of the points $P_1', P_2', \ldots$. Thus $\sum_i \sum_j \mathfrak{U}(P_{ij} \mid \mathfrak{M})$ maps to the neighborhood $\mathfrak{U}(P'' \mid \mathfrak{M}'')$.

THEOREM III. *Let $\mathfrak{M}$ be a bounded closed point set of a Euclidean space whose points are divided into equivalence classes. Let $\mathfrak{M}$ be mapped continuously onto a point set $\mathfrak{M}'$ in another Euclidean space in such a way that equivalent points of $\mathfrak{M}$ map to the same point of $\mathfrak{M}'$ and nonequivalent points of $\mathfrak{M}$ map to distinct* points of $\mathfrak{M}'$; then $\mathfrak{M}'$ arises from $\mathfrak{M}$ by means of identification of equivalent points.*

Proof. From Theorem II one needs only to show that if $P_1, P_2, \ldots$ are all of the points which map to P' and if $\mathfrak{U}(P_i \mid \mathfrak{M})$ is an arbitrary neighborhood of P_i, then the image of the set union $\mathfrak{U}(P_1 \mid \mathfrak{M}) + \mathfrak{U}(P_2 \mid \mathfrak{M}) + \cdots$ is a neighborhood $\mathfrak{U}(P' \mid \mathfrak{M}')$. Obviously this image is a subset $\mathfrak{N}'$ of $\mathfrak{M}'$ such that $\mathfrak{N}'$ contains P'. If we were to assume that $\mathfrak{N}'$ were not a neighborhood of P', then in each ε-neighborhood of P' there would exist points of $\mathfrak{M}'$ which do not belong to $\mathfrak{N}'$. In particular, corresponding to each ε of the sequence $1/2, 1/3, \ldots, 1/i, \ldots$ there would then exist a point Q_i' which would belong to the $(1/i)$-neighborhood of P' but not to $\mathfrak{N}'$. Let $Q_1, Q_2, \ldots$ be arbitrary points in the preimage of the sequence $Q_1', Q_2', \ldots$ such that Q_i' is the image of Q_i. Then because $\mathfrak{M}$ is bounded and closed the sequence $Q_1, Q_2, \ldots$ will have at least one point of accumulation, Q. Each neighborhood of the image point Q' of Q will then contain infinitely many points of the sequence $Q_1', Q_2', \ldots$ because the mapping is continuous. Since P' is the only accumulation point of this sequence, it follows that $Q' = P'$.

---

*This is not of course always possible, for an arbitrary division into equivalence classes.

Therefore Q is one of the points $P_1, P_2, \ldots$. Suppose, for example, $Q = P_1$. Then infinitely many of the points $Q_1, Q_2, \ldots$ lie in $\mathfrak{U}(P_1 \mid \mathfrak{M})$ and thus infinitely many of the points $Q'_1, Q'_2, \ldots$ lie in $\mathfrak{N}'$. This is a contradiction; thus $\mathfrak{N}'$ is in fact a neighborhood of P'.

We are now in a position to answer the question: When does the continuity of two out of the three mappings coupled by means of the relation $\psi\varphi = \chi$ (first φ, then ψ!) imply continuity of the third mapping? As noted earlier, χ is continuous whenever both φ and ψ are continuous. In contrast, the continuity of ψ and χ does not imply that of φ, since, for example, χ will be continuous for an arbitrary choice of φ when ψ transforms all points to a single point. It follows from Theorem III, however, that the continuity of φ and χ will imply the continuity of ψ under certain conditions. We have

THEOREM IV. *Let a bounded closed set $\mathfrak{M}$ of a Euclidean space be mapped into another Euclidean space by means of a continuous mapping φ. Subsequently map the image $\mathfrak{M}'$ into an arbitrary neighborhood space by means of a mapping ψ (which we do not assume in advance to be continuous). If $\mathfrak{M}''$ is the image of $\mathfrak{M}'$ and the mapping $\psi\varphi = \chi$ is a continuous mapping of $\mathfrak{M}$ onto $\mathfrak{M}''$, then ψ is also continuous.*

Proof. Let P' be an arbitrary point of $\mathfrak{M}'$ and let $P'' = \psi(P')$ be its image in $\mathfrak{M}''$. Let $P_1, P_2, \ldots$ be all points of $\mathfrak{M}$ which are mapped by φ to P'. Because of the continuity of χ there will exist, for a given neighborhood $\mathfrak{U}(P'' \mid \mathfrak{M}'')$, neighborhoods $\mathfrak{U}(P_1 \mid \mathfrak{M}), \mathfrak{U}(P_2 \mid \mathfrak{M}), \ldots$, all of which are mapped into $\mathfrak{U}(P'' \mid \mathfrak{M}'')$. If one now declares that all points of $\mathfrak{M}$ which transform to the same point of $\mathfrak{M}'$ are equivalent, then from Theorem III $\mathfrak{M}'$ arises from $\mathfrak{M}$ by means of identification of equivalent points. The set union $\mathfrak{U}(P_1 \mid \mathfrak{M}) + \mathfrak{U}(P_2 \mid \mathfrak{M}) + \cdots$ thus transforms to a neighborhood $\mathfrak{U}(P' \mid \mathfrak{M}')$ which, in turn, maps into $\mathfrak{U}(P'' \mid \mathfrak{M}'')$. That is, ψ is continuous at P' and is therefore continuous at each point of $\mathfrak{M}'$.

In Theorem IV, if φ is a one-to-one mapping which is continuous in the direction $\mathfrak{M} \to \mathfrak{M}'$ and one sets $\psi = \varphi^{-1}$, then $\psi\varphi$ is the identity mapping of $\mathfrak{M}$ onto itself and is therefore continuous. We have thus proven

THEOREM V. *A one-to-one continuous mapping of a bounded closed subset of a Euclidean space is a homeomorphism.*

EXAMPLE 1. Let the set $\mathfrak{M}$ be a disk, including its boundary circle. When all points of the boundary circle are identified the 2-sphere results. For one can map the disk onto the 2-sphere continuously so that the map is one-to-one except on the boundary circle. Map the disk radii, directed from the disk center point to the disk boundary, onto the sphere meridians of longitude directed from south pole to north pole. The boundary point of a radius will always map to the north pole. From Theorem III the 2-sphere arises from the disk by identifying all points which map to the north pole.

EXAMPLE 2. In the set of all lines passing through a point in Euclidean 3-space declare that a neighborhood of a line g is the subset of all lines which belong to a circular cone about this line (and also any set of lines containing the subset of all lines belonging to such a cone). A neighborhood space homeomorphic to this neighborhood space will arise from the 2-sphere by means of identification of diametrically opposite points of the sphere. This gives the neighborhood space introduced in §2, called the "projective plane."

9. n-Simplexes

Let $n + 1$ linearly independent points

$$P_0, P_1, \ldots, P_n$$

be given in the m-dimensional Euclidean space, $\Re^m$; $0 \leq n \leq m$. [We say that $n + 1$ points are linearly independent if they do not lie in an $(n - 1)$-dimensional linear subspace.*] Let the coordinates of the point P_i be

$$p_{i1}, p_{i2}, \ldots, p_{im} \qquad (i = 0, 1, \ldots, n). \tag{1}$$

Assign a mass

$$\mu_i \geq 0 \tag{2}$$

to each point P_i in such a way that the sum of all the assigned masses is equal to 1:

$$\mu_0 + \mu_1 + \cdots + \mu_n = 1. \tag{3}$$

These masses then determine a centroid X, that is, a point which has the coordinates

$$x_1 = \sum_0^n \mu_i p_{i1}, \qquad x_2 = \sum_0^n \mu_i p_{i2}, \qquad \ldots, \qquad x_m = \sum_0^n \mu_i p_{im}. \tag{4}$$

We call $\mu_0, \mu_1, \ldots, \mu_n$ the *barycentric coordinates* of X; they are constrained by the conditions (2) and (3). For each choice of $\mu_0, \mu_1, \ldots, \mu_n$ satisfying (2) and (3) one obtains a centroid X; as $\mu_0, \mu_1, \ldots, \mu_n$ are allowed to vary over all real numbers satisfying (2) and (3), a set of such centroid points will be generated in $\Re^m$. This set of points is said to constitute a "rectilinear" or "geometric" *n-simplex*, $\mathfrak{E}^n$ in the Euclidean space $\Re^m$. Accordingly, $\mathfrak{E}^n$ is a closed subset of the Euclidean space. The points $P_0, P_1, \ldots, P_n$ are called the *vertices* of the n-simplex. The simplex is fully determined by specifying its vertices, and can be described by its vertices as well as by a single symbol, $\mathfrak{E}^n$.

Introduce a parallel** coordinate system $\xi_1, \xi_2, \ldots, \xi_m$ to replace the coordinate system $x_1, x_2, \ldots, x_m$ by using a linear transformation with nonvanishing determinant. Let $\pi_{i1}, \pi_{i2}, \ldots, \pi_{im}$ be the coordinates of P_i in the new coordinate system. Then the two sides of (4) transform in the same

---

*Cf. Schreier and Sperner [1] for the fundamental concepts of affine geometry which are not explicitly explained here.

**Editor's note*: The term "parallel" is used very loosely here; the intended meaning is "a new coordinate system obtained from the old one by a linear transformation" (not by a translation).

way, and one obtains the equations

$$\xi_1 = \sum_{i=0}^{n} \mu_i \pi_{i1}, \qquad \xi_2 = \sum_{i=0}^{n} \mu_i \pi_{i2}, \qquad \cdots, \qquad \xi_m = \sum_{i=0}^{n} \mu_i \pi_{im}. \qquad (5)$$

Thus the coordinates of the centroid are expressed in the same way by barycentric coordinates in all parallel coordinate systems.

In particular, one can choose the coordinate system $\xi_1, \xi_2, \ldots, \xi_m$ so that $\xi_1, \xi_2, \ldots, \xi_n$ are parallel coordinates in the linear subspace $\mathfrak{L}^n$ to which $\mathfrak{E}^n$ belongs. In that case $\pi_{i,n+1}, \pi_{i,n+2}, \ldots, \pi_{im}$ are all equal to zero and the system (5) becomes

$$\xi_1 = \sum_{i=0}^{n} \mu_i \pi_{i1}, \qquad \xi_2 = \sum_{i=0}^{n} \mu_i \pi_{i2}, \qquad \cdots, \qquad \xi_n = \sum_{i=0}^{n} \mu_i \pi_{in}. \qquad (6)$$

One can specialize even further and can take the vectors going from P_0 outward to the vertices $P_1, P_2, \ldots, P_n$ as basis vectors of the coordinate system, so that $\pi_{ii} = 1$ for $i = 1, 2, \ldots, n$ and all other π_{ij} are equal to zero, so that the system of equations

$$\xi_1 = \mu_1, \qquad \xi_2 = \mu_2, \qquad \cdots, \qquad \xi_n = \mu_n \qquad (7)$$

results. In this coordinate system the barycentric coordinates $\mu_1, \mu_2, \ldots, \mu_n$ coincide with the parallel coordinates. The points in $\mathfrak{E}^n$ satisfy the inequalities

$$\xi_1 \geq 0, \qquad \xi_2 \geq 0, \qquad \cdots, \qquad \xi_n \geq 0,$$

$$\xi_1 + \xi_2 + \cdots + \xi_n \leq 1$$

in this case.

The simplex, taken as a subset of the Euclidean space, is itself a neighborhood space. An ε-neighborhood of a point of the simplex consists of those points which are common to the simplex and to an ε-ball described about that point.

The 0-simplex is a single point, the 1-simplex is a line segment, the 2-simplex is a triangle, and the 3-simplex is a tetrahedron.

Those points whose nth barycentric coordinate is equal to zero, that is $\mu_n = 0$, constitute an $(n-1)$-simplex, the $(n-1)$-dimensional face $\mathfrak{E}_n^{n-1}$ of $\mathfrak{E}^n$, which lies opposite to the point P_n. This is so because formulas (2) to (4) hold for these point sets when one replaces the index n by $n-1$. In the same way, the face $\mathfrak{E}_i^{n-1}$ lies opposite to the vertex P_i.

A *k-dimensional face* of the simplex $\mathfrak{E}^n$ is a point set for which $n-k$ barycentric coordinates are equal to zero and only the remaining $k+1$ coordinates vary subject to (2) and (3). A k-dimensional face is also a simplex $\mathfrak{E}^k$. In particular, the 1-dimensional face simplexes are called the *edges* of $\mathfrak{E}^n$. The vertices of a k-dimensional face are also vertices of $\mathfrak{E}^n$ and, conversely, each arbitrary set of $k+1$ vertices of $\mathfrak{E}^n$ spans a k-dimensional face. There then exist $\binom{m+1}{k+1}$ k-dimensional faces of $\mathfrak{E}^n$.

The simplex $\mathfrak{S}^n$ consists of the totality of line segments which connect points of the face $\mathfrak{S}_n^{n-1}$ with the opposite vertex P_n. This is so because one can find the centroid of $n + 1$ mass points by first finding the centroid of the masses $\mu_0, \mu_1, \ldots, \mu_{n-1}$ (this is a point of $\mathfrak{S}_n^{n-1}$) and then finding the centroid of a mass $\mu_0 + \mu_1 + \cdots + \mu_{n-1}$ concentrated at this point and the mass μ_n at P_n. The desired centroid lies on the line segment connecting the latter two points. Each point of the connecting segment is a point of $\mathfrak{S}^n$ since one can let the mass μ_n decrease from 1 to 0 without changing the ratios of the remaining n masses; during this decrease the centroid of all the masses moves along the line segment connecting P_n to the point on the opposite face determined by the fixed mass ratio.

When one connects all points of a point set $\mathfrak{M}$ of a Euclidean space with a fixed point P by means of straight line segments, then one calls this procedure *projection to the point set $\mathfrak{M}$ from the projection center P*. The totality of connecting line segments is called the *projection cone*. The *n*-simplex then arises by means of projection to one of its $(n - 1)$-dimensional faces from the vertex opposite to that face. One can generate the *n*-simplex $\mathfrak{S}^n$ by means of successive projections, starting with an 0-simplex. One first obtains an edge by projection to a vertex P_0 from another vertex, for example P_1. By projection to that edge from a new vertex P_2 one obtains a 2-simplex, and so forth until $\mathfrak{S}^n$ is finally obtained by means of a projection to an $(n - 1)$-dimensional face from the last vertex P_n not previously used as a projection center.

Convexity is an important property of an *n*-simplex. A *convex region* of a Euclidean space is defined to be a closed bounded set such that for each pair of points of the set, all points of the line segment connecting them are also contained in the set. The *dimension* of a convex region is defined to be the dimension of the lowest dimensional linear space, $\mathfrak{L}^r$ which contains the convex region. The convex region possesses interior points relative to $\mathfrak{L}^r$ since it contains $r + 1$ linearly independent points and thereby contains a whole *r*-simplex which, in turn, contains interior points relative to $\mathfrak{L}^r$. We call the *set of boundary points relative to $\mathfrak{L}^r$* the *boundary of the convex region* and call the *interior points relative to $\mathfrak{L}^r$* the *inner points of the convex region*. Later, in Chapter V, we shall be able to show that boundary and inner points of a convex region differ from one another with regard to intrinsic topological properties, independent of any embedding into a Euclidean space. One should take care not to confuse inner points with interior points of a convex region relative to an arbitrary Euclidean embedding space. Inner points are interior points only if the dimension of the convex region is the same as that of the embedding space. Otherwise, each point of the convex region is a boundary point of the region relative to the embedding space. A 0-dimensional convex region consists of a single inner point.

The projection cone to a convex region $\mathfrak{B}$ from a point is again a convex region. Two arbitrary points Q_1 and Q_2 of the projection cone can always be projected into certain image points in $\mathfrak{B}$ (not necessarily uniquely determined); the latter points can be connected by a line segment belonging entirely to $\mathfrak{B}$, and the projection cone to this segment will contain the line segment $(Q_1 Q_2)$.

Because the n-simplex $\mathfrak{E}^n$ arises by means of a series of projections from a point P_0, which is a 0-dimensional convex region, it follows that $\mathfrak{E}^n$ is itself a convex region and, in fact, is n-dimensional. It is, moreover, the smallest convex region which contains the $n + 1$ vertices of $\mathfrak{E}^n$. For this reason one says that it is the *convex hull* of the $n + 1$ points.

As a convex region the n-simplex has a definite boundary, which obviously consists of all of its faces. The remaining points are inner points. A 0-simplex has no boundary; it consists of a single inner point.

Every simplex has a definite *midpoint*. This is the point having barycentric coordinates

$$\mu_0 = \mu_1 = \cdots = \mu_n \quad (= 1/(n + 1)).$$

A *linear* (affine) *mapping* of an n-simplex $\mathfrak{E}^n$ onto an r-simplex $'\mathfrak{E}^r$ can be obtained by mapping the vertices of $\mathfrak{E}^n$ to the vertices of $'\mathfrak{E}^r$. That is, one assigns a vertex of $'\mathfrak{E}^r$ to each vertex of $\mathfrak{E}^n$ and omits no vertex of $'\mathfrak{E}^r$ in the procedure; the masses of $\mathfrak{E}^n$ are placed at the image vertices and the centroid of $\mathfrak{E}^n$ is assigned to the image centroid. Each point of $\mathfrak{E}^n$ is thereby given a uniquely determined image point in $'\mathfrak{E}^r$. This linear mapping is determined uniquely by the mapping of the vertices.

If $r < n$, the mapping of $\mathfrak{E}^n$ onto $'\mathfrak{E}^r$ is said to be *degenerate*. In this case at least two vertices of $\mathfrak{E}^n$ will collapse into some vertex of $'\mathfrak{E}^r$. On the other hand, if $n = r$, the mapping is topological. Each face of $\mathfrak{E}^n$ is mapped linearly onto the face of $'\mathfrak{E}^r$ which is "spanned" by the images of its vertices. Therefore if two vertices collapse into one vertex, the whole edge spanned by them will collapse into this vertex.

To describe the linear mapping of $\mathfrak{E}^n$ onto $'\mathfrak{E}^r$ analytically, choose parallel coordinate systems $\xi_1, \xi_2, \ldots, \xi_n$ and $'\xi_1, '\xi_2, \ldots, '\xi_r$ respectively, in the linear spaces $\mathfrak{L}^n$ and $'\mathfrak{L}^r$ to which $\mathfrak{E}^n$ and $'\mathfrak{E}^r$ belong. Let

$$\pi_{ij} \quad (i = 0, 1, \ldots, n, j = 1, 2, \ldots, n)$$

be the coordinates of the vertices $P_0, P_1, \ldots, P_n$ of $\mathfrak{E}^n$ and let

$$'\pi_{ik} \quad (i = 0, 1, \ldots, n, k = 1, 2, \ldots, r)$$

be the coordinates of the image points $'P_0, 'P_1, \ldots, 'P_n$; several of these points may, of course, coincide. The coordinates of the centroid of the masses $\mu_0, \mu_1, \ldots, \mu_n$ placed at $P_0, P_1, \ldots, P_n$ are, from Eq. (4),

$$\xi_1 = \sum_{i=0}^{n} \mu_i \pi_{i1}, \quad \xi_2 = \sum_{i=0}^{n} \mu_i \pi_{i2}, \quad \ldots, \quad \xi_n = \sum_{i=0}^{n} \mu_i \pi_{in} \tag{8}$$

and the centroid of the masses $\mu_0, \mu_1, \ldots, \mu_n$ located at the image points $'P_0, 'P_1, \ldots, 'P_n$ correspondingly has coordinates

$$'\xi_1 = \sum_{i=0}^{n} \mu_i \,'\pi_{i1}, \quad '\xi_2 = \sum_{i=0}^{n} \mu_i \,'\pi_{i2}, \quad \ldots, \quad '\xi_r = \sum_{i=0}^{n} \mu_i \,'\pi_{ir}. \tag{9}$$

If one solves the system of equations (8) taken together with the equation

$$\mu_0 + \mu_1 + \cdots + \mu_n = 1 \tag{3}$$

relating the masses $\mu_0, \mu_1, \ldots, \mu_n$ (which is always possible because the points $P_0, P_1, \ldots, P_n$ are independent) and sets their values into (9), one gets a system of linear equations having the form

$$'\xi_1 = \sum_{j=1}^{n} \alpha_{1j}\xi_j, \quad '\xi_2 = \sum_{j=1}^{n} \alpha_{2j}\xi_j, \quad \ldots, \quad '\xi_r = \sum_{j=1}^{n} \alpha_{rj}\xi_j.$$

The linear mapping of $\mathfrak{E}^n$ onto $'\mathfrak{E}^r$ is, therefore, given by a uniquely determined linear mapping of $\mathfrak{L}^n$ onto $'\mathfrak{L}^r$.

A simplex becomes oriented when one specifies a particular ordering of its vertices. Two orderings of the vertices which agree up to an even permutation* of the vertices determine the same orientation. An oriented simplex E^n is given by means of the ordering of its vertices $P_0, P_1, \ldots, P_n$. We denote it by

$$E^n = +(P_0 P_1 \cdots P_n).$$

Upon transposing P_0 and P_1 the orientation of the simplex reverses and we denote the simplex having reverse orientation by

$$-E^n = -(P_0 P_1 \cdots P_n) = +(P_1 P_0 \cdots P_n).$$

We use a Latin letter to distinguish an oriented simplex from a nonoriented simplex. When we wish to denote a nonoriented simplex by its vertices we write

$$\mathfrak{E}^n = (P_0 P_1 \cdots P_n),$$

without a sign in front. Our definition of orientation almost suggests itself, because a linear mapping of the n-simplex onto itself which is determined by a permutation of its vertices will have a positive or negative determinant, respectively, according to whether the permutation is even or odd. To see this we note that each permutation of the $n + 1$ vertices can be obtained by means of a succession of transpositions of two vertices. A linear mapping which permutes only two vertices, let us say P_1 and P_2, has a negative determinant. We have only to choose P_0 as the initial point and choose the edges directed towards other vertices as basis vectors, and in this coordinate system the

---

*See, for example, van der Waerden [3, Vol. I, p. 24].

mapping will take the form

$$x_1' = x_2, \qquad x_2' = x_1, \qquad x_3' = x_3, \qquad \ldots, \qquad x_n' = x_n$$

and therefore will have determinant -1. The reasoning breaks down but the conclusion also remains true for the trivial case $n = 1$. Accordingly, for example, all even permutations of the regular tetrahedron are geometrically distinguishable from the odd permutations by the fact that they can be effected by rigid rotations about the tetrahedron midpoint.

To simplify matters formally we also orient the 0-simplexes. Here the orientation consists only of assigning a plus or minus sign to the simplex in question.

For the case $n = 1$ the orientation $+(P_0 P_1)$ determines a traversal of the segment for which P_0 serves as "initial vertex" and P_1 serves as "final vertex." An oriented 1-simplex can then be described as a line segment with an arrowhead upon it, directed from the initial point to the final point.

In the case $n = 2$, the ordering $+(P_0 P_1 P_2)$ and its even permutations $+(P_1 P_2 P_0)$ and $+(P_2 P_0 P_1)$ describe one and the same cyclic traversal of edges, thus a definite traversal sense is given to the triangle. The orientation of a triangle can therefore be indicated by a circular arrow drawn inside it.

In the case $n = 3$ the orientation $+(P_0 P_1 P_2 P_3)$ determines a screw sense. A screw which bores into the tetrahedron from the vertex P_0 and turns according to the ordering of the other three vertices will be a right-handed or left-handed screw according to the orientation of the tetrahedron.

Up to now we have considered simplexes of a Euclidean space and their linear mappings. If we now map such a simplex e^n topologically onto a neighborhood space $\mathfrak{M}$ (so that each point of $\mathfrak{M}$ becomes an image point), then we can also call the image $\mathfrak{E}^n$ a simplex, or more precisely a *topological simplex*, and we call e^n the *preimage* of $\mathfrak{E}^n$. If one regards the same neighborhood space $\mathfrak{M}$ as the image of another preimage, $\bar{e}^n$, *we consider the topological simplexes $\mathfrak{E}^n$ and $\overline{\mathfrak{E}}^n$ to be the same if and only if e^n can be mapped linearly onto $\bar{e}^n$ so that corresponding points have the same image point in $\mathfrak{M}$.*

The following concepts can then be carried over to a topological simplex from its preimage: edge, i-dimensional face, midpoint, straight line segment connecting two points, boundary, inner point. One cannot, however, speak of the distance between two points, since this changes during a linear mapping of two preimages onto one another. One can also consider the rectilinear simplexes of Euclidean spaces to be topological simplexes; here, preimage and image coincide and the mapping is the identity mapping. When one speaks simply of "a simplex" one refers always to a topological simplex, which of course can also be a rectilinear simplex in particular circumstances.

If, for example, we project the surface of a tetrahedron from the tetrahedron midpoint onto a circumscribed 2-sphere, then the four rectilinear 2-simplexes of the tetrahedron transform to four topological simplexes of the 2-sphere (spherical triangles).

Exactly as for a rectilinear simplex, one can orient a topological simplex by giving an ordering of its vertices and one can do this in two distinct ways.

It is also clear what is meant by *a linear mapping of two topological simplexes* $\mathfrak{E}^n$ and $'\mathfrak{E}^n$. This is a topological mapping of $\mathfrak{E}^n$ onto $'\mathfrak{E}^n$ which, when viewed as a mapping between the corresponding rectilinear simplexes e^n and $'e^n$ is a linear mapping of rectilinear simplexes.

10. Simplicial Complexes

Complexes are special neighborhood spaces which are constructed out of simplexes. In the future we will deal exclusively with such spaces. We shall study complexes in general in Chapters II–V, VII, VIII and XI and we shall study certain special complexes, i.e., manifolds, in Chapters VI, IX, and X. Complexes will be the basic concept in all investigations which follow.

A *complex* is a neighborhood space which can be simplicially decomposed. Simplicial decomposition of a neighborhood space $\mathfrak{R}$ is defined as follows: Let either a finite number or a countable infinity of simplexes (that is, topological images of rectilinear simplexes), having dimensions 0 through n, lie in $\mathfrak{R}$. Along with each simplex, all of its faces should belong to this set of simplexes. The simplexes provide a *simplicial decomposition* of the space $\mathfrak{R}$ if the following four conditions are fulfilled:

(k1) Each point belongs to at least one simplex.

(k2) Each point belongs to only finitely many simplexes.

(k3) Two simplexes are either disjoint, or one of them is a face of the other, or they have a common face consisting of the set theoretic intersection of the two simplexes.

[The definition of equality of topological simplexes is to be kept in mind here. In order for two topological 2-simplexes of $\mathfrak{R}$ to have a face in common it is not sufficient that their preimage triangles map to the same point set of $\mathfrak{R}$ under the topological mapping. Rather, the mapping of the preimage triangles into $\mathfrak{R}$ must be such that corresponding points of the two linear preimage faces (which are linearly related to one another and not merely related by an arbitrary topological relation) have the same image point in $\mathfrak{R}$.]

(k4) If P is a point of the neighborhood space $\mathfrak{R}$ and if $\mathfrak{U}(P \mid \mathfrak{E}_1)$, $\mathfrak{U}(P \mid \mathfrak{E}_2), \ldots, \mathfrak{U}(P \mid \mathfrak{E}_r)$ are neighborhoods of P in the simplexes $\mathfrak{E}_1$, $\mathfrak{E}_2, \ldots, \mathfrak{E}_r$, respectively, where these simplexes comprise all of the simplexes of the simplicial decomposition* which contain P, then the set theoretic union $\mathfrak{U}(P \mid \mathfrak{E}_1) + \mathfrak{U}(P \mid \mathfrak{E}_2) + \cdots + \mathfrak{U}(P \mid \mathfrak{E}_r)$ is also a neighborhood $\mathfrak{U}(P \mid \mathfrak{R})$ of P in $\mathfrak{R}$.

---

*The dimensions of the simplexes $\mathfrak{E}_1, \mathfrak{E}_2, \ldots, \mathfrak{E}_r$ lie between 0 and n. We have not indicated the dimension index, which is always superscripted.

If the neighborhood space $\Re$ can be divided into simplexes in this way, then it is called a *complex*. If we select, out of the various possible ways to decompose a complex into simplexes, one such way, then we call the complex divided into simplexes in this particular way a *simplicial complex* or, more precisely, a simplicially decomposed complex or a complex with a simplicial decomposition. It is customary to designate two simplexes of $\Re$ as being *incident* if one of them is a face of the other. A simplicial complex is said to be *n-dimensional* if it contains at least one *n*-simplex but none of higher dimension.

We shall see in the course of the development that practically all topologically important figures are complexes. This is indicated by the following lemma:

LEMMA. *A point set $\Re$ of a Euclidean space $\Re^m$ which consists of a finite or countably infinite number of geometric simplexes of dimensions 0 through n is a simplicial complex if, in addition to condition* (k3) (which states that the simplexes do not overlap or penetrate one another) *the following condition is fulfilled: each point of $\Re$ has a neighborhood which has points in common with only a finite number of simplexes.*

Proof. As a subset of the Euclidean space, the point set $\Re$ is a neighborhood space. The given simplexes and their faces will form a simplicial decomposition of $\Re$. Since the conditions (k1) and (k2) are obviously satisfied and condition (k3) is assumed to hold, it remains only to prove (k4). The neighborhood $\mathfrak{U}(P \mid \mathfrak{E}_i)$ mentioned in condition (k4) includes a certain ε_i-neighborhood of P relative to $\mathfrak{E}_i$; it is formed by all points of $\mathfrak{E}_i$ which have distance less than ε_i from P in $\Re^m$. If one chooses η to be a positive number smaller than $\varepsilon_1, \varepsilon_2, \ldots, \varepsilon_r$ and also so small that the η-neighborhood of P has points in common only with the simplexes $\mathfrak{E}_1, \mathfrak{E}_2, \ldots, \mathfrak{E}_r$ (which is always possible according to the conditions of the lemma), then this η-neighborhood of P relative to $\Re$ is a subset of $\mathfrak{U}(P \mid \mathfrak{E}_1) + \mathfrak{U}(P \mid \mathfrak{E}_2) + \cdots + \mathfrak{U}(P \mid \mathfrak{E}_r)$ and therefore this set union is also a neighborhood of P relative to $\Re$ and thus satisfies condition (k4).

EXAMPLE 1. The 2-sphere is a complex. From the last lemma, the tetrahedron with plane faces in Euclidean 3-space is a simplicial complex. It can be mapped topologically onto a 2-sphere circumscribed about its midpoint, by means of a projection from its midpoint. The topological image gives a simplicial decomposition of the 2-sphere. One can get another simplicial decomposition of the 2-sphere from the octahedron or icosahedron. For this see §14.

EXAMPLE 2. The Euclidean plane is a complex, since one can tile it with equilateral triangles.

EXAMPLE 3. An individual triangle of the Euclidean plane is also a complex, of course, but its decomposition into infinitely many triangles as shown in Fig. 32, for example, is not a simplicial decomposition. Condition (k4) is violated at the vertex point P. Here, the set union $\mathfrak{U}(P \mid \mathfrak{E}_1) + \mathfrak{U}(P \mid \mathfrak{E}_2) + \cdots + \mathfrak{U}(P \mid \mathfrak{E}_r)$ is not a neighborhood $\mathfrak{U}(P \mid \Re)$. It consists only of the point P. Aside from the simplex P there exists no simplex of the decomposition to which P belongs.

Additional examples may be found in §14.

For use later we indicate how one can construct a simplicial complex from its individual simplexes by means of identifications. Let us assume that a preimage is assigned to each simplex of the decomposition, the preimages being located in a Euclidean space of sufficiently high dimension. These preimages may all be disjoint from one another. Taken together they form a neighborhood space $\mathfrak{M}$ which is mapped continuously onto $\mathfrak{K}$, since each individual simplex is mapped topologically onto a subset of $\mathfrak{K}$. If one now declares that all points having the same image point in $\mathfrak{K}$ are equivalent and identifies them, then one obtains exactly the simplicial complex $\mathfrak{K}$. For if one selects arbitrary neighborhoods in $\mathfrak{M}$ of those preimage points which map to a given point in $\mathfrak{K}$, then, from (k4), the image of the set theoretic union of these neighborhoods is a neighborhood of the image point. Thus, by Theorem II of §8, $\mathfrak{K}$ arises from $\mathfrak{M}$ by means of identification of equivalent points.

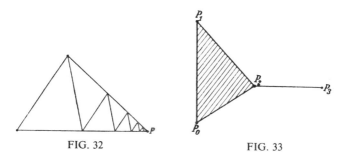

FIG. 32   FIG. 33

11. The Schema of a Simplicial Complex

Two simplexes of the simplicial decomposition of a complex never have all vertices identical, because of (k3). Because of this property one can describe a simplicial complex by means of its vertices and a directory listing the simplexes which they span. Such a list is called a *schema* of the complex. The schema of the complex illustrated in Fig. 33, consisting of a triangle and a line segment, then reads

$$(P_0 P_1 P_2); (P_0 P_1); (P_1 P_2); (P_2 P_0); (P_2 P_3); (P_0); (P_1); (P_2); (P_3).$$

Another type of scheme, the incidence matrices, will be used later to describe complexes.

If two simplicial complexes $\mathfrak{K}$ and $\mathfrak{K}'$ have the same schema, then they are homeomorphic. More precisely, they can be mapped topologically onto one another so that the simplexes of one map linearly to those of the other. In order to effect this topological mapping we notice first that a point P of $\mathfrak{K}$ is an inner point of exactly one simplex of the simplicial decomposition. For by (k1) P belongs to at least one simplex and is therefore an inner point of this simplex or of one of its faces. From (k3) there cannot exist two simplexes such that P is an inner point of each simplex. If $\mathfrak{E}$ is the simplex of $\mathfrak{K}$ that has

P as an inner point, then one looks for the vertices of $\mathfrak{K}'$ which correspond to the vertices of $\mathfrak{E}$. These are the vertices of a simplex $\mathfrak{E}'$ which is mapped linearly onto $\mathfrak{E}$ by means of the assignment of vertices. The image point P' of P is the image of P under the desired mapping of $\mathfrak{K}$ onto $\mathfrak{K}'$. We have now produced a one-to-one mapping of $\mathfrak{K}$ onto $\mathfrak{K}'$. The mapping is also topological. For if $\mathfrak{U}(P' \mid \mathfrak{K}')$ is a neighborhood of P' in $\mathfrak{K}'$ and $\mathfrak{E}'_1, \mathfrak{E}'_2, \ldots, \mathfrak{E}'_r$ are the simplexes to which P' belongs, then the intersection of $\mathfrak{U}(P \mid \mathfrak{K}')$ with the simplexes $\mathfrak{E}'_i$ $(i = 1, 2, \ldots,$ or $r)$ is a neighborhood $\mathfrak{U}(P' \mid \mathfrak{E}'_i)$ of P' in the simplex $\mathfrak{E}'_i$, according to the definition of neighborhoods in a subset of a neighborhood space. Since the simplex $\mathfrak{E}_i$ has now been mapped topologically onto $\mathfrak{E}'_i$ by means of the linear mapping which was constructed, then the neighborhood $\mathfrak{U}(P' \mid \mathfrak{E}'_i)$ corresponds to a neighborhood $\mathfrak{U}(P \mid \mathfrak{E}_i)$ in $\mathfrak{E}_i$. By (k4) the set union $\mathfrak{U}(P \mid \mathfrak{E}_1) + \mathfrak{U}(P \mid \mathfrak{E}_2) + \cdots + \mathfrak{U}(P \mid \mathfrak{E}_r)$ is then a neighborhood of P in $\mathfrak{K}$. Its image in $\mathfrak{K}'$ belongs to $\mathfrak{U}(P' \mid \mathfrak{K}')$. Thus the mapping is continuous in the direction $\mathfrak{K} \to \mathfrak{K}'$ and since $\mathfrak{K}$ was not distinguished *a priori* from $\mathfrak{K}'$, the mapping is also continuous in the reverse direction and is, hence, topological.

We will now discuss the question of when a set $\mathfrak{W}$ having either a finite number or a countable infinity of elements, in which certain "distinguished" subsets are defined, can be the schema of an n-dimensional simplicial complex, in the sense that the elements of $\mathfrak{W}$ are in one-to-one correspondence with the vertices of the complex and the distinguished subsets are in one-to-one correspondence with the sets of vertices of simplexes. The following three conditions are clearly necessary:

(Sch1) Each subset of a distinguished subset is again a distinguished subset, since all face simplexes of a simplex of a simplicial complex also belong to the complex.

(Sch2) The number of elements in a distinguished subset is less than or equal to $n + 1$.

(Sch3) Each element occurs in only a finite number of distinguished subsets, because of (k2).

To show that these are also sufficient conditions we will construct the corresponding geometric complex in the $(2n + 1)$-dimensional Euclidean space $\mathfrak{R}^{2n+1}$. With this goal in mind we establish a one-to-one relation between the elements of $\mathfrak{W}$ and a set of points in $\mathfrak{R}^{2n+1}$ having the same cardinality as $\mathfrak{W}$. These points must satisfy two conditions:

(1) They have no point of accumulation.

(2) Each set of $k + 1$ points span a k-dimensional linear space, for $k \leq 2n + 1$; that is, the smallest linear space to which they belong is k-dimensional.

These conditions can always be fulfilled, for if one chooses the first r points $P_1, P_2, \ldots, P_r$ so that condition (2) is satisfied, then one can choose the point P_{r+1} so that its x_1-coordinate is at least 1 greater than any of the

x_1-coordinates of $P_1, P_2, \ldots, P_r$ and in addition so that P_{r+1} belongs to none of the linear spaces which are spanned by each k of the points $P_1, P_2, \ldots, P_r$ ($k \leqq 2n + 1$). The latter condition can be fulfilled because the finitely many linear spaces to be avoided have at most the dimension $2n$. In that case (2) is also satisfied for the points $P_1, P_2, \ldots, P_{r+1}$. It then follows that (2) holds for the entire sequence. Condition (1) holds because the x_1-coordinates increase without bound in case the point set is infinite.

We now let the p-simplex spanned by associated points of the sequence correspond to a distinguished subset of $p + 1$ elements in $\mathfrak{W}$. We thus obtain a finite or infinite set $\mathfrak{K}$ of geometric simplexes of dimensions 0 through n in $\mathfrak{R}^{2n+1}$. From the lemma of §10 it easily follows that $\mathfrak{K}$ is a complex. Let us consider two simplexes $\mathfrak{E}^p$ and $\mathfrak{E}^q$ of $\mathfrak{K}$. If $k + 1$ is the number of vertices which belong to at least one of the two simplexes $\mathfrak{E}^p$ and $\mathfrak{E}^q$, then $k + 1 \leqq 2n + 2$ because $p \leqq n$ and $q \leqq n$. As a consequence of (2) these $k + 1$ vertices span a k-simplex $\mathfrak{E}^k$, of which $\mathfrak{E}^p$ and $\mathfrak{E}^q$ are faces. Therefore $\mathfrak{E}^p$ and $\mathfrak{E}^q$ are either disjoint or have a common face as their intersection. We must also show that there exists a neighborhood of each point Q of $\mathfrak{K}$ such that this neighborhood has points in common with only finitely many simplexes. We circumscribe a ball of arbitrary radius $\varepsilon > 0$ about Q. Let $\bar{x}_1$ be the x_1-coordinate of Q. In $\mathfrak{K}$ there exist only finitely many vertices having x_1-coordinate smaller than $\bar{x}_1 + \varepsilon$. From (Sch3) there are only finitely many simplexes of $\mathfrak{K}$ incident with these vertices. Only these simplexes can protrude into the ε-ball about Q. Thus, from the lemma, $\mathfrak{K}$ is a complex.

It follows immediately from our construction of $\mathfrak{K}$ that $\mathfrak{K}$ has exactly the schema of $\mathfrak{W}$. We have, at the same time, proved the theorem:

One can embed each n-dimensional complex geometrically in a $(2n + 1)$-dimensional Euclidean space; that is, for each n-dimensional complex there is a geometric complex in $\mathfrak{R}^{2n+1}$ having the same schema.

The schema of a simplicial complex gives a better overall view of the structure of possible simplicial complexes than the original, more complicated definition. It also opens up to us the possibility of another approach to the study of topology, the strictly combinatorial method. We will not pursue the strictly combinatorial method in this book. In that approach the schema is not merely an aid to the description of the complex. On the contrary, the schema itself defines the simplicial complex. Other simplicial complexes which are related to the original simplicial complex can be obtained from it by means of certain combinatorial alterations (combinatorial subdivision for example). Simplicial mappings* then take the place of homeomorphism. The detailed development of this way of thinking will be illustrated by an example that we will use occasionally to study surface topology (§37). In strict combinatorial topology the individual simplex is a distinguished set of finitely many vertices. These are not bound together by "continuous space sauce." The (at most countably many) vertices of the schemas and their distinguished subsets are the objects under study and simplicial mappings are the relations,

---

* *Editor's note*: The concept of "elementary relatedness" appears in the German original. We have translated this very freely here because we felt that usage has changed sufficiently to render the original text meaningless to a reader in 1978. Later, in §37, the concept of "elementary relatedness" will appear again; however, there it will be defined with some care, and we shall adhere to the original text.

whereas for us the (continuously many) points and their neighborhoods are the objects and continuous mappings are the relations.[10]

12. Finite, Pure, Homogeneous Complexes

A simplicial complex $\mathfrak{K}$ is said to be *finite* or *infinite*, respectively, according to whether it is built of finitely many or infinitely many simplexes. In a finite complex each infinite point set has at least one accumulation point; since only finitely many simplexes are available there must exist a simplex in which infinitely many points of the set lie. On the other hand, if $\mathfrak{K}$ is infinite then the set of midpoints of all the simplexes gives an example of an infinite set without an accumulation point. For if P is an arbitrary point of $\mathfrak{K}$, then the set theoretic union of the simplexes $\mathfrak{E}_1, \mathfrak{E}_2, \ldots, \mathfrak{E}_r$ to which P belongs is a neighborhood of P, from (k4), and this neighborhood contains only finitely many points of that set, namely the midpoints of the simplexes $\mathfrak{E}_1, \mathfrak{E}_2, \ldots, \mathfrak{E}_r$ and their faces. Thus the concepts of "finite" and "infinite" are recognized to be topological invariants.* *A finite simplicial complex cannot be homeomorphic to an infinite simplicial complex.* If a complex admits a simplicial decomposition into finitely many simplexes, then it has no simplicial decomposition into infinitely many simplexes (cf. Example 3 of §10). We are thus justified in speaking simply of finite and infinite complexes, omitting the word "simplicial." As an example, the 2-sphere is a finite complex; when one point is removed it becomes an infinite complex, the Euclidean plane, as we saw in §6.

If one selects an arbitrary nonempty set of simplexes in a simplicial complex, then these simplexes together with their faces form a *subcomplex* $\mathfrak{K}_1$ of $\mathfrak{K}$, which is obviously again a simplicial complex.

A subcomplex $\mathfrak{K}_1$ is said to be *isolated* (in $\mathfrak{K}$) if each simplex of $\mathfrak{K}$ which has a simplex of $\mathfrak{K}_1$ as a face likewise belongs to $\mathfrak{K}_1$. $\mathfrak{K}$ can be decomposed in a unique way into a definite maximal number τ (which may be ∞), of isolated disjoint subcomplexes. If $\tau = 1$, then $\mathfrak{K}$ is said to be *connected*. A connected simplicial complex is obviously also characterized by the fact that one can join any two vertices, let us say P and Q, by an edge path; that is, by a sequence of oriented edges, the first having P as its initial point and the last having Q as its final point, also the final point of an arbitrary edge coinciding with the initial point of the succeeding edge.

One can characterize an isolated subcomplex in a topologically invariant manner as a nonempty subset of $\mathfrak{K}$ which is simultaneously open and closed or, what is the same, as a nonempty subset without boundary.

Proof. It is clear that an isolated subcomplex is a nonempty subset without boundary. Conversely, let $\mathfrak{N}$ be an arbitrary nonempty subset of $\mathfrak{K}$ without

---

* *Editor's note*: What has just been observed is that a finite simplicial complex is compact while an infinite simplicial complex is not compact.

boundary. If P is a point of $\mathfrak{N}$ and $\mathfrak{E}^i$ is a simplex of the simplicial decomposition of $\mathfrak{R}$ such that P lies in $\mathfrak{E}^i$, then all of $\mathfrak{E}^i$ is contained in $\mathfrak{N}$. For suppose that P' is a point of $\mathfrak{E}^i$ which does not belong to $\mathfrak{N}$. We then parameterize the line segment connecting P to P' (linearly) with a parameter s varying from 0 to 1 and form the least upper bound $\bar{s}$ of the parameter values of all points belonging to $\mathfrak{N}$. In each neighborhood of the point $\bar{P}$ determined by the parameter value $\bar{s}$ there are points which do not belong to $\mathfrak{N}$ as well as points which belong to $\mathfrak{N}$. That is, $\bar{P}$ is a boundary point of $\mathfrak{N}$, in contradiction to our assumption. Thus $\mathfrak{N}$ is a subcomplex $\mathfrak{R}_1$ of $\mathfrak{R}$ which contains along with each point P all of the simplexes of $\mathfrak{R}$ on which P lies. Thus $\mathfrak{R}_1$ is also isolated.

Because of this topological characterization of an isolated subcomplex there also follows *the topological invariance of the concept "connected" and of the integer τ.*

An n-dimensional simplicial complex is said to be *pure* if each k-simplex ($k < n$) is a face of at least one n-dimensional simplex. Otherwise it is said to be *impure*. By the *boundary* of a pure n-dimensional simplex we mean the totality of $(n-1)$-simplexes which are incident with an odd number of n-simplexes.

As an example, suppose that a pure complex consists of four plane triangles of Euclidean 3-space having one edge in common, and that these triangles are located like the pages of an open book, spreading out fanwise from the book's spine. The boundary is then formed by all of the triangle edges with the exception of the common edge. The present definition of boundary is in agreement with the definition of the boundary of a simplex given earlier in §9, since an individual simplex is a pure complex. The boundary of a convex region, defined in §9, as well as that of the closed n-ball, defined in §14, will also be included in the present definition as soon as we have shown that the convex region and the closed n-ball are pure complexes (§14).

In Chapter V we shall develop methods to prove that the concepts "n-dimensional, pure, boundary" are topological invariants. These later results are mentioned to justify the fact that we occasionally refer to a complex $\mathfrak{R}$ by using the notation $\mathfrak{R}^n$, indicating its dimension with a superscript. At present we note only that the dimension 0 is a topological invariant. A 0-dimensional complex consists exclusively of isolated points. Each point is its own neighborhood. This is not true in higher dimensions.

Among the complexes, the *homogeneous complexes* deserve special mention. An n-dimensional complex is said to be *homogeneous* if to each of its points there corresponds a neighborhood of that point which can be mapped topologically onto the interior of an n-dimensional ball. The interior of the n-ball is formed by those points of Euclidean n-space whose coordinates satisfy the inequality $x_1^2 + x_2^2 + \cdots + x_n^2 < 1$. As an example, the 2-sphere is a homogeneous 2-dimensional complex. In contrast, the closed disk is a nonhomogeneous complex because the boundary points do not satisfy the homogeneity condition. This will be shown in Chapter V. Likewise, the edge complex of the tetrahedron is a nonhomogeneous 1-dimensional complex.

Although the question of whether a particular complex is homogeneous is easily determined intuitively in low dimensions (for example, Euclidean 3-space is clearly a homogeneous complex), it is not so accessible to mathematical treatment in more than three dimensions (§68). Therefore we do not place it at the center of our investigations.

13. Normal Subdivision

Our ultimate goal, to read off properties of the complex which are independent of the choice of decomposition from a given simplicial decomposition of a complex, cannot be reached by a procedure of classifying the various possible decompositions of the complex. The multiplicity of decompositions makes it difficult to obtain an overall view. To approach our goal it will be necessary, however, to learn how to derive further decompositions of a complex, starting from a given simplicial decomposition. It is only after we have obtained a sufficiently fine decomposition that we will be able to compare it to another arbitrary decomposition or to the complex considered merely as a point set (Chapter V).

Subdividing a simplicial complex consists of decomposing each simplex into smaller subsimplexes in such a way that a simplicial complex again results. Later, we shall need only one particular type of subdivision, the so-called *normal subdivision*. It is obtained in the following way. One first subdivides each edge of the simplicial complex through its midpoint, decomposing the edge into two 1-simplexes. This provides a subdivision of the boundary 1-simplexes of each 2-simplex of the simplicial complex. Project these subdivided boundary 1-simplexes linearly from the midpoint of each 2-simplex. We have already defined the midpoint of a topological simplex by use of rectilinear preimage simplexes (§9); the projection of our subdivided boundary edge is first defined in the preimage simplex and subsequently carried over to the topological simplex. Each 2-simplex then decomposes into six 2-simplexes, because the projection cone of a 1-simplex is a 2-simplex. Proceeding further, one projects the subdivided 2-simplex boundary faces from the midpoint of each 3-simplex. From each 3-simplex one obtains a set of 4! 3-simplexes and, additionally, new 2-simplexes and 1-simplexes and also a new 0-simplex, the projection center. One proceeds in this way up to the *n*-simplexes. Figure 34 shows a normally subdivided 2-simplex.

The normal subdivision of a simplicial complex $\mathfrak{K}$ is again a simplicial complex. This follows from the lemma of §10 if one embeds $\mathfrak{K}$ in a Euclidean space and performs the subdivision in that Euclidean space.

FIG. 34

Since the normal subdivision is known to be a simplicial complex, one can again form its normal subdivision. Proceeding in this way, one can arrive at a g-fold normal subdivision. By choosing g sufficiently large, one can make the subsimplexes arbitrarily small. More precisely, we have the following lemma.

LEMMA. *A subsimplex of the g-fold normal subdivision of a geometric n-simplex $\mathfrak{E}^n$ will be (properly) contained in a simplex similarly positioned to $\mathfrak{E}^n$ and diminished in size by a factor $[n/(n+1)]^g$.*

Proof. We will need only to prove the particular case $g = 1$ since the lemma will follow directly by a g-fold application of this special case. If $\mathbf{E}^n$ is the subsimplex of $\mathfrak{E}^n = (P_0 P_1 \cdots P_n)$ under consideration, then there is no loss of generality if one assumes that its vertices $M_0, M_1, \ldots, M_n$ are just the midpoints of the faces $(P_0), (P_0 P_1), \ldots, (P_0 P_1 \cdots P_n)$. The barycentric coordinates of M_i are (§9)

$$\mu_0 = \mu_1 = \cdots = \mu_i = 1/(i+1), \qquad \mu_{i+1} = \mu_{i+2} = \cdots = \mu_n = 0. \quad (1)$$

In the parallel coordinate system $\xi_1, \xi_2, \ldots, \xi_n$ whose basis vectors are directed from P_0 to the vertices $P_1, P_2, \ldots, P_n$ the coordinates $\xi_1, \xi_2, \ldots, \xi_n$ coincide, respectively, in that order, with the barycentric coordinates $\mu_1, \mu_2, \ldots, \mu_n$ (§9). $\mathfrak{E}^n$ is determined by means of the inequalities

$$\xi_1 \geq 0, \quad \xi_2 \geq 0, \quad \ldots, \quad \xi_n \geq 0,$$
$$\xi_1 + \xi_2 + \cdots + \xi_n \leq 1. \quad (2)$$

Let us look at the simplex $'\mathfrak{E}^n$ which is positioned similarly to $\mathfrak{E}^n$ and reduced in size by a factor $n/(n+1)$:

$$\xi_1 \geq 0, \quad \xi_2 \geq 0, \quad \ldots, \quad \xi_n \geq 0,$$
$$\xi_1 + \xi_2 + \cdots + \xi_n \leq n/(n+1). \quad (3)$$

Since the coordinates (1) of M_i ($i = 0, 1, \ldots, n$) satisfy the inequalities (3), then the vertices of $\mathbf{E}^n$ belong to $'\mathfrak{E}^n$ and, because of its convexity, $\mathbf{E}^n$ itself belongs to $'\mathfrak{E}^n$.

Problem

Let a neighborhood space be divided into simplexes so that (k1), (k2) and (k4) are satisfied and two simplexes are either disjoint, identical or possess one or more common faces. Show that one obtains a simplicial complex by normally subdividing all simplexes. (Prove by induction over the dimension of the complex that (k3) of §10 also holds for the normal subdivision; that is, if two simplexes are not disjoint or identical, they can only intersect along a common face.*

14. Examples of Complexes

The concept of a complex is of great generality. To indicate the breadth of the concept we shall prove that several important and well-known neighborhood spaces are complexes. This will also provide material for illustrative examples later in the text.

The Closed n-Ball

The set of points of Euclidean n-space $\mathfrak{R}^n$ whose coordinates satisfy the inequality

$$x_1^2 + x_2^2 + \cdots + x_n^2 \leqq 1 \qquad (n \geqq 1) \tag{1}$$

is called a *standard unit ball* of the Euclidean space. A topological image of the standard unit ball is called a *closed n-ball*.** In the case $n = 2$ it is also called a *closed disk*. The points for which the equality sign is valid in Eq. (1) form the boundary of the standard unit ball and (after a subsequent topological mapping) *the boundary of the closed n-ball* (cf. §9). Topological invariance of the boundary will be established in Chapter V.

Each convex n-dimensional region $\mathfrak{B}$ in $\mathfrak{R}^n$ is a closed n-ball.

To prove this we circumscribe a standard ball $\mathfrak{k}$ about an inner point O of $\mathfrak{B}$ in such a way that the whole ball lies in $\mathfrak{B}$. Let $\bar{\mathfrak{k}}$ be the boundary of the ball (Fig. 35). We first show that a ray directed outwards from O intersects the boundary $\bar{\mathfrak{B}}$ of $\mathfrak{B}$ in exactly one point. Because $\mathfrak{B}$ is bounded, the ray intersects at least one boundary point of $\mathfrak{B}$. As the boundary of $\mathfrak{B}$ is a closed subset of $\mathfrak{R}^n$ (§7), then among those points of the ray which lie in $\bar{\mathfrak{B}}$ there will exist a point of maximum distance from O. Call this point $\bar{P}$. All points of $O\bar{P}$ with the exception of $\bar{P}$ are inner points of $\mathfrak{B}$. To see this we join $\bar{P}$ by line segments to all points of $\mathfrak{k}$. Since $\mathfrak{B}$ is convex, all of the connecting line segments lie in $\mathfrak{B}$. One can therefore construct a ball about each point of the segment $O\bar{P}$ such that the ball lies entirely in $\mathfrak{B}$. This is the ball located in similitude to $\mathfrak{k}$, with respect to $\bar{P}$ as center of projection (Fig. 35).

Let the intersection point of $O\bar{P}$ with $\bar{\mathfrak{k}}$ be called $\bar{p}$. A one-to-one mapping of $\bar{\mathfrak{B}}$ onto $\bar{\mathfrak{k}}$ is produced by making the correspondence $\bar{P} \longleftrightarrow \bar{p}$. This mapping is continuous in the direction $\bar{p} \to \bar{P}$. For if the mapping were discontinuous at the point $\bar{p}$, then for a given neighborhood $\mathfrak{U}(\bar{P} \mid \mathfrak{B})$ there would exist a

---

Editor's note: We have taken the liberty of altering the term "inzident" in the original German text to "identical," because the problem seemed unclear to us any other way.

**Editor's note*: The term *n-dimensional element* is used in the older German literature. The term *elementary surface patch* (Elementarflachenstuck) describes the case $n = 2$.

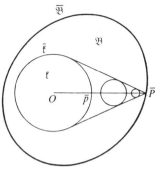

FIG. 35

sequence of points $\bar{p}_1, \bar{p}_2, \ldots$ of $\bar{\mathfrak{k}}$ converging to $\bar{p}$ such that all points of the image sequence $\bar{P}_1, \bar{P}_2, \ldots$ would lie outside $\mathfrak{U}(\bar{P} \mid \bar{\mathfrak{B}})$. This sequence of image points has an accumulation point on $\bar{\mathfrak{B}}$ which must lie on the ray $O\bar{P}$ and can therefore only be the point $\bar{P}$. This would imply that almost all points of $\bar{P}_1, \bar{P}_2, \ldots$ lie in $\mathfrak{U}(\bar{P} \mid \bar{\mathfrak{B}})$, which contradicts our assumption of discontinuity at $\bar{p}$.

The mapping is also continuous in the direction $\bar{P} \to \bar{p}$, from Theorem V, §8, and is therefore topological.

One can extend the mapping from $\bar{\mathfrak{B}}$ onto $\bar{\mathfrak{k}}$ to a mapping of the full region $\mathfrak{B}$ onto the closed ball $\mathfrak{k}$ by mapping the segment $O\bar{P}$ linearly onto the segment $O\bar{p}$.

Since the geometric n-simplex is a convex region of Euclidean n-space $\mathfrak{R}^n$, it can then be mapped topologically onto the closed n-ball. We have thus proved that *the closed n-ball is a pure complex*.

The *n*-Sphere $\mathfrak{S}^n$

The set of points of Euclidean $(n + 1)$-space $\mathfrak{R}^{n+1}$ whose coordinates satisfy the equation

$$x_1^2 + x_2^2 + \cdots + x_{n+1}^2 = 1 \qquad (n \geqq 0)$$

is called the n-dimensional unit sphere of $\mathfrak{R}^{n+1}$. A topological image of this set is called the *n-sphere or n-dimensional spherical space* and is denoted by $\mathfrak{S}^n$. The 0-sphere is a pair of points.

The boundary of an $(n + 1)$-dimensional convex region $\mathfrak{B}^{n+1}$ is an n-sphere, for we have just presented a topological mapping of $\mathfrak{B}^{n+1}$ onto the $(n + 1)$-dimensional standard ball such that the boundary of $\mathfrak{B}^{n+1}$ transforms to the boundary of the ball. It follows immediately that the n-sphere is homeomorphic to the boundary of an $(n + 1)$-simplex. This provides the simplest simplicial decomposition of the n-sphere. For the case $n = 2$ this decomposition is formed by the four triangles of a tetrahedron.

On occasion we shall require another simplicial decomposition of $\mathfrak{S}^n$

besides the tetrahedral decomposition. We will require the generalization of the octahedral subdivision of the 2-sphere to dimension n.

Let

$$\mathfrak{v}_1, \mathfrak{v}_2, \ldots, \mathfrak{v}_{n+1}$$

be unit vectors directed from the origin O of $\mathfrak{R}^{n+1}$ to the points

$$(1, 0, 0, \ldots, 0), (0, 1, 0, \ldots, 0), \ldots, (0, 0, 0, \ldots, 1).$$

The vectors

$$\varepsilon_1\mathfrak{v}_1, \varepsilon_2\mathfrak{v}_2, \ldots, \varepsilon_{n+1}\mathfrak{v}_{n+1} \qquad (\varepsilon_k = \pm 1) \tag{2}$$

"span" an $(n + 1)$-simplex, namely the $(n + 1)$-simplex whose vertices are 0 and the endpoints of the vectors (2). There exist 2^{n+1} sign combinations in (2) and, therefore, the same number of simplexes. Each pair of these simplexes either intersects in a face which is spanned by its common vectors or else has only the point 0 in common. In the latter case the vectors of one of the $(n + 1)$-simplexes are directed opposite to those of the other simplex. The 2^{n+1} simplexes form a simplicial complex, from the lemma of §10. As is easily seen, it is convex and is therefore a closed $(n + 1)$-ball. Its boundary, an n-sphere, is formed by 2^{n+1} n-simplexes whose vertices are the endpoints of the set of vectors (2). This is the "octahedral" decomposition of the n-sphere. It is center symmetric, in contrast to the tetrahedral decomposition, that is, it transforms to itself by the interchange of diametrically opposite points, that is, by the mapping

$$x_1' = -x_1, \qquad x_2' = -x_2, \qquad \ldots, \qquad x_{n+1}' = -x_{n+1}.$$

The n-sphere is decomposed into two n-hemispheres by the hyperplane $x_{n+1} = 0$. These n-hemispheres are closed n-balls, as one can recognize when one projects them parallel to the x_{n+1} axis into the hyperplane $x_{n+1} = 0$. The n-sphere thus arises from two closed n-balls by mapping their boundaries topologically onto one another and identifying corresponding points.[11]

Another way to generate the n-sphere is to identify those boundary points of the closed n-ball

$$x_1^2 + x_2^2 + \cdots + x_n^2 \leqq 1$$

which are located mirror symmetrically with respect to the equatorial hyperplane $x_n = 0$, for the equatorial hyperplane decomposes the closed n-ball into two n-balls whose boundary spheres are to be identified.

The *punctured n-sphere* is the n-sphere with one of its points removed. The punctured n-sphere is homeomorphic to Euclidean n-space, $\mathfrak{R}^n$. This can be seen by noting that the formulas of stereographic projection (§7) remain valid for arbitrary dimension n. It is therefore possible to artificially make the stereographic projection into a one-to-one correspondence by adding an improper (or infinitely distant) point to $\mathfrak{R}^n$, the image of the center of projection, and by choosing as its neighborhoods the images of the neighborhoods of the projection center on the n-sphere. These neighborhoods are point sets in $\mathfrak{R}^n$ which contain all points outside of a sufficiently large n-ball and of course also contain the improper point itself. Putting this in another way, the n-sphere is obtained from Euclidean n-space by the addition of a single improper point.

The Simplicial Star $\mathfrak{S}t^n$

Given a finite $(n-1)$-dimensional simplicial complex $\mathfrak{R}^{n-1}$ one can obtain from it an n-dimensional simplicial complex called a *simplicial star* $\mathfrak{S}t^n$. The simplicial star will often be used later in the text. The schema of this complex is obtained from that of $\mathfrak{R}^{n-1}$ by choosing an additional vertex O and joining each i-simplex $\mathfrak{E}_i = (P_0 P_1 \cdots P_i)$ with O to generate an $(i+1)$-simplex $(O P_0 P_1 \cdots P_i)$. The point O is called the *center point* and the complex $\mathfrak{R}^{n-1}$ is called the *outer boundary* of the simplicial star. (Note that the outer boundary of a simplicial star is not the same as the boundary of a simplex or a convex region.) As an example, all simplexes of a simplicial complex which have a vertex in common form a simplicial star. Figure 36 shows a 2-dimensional simplicial star; the simplexes of the outer boundary $\mathfrak{R}'$ appear in heavy print.

Each convex n-dimensional region of Euclidean n-space whose boundary is a complex $\mathfrak{S}^{n-1}$ consisting of geometric $(n-1)$-simplexes may be used to construct an n-dimensional simplicial star. One needs only to project the geometric simplexes from an inner point O of the convex region. From the lemma of §10 concerning complexes in Euclidean spaces, an n-dimensional complex will be generated by this procedure and it is an $\mathfrak{S}t^n$ because its schema is that of an $\mathfrak{S}t^n$.

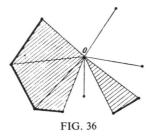

FIG. 36

n-Dimensional Projective Space $\mathfrak{P}^n$

The lines passing through the origin O of Euclidean $(n+1)$-space $\mathfrak{R}^{n+1}$ $(n \geqslant 2)$ form a complex $\mathfrak{P}^n$, *n-dimensional projective space*. The lines are to be regarded as "points" of a neighborhood space, where one takes as a neighborhood of a line g the subset of lines passing through the points of a neighborhood in $\mathfrak{R}^{n+1}$ of some arbitrary point of g. In particular, the lines which join O with all points of an ε-ball neighborhood of a point of g other than O form a special neighborhood of g, a "circular cone" about g. A line g will intersect the unit n-sphere $\mathfrak{S}^n$ of $\mathfrak{R}^{n+1}$ in two diametrically opposite points, P_1 and P_2. *One obtains a projective space when one declares each point of the set* $\mathfrak{M} = \mathfrak{S}^n$ *to be equivalent to the point diametrically opposite to it.* Pairs of diametrically opposite points are thus in one-to-one correspondence with the lines which comprise $\mathfrak{P}^n$. To see that $\mathfrak{P}^n$ is generated by means of this pairwise identification of points, one has now only to show that the set theoretic union of two arbitrary neighborhoods, $\mathfrak{U}(P_1 \mid \mathfrak{M})$ and $\mathfrak{U}(P_2 \mid \mathfrak{M})$, is transformed by this pair association to a neighborhood of g and that one obtains all neighborhoods of g in this manner. We leave the proof to the reader.

One may also obtain $\mathfrak{P}^n$ *from an n-ball by identifying diametrically opposite points on its boundary* [*which is an* $(n-1)$-*sphere*]. We described this representation of $\mathfrak{P}^n$ in §2 for the case $n = 2$; clearly it can be extended to the case of arbitrary dimension.

The points which lie on a given line passing through the origin of $\mathfrak{R}^{n+1}$ have (real) coordinates

$$x_1, x_2, \ldots, x_{n+1}$$

which differ only by a common factor. One can then place such a line (and therefore also projective points) in one-to-one correspondence with $(n+1)$-tuples of $x_1, x_2, \ldots, x_{n+1}$, which are determined up to a nonzero multiplicative factor. The $(n+1)$-tuple consisting exclusively of zeros is to be omitted [since no line corresponds to this $(n+1)$-tuple]. A neighborhood of such a "projective point" $\bar{x}_1 : \bar{x}_2 : \cdots : \bar{x}_{n+1}$ is formed by all projective points $x_1 : x_2 : \cdots : x_{n+1}$ for which

$$|x_1 - \bar{x}_1| < \varepsilon, \qquad \ldots, \qquad |x_{n+1} - \bar{x}_{n+1}| < \varepsilon,$$

for fixed values of $\bar{x}_1, \bar{x}_2, \ldots, \bar{x}_{n+1}$. These projective points correspond to all lines which are directed from the origin O of $\mathfrak{R}^{n+1}$ toward the points of a cube neighborhood in $\mathfrak{R}^{n+1}$ of the point $(\bar{x}_1, \bar{x}_2, \ldots, \bar{x}_{n+1})$. The size of such a neighborhood depends not only upon the projective point $\bar{x}_1 : \bar{x}_2 : \cdots : \bar{x}_{n+1}$ and the choice of ε, but also upon the choice of the common factors of the $\bar{x}_i$.

One obtains a simplicial decomposition of $\mathfrak{P}^n$ by normally subdividing the octahedral decomposition of the n-sphere $\mathfrak{S}^n$ given earlier and identifying diametrically opposite simplexes. One requires that the identifications be performed on the normal subdivision rather than on the octahedral decomposition itself because condition (k3) for complexes would otherwise be violated. Two arbitrary n-simplexes of the octahedral decomposition of $\mathfrak{S}^n$ would have the same vertices if the diametral point identification were performed directly on the octahedral subdivision (cf. the problem at the end of §13).

Topological Products

Given two neighborhood spaces $\mathfrak{A}$ and $\mathfrak{B}$ one can derive a new neighborhood space, the *topological product* $\mathfrak{A} \times \mathfrak{B}$, by means of the following definition: A point $A \times B$ of the topological product $\mathfrak{A} \times \mathfrak{B}$ corresponds to a point pair consisting of the point A of $\mathfrak{A}$ and the point B of $\mathfrak{B}$. A neighborhood of the point $\bar{A} \times \bar{B}$ is formed by all points $A \times B$ for which A lies in a neighborhood of $\bar{A}$ in $\mathfrak{A}$ and B lies in a neighborhood of $\bar{B}$ in $\mathfrak{B}$; in addition each set containing such a set is also declared to be a neighborhood of $\bar{A} \times \bar{B}$.

As an example, the Euclidean plane is the topological product of two real number lines, the x_1-axis and the x_2-axis.

More generally, if $\mathfrak{A}$ and $\mathfrak{B}$ are point sets of the Euclidean spaces $\mathfrak{R}^a$ and $\mathfrak{R}^b$, respectively, then $\mathfrak{A} \times \mathfrak{B}$ can be regarded as a point set of the Euclidean space $\mathfrak{R}^{a+b} = \mathfrak{R}^a \times \mathfrak{R}^b$. To see this let

$$x_1, x_2, \ldots, x_a \qquad (A)$$

be the coordinates of a point A of $\mathfrak{A}$ and let

$$x_{a+1}, x_{a+2}, \ldots, x_{a+b} \qquad (B)$$

be the coordinates of a point B of $\mathfrak{B}$. We then choose as the point $A \times B$ the point having coordinates

$$x_1, x_2, \ldots, x_a, x_{a+1}, \ldots, x_{a+b}. \qquad (A \times B)$$

Obviously the set of points $A \times B$, considered as a subset of $\mathfrak{R}^{a+b}$, is a neighborhood space homeomorphic to the topological product $\mathfrak{A} \times \mathfrak{B}$.

Problem

If $\mathfrak{A}$ and $\mathfrak{B}$ are convex regions of dimension α and β in the Euclidean spaces $\mathfrak{R}^a$ and $\mathfrak{R}^b$, respectively, then the topological product, regarded (as above) as a subset of the Euclidean space $\mathfrak{R}^{a+b}$, is a convex region having dimension $\alpha + \beta$; its boundary is formed by the topological product of the boundary of $\mathfrak{A}$ with $\mathfrak{B}$ taken in set theoretic union with the topological product of the boundary of $\mathfrak{B}$ with $\mathfrak{A}$.

THEOREM. *The topological product of two complexes is again a complex.*

Proof. Embed a simplicial decomposition of the complex $\mathfrak{A}$ into a Euclidean space $\mathfrak{R}^a$ and embed a simplicial decomposition of $\mathfrak{B}$ into $\mathfrak{R}^b$; this is possible from §11. Then $\mathfrak{A} \times \mathfrak{B}$ is a point set of the Euclidean space $\mathfrak{R}^a \times \mathfrak{R}^b$ and is in fact the set theoretic union of those sets consisting of the product of a simplex of $\mathfrak{A}$ with a simplex of $\mathfrak{B}$. Each such product, $\mathfrak{E}^\alpha \times \mathfrak{E}^\beta$ is a convex region of dimension $\alpha + \beta$, whose boundary is formed by the $(\alpha + \beta - 1)$-dimensional regions $\mathfrak{E}_\mu^{\alpha-1} \times \mathfrak{E}^\beta$ and $\mathfrak{E}^\alpha \times \mathfrak{E}_\nu^{\beta-1}$ (cf. the preceding problem). Here $\mathfrak{E}_\mu^{\alpha-1}$ denotes a face of $\mathfrak{E}^\alpha$ and $\mathfrak{E}_\nu^{\beta-1}$ denotes a face of $\mathfrak{E}^\beta$. The simplicial decomposition of $\mathfrak{A} \times \mathfrak{B}$ in $\mathfrak{R}^{a+b}$ can be constructed by means of a sequence of projections. One first divides each 1-dimensional region (straight line segment) through its midpoint. After the $(k - 1)$-dimensional regions have been simplicially decomposed one projects the boundary of a k-dimensional region from an inner point, obtaining the simplicial decomposition of the k-dimensional region. One can simplicially decompose regions of any dimension in this way. The set of simplexes obtained in this way satisfies the necessary conditions for the lemma of §10 and one thus has a simplicial decomposition of $\mathfrak{A} \times \mathfrak{B}$.

For use later we prove the following lemma:

LEMMA. *Let the topological product $\mathfrak{K}^n \times \mathfrak{t}$ of a complex $\mathfrak{K}^n$ and the unit interval $0 \leqq t \leqq 1$ be mapped into a simplicial complex $\mathfrak{K}^m$ by a mapping f as follows: For each point P of $\mathfrak{K}^n$ let $f(P \times 0)$ and $f(P \times 1)$ belong to the same simplex of $\mathfrak{K}^m$ and let the interval $P \times \mathfrak{t}$ transform linearly to the segment joining $f(P \times 0)$ and $f(P \times 1)$.*

If the restriction of f to the subsets $\mathfrak{K}^n \times 0$ and $\mathfrak{K}^n \times 1$ is continuous, then f is continuous everywhere.

Proof. We regard $\mathfrak{K}^n$ as a subset of a Euclidean space $\mathfrak{R}^a$ having coordinates $x_1, x_2, \ldots, x_a$. $\mathfrak{K}^n \times \mathfrak{t}$ is then a subset of the Euclidean space $\mathfrak{R}^a \times \mathfrak{R}^1$, where $\mathfrak{R}^1$ denotes the real number line with coordinate t. Likewise we regard $\mathfrak{k}^m$ as being embedded geometrically in a Euclidean space $\mathfrak{R}^b$ having coordinates $y_1, y_2, \ldots, y_b$. The mapping is then accomplished by means of the mapping functions

$$y_i(x_1, x_2, \ldots, x_a, t) = y_i(x_1, x_2, \ldots, x_a, 0) \cdot (1 - t)$$
$$+ y_i(x_1, x_2, \ldots, x_a, 1) \cdot t \qquad (i = 1, 2, \ldots, b).$$

Since $y_i(x_1, x_2, \ldots, x_a, 0)$ and $y_i(x_1, x_2, \ldots, x_a, 1)$ are continuous functions of $x_1, x_2, \ldots, x_a$ by assumption (cf. §6), it follows that $y_i(x_1, x_2, \ldots, x_a, t)$ is continuous in $x_1, x_2, \ldots, x_a, t$; hence f is continuous.

Problems

1. Show that $\mathfrak{S}^n$ is generated by identification of all boundary points of the closed n-ball.

2. If one assigns the point A of $\mathfrak{A}$ to each point $A \times B$ of the topological product $\mathfrak{A} \times \mathfrak{B}$, then this is a continuous mapping of $\mathfrak{A} \times \mathfrak{B}$ onto $\mathfrak{A}$.

3. A mass point is constrained to move with constant speed on (a) a torus, (b) a 2-sphere. Show that the phase space is (a) the topological product of the torus and the circle $\mathfrak{S}^1$; (b) projective 3-space $\mathfrak{P}^3$.[12]

4. Prove: The unit 3-sphere of Euclidean 4-space $\mathfrak{R}^4$, which has the equation

$$x_1^2 + x_2^2 + x_3^2 + x_4^2 = 1, \tag{3}$$

can be decomposed into two solid tori (topological product of the closed disk and the circle):

$$x_1^2 + x_2^2 \leqq x_3^2 + x_4^2 \quad \text{and} \quad x_1^2 + x_2^2 \geqq x_3^2 + x_4^2.$$

These solid tori are congruent, i.e., they can be transformed to one another by a rigid Euclidean motion about the midpoint of the sphere. The common boundary surface, whose points satisfy Eq. (3) and also

$$x_1^2 + x_2^2 = x_3^2 + x_4^2 \tag{4}$$

can be transformed, by a stereographic projection, to a torus of revolution. The stereographic projection is defined by projecting from the north pole $(0, 0, 0, 1)$ to the equatorial hyperplane $x_4 = 0$. The hyperplane is taken to have Cartesian coordinates ξ_1, ξ_2, ξ_3 and the torus of

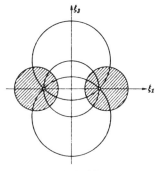

FIG. 37

revolution has the ξ_3-axis as its axis of rotation. One sees in this way that the 3-sphere (which is obtained by adding the image of the north pole to the equatorial hyperplane) may be decomposed into the union of two solid tori. The "core" of the first solid torus is the unit circle in the (ξ_1, ξ_2)-plane, and the core of the second solid torus is the ξ_3-axis plus the point at infinity (Fig. 37).

5. Show that the torus of Problem 4 can be mapped topologically onto itself by means of a rigid motion in spherical 3-space in such a way that longitude and meridian circles are interchanged (cf. §1 and the problem in §46).

6. When one identifies diametrically opposite points of the unit 3-sphere, transforming it to projective 3-space $\mathfrak{P}^3$, then the torus mentioned above becomes an hyperboloid of one sheet which is defined in projective coordinates $x_1 : x_2 : x_3 : x_4$ by Eq. (4). It is homeomorphic to the torus and likewise decomposes $\mathfrak{P}^3$ into two solid tori. Both projective and spherical 3-space can therefore be decomposed, by means of a torus, into two solid tori. (These solid tori can be transformed to one another projectively or by spherical rigid motions respectively.) Carry the proof through in detail.

7. Prove: *Each nonempty open subset of Euclidean n-space is an infinite complex.* (Scoop out the subset by means of *n*-dimensional cubes which are continually made finer and subdivide these simplicially.)

CHAPTER THREE

HOMOLOGY GROUPS

We have already become acquainted with the concept of a complex and its simplicial decomposition. All of the investigations which follow are directed toward the unsolved problem of completely classifying complexes and investigating their properties. By the "properties" of complexes we mean those properties which are independent of the individual simplicial decomposition and are preserved by topological mappings. In spite of this, we shall begin by investigating a fixed simplicial decomposition of a complex. In the next chapter we shall prove that important properties learned by studying the simplicial decomposition, namely, the homology groups, are independent of the particular simplicial decomposition and are topologically invariant.

The considerations in this chapter are purely combinatorial. No use is made of the concepts of neighborhood or continuity. Rather, it suffices to describe an n-dimensional complex simply by presenting its schema, that is, a set of "vertices" in which certain distinguished subsets are declared to be "simplexes" and for which the conditions (Sch1) through (Sch3) of §11 are satisfied. For simplicity, we shall only deal with finite complexes.

15. Chains

A simplicial k-dimensional chain or *simplicial k-chain* in the finite n-dimensional simplicial complex $\mathfrak{R}^n$ consists of a collection of k-simplexes of $\mathfrak{R}^n$, each being assigned a definite orientation and a definite positive multiplicity.

The k-chain 0 is defined to be the k-chain which consists of 0 k-simplexes. For $k > n$ the k-chain 0 is the only k-chain. Each oriented k-simplex, counted with multiplicity 1, is a k-chain.

To say that an oriented simplex E^k appears in a k-chain with multiplicity a is the same as saying that the oppositely oriented simplex $-E^k$ appears in the chain with multiplicity $-a$. If a k-simplex does not appear in a k-chain one says that it occurs with multiplicity 0.

We now define the sum of two chains U^k and V^k. It is again a k-chain, $U^k + V^k$, which is obtained in the following manner: If an oriented simplex

E_κ^k appears in U^k with multiplicity u_κ and appears in V^k with multiplicity v_κ, then E_κ^k appears in the chain $U^k + V^k$ with multiplicity $u_\kappa + v_\kappa$. If one regards all k-simplexes of $\mathfrak{R}^n$ as having been given fixed orientation and denotes these simplexes by E_κ^k ($\kappa = 1, 2, \ldots, \alpha^k$), then a k-chain is uniquely represented by an α^k-dimensional integer-valued vector

$$(u_1, u_2, \ldots, u_{\alpha^k}). \qquad (U^k)$$

Two chains U^k and V^k are added by adding their corresponding vectors.* But this implies that the k-chains form an Abelian group,** in particular, an α^k-dimensional lattice $\mathfrak{T}^k$. The oriented simplexes $E_1^k, E_2^k, \ldots, E_{\alpha^k}^k$ can be used as a basis for the group (§86). We then have

$$U^k = u_1 E_1^k + u_2 E_2^k + \cdots + u_{\alpha^k} E_{\alpha^k}^k,$$

where, as always, we use additive notation when dealing with Abelian groups. As a rule of computation let us note that when $mU^k = 0$ and $m \neq 0$, it follows that $U^k = 0$, because all elements of $\mathfrak{T}^k$ have infinite order.

16. Boundary, Closed Chains

If $\mathfrak{E}^{k-1}$ is a face of the oriented simplex E^k, then the orientation of E^k will at the same time determine a particular *"induced orientation"* of $\mathfrak{E}^{k-1}$ according to the following rule: If P_0 is the vertex of E^k lying opposite the face $\mathfrak{E}^{k-1}$ and $E^k = \varepsilon(P_0 P_1 \cdots P_k)$, then the orientation induced in $\mathfrak{E}^{k-1}$ is given by $\varepsilon(P_1 P_2 \cdots P_k)$. The *boundary of an oriented simplex* E^k is defined to be the $(k-1)$-chain formed by the $(k-1)$-dimensional faces of E^k in which each face simplex appears with the induced orientation and with multiplicity $1.^\dagger$ Let $E_0^{k-1}, E_1^{k-1}, \ldots, E_k^{k-1}$ denote the $(k-1)$-dimensional face simplexes of E^k, with each E_i^{k-1} given an arbitrary orientation. Then the boundary of E^k is the $(k-1)$-chain

$$\mathfrak{R}\partial E^k = \varepsilon \sum_{i=0}^{k} (-1)^i (P_0 P_1 \cdots P_{i-1} P_{i+1} \cdots P_k) = \sum_{i=0}^{k} \varepsilon_i E_i^{k-1}.$$

Here $\varepsilon_i = 1$ or $\varepsilon_i = -1$, respectively, according to whether the arbitrarily oriented simplex E_i^{k-1} does or does not possess the induced orientation. We use the symbol $\mathfrak{R}\partial$ to denote a boundary.

---

*The superscript k on α^k indicates dimensionality, as do all superscripts, and is not to be interpreted as an exponent.

**The concepts used in the treatment to follow, namely, vector, lattice, Abelian group, direct sum, etc., are defined in Chapter XII.

†One should take care to distinguish between the boundary (chain) of an oriented simplex and the boundary of a nonoriented simplex (more generally, of a pure complex; see §12). The former is a linear combination of oriented simplexes, the latter a point set.

The boundary of an arbitrary k-chain $U_k = \sum_{\kappa=1}^{\alpha^k} u_\kappa E_\kappa^k$ is defined to be the sum of the boundaries of the individual k-simplexes:

$$\Re\partial U^k = \Re\partial \sum_{\kappa=1}^{\alpha^k} u_\kappa E_\kappa^k = \sum_{\kappa=1}^{\alpha^k} u_\kappa \Re\partial E_\kappa^k.$$

Accordingly, we have the rules of computation:

$$\Re\partial(U^k \pm V^k) = \Re\partial U^k \pm \Re\partial V^k, \tag{1}$$

$$\Re\partial m U^k = m\Re\partial U^k, \tag{2}$$

also,

$$\Re\partial m U^k = 0 \text{ and } m = 0 \quad \text{imply} \quad \Re\partial U^k = 0, \tag{3}$$

for according to (2) we have $\Re\partial m U^k = m\Re\partial U^k$ and from this it follows that $\Re\partial U^k = 0$.

The boundary of a 0-dimensional chain is defined to be the integer 0.

A k-chain is said to be a *cycle* or is said to be *closed* if its boundary vanishes; that is, the boundary of the k-chain is the $(k-1)$-chain 0 in the case $k \neq 0$, or the integer 0 in the case $k = 0$. Thus every 0-chain is closed.

Because of (1), the sum and the difference of two closed chains are again closed chains. The closed k-chains therefore form a sublattice $\mathfrak{G}^k$ of the lattice $\mathfrak{T}^k$ of all k-chains. In the case $k = 0$, $\mathfrak{G}^k$ coincides with $\mathfrak{T}^k$.

For the boundary chains we have the important lemma:

Every boundary chain is closed.

Proof. Since each $(k-1)$-dimensional boundary chain is the sum of boundary chains of oriented k-simplexes, it suffices to show that the boundary of an oriented k-simplex E^k is closed. We may assume here that $k > 1$ since the boundary of a 1-simplex is closed, as is every 0-chain. Let

$$E^k = (P_0 P_1 P_2 \cdots P_k)$$

and let $\mathfrak{G}^{k-2}$ be a given $(k-2)$-simplex which is incident with E^k. It is no restriction to assume that exactly $P_2, P_3, \ldots, P_k$ are its vertices, since one can bring any $k-1$ vertices whatever into last place in the symbol $(P_0 P_1 P_2 \cdots P_k)$ by means of an even permutation. From the definition of a boundary we have

$$\Re\partial E^k = (P_1 P_2 \cdots P_k) - (P_0 P_2 \cdots P_k) + \cdots$$

and therefore

$$\Re\partial \Re\partial E^k = (P_2 \cdots P_k) - (P_2 \cdots P_k) + \cdots.$$

The only simplexes in $\Re\partial \Re\partial E^k$ which contain the vertices $P_2, \ldots, P_k$ and only these vertices, are the two simplexes above. Since they cancel each other, the simplex $(P_2 \cdots P_k)$ does not appear in $\Re\partial \Re\partial E^k$. Thus $\Re\partial \Re\partial E^k = 0$.

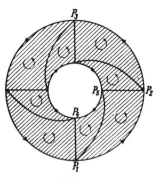

FIG. 38

EXAMPLE. Let $\Re^n$ be the complex illustrated in Fig. 38, a simplicial decomposition of the annulus. The triangles, each taken with multiplicity 1, form a 2-chain U^2, where the triangle orientations are indicated by means of circular arrows. We traverse all of the edges, in order to determine the boundary of U^2. The edge (P_1P_5) will not appear in $\Re\partial U^2$ because the only triangles bordering on it, $+(P_4P_1P_5)$ and $+(P_2P_5P_1)$, induce opposite orientations, namely, the respective orientations $+(P_1P_5)$ and $+(P_5P_1)$. In like manner the edge (P_1P_4) or any other intermediate edge will not appear in $\Re\partial U^2$. On the other hand, the edge (P_1P_2) will be given the orientation $+(P_1P_2)$ from the only bordering triangle, $+(P_5P_1P_2)$. Thus $+(P_1P_2)$ appears in the boundary of U^2 with multiplicity 1. $\Re\partial U^2$ will then consist of all of the edges of the inner and outer contours, where each edge is counted once and is taken with the orientation of the arrow. $\Re\partial U^2$ is in fact closed. The edges of the inner and of the outer contours, taken separately, form closed chains, B_i^1 and B_a^1. For example, the orientation $+(P_2)$ is induced at the vertex P_2 by the edge $+(P_1P_2)$ while the orientation $-(P_2)$ is induced at P_2 by the edge $+(P_2P_3)$.

17. Homologous Chains

Although every boundary chain on $\Re^n$ is closed, not every closed chain is a boundary chain. As an example, the outer contour B_a^1 of the annulus which was just introduced, does not bound. Whenever a closed k-chain U^k is the boundary of a $(k+1)$-chain U^{k+1} on $\Re^n$, then U^k is said to be bounding or *null homologous*; in symbols

$$U^k \sim 0.$$

More precisely, one must say $U^k \sim 0$ on $\Re^n$, which indicates that U^{k+1} lies on $\Re^n$. Correspondingly, if $\mathfrak{k}$ is a subcomplex of $\Re^n$ to which U^{k+1} belongs, one says that

$$U^k \sim 0 \qquad \text{on} \quad \mathfrak{k}.$$

If $U^k \sim 0$ on $\mathfrak{k}$, then also $U^k \sim 0$ on $\Re^n$, but the converse is not necessarily true.

More generally, two closed or nonclosed chains are *homologous* to one another if they differ by a null homologous chain, that is, if their difference is null homologous:

$$U^k \sim V^k \qquad \text{iff} \qquad U^k - V^k \sim 0.$$

A relation $U^k \sim V^k$ is called a *homology*.

Because null homologous chains are closed, it follows from $U^k \sim V^k$ that $\Re\partial(U^k - V^k) = 0$, that is, $\Re\partial U^k = \Re\partial V^k$; thus homologous chains have the same boundary. If $U^k \sim 0$ and $V^k \sim 0$, so that

$$U^k = \Re\partial U^{k+1}, \qquad V^k = \Re\partial V^{k+1},$$

then from Eq. (1) in §16,

$$U^k \pm V^k = \Re\partial(U^{k+1} \pm V^{k+1}),$$

so that $U^k \pm V^k \sim 0$. *The sum and the difference of null homologous chains are again null homologous chains.* Thus the null homologous k-chains form a sublattice, $\mathfrak{N}^k$ of the lattice $\mathfrak{G}^k$ of all closed k-chains. One can therefore add homologies, subtract them, and multiply them by an integer. On the contrary, as we will see immediately in an example, it is not permitted to divide them by a common factor. *Thus in general it does not follow that $mU^k \sim 0$ implies $U^k \sim 0$.*

We have, furthermore: If $U^k \sim V^k$ and $V^k \sim W^k$, then $U^k \sim W^k$. That is, if $U^k - V^k \sim 0$ and $V^k - W^k \sim 0$, then also their sum $U^k - W^k \sim 0$. One can then divide all of the closed k-chains into equivalence classes of homologous chains, the so-called *homology classes*. As an example, the null homologous k-chains form such a homology class.

EXAMPLE 1. We shall determine the 1-dimensional homology classes of the annulus (Fig. 38). We use the notation and results of the previous section; we had

$$\Re\partial U^2 = B_a^1 + B_i^1;$$

therefore

$$B_a^1 + B_i^1 \sim 0 \qquad \text{or} \qquad B_a^1 \sim - B_i^1,$$

so that B_a^1 and $- B_i^1$ belong to the same homology class.

Each arbitrary closed 1-chain U^1 is homologous to a multiple of B_i^1. Suppose that U^1 contains an edge of the outer contour, for example. One can replace this edge by the two other sides of the adjacent triangle, by adding the boundary Δ^1 of this triangle, oriented suitably, to U^1. Since $\Delta^1 \sim 0$, this replacement transforms U^1 to a chain homologous to U^1. There thus exists a chain $'U^1$ homologous to U^1, which no longer contains any edge of the outer contour. If a "diagonal" edge appears in $'U^1$, for example, $+(P_1P_5)$, one can likewise replace this by the edges $+(P_1P_4) + (P_4P_5)$, and, if one proceeds in the same manner with all of the diagonal edges appearing in $'U^1$, one obtains a chain $''U^1$ which contains neither outer nor diagonal edges, is closed, and is homologous to U^1. But in that case $''U^1$ can only contain edges of the inner contour. For if a radial edge were to appear, for example, (P_5P_2), then due to the fact that $''U^1$ is closed the orientations induced in the vertex P_2 would have to cancel and since no edges of $''U^1$ are incident with P_2 aside from (P_5P_2), then (P_5P_2) would have to appear with multiplicity zero in U^1. Thus $''U^1$ lies entirely on the inner contour and, due to the fact that it is closed, must contain all edges of the inner contour equally often. Thus we have shown that

$$''U^1 = mB_i^1,$$

where m is an integer.

We have proven in this way that each closed 1-chain is homologous to one of the chains

$$0, \pm B_i^1, \pm 2B_i^1, \pm 3B_i^2, \ldots.$$

No two of these chains are homologous to one another. For suppose that

$$m'B_i^1 \sim m''B_i^1;$$

then $(m' - m'')B_i^1 = mB_i^1 \sim 0$. It would follow from this that $m = 0$ and thus $m' = m''$. For if the chain mB_i^1 is the boundary of a 2-chain, then any triangle adjacent to the inner boundary will appear in the 2-chain with multiplicity m. But since no intermediate edges can appear in the boundary of the 2-chain, then all other triangles must appear with multiplicity m; thus the 2-chain is equal to mU^2. The boundary of this is $m(B_a^1 + B_i^1)$, which is equal to mB_i^1 only if $m = 0$. This shows that there exist infinitely many homology classes and the multiples of B_i^1 are their representatives.

EXAMPLE 2. Let $\mathfrak{R}^n$ be the projective plane $\mathfrak{P}^2$ with the simplicial decomposition shown in Fig. 39. The disk shown in Fig. 39 transforms to the projective plane, as discussed in §2, when one identifies diametrically opposite edges and vertices. With this stipulation the figure is nothing other than an abbreviated description of the schema of the simplicial complex which represents the projective plane. Thus, for example, the two points P of the figure are one and the same point in the projective plane and the 1-chain A^1 which consists of the four oriented vertices of the upper semicircle, counted once, is the same as the chain formed from the four edges of the lower semicircle. All triangles are oriented so that for each intermediate edge of the disk opposite orientations are induced by the two triangles adjacent to that edge. One then obtains a 2-chain U^2 having $\mathfrak{R}\partial U^2 = +2A^1$ for the orientations chosen in the figure. Thus $2A^1 \sim 0$. The most general 2-chain which has a multiple of A^1 as a boundary is a multiple of U^2, for example, kU^2. For an intermediate edge of the disk cannot appear in the boundary of this 2-chain; thus all triangles appear equally often. But the boundary of kU^2 is $2kA^1$. This means that A^1, which is a closed 1-chain of the projective plane (a projective line), is not ~ 0 but that $2A^1 \sim 0$. Thus $A^1 \sim 0$ does not follow from $2A^1 \sim 0$.

An arbitrary closed 1-chain U^1 is either ~ 0 or $\sim A^1$. That is, there exist only two 1-dimensional homology classes. If the chain U^1 does not already lie on the boundary of the disk of the figure one can, as for the case of the annulus, push it step by step to the rim, by replacing each intermediate edge with two edges of a suitably oriented adjacent triangle. This results in a 1-chain $'U^1 \sim U^1$, lying on the periphery of the disk. This 1-chain can only be a multiple of A^1, since it is closed in $\mathfrak{P}^2$. Since an even multiple of A^1 is null homologous, $'U^1$ and U^1 are either ~ 0 or $\sim A^1$.

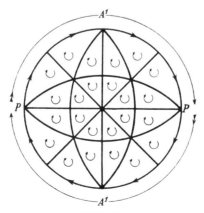

FIG. 39

18. Homology Groups

In group theoretic terms the homology classes are nothing other than the residue classes of the lattice $\mathfrak{G}^k$ of closed k-chains relative to the sublattice $\mathfrak{N}^k$ of null homologous k-chains. Two closed chains U^k and V^k are homologous if and only if their difference is null homologous, that is, the two chains belong to the same residue class of $\mathfrak{N}^k$ in $\mathfrak{G}^k$. Thus the k-dimensional homology classes themselves form an Abelian group if one defines the sum of two homology classes H_1^k and H_2^k to be the homology class determined by the sum of a chain from H_1^k and a chain from H_2^k (elementwise addition). This group is the factor group $\mathfrak{G}^k/\mathfrak{N}^k$. It is called the kth *homology group* $\mathfrak{H}^k$ of the simplicial complex $\mathfrak{K}^n$.

As long as we are dealing with a particular simplicial decomposition of the complex, it is not self-evident why we pay attention to the factor group of the lattices of the closed chains by the null homologous chains and not, for example, all k-chains by the closed k-chains. The reason is that the homology groups are a topologically invariant property of the complex. They are independent of the choice of simplicial decomposition. Their topological invariance is intuitively reasonable, as will be illustrated by the examples of the annulus, the projective plane, and the torus, It is not, however, easy to prove this. The proof will in fact occupy the whole of the next chapter.

The groups $\mathfrak{T}^k$, $\mathfrak{G}^k$, and $\mathfrak{N}^k$ are lattices, that is, free Abelain groups having finitely many generators. In contrast, the homology group $\mathfrak{H}^k$ will in general possess nonzero elements of finite order. On the other hand, $\mathfrak{H}^k$ will still have only finitely many generators, since it is a factor group of the finitely generated group $\mathfrak{G}^k$. Thus $\mathfrak{H}^k$ is the direct sum of finitely many cyclic groups, namely, p^k free cyclic groups, and a number ρ^k of cyclic groups of finite orders $c_1^k, c_2^k, \ldots, c_{\rho^k}^k$, where each c^k is a divisor of the preceding one (§86).[*] The integer p^k and the orders $c_1^k, c_2^k, \ldots, c_{\rho^k}^k$ are determined uniquely by the group $\mathfrak{H}^k$ and are characteristic of the group. The integer p^k is called the kth *Betti number* and $c_1^k, c_2^k, \ldots, c_{\rho^k}^k$ are called the *k-dimensional torsion coefficients* of the simplicial complex $\mathfrak{K}^n$. The name "torsion coefficient" will be justified later by the example of the lens spaces (§61).

If one selects a generating element from each of the $\rho^k + p^k$ cyclic groups whose direct sum is $\mathfrak{H}^k$ and, further, selects a representative k-chain from each of the generating elements, which in fact are homology classes, one obtains $\rho^k + p^k$ closed k-chains

$$A_1^k, \ldots, A_{\rho^k}^k, B_1^k, \ldots, B_{p^k}^k. \tag{1}$$

Each arbitrary closed chain U^k is then homologous to a linear combination

$$U^k \sim \xi_1 A_1^k + \xi_2 A_2^k + \cdots + \xi_{\rho^k} A_{\rho^k}^k + \eta_1 B_1^k + \eta_2 B_2^k + \cdots + \eta_{p^k} B_{p^k}^k. \tag{2}$$

[*] The superscript k denotes dimension and is not an exponent.

Since $c_\mu^k A_\mu^k \sim 0$ one can arrange that $0 \leqq \xi_\mu < c_\mu^k$. On the other hand, the η_ν can vary without bound. With this normalization the coefficients ξ_μ, η_ν are clearly determined by U^k. One calls such a system of $\rho^k + p^k$ closed k-chains (1) a *k-dimensional homology basis*. It is important that the kth homology group is not only an abstract group determined by its Betti number and torsion coefficients, but also that it is realizable by a homology basis.

The appearance of a torsion coefficient c_μ^k indicates topologically the existence of a closed k-chain which becomes homologous to 0 only after a c_μ^k-fold repetition.

The Betti number p^k gives the maximum number of homologously independent closed k-chains. Finitely many k-chains $U_1^k, \ldots, U_r^k$ are said to be *homologously independent* if no homology of the form

$$t_1 U_1^k + t_2 U_2^k + \cdots + t_r U_r^k \sim 0$$

holds them, unless all the coefficients vanish. If, on the other hand, there exists a nontrivial homology involving them, they are said to be *homologously dependent*. In the homology basis (1) each of the chains A_μ^k is itself homologously dependent, since $c_\mu^k A_\mu^k \sim 0$; on the other hand, the chains $B_1^k, B_2^k, \ldots, B_{p^k}^k$ are homologously independent because

$$\eta_1 B_1^k + \eta_2 B_2^k + \cdots + \eta_{p^k} B_{p^k}^k$$

is homologous to 0 only when all of the η vanish. No more than p^k homologously independent k-chains can exist, since by (2) an appropriate multiple, for example, the c_1^k-fold multiple, of an arbitrary closed k-chain will be homologous to a linear combination of the p^k chains $B_1^k, B_2^k, \ldots, B_{p^k}^k$.

The 0th homology group of a connected simplicial complex is always the free cyclic group.

One can choose an oriented 0-simplex, for example E_1^0, as a homology basis. Consider all of the 0-simplexes $E_1^0, E_2^0, \ldots, E_{\alpha^0}^0$ to be oriented with the $+$ sign. If

$$U^0 = u_1 E_1^0 + u_2 E_2^0 + \cdots + u_{\alpha^0} E_{\alpha^0}^0$$

is an arbitrary 0-chain, then one can connect the vertices E_1^0 and E_ν^0 by an edge path, since $\mathfrak{R}^n$ is connected, and one can orient its edges so that successive edges induce opposite orientations at a vertex. By doing this the edge path becomes a 1-chain U^1 and we have

$$\mathfrak{R} \partial U^1 = \pm (E_1^0 - E_\nu^0);$$

hence $E_\nu^0 \sim E_1^0$. Thus U^0 is homologous to the $(u_1 + \cdots + u_{\alpha^0})$-fold multiple of E_1^0. No nonzero multiple of E_1^0 is homologous to 0. For at the boundary of a 1-simplex one vertex occurs with multiplicity $+1$ and the other vertex with multiplicity -1. The sum of these multiplicities is 0. This is also the case for the boundary of an arbitrary 1-chain. That is, the sum of the vertex multiplicities is 0 for each null homologous 0-chain. Consequently, E_1^0 is a homology basis and the 0th homology group is the free group with one generator.

The 0-dimensional homology classes can be characterized more simply by use of the concept of the *sum of the coefficients of a* 0-*chain*, as follows: we define the sum of the coefficients of a 0-chain to be the sum of the multiplicities of all of the 0-simplexes of the chain when oriented with the $+$ sign. For example, in the chain U^0 the sum of the coefficients in $u_1 + \cdots + u_{\alpha^0}$. One can then say: In a connected complex two 0-chains are homologous to one another if and only if the sum of the coefficients is the same for both chains.

One can prove easily that the 0th homology group is the free Abelian group of τ generators if $\Re^n$ consists of τ isolated subcomplexes. *Thus there exist no* 0-*dimensional torsion coefficients.*

For $k > n$ the homology group $\mathfrak{H}^k$ of $\Re^n$ consists of the null element alone. For there exists just one kth homology class. It contains only the single k-chain 0.

For $k = n$ there exist just as many homology classes as closed n-chains. That is, it always follows from $U^n \sim V^n$ that $U^n = V^n$, for there exists no $(n + 1)$-chain other than the $(n + 1)$-chain 0. One calls a sequence of k-chains $U_1^k, \ldots, U_r^k$ *linearly dependent* or *linearly independent* respectively according to whether an equation

$$t_1 U_1^k + t_2 U_2^k + \cdots + t_r U_r^k = 0$$

does or does not hold, where not all coefficients are zero. One can then say: For $k = n$ linear dependence is the same as homologous dependence. The nth Betti number p^n is therefore equal to the maximum number of linearly independent closed n-chains. *There are no n-dimensional torsion coefficients.* From $mU^u \sim 0$ it follows that $mU^n = 0$ and when $U^n \neq 0$, then $m = 0$.

Let us mention briefly how one defines homology groups for infinite complexes. A chain is again an aggregate of finitely many simplexes. Boundary, closedness, and homology are defined as for finite complexes. The elements of the homology group for the dimension k are again the homology classes of the k-dimensional closed k-chains. On the other hand, the homology groups of an infinite complex will not in general have a finite number of generators, so that they can not be characterized by Betti numbers and torsion coefficients.

19. Computation of the Homology Groups in Simple Cases

1. In the annulus (§16) each 1-chain U^1 is homologous to a multiple of the chain B_i^1 and no multiple of B_i^1 is ~ 0. The homology group $\mathfrak{H}^1$ is therefore the free cyclic group, and B_i^1 is a homology basis. As we have

already shown, $\mathfrak{H}^0$ is the free cyclic group; $\mathfrak{H}^2$ consists only of the zero element, since there exist no closed 2-chains other than 0. The Betti numbers are

$$p^0 = 1, \qquad p^1 = 1, \qquad p^2 = 0.$$

Torsion coefficients are absent in all dimensions.

2. The projective plane (§17). There are two 1-dimensional homology classes; $\mathfrak{H}^1$ is therefore the group of order 2, and the chain A^1 (the projective line) forms a homology basis. Since there are no closed 2-chains, $\mathfrak{H}^2$ consists of the zero element:

$$p^0 = 1, \qquad p^1 = 0, \qquad p^2 = 0.$$

There is one 1-dimensional torsion coefficient, having the value 2.

3. Let a simplicial decomposition of the torus be given by the square shown in Fig. 40. Opposite sides of the square are to be identified (cf. §1). The edges belonging to the decomposition can be classified as either boundary or inner edges. Boundary edges are those that lie on either of the sides a and b of the square. The inner edges can be classified as either vertical, diagonal or horizontal edges. To determine the homology groups we proceed as in the case of the annulus:

I. For each closed 1-chain U^1 there exists a 1-chain homologous to it, composed entirely of boundary edges. Each inner vertical or diagonal edge can be replaced by the other two sides of the adjacent triangle lying to its right, namely, by addition of a null homologous boundary chain of this triangle. Each inner horizontal edge can be replaced by the other two edges of the triangle lying underneath it. After finitely many such steps we arrive at a homologous chain $'U^1 \sim U^1$ which lies on the rim of the square. Since $'U^1$ is closed, all oriented edges of the square side a must occur equally often, for example, α-fold, and likewise all oriented edges of side b must occur equally often, for example, β-fold.

Let us call the closed 1-chain arising from all of the singly counted edges of the square side a (with orientation as shown in the figure) a^1, and let us correspondingly introduce the 1-chain b^1, so that we have

$$U^1 \sim {}'U^1 = \alpha a^1 + \beta b^1.$$

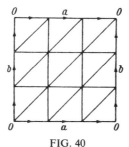

FIG. 40

The homology classes of the closed chains a^1 and b^1 are then the generators of the homology group for the dimension 1.

II.   One has $\alpha a^1 + \beta b^1 \sim 0$ if and only if $\alpha = \beta = 0$. To prove this we orient the triangles of the square so that opposite orientations are induced in all inner edges by the two adjoining triangles. The 2-chain U^2 which consists of all these oriented triangles then has no boundary at all because opposite sides of the square are to be regarded as identical on the torus. Now assume that the chain $\alpha a^1 + \beta b^1 \sim 0$; then this 1-chain is the boundary of a 2-chain which contains all triangles equally often, since the inner edges must cancel out. That 2-chain is therefore a multiple of U^2 and, likewise, is thus closed. Thus $\alpha a^1 + \beta b^1 = 0$, and $\alpha = \beta = 0$ because the chains a^1 and b^1 have no common 1-simplex. This means that the homology group $\mathfrak{H}^1$ is the free Abelian group having two generators, and a^1 and b^1 (the meridian and longitude circles of the torus) form a homology basis of $\mathfrak{H}^1$. $\mathfrak{H}^0$ and $\mathfrak{H}^2$ are free cyclic groups. One then has

$$p^0 = 1, \qquad p^1 = 2, \qquad p^2 = 1.$$

There are no torsion coefficients.

If we had decomposed the square of Fig. 40 by more than three vertical and horizontal edges, and the corresponding diagonal edges, we obviously would have obtained the same homology groups for this simplicial complex. One can also prove that one gets the same result for any other not-too-complicated simplicial decomposition of the torus. One would then suspect that the homology groups are independent of the chosen decomposition of the torus and that they are in fact determined by the torus as a neighborhood space. But until we have completed the proof of this (Chapter IV) we must allow the possibility that there exist decompositions which lead to a result different from that obtained with the present decomposition.

4.   The simplicial star. Let $\mathfrak{K}^n$ be a simplicial star with center point O and outer boundary $\mathfrak{K}^{n-1}$ and let U^k be an arbitrary closed k-chain on $\mathfrak{K}^n$ ($k \leqslant 0$). One can transform U^k to a homologous chain $'U^k$ in which no k-simplex of $\mathfrak{K}^{n-1}$ appears. For example, if the simplex $E^k = \varepsilon(P_0 P_1 \cdots P_k)$ of $\mathfrak{K}^{n-1}$ appears in U^k with multiplicity u, then one subtracts the u-fold boundary of the simplex

$$E^{k+1} = \varepsilon(O P_0 P_1 \cdots P_k)$$

from U^k and one obtains a chain homologous to U^k such that E^k no longer appears. In this manner one removes in sequence all simplexes of U^k which lie on $\mathfrak{K}^{n-1}$ and one finally obtains a chain $'U^k \sim U^k$ whose k-simplexes are all incident with O. But in that case $'U^k = 0$. For if a simplex $'E^k$ appears in $'U^k$ with multiplicity $'u$, then the face $'E^{k-1}$ of $'E^k$ lying opposite to vertex O will likewise appear with multiplicity $'u$ in the boundary of $'U^k$. This is because $'E^k$ is the only k-simplex in $\mathfrak{K}^n$ that has $'E^{k-1}$ as a face and O as a vertex. Thus $'u = 0$ follows from $\Re \partial' U^k = 0$. We have then shown that for $k > 0$ each closed k-chain on a simplicial star is null homologous,

$$p^0 = 1, \qquad p^1 = \cdots = p^n = 0.$$

There are no torsion coefficients.

5. The n-simplex. Since an n-simplex can be regarded as a simplicial star whose center point is a vertex and whose outer boundary is the opposing face, all homology groups of an n-simplex except the 0th group are trivial.

6. The n-sphere. We use the tetrahedral decomposition (§14). Accordingly, the n-sphere $\mathfrak{S}^n$ is the boundary of an $(n + 1)$-simplex, $\mathfrak{E}^{n+1}$. From (5), a closed chain U^k on $\mathfrak{S}^n$ is null homologous on $\mathfrak{E}^{n+1}$ when $k > 0$; that is, U^k is the boundary of a chain U^{k+1}. But U^{k+1} itself lies on $\mathfrak{S}^n$, except when $k + 1 = n + 1$. Thus for $0 < k < n$, $U^k \sim 0$ on $\mathfrak{S}^n$. In the case $k = n$, U^{k+1} must be equal to a multiple of the chain which consists of the oriented simplex E^{n+1}, that is,

$$U^{k+1} = uE^{n+1},$$

because E^{n+1} is the only $(n + 1)$-simplex which occurs, thus

$$U^k = u\Re \partial E^{n+1}.$$

Since $u\Re \partial E^{n+1}$ is not ~ 0 on $\mathfrak{S}^n$ when $u \neq 0$, for $\mathfrak{S}^n$ is only n-dimensional, we have thus shown that the homology groups of the boundary of an $(n + 1)$-simplex all consist of the null element, except for the 0th and nth groups, which are free cyclic groups:

$$p^0 = 1, \qquad p^1 = p^2 = \cdots = p^{n-1} = 0, \qquad p^n = 1,$$

and torsion coefficients are absent.

More difficult examples, with arbitrarily many torsion coefficients of arbitrary value, can be found among the 3-dimensional manifolds discussed in Chapter IX, for example, in §61 and in §62, Problem 4.

20. Homologies with Division

In addition to the lattices $\mathfrak{T}^k$, $\mathfrak{G}^k$, $\mathfrak{N}^k$, we shall also consider the lattice $\mathfrak{D}^k$, which consists of all k-chains for which a nonzero multiple of the chain belongs to $\mathfrak{N}^k$. The chains which have this property obviously form a lattice. For if cU^k and dV^k belong to $\mathfrak{N}^k (c, d \neq 0)$, then so does $cd(U^k \pm V^k)$. That is, the sum and the difference of two chains U^k and V^k also belong to $\mathfrak{D}^k$. $\mathfrak{N}^k$ is a sublattice of $\mathfrak{D}^k$ and $\mathfrak{D}^k$ in turn is a sublattice of $\mathfrak{G}^k$. For the chain cU^k belonging to $\mathfrak{N}^k$ is closed, as are all chains of $\mathfrak{N}^k$, and it follows that U^k is closed by formula (3) of §16. One then has the following sequence of lattices

$$\mathfrak{T}^k, \mathfrak{G}^k, \mathfrak{D}^k, \mathfrak{N}^k$$

in which each one contains the lattice following.

The chains of $\mathfrak{D}^k$ are said to be *division–null homologous* (symbolically: ≈ 0). This signifies that there exists an integer $c \neq 0$ such that $cU^k \sim 0$. One also calls division–null homologous chains "boundary divisors." Two arbitrary chains T_1^k and T_2^k of $\mathfrak{T}^k$ are said to be *division-homologous*,

$$T_1^k \approx T_2^k,$$

if their difference is ≈ 0. Such a relation is called a division homology.

The fact that one may add, subtract, and multiply division homologies by integers is only another expression of the lattice property of $\mathfrak{D}^k$. In contrast to ordinary homologies one can also divide them by a nonzero factor, and hence their name. The relation

$$bW^k \approx 0 \tag{1}$$

indicates that $c(bW^k) \sim 0$ for some appropriate $c \neq 0$. When, however, $b \neq 0$ this relation has the same meaning as

$$W^k \approx 0. \tag{2}$$

Thus (2) follows from (1).

We now examine the decomposition of $\mathfrak{D}^k$ into residue classes relative to the sublattice $\mathfrak{N}^k$. These residue classes are homology classes which have finite order, when considered as elements of the homology group $\mathfrak{H}^k = \mathfrak{G}^k / \mathfrak{N}^k$. This is so because, for each chain of $\mathfrak{D}^k$, there will exist a nonzero multiple which belongs to $\mathfrak{N}^k$. Since, conversely, each chain having a multiple which is null homologous lies in $\mathfrak{D}^k$, then $\mathfrak{D}^k$ consists of exactly the homology classes of finite order. Expressed in another way, the factor group $\mathfrak{D}^k / \mathfrak{N}^k$ is the subgroup of the homology group which is formed by elements of finite order. This subgroup is called the *torsion group* for the dimension k. Its order is equal to the product of the k-dimensional torsion coefficients.

If, on the other hand, one decomposes $\mathfrak{G}^k$ relative to the sublattice $\mathfrak{D}^k$, then there results a division of all closed chains into classes of division-homologous chains. The factor group $\mathfrak{G}^k / \mathfrak{D}^k$ is called the *Betti group*[13] for the dimension k. Based upon the group theoretic relation (§83) which states that

$$\mathfrak{G}^k / \mathfrak{D}^k \quad \text{is isomorphic to} \quad (\mathfrak{G}^k / \mathfrak{N}^k) / (\mathfrak{D}^k / \mathfrak{N}^k),$$

one can also regard the Betti group as the factor group of the homology group $\mathfrak{H}^k = \mathfrak{G}^k / \mathfrak{N}^k$ *relative to the torsion group* $\mathfrak{D}^k / \mathfrak{N}^k$. Since $\mathfrak{H}^k$ is the direct sum of the torsion group and a free Abelian group of p^k-generators then *the Betti group is itself a free Abelian group of p^k-generators*. Introduction of the lattice $\mathfrak{D}^k$ thus allows a decomposition of the homology group into a finite part, the torsion group, and an infinite part, the Betti group. It is to be noted, of course, that the Betti group is not to be spoken of as a subgroup but as a factor group of the homology group.

The homology basis now consists of a torsion basis and a Betti basis. One represents the torsion group as a direct sum of cyclic groups of orders

$c_1^k, c_2^k, \ldots, c_{\rho^k}^k$ where the c_ν^k denote the k-dimensional torsion coefficient; one selects a generating element (that is, a particular homology class) from each of these cyclic groups and chooses a representative chain from each of these classes. These ρ^k chains form a *torsion basis*.

Correspondingly, if one represents the Betti group as a direct sum of p^k free cyclic groups, selects a generating element from each of these groups (this is a class of chains division-homologous to one another), and chooses a representative chain from each of these classes, these p^k chains form a *Betti basis*. Each arbitrary closed chain is then division-homologous to a linear combination of the p^k basis chains. In the homology basis

$$A_1^k, A_2^k, \ldots, A_{\rho^k}^k, B_1^k, B_2^k, \ldots, B_{p^k}^k$$

of Section 18 the A^k form a torsion basis, and the B^k form a Betti basis. The transformation from one Betti basis

$$B_1^k, B_2^k, \ldots, B_{p^k}^k$$

to another,

$$'B_1^k, 'B_2^k, \ldots, 'B_{p^k}^k$$

is accomplished by means of an integer-valued unimodular transformation

$$'B_\mu^k \approx \sum_{\nu=1}^{p^k} b_{\mu\nu} B_\nu^k \qquad (\mu = 1, 2, \ldots, p^k).$$

In the case of the torus (§19, Example 3) the torsion basis is empty so the torsion group consists of the null element; the chains a^1 $(= B_1^1)$ and b^1 $(= B_2^1)$ form a Betti basis; these chains are the meridian circle and the longitude circle. In the case of the projective plane (§19, Example 2) the Betti basis is empty, so the Betti group consists of the null element; the chain A^1, the projective line, forms a torsion basis.

We now collect together and summarize the various groups and lattices encountered so far:

$\mathfrak{T}^k$	the lattice of all k-chains,
$\mathfrak{G}^k$	the lattice of closed k-chains,
$\mathfrak{D}^k$	the lattice of division-null homologous k-chains,
$\mathfrak{N}^k$	the lattice of null homologous k-chains,
$\mathfrak{H}^k = \mathfrak{G}^k/\mathfrak{N}^k$	the homology group,
$\mathfrak{D}^k/\mathfrak{N}^k$	the torsion group,
$\mathfrak{G}^k/\mathfrak{D}^k$ or $\mathfrak{H}^k/(\mathfrak{D}^k/\mathfrak{N}^k)$	the Betti group.

21. Computation of Homology Groups from the Incidence Matrices

We have previously defined the homology groups of an arbitrary simplicial complex. We now derive a general procedure allowing their computation, at least for a finite complex. Toward this goal we assume that all simplexes are given with fixed orientations (the 0-simplex with the $+$ sign) and are denoted by

$$E_\kappa^k \qquad (\kappa = 1, 2, \ldots, \alpha^k, k = 0, 1, \ldots, n).$$

For $0 \leqq k \leqq n - 1$ one has the boundary relations

$$\Re \partial E_\lambda^{k+1} = \sum_{\kappa=1}^{\alpha^k} \varepsilon_{\kappa\lambda}^k E_\kappa^k \qquad (\lambda = 1, 2, \ldots, \alpha^{k+1}). \tag{1}$$

The matrix

$$(\varepsilon_{\kappa\lambda}^k) = \mathbf{E}^k \qquad (k = 0, 1, \ldots, n-1)$$

can be regarded as a table of rows and columns. The left entries are the oriented simplexes of dimension k, while the upper entries are the oriented simplexes of dimension $k + 1$. The intersection of the κth row and λth column is the coefficient $\varepsilon_{\kappa\lambda}^k$ with which the oriented simplex E_κ^k appears in the boundary chain of the simplex E_λ^{k+1}. The coefficient $\varepsilon_{\kappa\lambda}^k$ is $+1$, -1, or 0, respectively, depending upon whether the oriented simplex E_κ^k has the orientation induced by E_λ^{k+1}, the opposite orientation, or is not at all incident with E_λ^{k+1}. The matrix $\mathbf{E}^k$ is called the *incidence matrix*, for the dimension k, of the simplicial decomposition of the complex.[14]

The incidence matrices $\mathbf{E}^0$ and $\mathbf{E}^1$ of a simplicial complex consisting of a single 2-simplex $\mathfrak{E}^2$, assigned the orientations of Fig. 41, are

$\mathbf{E}^0$	E_1^1	E_2^1	E_3^1
E_1^0	0	-1	-1
E_2^0	$+1$	0	$+1$
E_3^0	-1	$+1$	0

$\mathbf{E}^1$	E^2
E_1^1	$+1$
E_1^2	$+1$
E_1^3	-1

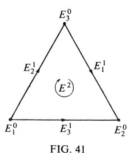

FIG. 41

The simplicial complex is completely determined by the incidence matrices. For by means of them all of the $(k - 1)$-dimensional faces of a simplex E_κ^k are determined, from each face the $(k - 2)$-dimensional faces are determined, and so forth until the vertices of E_κ^k are finally determined. Thus the schema of the complex is given and, thereby, the complex itself. It must then be possible to read all of the properties of a simplicial complex from the incidence matrices. This is true in particular for the homology groups, which we shall now obtain from the incidence matrices.

If $k > 0$ in Eq. (1), then taking the boundary of both sides yields

$$\mathfrak{R}\partial\,\mathfrak{R}\partial E_\lambda^{k-1} = \sum_{\kappa=1}^{\alpha^k} \varepsilon_{\kappa\lambda}^k\,\mathfrak{R}\partial E_\kappa^k = \sum_{\kappa=1}^{\alpha^k} \varepsilon_{\kappa\lambda}^k \sum_{i=1}^{\alpha^{k-1}} \varepsilon_{i\kappa}^{k-1} E_i^{k-1}$$

$$= \sum_{i=1}^{\alpha^{k-1}} \left(\sum_{\kappa=1}^{\alpha^k} \varepsilon_{i\kappa}^{k-1}\varepsilon_{\kappa\lambda}^k \right) E_i^{k-1} = 0$$

since the boundary of a closed chain is zero. The individual coefficients are all zero because the E_i^{k-1} are linearly independent:

$$\sum_{\kappa=1}^{\alpha^k} \varepsilon_{i\kappa}^{k-1}\varepsilon_{\kappa\lambda}^k = 0 \qquad (i = 1, 2, \ldots, \alpha^{k-1}, \lambda = 1, 2, \ldots, \alpha^{k+1}),$$

a system of equations which we can replace, using the algebra of matrices, by the single matrix equation

$$\mathbf{E}^{k-1}\mathbf{E}^k = 0 \qquad (k = 1, 2, \ldots, n-1). \tag{2}$$

These equations are the arithmetic expression of the fact that each boundary chain is closed, for they merely say that the boundary of the boundary of each $(k + 1)$-simplex vanishes.

The oriented simplexes

$$E_1^k, E_2^k, \ldots, E_{\alpha^k}^k$$

form a basis for the lattice of all k-chains $\mathfrak{X}^k$. In each dimension $k = 0, 1, \ldots, n$ we now introduce a new basis

$$U_1^k, U_2^k, \ldots, U_{\alpha^k}^k$$

to replace it. In place of the boundary relations (1) there will hold the boundary relations

$$\mathfrak{R}\partial U_\lambda^{k+1} = \sum_{\kappa=1}^{\alpha^k} {}'\varepsilon_{\kappa\lambda}^k U_\kappa^k, \tag{1'}$$

to give the new matrices $'\mathbf{E}^k$. Since, as before, the boundary of a chain boundary vanishes, equations

$$'\mathbf{E}^{k-1}, '\mathbf{E}^k = 0$$

hold. We now attempt to find a new basis for each of the lattices $\mathfrak{X}^k$ $(k = 0, 1, \ldots, n)$ so that the new matrices take the simplest possible form, the normal form $\mathbf{H}^k$. We carry out the transformation from the original bases of the $\mathbf{E}^k$ to the final bases stepwise. Each step consists of changing only one basis chain for a particular dimension. The change is made by means of one of the following elementary transformations:

(a)   replacement of E_τ^t by $U_\tau^t = E_\tau^t + E_\nu^t$ $(\tau \neq \nu)$;

(b)   replacement of E_τ^t by $U_\tau^t = - E_\tau^t$.

After a transformation (a) all chains U_κ^k coincide with the old E_κ^k, except for U_τ^t. Only the incidence matrices $\mathbf{E}^{t-1}$ and $\mathbf{E}^t$ are thereby changed. In $\mathbf{E}^{t-1}$ the change is restricted to the τth column. To find this we form

$$\mathcal{R}\partial U_\tau^t = \mathcal{R}\partial(E_\tau^t + E_\nu^t) = \sum_{\sigma=1}^{\alpha^{t-1}} \left(\varepsilon_{\sigma\tau}^{t-1} + \varepsilon_{\sigma\nu}^{t-1} \right) U_\sigma^{t-1} = \sum_{\sigma=1}^{\alpha^{t-1}} {}'\varepsilon_{\sigma\tau}^{t-1} U_\sigma^{t-1}$$

giving,

$$'\varepsilon_{\sigma\tau}^{t-1} = \varepsilon_{\sigma\tau}^{t-1} + \varepsilon_{\sigma\nu}^{t-1}.$$

Thus, to obtain $'\mathbf{E}^{t-1}$ one must add the νth column of $\mathbf{E}^{t-1}$ to the τth column. In the left entry of the matrix $\mathbf{E}^t$ one replaces E_τ^t by $U_\tau^t = E_\tau^t + E_\nu^t$. Upon forming

$$\mathcal{R}\partial U_\lambda^{t+1} = \mathcal{R}\partial E_\lambda^{t+1} = \cdots + \varepsilon_{\tau\lambda}^t E_\tau^t + \cdots + \varepsilon_{\nu\lambda}^t E_\nu^t + \cdots$$

$$= \cdots + \varepsilon_{\tau\lambda}^t (E_\tau^t + E_\nu^t) + \cdots + (\varepsilon_{\nu\lambda}^t - \varepsilon_{\tau\lambda}^t) E_\nu^t + \cdots$$

$$= \sum_{\mu=1}^{\alpha^t} {}'\varepsilon_{\mu\lambda}^t U_\mu^t,$$

one recognizes that

$$'\varepsilon_{\mu\lambda}^t = \varepsilon_{\mu\lambda}^t \qquad \text{for} \quad \mu \neq \nu,$$

$$'\varepsilon_{\nu\lambda}^t = \varepsilon_{\nu\lambda}^t - \varepsilon_{\tau\lambda}^t.$$

That is, $'\mathbf{E}^t$ is obtained from $\mathbf{E}^t$ by subtracting the τth row from the νth row.

An elementary transformation (b) changes the sign of the τth column of $\mathbf{E}^{t-1}$ and the τth row of $\mathbf{E}^t$.

Just as we used an elementary transformation to transform the basis consisting of the oriented simplexes E to a basis consisting of the U, we can apply an elementary transformation to the basis consisting of chains U and transform it to a basis consisting of chains V. The matrices $'\mathbf{E}^k$ will be subjected to the same kind of operations as those just performed on the matrices $\mathbf{E}^k$. In particular, one can perform a row addition in $\mathbf{E}^t$ by using the transformations (a) and (b). One changes the sign of the τth row of $\mathbf{E}^t$, applies the transformation (a), and then once again reproduces the original sign of the τth row.

The elementary transformations (a) and (b) of chains can be used to effect elementary matrix operations, that is, row and column addition, and multiplication of a row or column by -1.* One can bring an integer-valued matrix to the normal form of §87, by using such transformations. In this normal form, the invariant factors are in the main diagonal beginning from the upper left, and their number is equal to the rank of the matrix; all other elements are 0.

We begin the normalization process for the incidence matrices by bringing the matrix $\mathbf{E}^0$ into the normal form.** Using appropriate elementary transformations we rearrange the invariant factors: they will no longer stand in the main diagonal, upper left (Fig. 42). Instead they will cut off the upper right-hand corner of the matrix $\mathbf{E}^0$ (Fig. 43). Their number is equal to the rank γ^0 of $\mathbf{E}^0$. While doing this, row rearrangements are simultaneously carried out on $\mathbf{E}^1$, since each left entry of $\mathbf{E}^1$ must agree with the corresponding upper entry of $\mathbf{E}^0$. As before, $\mathbf{E}^0\mathbf{E}^1 = 0$, so the last γ^0 rows of the rearranged matrix $\mathbf{E}^1$ must consist only of zeros. The upper $\alpha^1 - \gamma^0$ rows of the matrix $\mathbf{E}^1$ must be given further manipulation. Their rearrangements correspond to column rearrangements of $\mathbf{E}^0$, which now may be permitted to occur only in the first $\alpha^1 - \gamma^0$ columns, since the latter consist only of zeros and the values of the ε_{ik}^0 will be unchanged. $\mathbf{E}^0$ will then have assumed its final form, which we call $\mathbf{H}^0$. By performing column rearrangements of $\mathbf{E}^1$, which have no effect on the matrix $\mathbf{H}^0$, $\mathbf{E}^1$ is brought into the normal form $\mathbf{H}^1$ in which the invariant factors have positions corresponding to those in $\mathbf{H}^0$. The remaining matrices up to and including $\mathbf{E}^{n-1}$ are rearranged stepwise in like manner.

The incidence matrix $\mathbf{E}^k$ has now been transformed to its *normal form* $\mathbf{H}^k$. It contains only zeros, except on a segment cutting off the upper right-hand

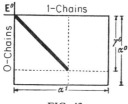

FIG. 42

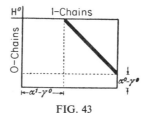

FIG. 43

---

*The elementary operations (a) and (b) are, by the way, special cases of integer-valued unimodular transformations which transform the E^t basis to a new U^t basis. It is generally true that, when σ is held fixed, the rows of variables $\varepsilon_{\sigma\tau}^{t-1}$ transform cogrediently in the index τ to the E_τ^t; when λ is held fixed, the rows of variables $\varepsilon_{\tau\lambda}^t$ transform contragrediently, because of formula (1). Since we make no general use of integer-valued unimodular transformations of chains, we can restrict ourselves here to the operations (a) and (b). We refer to Section 71 for the general case. The fact that each integer-valued unimodular transformation can be realized by a sequence of transformations (a) and (b) will not be used, and we therefore omit its proof.

**The fact that matrices $\mathbf{E}^0, \mathbf{E}^1, \ldots, \mathbf{E}^n$ are incidence matrices of a simplicial complex plays no role in the normalization process. We make use only of the relation $\mathbf{E}^{k-1}\mathbf{E}^k = 0$.

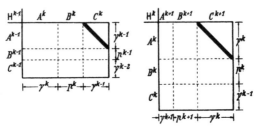

FIG. 44

corner (Fig. 44). Along this segment there are first ρ^k integers different than one: $c_1^k, c_2^k, \ldots, c_{\rho^k}^k$, each of which is a divisor of the integer preceding and which we recognize to be the k-dimensional torsion coefficients; these are followed by $\gamma^k - \rho^k$ ones.

Having normalized the incidence matrices, our goal is now easily reached. The lattices $\mathfrak{T}^k$, $\mathfrak{G}^k$, $\mathfrak{N}^k$ of all k-chains, closed k-chains and null homologous k-chains, respectively, are determined as follows.

Among the k-chains in the upper entries of $\mathbf{H}^{k-1}$ ($k = 1, 2, \ldots, n$) the last γ^{k-1} chains have nonzero boundary, since the invariant factors c_1^{k-1}, $c_2^{k-1}, \ldots, c_{\gamma^{k-1}}^{k-1}$ stand in the last γ^{k-1} columns of $\mathbf{H}^{k-1}$. Let us call these chains

$$C_1^k, C_2^k, \ldots, C_{\gamma^{k-1}}^k.$$

Let us call the first γ^k-chains in the left entry of $\mathbf{H}^k$

$$A_1^k, A_2^k, \ldots, A_{\gamma^k}^k.$$

They are division–null homologous (≈ 0). For the invariant factors $c_1^k, c_2^k, \ldots, c_{\gamma^k}^k$ are in the last γ^k columns of $\mathbf{H}^k$, so one has the boundary relations

$$\mathfrak{R} \partial C_\lambda^{k+1} = c_\lambda^k A_\lambda^k \qquad (\lambda = 1, 2, \ldots, \gamma^k, k = 0, 1, \ldots, n - 1). \tag{3}$$

The chains C^k and A^k will not appear for every value of k from 0 through n. For example, the C^0 are missing for $k = 0$ since every 0-chain is closed, and the A^n are missing for $k = n$ since no $(n + 1)$-simplexes exist and thus there exist no division–null homologous n-chains other than the n-chain 0. It is clear that a chain A^k cannot be a C^k-chain since the A^k, as division–null homologous chains, are closed while the C^k are not. On the other hand, the α^k basis chains will not in general be exhausted by the chains A^k and C^k. Call the remaining chains

$$B_1^k, B_2^k, \ldots, B_{p^k}^k.$$

They, like the A^k, are closed, but are not division–null homologous and they number

$$p^k = \alpha^k - \gamma^k - \gamma^{k-1} \tag{4}$$

for $0 < k < n$. For $k = 0$, $p^0 = \alpha^0 - \gamma^0$, and for $k = n$, $p^n = \alpha^n - \gamma^{n-1}$. If one specifies as a convention that $\gamma^{-1} = \gamma^n = 0$, then formula (4) is valid for all k from 0 through n.

Because the upper entries of $\mathbf{H}^{k-1}$ represent a basis of the lattice $\mathfrak{T}^k$ of all k-chains ($k = 1, 2, \ldots, n$), each k-chain can be written in the form

$$V^k = \sum_{\lambda=1}^{\gamma^k} x_\lambda^k A_\lambda^k + \sum_{\mu=1}^{p^k} y_\mu^k B_\mu^k + \sum_{\nu=1}^{\gamma^{k-1}} z_\nu^k C_\nu^k.$$

Since the chains A_λ^k and B_μ^k are closed V^k is closed only if

$$\sum_{\nu=1}^{\gamma^{k-1}} z_\nu^k \Re \partial C_\nu^k = \sum_{\nu=1}^{\gamma^{k-1}} z_\nu^k c_\nu^{k-1} A_\nu^{k-1} = 0,$$

which is the case only when all z_ν^k are equal to 0, since the chains A_ν^{k-1} are linearly independent. Also, in the case $k = 0$, there is no chain C_ν^0, and each closed chain is a linear combination of the chains A_λ^0 and B_λ^0.

Thus, for $k = 0, 1, \ldots, n$, the lattice $\mathfrak{G}^k$ of closed k-chains will be spanned by the γ^k-chains A_λ^k and the p^k chains B_μ^k.

We now look at the sublattice $\mathfrak{N}^k$ of null homologous k-chains. In order for a closed k-chain

$$V^k = \sum_{\lambda=1}^{\gamma^k} x_\lambda^k A_\lambda^k + \sum_{\mu=1}^{p^k} y_\mu^k B_\mu^k \qquad (k = 0, 1, \ldots, n-1) \qquad (5)$$

to be null homologous it must be the boundary of a $(k+1)$-chain, and since the $(k+1)$-chains A_ρ^{k+1} and B_σ^{k+1} are closed, it must be the boundary of a linear combination of the chains C_λ^{k+1}. It must therefore be of the form

$$V^k = \Re \partial \sum_{\lambda=1}^{\gamma^k} z_\lambda^{k+1} C_\lambda^{k+1} = \sum_{\lambda=1}^{\gamma^k} z_\lambda^{k+1} c_\lambda^k A_\lambda^k. \qquad (6)$$

Because of the linear independence of the chains A_λ^k and B_μ^k, the right-hand sides of Eqs. (5) and (6) must coincide coefficient for coefficient; thus we must have

$$x_\lambda^k = z_\lambda^{k+1} c_\lambda^k, \qquad y_\mu^k = 0.$$

Accordingly, the k-chain (5) is homologous to 0 if and only if the coefficients satisfy the conditions

$$x_\lambda^k \equiv 0 \pmod{c_\lambda^k}, \qquad y_\mu^k = 0. \qquad (7)$$

These equations also hold for $k = n$ because the chain A^n is lacking and the only null homologous n-chain is the n-chain 0. Thus two closed k-chains V^k and $'V^k$ belong to the same homology class ($k = 0, 1, \ldots, n$) if and only if

$$x_\lambda^k \equiv {'x_\lambda^k} \pmod{c_\lambda^k} \qquad \text{and} \qquad y_\mu^k = {'y_\mu^k}. \qquad (8)$$

When one reduces the coefficients x_λ^k appearing in (5) modulo c_λ^k to the interval

$$0 \leq \xi_\lambda^k < c_\lambda^k, \tag{9}$$

one obtains a homologous chain

$$V^k \sim \sum_{\lambda=1}^{\rho^k} \xi_\lambda^k A_\lambda^k + \sum_{\mu=1}^{p^k} \eta_\mu^k B_\mu^k. \tag{10}$$

In this chain all A_λ^k are missing for which $c_\lambda^k = 1$, because of the normalization condition (9), so that the first sum only runs from 1 through ρ^k; here ρ^k denotes the number of invariant factors of the matrix $\mathbf{E}^k$ which are different than unity. The coefficients ξ_λ^k and η_μ^k are unambiguously determined by V^k, because of (8).

In other words: The cyclic subgroups of the homology group $\mathfrak{H}^k$ which are generated by the homology classes of the chains $A_1^k, A_2^k, \ldots, A_{\rho^k}^k$ have orders $c_1^k, c_2^k, \ldots, c_{\rho^k}^k$, from (7) and (8); those generated by the homology classes of the chains $B_1^k, B_2^k, \ldots, B_{p^k}^k$ are free Abelian groups. Due to (10) each element of the homology group can be unambiguously represented as a sum of elements of these subgroups, one element occurring in the sum for each subgroup. $\mathfrak{H}^k$ is thereby the direct sum of these subgroups; $A_1^k, A_2^k, \ldots, A_{\rho^k}^k$ together with $B_1^k, B_2^k, \ldots, B_{p^k}^k$ form a homology basis; the chains $B_1^k, B_2^k, \ldots, B_{p^k}^k$ form a Betti basis for the dimension k. The integer $p^k = \alpha^k - \gamma^k - \gamma^{k-1}$ is the Betti number and $c_1^k, c_2^k, \ldots, c_{\rho^k}^k$ are the k-dimensional torsion coefficients. The homology groups have therefore been determined and we have proved the following theorem:

THEOREM. *If α^k is the number of k-simplexes in the simplicial complex $\mathfrak{R}^n$, and if γ^k is the rank of the incidence matrix $\mathbf{E}^k$, then the Betti number for the dimension k is*

$$p^k = \alpha^k - \gamma^k - \gamma^{k-1}$$

$(\gamma^{-1} = \gamma^n = 0)$, *and the torsion coefficients for the dimension k are those invariant factors of $\mathbf{E}^k$ which differ from 1; they are missing for $k = n$.*

We sometimes call the Betti numbers and torsion coefficients the *numerical invariants* of the complex, to distinguish them from other invariant mathematical objects associated with the complex, such as, for example, the fundamental group, which will be derived later.

22. Block Chains

While the determination of the homology groups from the incidence matrices is always possible in principle, the actual computation may be quite involved. For example, if one starts from the simplicial decomposition of the torus given in Fig. 40 one has $\alpha^0 = 9$, $\alpha^1 = 27$, $\alpha^2 = 18$; one must perform computations with matrices having 27 rows, which are tedious even to write down, let alone to transform into normal form.

A simpler procedure for computation of the numerical invariants and homology groups is clearly desirable. We can find such a procedure. We choose as terms of the k-chains not individual simplexes, as was done previously, but whole chains, so-called blocks. Just as we assembled individual simplexes together to form k-chains, we build block chains from finitely many blocks and use these block chains as a foundation for the construction of the homology groups.

For each dimension $k = 0, 1, \ldots, n$ we select a finite set of simplicial chains

$$Q_1^k, Q_2^k, \ldots, Q_{\bar{\alpha}^k}^k$$

which we call *k-blocks* and which satisfy the following conditions:

(Bl1) $Q_1^k, Q_2^k, \ldots, Q_{\bar{\alpha}^k}^k$ are linearly independent; that is,

$$t_1 Q_1^k + t_2 Q_2^k + \cdots + t_{\bar{\alpha}^k} Q_{\bar{\alpha}^k}^k = 0$$

implies $t_1 = t_2 = \cdots = t_{\bar{\alpha}^k} = 0$.

This condition will always be satisfied, for example, when no two blocks Q_i^k and Q_j^k have a k-simplex in common.

A linear combination of blocks is called a *block chain*. Because the blocks are linearly independent, two block chains $\sum t_\mu Q_\mu^k$ and $\sum' t_\mu Q_\mu^k$ will be equal if and only if $t_1 = {'}t_1, t_2 = {'}t_1, \ldots, t_{\bar{\alpha}^k} = {'}t_{\bar{\alpha}^k}$. Since the block chains are simplicial chains, each block chain will have a well-defined boundary and there will exist closed and null homologous block chains.

We require further of the blocks:

(Bl2) The boundary of a $(k + 1)$-dimensional block chain is a k-dimensional block chain.

Obviously it suffices to require that the boundary of each $(k + 1)$-block be a k-block chain:

$$\Re \partial Q_\lambda^{k+1} = \sum_{\kappa=1}^{\bar{\alpha}^k} \bar{\varepsilon}_{\kappa\lambda}^k Q_\kappa^k \qquad (\lambda = 1, 2, \ldots, \bar{\alpha}^{kH}). \tag{1}$$

(Bl3) For each closed simplicial k-chain there exists a homologous block chain.

(Bl4) If a k-block chain is null homologous, that is, the boundary of a simplicial $(k + 1)$-chain, it is also the boundary of a $(k + 1)$-dimensional block chain.

The conditions (Bl1) through (Bl4) are consistent, because the oriented simplexes of $\Re^n$ form a particular block system for which the four conditions are obviously satisfied.

For the computation of the homology groups it now suffices to use the coarse structured block chains instead of the fine structured simplicial chains. The computational procedure is exactly the same as the procedure used earlier, only one begins not with incidence matrices of the simplicial complex, but with block incidence matrices

$$\left(\bar{\varepsilon}_{\kappa\lambda}^{k} \right) = \overline{\mathbf{E}}^{k}$$

which are given by the boundary relations (1). Nevertheless, we will repeat the derivation in order to indicate at which points conditions (Bl1) through (Bl4) are used.

If one takes the boundary on both sides of (1), for $k > 0$, the $(k - 1)$-chain 0 must result since the boundary of a boundary vanishes. Thus

$$\sum_{\kappa=1}^{\bar{\alpha}^{k}} \sum_{i=1}^{\bar{\alpha}^{k-1}} \bar{\varepsilon}_{\kappa\lambda}^{k} \bar{\varepsilon}_{i\kappa}^{k-1} Q_{i}^{k-1} = 0$$

which, due to the linear independence of the Q_{i}^{k-1}, is equivalent to

$$\sum_{\kappa=1}^{\bar{\alpha}^{k}} \bar{\varepsilon}_{i\kappa}^{k-1} \bar{\varepsilon}_{\kappa\lambda}^{k} = 0 \qquad \text{or} \qquad \overline{\mathbf{E}}^{k-1} \overline{\mathbf{E}}^{k} = 0.$$

By carrying out unimodular transformations upon the blocks Q^{k} which correspond to the entries of the matrices, we bring the matrices $\overline{\mathbf{E}}^{k}$ simultaneously to the normal form $\overline{\mathbf{H}}^{k}$, as was done in §21.

The block chains which correspond to the entries of the matrices $\overline{\mathbf{H}}^{k}$ are always linearly independent; for each dimension they decompose to three types:

1.   The chains $\overline{A}_{\lambda}^{k}$ $(\lambda = 1, 2, \ldots, \bar{\gamma}^{k})$; they are division–null homologous.
2.   The chains $\overline{B}_{\mu}^{k}$ $(\mu = 1, 2, \ldots, \bar{p}^{k})$; these chains are also closed, but no nonzero multiple is a boundary of a $(k + 1)$-block chain.
3.   The chains $\overline{C}_{\nu}^{k}$ $(\nu = 1, 2, \ldots, \bar{\gamma}^{k-1})$; these chains are not closed and we have

$$\mathcal{R} \partial \overline{C}_{\nu}^{k} = \bar{c}_{\nu}^{k-1} \overline{A}_{\nu}^{k-1}.$$

Here $\bar{\gamma}^{k}$ is the rank of $\overline{\mathbf{E}}^{k}$, $\bar{c}_{\nu}^{k}$ are the invariant factors of $\overline{\mathbf{E}}^{k}$, and $\bar{p}^{k} = \bar{\alpha}^{k} - \bar{\gamma}^{k} - \bar{\gamma}^{k-1}$ (with $\bar{\gamma}^{-1} = 0$ and $\bar{\gamma}^{n} = 0$). It now follows, as earlier: The most general closed k-block chain is a linear combination of the $\overline{A}_{\lambda}^{k}$ and the $\overline{B}_{\mu}^{k}$. The null homologous k-block chains, which due to (Bl4) are also boundaries of $(k + 1)$-clock chains, are the chains $\bar{c}_{\kappa}^{k} \overline{A}_{\kappa}^{k}$ and the linear combinations which result from them. Each closed block chain, and hence [from (Bl3)] each closed simplicial chain, is thereby homologous to one and only one linear combination

$$\xi_{1} \overline{A}_{1}^{k} + \xi_{2} \overline{A}_{2}^{k} + \cdots + \xi_{\bar{\gamma}^{k}} \overline{A}_{\bar{\gamma}^{k}}^{k} + \eta_{1} \overline{B}_{1}^{k} + \eta_{2} \overline{B}_{2}^{k} + \cdots + \eta_{\bar{p}^{k}} \overline{B}_{\bar{p}^{k}}^{k},$$

where $0 \leqq \xi_\lambda < \bar{c}_\lambda^k$ and the range of the η_μ is not restricted. Consequently, those $\bar{A}^k$ for which the associated invariant factor is $\neq 1$, together with the $\bar{B}^k$ form a homology basis. The $\bar{B}^k$ alone form a Betti basis for the dimension k. The Betti number is therefore

$$p^k = \bar{p}^k = \bar{\alpha}^k - \bar{\gamma}^k - \bar{\gamma}^{k-1} \qquad (2)$$

and the torsion coefficients are those invariant factors $c_\nu^k = \bar{c}_\nu^k$ of $\bar{\mathbf{E}}^k$ which are not equal to 1.

EXAMPLE 1. One can use the following chains as a block system on the torus (§19): the vertex O, the 1-chains a^1 and b^1 (meridian and latitude circles), and the 2-chain U^2 which is formed by the coherently oriented 2-simplexes. (Bl1) is satisfied because the k-blocks have no common k-simplex. (Bl2) is satisfied because they are closed chains. (Bl3) is satisfied for $k = 0$ because any pair of 0-simplexes of a connected complex is homologous; it is satisfied for $k = 1$ because we have shown that each closed 1-chain $\sim \alpha a^1 + \beta b^1$; it is satisfied for $k = 2$ because all closed 2-chains are multiples of the block chain U^2 consisting of all 2-simplexes. Finally, (Bl4) is satisfied, since for $k = 0$ the only null homologous 0-block chain is the block chain 0; for $k = 1$ we proved under (II) of §19 that the only null homologous 1-chain composed of the blocks a^1 and b^1 is the 1-chain 0; there are no null homologous 2-chains. The block matrices then read

$\bar{\mathbf{E}}^0$	a^1	b^1
O	0	0

$\bar{\mathbf{E}}^1$	U^2
a^1	0
b^1	0

From these matrices one can verify that the homology groups are as computed earlier.

EXAMPLE 2. A block system for the octahedral decomposition of the 2-sphere (§14) is formed by two diametrically opposite vertices of the equatorial circle, the two semicircles joining them (each consisting of two 1-simplexes) on the equatorial circle, and the two hemispherical surfaces into which the sphere is bisected by the equatorial circle.

From this block system one can obtain a block system for the projective plane as follows. After making a normal subdivision, one identifies diametrically opposite blocks of the 2-sphere to form a single block of the projective plane. The resulting block system has exactly one block for each dimension. This procedure can be extended to the case of arbitrary dimension, and one can thereby compute the block incidence matrices and hence the homology groups of projective n-space $\mathfrak{P}^n$. We will not perform this calculation, because it is somewhat complicated to prove that we are in fact dealing with a block system. We shall later find a simpler way to obtain these homology groups (§31).

The concept of the block chain will be a significant tool in the theory of manifolds, for example in §41, §61, and §67.

23. Chains mod 2, Connectivity Numbers, Euler's Formula

The concepts of chain and boundary and the related developments of the previous sections depend in an essential way upon the concept of the orientation of a simplex. One can construct an analogous theory of chains which disregards all orientations. Just as the homology groups and Betti numbers were obtained by means of chains of oriented simplexes, the connection groups and connectivity numbers will be obtained by means of chains of nonoriented simplexes, also referred to as chains mod 2. It will later prove to be the case that the theories and methods valid for oriented chains carry over immediately to chains mod 2. Admittedly, connection groups are of limited importance, in the sense that they can be derived from homology groups. Occasionally, however, because they possess a greater generality, they allow general statements to be made in cases where the homology groups fail. For example, they will be used later to extend theorems such as the duality theorem, which holds only in orientable manifolds for ordinary chains, to nonorientable manifolds.[15]

In the treatment which follows we shall disregard the orientation of the simplexes or, what amounts to the same thing, we shall identify each oriented simplex with itself, oppositely oriented. In this case the double of a chain U^k, $U^k + U^k$, is the same as the chain $U^k - U^k = 0$. It necessarily follows that two chains

$$U^k = u_1 E_1^k + u_2 E_2^k + \cdots + u_{\alpha^k} E_{\alpha^k}^k$$

and

$$'U^k = 'u_1 E_1^k + 'u_2 E_2^k + \cdots + 'u_{\alpha^k} E_{\alpha^k}^k$$

are regarded as being the same if their corresponding coefficients differ by even integers, that is, $u_\nu = 'u_\nu \pmod 2$ for $\nu = 1, 2, \ldots, \alpha^k$. One also calls the chains *congruent mod* 2. One can associate to each class of mod 2 congruent chains U^k, $'U^k$ a unique k-dimensional subcomplex. A nonoriented simplex $\mathfrak{E}^k$ will belong to this subcomplex if and only if the associated oriented simplex E^k occurs in U^k, and hence also in $'U^k$, with odd multiplicity.* This subcomplex is called the subcomplex belonging to U^k (and likewise to $'U^k$). (It does not necessarily consist of all of the simplexes appearing in U^k!) A "chain mod 2" is now nothing other than a k-dimensional pure subcomplex. The reason that we choose the terminology "chain mod 2" and not "subcomplex" is because later, when we study singular chains mod 2, the term subcomplex will be inappropriate. Nevertheless, we shall denote chains mod 2, like complexes, with German letters.

We now define the sum and the boundary of chains mod 2. For this, we begin with ordinary chains.

If

$$W^k = U^k + V^k,$$

and $\mathfrak{W}^k$, $\mathfrak{U}^k$, and $\mathfrak{V}^k$ denote the associated chains mod 2, then we define as the *sum* $\mathfrak{U}^k + \mathfrak{V}^k$ the chain mod 2

$$\mathfrak{W}^k = \mathfrak{U}^k + \mathfrak{V}^k.$$

According to this definition, $\mathfrak{U}^k + \mathfrak{V}^k$ *consists of all simplexes* which appear in $U^k + V^k$ with odd multiplicity and therefore appear with odd multiplicity in

---

*Here, in contrast to §12, it is necessary to consider the "empty subcomplex" to be among the subcomplexes.

exactly one of the chains U^k or V^k, that is, of all nonoriented simplexes *which appear in exactly one of the chains mod* 2, $\mathfrak{U}^k$ *or* $\mathfrak{B}^k$. On the contrary, if a simplex $\mathfrak{S}^k$ belongs to $\mathfrak{U}^k$ as well as $\mathfrak{B}^k$, then it no longer appears in the sum.

We saw earlier that each chain can be represented by a *vector*, the components of which are the multiplicities of the simplexes appearing in the chain. In this way, chain addition was reduced to vector addition. Vector representation is also possible for the chains mod 2. In the latter case the components of the vectors are not integers, but are residue classes mod 2. We shall indicate the residue classes of the even and of the odd integers by the symbols $\breve{0}$ and $\breve{1}$ respectively. We then have the rules of computation

$$\breve{0} + \breve{0} = \breve{0}, \quad \breve{0} + \breve{1} = \breve{1}, \quad \breve{1} + \breve{1} = \breve{0},$$

$$\breve{0} \cdot \breve{0} = \breve{0}, \quad \breve{0} \cdot \breve{1} = \breve{0} \quad \breve{1} \cdot \breve{1} = \breve{1}.$$

The computation rule $\breve{1} + \breve{1} = \breve{0}$ states, for example, that the sum of two odd numbers is an even number. A chain mod 2, $\mathfrak{U}^k$, can then be written as a vector

$$\mathfrak{U}^k = (\breve{u}_1, \breve{u}_2, \ldots, \breve{u}_{\alpha^k}),$$

where $\breve{u} = \breve{1}$ or $\breve{0}$, according to whether the simplex $\mathfrak{S}^k$ does or does not appear, respectively, in $\mathfrak{U}^k$. The sum of $\mathfrak{U}^k$ and

$$\mathfrak{B}^k = (\breve{v}_1, \breve{v}_2, \ldots, \breve{v}_{\alpha^k})$$

is given by

$$\mathfrak{U}^k + \mathfrak{B}^k = (\breve{u}_1 + \breve{v}_1, \breve{u}_2 + \breve{v}_2, \ldots, \breve{u}_{\alpha^k} + \breve{v}_{\alpha^k}).$$

In particular, the simplexes $\mathfrak{S}_1^k, \mathfrak{S}_2^k, \ldots, \mathfrak{S}_{\alpha^k}^k$ correspond to the unit vectors

$$\mathfrak{S}_1^k = (\breve{1}, \breve{0}, \ldots, \breve{0}), \quad \mathfrak{S}_2^k = (\breve{0}, \breve{1}, \ldots, \breve{0}), \quad \ldots, \quad \mathfrak{S}_{\alpha^k}^k = (\breve{0}, \breve{0}, \ldots, \breve{1}),$$

so that one can also write for $\mathfrak{U}^k$,

$$\mathfrak{U}^k = \breve{u}_1 \mathfrak{S}_1^k + \breve{u}_2 \mathfrak{S}_2^k + \cdots + \breve{u}_{\alpha^k} \mathfrak{S}_{\alpha^k}^k. \tag{1}$$

We will adopt the convention that one can simply write $\mathfrak{S}^k$ instead of $\breve{1}\mathfrak{S}^k$ and one can simply omit $\breve{0}\mathfrak{S}^k$. If all $\breve{u} = \breve{0}$ we write $\mathfrak{U}^k = 0$.

We have already agreed that the boundary of an oriented simplex E^k is the sum of its $(k-1)$-dimensional faces, furnished with the induced orientations. Since we dispense with all orientations in the theory of chains mod 2, we will have to regard $\mathfrak{R} \partial \mathfrak{S}^k$ as the chain mod 2 formed from all of the nonoriented faces of $\mathfrak{S}^k$:

$$\mathfrak{R} \partial \mathfrak{S}_\kappa^k = \sum_{i=1}^{\alpha^{k-1}} \breve{\varepsilon}_{i\kappa}^{k-1} \mathfrak{S}_i^{k-1}; \tag{2}$$

here $\breve{\varepsilon}_{i\kappa}^{k-1} = \breve{1}$ or $\breve{0}$ according to whether $\mathfrak{S}_i^{k-1}$ is or is not incident, respectively, with $\mathfrak{S}_\kappa^k$. The boundary of an arbitrary chain mod 2, (1), is then defined as the sum of the boundaries of the individual simplexes,

$$\mathfrak{R}\partial\mathfrak{U}^k = \sum_{\kappa=1}^{\alpha^k} \check{u}_\kappa \mathfrak{R}\partial\mathfrak{E}^k_\kappa.$$

Consequently, *a* $(k-1)$-*simplex* $\mathfrak{E}^{k-1}_i$ *will belong to* $\mathfrak{R}\partial\mathfrak{U}^k$ *if and only if it is incident to an odd number of simplexes of* $\mathfrak{U}^k$. For 0-dimensional chains mod 2 the boundary is the number 0.

One can now carry over the concepts defined for ordinary chains to chains mod 2.

A chain mod 2 is said to be *closed* if its boundary vanishes;

$$\mathfrak{R}\partial\mathfrak{U}^k = 0.$$

As an example, in the projective plane (Fig. 39) the totality of triangles is a closed chain mod 2 because exactly two triangles are incident on each edge. On the other hand, as we know, there are no closed 2-chains on the projective plane. As another example, each pure unbounded subcomplex in the sense of §12 is a closed chain mod 2.

All 0-dimensional chains mod 2 are closed.

A chain mod 2, $\mathfrak{U}^k$, is said to be null homologous if it is the boundary of a $(k+1)$-chain mod 2. More generally, two (not necessarily closed) chains mod 2 are said to be *homologous* to one another if their difference is null homologous.

Every null homologous chain mod 2 is closed, since the boundary of an individual $(k+1)$-simplex is a closed chain mod 2.

One may now divide the closed chains mod 2 into homology classes. These homology classes form a group, $\check{\mathfrak{H}}^k$ when addition of two homology classes is defined by means of addition of two representatives. $\check{\mathfrak{H}}^k$ is obviously the analog of the homology group $\mathfrak{H}^k$ and is called the *kth connectivity group* of the simplicial complex $\mathfrak{R}^n$. For a finite complex $\mathfrak{R}^n$, $\check{\mathfrak{H}}^k$ is a finite group, because there are only finitely many distinct chains mod 2, namely, 2^{α^k}; thus there are only finitely many homology classes. Since $\mathfrak{U}^k + \mathfrak{U}^k = 0$ always, each element of $\check{\mathfrak{H}}^k$ is of order 2. $\check{\mathfrak{H}}^k$ is therefore the direct sum of finitely many, say q^k,* groups of order 2. The integer q^k is called the *kth connectivity number of* of $\mathfrak{R}^n$.**

A series of chains mod 2

$$\mathfrak{U}^k_1, \mathfrak{U}^k_2, \ldots, \mathfrak{U}^k_r \tag{3}$$

is said to be linearly independent if a linear equation

$$\check{\imath}_1\mathfrak{U}^k_1 + \check{\imath}_2\mathfrak{U}^k_2 + \cdots + \check{\imath}_r\mathfrak{U}^k_r = 0 \tag{4}$$

implies all $\check{\imath} = \check{0}$. If the chains mod 2 in (3) are also closed, then they are said

---

*The k is a superscript, not an exponent.
**The name will first be explained in the theorem of §41.

to be *homologously independent* if a homology

$$\check{i}_1 \mathfrak{U}_1^k + \check{i}_2 \mathfrak{U}_2^k + \cdots + \check{i}_r \mathfrak{U}_r^k \sim 0 \tag{5}$$

implies all $\check{i} = \check{0}$. Otherwise, one speaks of linearly or homologously dependent chains mod 2, respectively. As an example, r different k-simplexes are linearly independent.

One can obtain a system of homologously independent k-chains mod 2 by selecting the nonzero group element (that is, a homology class mod 2) from each of the q^k groups of order 2 whose direct sum is $\check{\mathfrak{H}}^k$ and then selecting a representative for each homology class. The selected chains are obviously homologously independent. Such a system of q^k chains mod 2 is called a k-dimensional *connectivity basis*. No more than q^k homologously independent chains mod 2 can exist. For if $\mathfrak{U}_1^k, \mathfrak{U}_2^k, \ldots, \mathfrak{U}_r^k$ are homologously independent, then among all of the possible linear combinations $\check{i}_1 \mathfrak{U}_1^k + \check{i}_2 \mathfrak{U}_2^k + \cdots + \check{i}_r \mathfrak{U}_r^k$, no two combinations can be homologous to one another. Thus there exist at least 2^r different homology classes. As there can only exist 2^{q^k} homology classes in all, since q^k is the order of $\check{\mathfrak{H}}^k$, it follows that $r \leqq q^k$. *The k-th connectivity number is therefore the maximum number of homologously independent k-chains mod* 2, and consequently represents the analog of the Betti number.

We previously computed the Betti number p^k from the incidence matrices. We can now derive the connectivity number q^k from the *incidence matrices mod* 2, which arise from the boundary relations (2):

$$\check{\mathbf{E}}^{k-1} = \left(\check{\varepsilon}_{i\kappa}^{k-1}\right) \qquad (i = 1, 2, \ldots, \alpha^{k-1}, \kappa = 1, 2, \ldots, \alpha^k, k = 1, 2, \ldots, n).$$

The procedure is almost word for word the same as before. Chains mod 2 replace the chains used previously and the elements of the incidence matrices are no longer integers but are the residue classes $\check{0}$ and $\check{1}$. The matrix equation $\mathbf{E}^{k-1}\mathbf{E}^k = 0$, upon which the normalization procedure essentially rests, is replaced by the equation

$$\check{\mathbf{E}}^{k-1}\check{\mathbf{O}}^k = \check{\mathbf{0}}.$$

Here $\check{\mathbf{O}}$ is an abbreviation for a matrix of α^{k-1} rows and α^{k+1} columns having all elements $\check{0}$. The second of the elementary transformations (a) and (b) of §21 is no longer needed, because replacement of a chain mod 2 by its negative is the same as the identity transformation, since $\check{1} = -\check{1}$. In the normal form $\check{\mathbf{H}}^k$ of the incidence matrix mod 2 $\check{\mathbf{E}}^k$ all elements are now equal to $\check{0}$ except for elements of a diagonal segment which cuts off the upper right-hand corner of $\check{\mathbf{H}}^k$; these elements are equal to $\check{1}$. Let their number be δ^k.

The chains mod 2 corresponding to the rows and columns of the matrices $\check{\mathbf{E}}^0, \check{\mathbf{E}}^1, \ldots, \check{\mathbf{E}}^{n-1}$ are, for each k, the α^k k-simplexes of $\mathfrak{R}^n$. These are linearly independent and each k-chain mod 2 is a linear combination of them. This property is obviously preserved under elementary transformations and will also hold for the chains mod 2 which stand in the entries of the normalized matrices $\check{\mathbf{H}}^0, \check{\mathbf{H}}^1, \ldots, \check{\mathbf{H}}^{n-1}$. For each dimension they decompose into three types, $\mathfrak{A}^k_\lambda, \mathfrak{B}^k_\mu$, and $\mathfrak{C}^k_\nu$. The $\mathfrak{C}^k_\nu$ ($\nu = 1, 2, \ldots, \delta^{k-1}$) are the last δ^k chains in the upper entry of $\check{\mathbf{H}}^{k-1}$. They are not closed, for an element $\check{1}$ appears in each column. The $\mathfrak{A}^k_\lambda$ ($\lambda = 1, 2, \ldots, \delta^k$) are the first δ^k chains in the left entry of $\check{\mathbf{H}}^k$. They are null homologous since one has

$$\mathfrak{R}\,\partial\mathfrak{C}^{k+1}_\lambda = \mathfrak{A}^k_\lambda \qquad (\lambda = 1, 2, \ldots, \delta^k). \tag{6}$$

The remaining chains mod 2 are called $\mathfrak{B}^k_\mu$ ($\mu = 1, 2, \ldots, \alpha^k\delta^k\delta^{k-1}$). They are closed but are not null homologous.[*]

The most general closed k-chain mod 2 is a linear combination of the $\mathfrak{A}^k_\lambda$ and the $\mathfrak{B}^k_\mu$. Since the $\mathfrak{A}^k_\lambda$ are ~ 0, it is then homologous to a linear combination

$$\sum_{\mu=1}^{\alpha^k - \delta^k - \delta^{k-1}} \check{\eta}_\mu \mathfrak{B}^k_\mu. \tag{7}$$

On the other hand, no such linear combination is ~ 0, since only the $\mathfrak{A}^k_\lambda$ appear in the boundary relations (6). Consequently the $\mathfrak{C}^k_\mu$ represent a k-dimensional connectivity basis, and the kth connectivity number is

$$q^k = \alpha^k - \delta^k - \delta^{k-1} \tag{8}$$

where we set $\delta^{-1} = \delta^n = 0$; δ^k is the rank of $\check{\mathbf{H}}^k$, that is, the number of rows in the subdeterminant having the largest number of rows which is not equal to $\check{0}$, that is, $= \check{1}$. This rank does not change under row and column additions. Thus δ^k is also the rank of the original incidence matrix mod 2, $\check{\mathbf{E}}^k$.

There is a relation between the Betti number p^k and the connectivity number q^k, which we shall now derive. Assume that the incidence matrix $\mathbf{E}^k$ has g^k even invariant factors, that is, $\gamma^k - g^k$ odd invariant factors (including those which have value 1). The diagonal form $\mathbf{H}^k$ of $\mathbf{E}^k$ will contain a $(\gamma^k - g^k)$-rowed subdeterminant of odd value, namely, the product of the $\gamma^k - g^k$ odd invariant factors, whereas all $(\gamma^k - g^k + 1)$-rowed subdeterminants are even. The same is true for $\mathbf{E}^k$, because the divisors of the determinant are invariant with respect to elementary transformations (Section 87). If one goes from $\mathbf{E}^k$ to the incidence matrix mod 2, $\check{\mathbf{E}}^k$, by replacing each even element with $\check{0}$ and each odd element with $\check{1}$, it follows that $\check{\mathbf{E}}^k$ has a $(\gamma^k - g^k)$-rowed subdeterminant of value $\check{1}$, while all $(\gamma^k - g^k + 1)$-rowed subdeterminants are equal to $\check{0}$. Consequently the rank δ^k of $\mathbf{E}^k$ is

$$\delta^k = \gamma^k - g^k.$$

---

[*]The chains mod 2 $\mathfrak{A}^k_\lambda$, $\mathfrak{B}^k_\mu$, and $\mathfrak{C}^k_\nu$ are not in general the chains mod 2 associated with the chains A^k_λ, B^k_μ, and C^k_ν introduced in §21. For example, if a chain C^k_ν lies above an even torsion coefficient in the normalized incidence matrix $\mathbf{H}^{k-1}$, then the associated chain mod 2 is closed and does not appear among the chains mod 2 $\mathfrak{C}^k_\nu$.

When we set this value into (8) it follows that

$$q^k = (\alpha^k - \gamma^k - \gamma^{k-1}) + g^k + g^{k-1}.$$

From §21, the expression in parentheses on the right is the Betti number p^k, giving

$$q^k = p^k + g^k + g^{k-1}, \tag{9}$$

where we set

$$g^{-1} = g^n = 0.$$

*The kth connectivity group is therefore determined by the homology groups of dimension k and $(k - 1)$.*These homology groups determine the Betti number p^k and the k-dimensional and $(k - 1)$-dimensional torsion coefficients and thus determine g^k and g^{k-1}, which, in turn, determine the connectivity number q^k and consequently the connectivity group.

The connectivity number q^k is never smaller than the Betti number p^k.

Block chains can also be defined for the chains mod 2 and they can be used for the computation of the connectivity groups. The conditions required to specify them correspond to the conditions (Bl1) through (Bl4). We leave the details as an exercise for the reader.

Among the complexes whose homology groups we have computed, only the projective plane (§19) possesses a torsion coefficient and, in fact, an even one. Here $g^1 = 1$, so that $q^1 = p^1 + g^1 = 1$, $q^2 = p^2 + g^1 = 1$, in agreement with the fact that there exists a chain mod 2 not homologous to zero of dimension 1 (a projective line), and one of dimension 2, formed by all of the triangles of the simplicial decomposition. In all of the other examples, the connectivity numbers coincide with the Betti numbers, due to the absence of torsion coefficients.

The Euler Characteristic

When one forms the alternating sums of connectivity numbers given in the formulas (9), and uses $g^{-1} = g^n = 0$ one obtains

$$\sum_{k=0}^{n} (-1)^k q^k = \sum_{k=0}^{n} (-1)^k p^k.$$

Then because of the formulas

$$p^k = \alpha^k - \gamma^k - \gamma^{k-1} \qquad (\S21),$$

and

$$p^k = \bar{\alpha}^k - \bar{\gamma}^k - \bar{\gamma}^{k-1} \qquad (\S22),$$

the right-hand side becomes

$$\sum_{k=0}^{n} (-1)^k \alpha^k = \sum_{k=0}^{n} (-1)^k \bar{\alpha}^k,$$

in which α^k is the number of k-simplexes of a simplicial decomposition and $\bar{\alpha}^k$ is the number of k-blocks of a block system. *One then has the relation*

$$-N = \sum_{k=0}^{n} (-1)^k p^k = \sum_{k=0}^{n} (-1)^k q^k = \sum_{k=0}^{n} (-1)^k \alpha^k = \sum_{k=0}^{n} (-1)^k \bar{\alpha}^k. \quad (10)$$

The number N is called the Euler characteristic of the simplicial complex $\Re^n$.

If we assume the result, first proved in the next chapter, that N is a topological invariant of $\Re^n$ and is independent of the particular simplicial decomposition, then formula (10) describes the *Euler polyhedral formula*, extended to complexes of arbitrary dimension. We shall check the formula for the case of the tetrahedron surface; the Betti numbers of this surface were found in Example 6, §19. The numbers of simplexes α^k can be counted directly. One gets

$$\sum_{k=0}^{2} (-1)^k p^k = 1 - 0 + 1 = \sum_{k=0}^{2} (-1)^k \alpha^k = 4 - 6 + 4 = 2 = -N;$$

thus the Euler characteristic of the tetrahedron is -2. One should compare this result with Chapter VI, §38 and §41.

Problems

1. Verify the formulas $\sum_{k=0}^{n} (-1)^k p^k = \sum_{k=0}^{n} (-1)^k \alpha^k$ in the examples of §19.

2. Show that the relation $N(\mathfrak{S}^n) = 2N(\mathfrak{P}^n)$ holds between the Euler characteristic $N(\mathfrak{S}^n)$ of the n-sphere and the Euler characteristic $N(\mathfrak{P}^n)$ of projective n-space.

3. Let $\Re^1$ be a 1-dimensional connected complex having α^0 0-simplexes and α^1 1-simplexes. With the help of the formula $-\alpha^0 \alpha^1 = -p^0 + p^1$, prove that one can remove exactly $-\alpha^0 + \alpha^1 + 1$ 1-simplexes from the complex (but no more) without disconnecting the resulting "edge complex" having the same vertices.

24. Pseudomanifolds and Orientability

We now turn our attention to a special class of complexes, the pseudomanifolds, which represent a first step, using methods available to us now, toward the manifolds.

A *closed pseudomanifold* is defined as follows:

(PM1) It is a pure, finite n-dimensional simplicial complex ($n \geq 1$); by "pure" we mean that each k-simplex is a face of at least one n-simplex (purity condition).

(PM2) Each $(n-1)$-simplex is a face of exactly two n-simplexes (nonbranching condition).

(PM3) Every two n-simplexes can be connected by means of a series of alternating n- and $(n-1)$-simplexes, each of which is incident with its successor (connectivity condition).

A closed pseudomanifold is said to be *orientable* if each of its n-simplexes can be oriented *coherently*, that is, oriented so that opposite orientations are induced in each $(n-1)$-simplex by the two adjoining n-simplexes. If no coherent orientation is possible, the pseudomanifold is said to be *nonorientable*.

A closed n-chain on an orientable and coherently oriented closed pseudomanifold is completely determined whenever one knows how often a single, arbitrarily chosen, oriented n-simplex appears in the chain. This is so because each of the n-simplexes adjoining this simplex must appear equally often for, from (PM3), one can reach each n-simplex by moving successively through adjoining simplexes; hence all n-simplexes must appear equally often. Consequently the nth homology group $\mathfrak{H}^n$ is the free cyclic group. In other words, the nth Betti number is $p^n = 1$. A basis for this group is one of the two chains which arise by virtue of the coherent orientation of the pseudomanifold. One can likewise show, for a nonorientable pseudomanifold, that there are no closed n-chains other than the chain 0, so that $\mathfrak{H}^n$ consists only of the null element; $p^n = 0$. One can thereby recognize whether a closed pseudomanifold is orientable by inspecting the nth homology group: *the necessary and sufficient condition for orientability is that the nth Betti number p^n have the value* 1.

On the other hand, *The nth connectivity number for both orientable and nonorientable closed pseudomanifolds is always $q^n = 1$.* For there exists exactly one nonvanishing closed n-chain mod 2, the collection of all n-simplexes.

We shall see later (§36) that the property of a complex being a pseudomanifold is a topologically invariant property. The proof will be carried through by reducing the defining properties (PM1) through (PM3) to homology properties (among which we include the connectivity numbers) and proving that the homology properties are topologically invariant. It is then important that one can replace the condition (PM3) by the condition $q^n = 1$. More precisely, the set of conditions

  (I)   (PM1), (PM2), (PM3)

is equivalent to

  (II)   (PM1), (PM2), $q^n = 1$.

We have just seen that $q^n = 1$ is a consequence of (I). Conversely (PM3) follows from (II): the n-dimensional chain mod 2, $\mathfrak{U}^n$, of all simplexes which can be connected by a series of alternating incident simplexes of dimensions n and $n-1$ with a given n-simplex $\mathfrak{E}^n$, is closed. For by (PM2) each $(n-1)$-simplex of $\mathfrak{R}^n$, and therefore also of $\mathfrak{U}^n$, is a face of exactly two n-simplexes. If there existed other n-simplexes besides those of $\mathfrak{U}^n$, then they would likewise form a closed n-chain mod 2; thus q^n would be at least $= 2$. This is a contradiction; thus $\mathfrak{U}^n$ must exhaust all of $\mathfrak{R}^n$.

One can also make several general statements about the $(n-1)$th homology group of a closed pseudomanifold.

THEOREM I. *The normal form $\mathbf{H}^{n-1}$ of the incidence matrix $\mathbf{E}^{n-1}$ has the form (1) for an orientable pseudomanifold $\mathfrak{K}^n$, and has the form (2) for a nonorientable pseudomanifold*:

$$\mathbf{H}^{n-1} = \begin{bmatrix} 0 & 1 & 0 & \cdots & 0 \\ 0 & 0 & 1 & \cdots & 0 \\ 0 & 0 & 0 & \cdots & 0 \\ \vdots & \vdots & \vdots & & \vdots \\ 0 & 0 & 0 & \cdots & 1 \\ 0 & 0 & 0 & \cdots & 0 \\ \vdots & \vdots & \vdots & & \vdots \\ 0 & 0 & 0 & \cdots & 0 \end{bmatrix} \tag{1}$$

$$\mathbf{H}^{n-1} = \begin{bmatrix} 2 & 0 & 0 & \cdots & 0 \\ 0 & 1 & 0 & \cdots & 0 \\ 0 & 0 & 1 & \cdots & 0 \\ \vdots & \vdots & \vdots & & \vdots \\ 0 & 0 & 0 & \cdots & 0 \\ 0 & 0 & 0 & \cdots & 1 \\ \vdots & \vdots & \vdots & & \vdots \\ 0 & 0 & 0 & \cdots & 0 \end{bmatrix} \tag{2}$$

In the case (2) its rank is equal to the number of columns; in the case (1) its rank is one less than the number of columns. Only in the case (2) does there exist an invariant factor other than 1 and it has the value 2.

Proof. Since each $(n-1)$-simplex of $\mathfrak{K}^n$ is incident with exactly two n-simplexes, there must be exactly two nonzero elements in each row of $\mathbf{E}^{n-1}$, each of value 1. The connectivity condition (PM3) states that whenever one divides the columns of $\mathbf{E}^{n-1}$ into two classes in any arbitrary manner, there will exist a row whose two ones are in columns belonging to different classes. After this, it is a purely arithmetic fact (§87) that there will only exist the two normal forms (1) and (2). The matrix (1) belongs to an orientable pseudomanifold because the n-chain in the upper entry of the first column is closed. In the case (2) such a closed chain does not appear; thus the pseudomanifold is nonorientable.

Since the invariant factors of $\mathbf{E}^{n-1}$ which differ from 1 are the $(n-1)$-dimensional torsion coefficients, the theorem is equivalent to the following theorem.

THEOREM II. *A closed pseudomanifold $\mathfrak{K}^n$ has no $(n-1)$-dimensional torsion coefficients if it is orientable; if it is nonorientable, it has exactly one, of value 2.*

In the nonorientable case there exists a closed $(n-1)$-chain U^{n-1}, uniquely determined up to null homologous chains, which is not itself null homologous but which is null homologous when doubled. One obtains such a chain U^{n-1} as follows. Let $E_1^n, E_2^n, \ldots, E_{\alpha_n}^n$ be arbitrarily oriented n-simplexes of $\Re^n$. The n-chain

$$U^n = E_1^n + E_2^n + \cdots + E_{\alpha_n}^n$$

has a nonzero boundary because $\Re^n$ is nonorientable. An $(n-1)$-simplex will occur in $\Re \partial U^n$ either 0 times or twice, according to whether the two adjoining n-simplexes induce opposite or like orientations in it, respectively. Thus $\Re \partial U^n$ is the double of a chain U^{n-1}. Therefore U^{n-1} when doubled is null homologous. If U^{n-1} itself were already the boundary of an n-chain V^n, then one would have

$$2U^{n-1} = \Re \partial 2V^n = \Re \partial U^n$$

and thus

$$\Re \partial (U^n - 2V^n) = 0.$$

Since, except for the n-chain 0, there exists no closed n-chain on a nonorientable pseudomanifold, it follows that $U^n = 2V^n$, in contradiction to the fact that U^n contains each n-simplex only once.

A *pseudomanifold with boundary* is defined by three properties. Two of these are (PM1) and (PM3); in place of (PM2) one has the condition $(\overline{PM}2)$: Each $(n-1)$-simplex is incident with at most two n-simplexes and there exists at least one $(n-1)$-simplex which is incident with only one n-simplex.

The boundary of a pseudomanifold with boundary, according to the definition of the boundary of a pure complex (footnote, §16), consists of the totality of those $(n-1)$-simplexes which are incident with a single n-simplex. All points and simplexes which do not lie on the boundary are called *inner* points and simplexes.

A pseudomanifold with boundary is said to be orientable if the n-simplexes can be oriented coherently, that is, so opposite orientations are induced in each inner $(n-1)$-simplex.

The annulus (§16) and the Möbius band (Fig. 45) are the simplest examples, respectively, of orientable and nonorientable pseudomanifolds with boundary. In Fig. 45 both sides c must be identified.

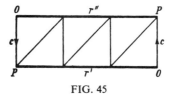

FIG. 45

Another example of a pseudomanifold with boundary is a simplicial star $\mathfrak{St}^n$, having center point O, whose outer boundary $\mathfrak{A}^{n-1}$ is a closed pseudomanifold. Properties (PM1) and (PM3) are satisfied for $\mathfrak{St}^n$ because they are satisfied for the outer boundary. $(\overline{PM2})$ is satisfied because an $(n-1)$-simplex of the outer boundary is incident with one n-simplex, while all other $(n-1)$-simplexes (which are incident with O) are incident with two n-simplexes. *If $\mathfrak{A}^{n-1}$ is orientable, then $\mathfrak{St}^n$ is also orientable, and conversely.* Let

$$OP_1 \cdots P_{n-1}$$

be an $(n-1)$-simplex of $\mathfrak{St}^n$ incident with O and let

$$E^n = +(OP_1 \cdots P_{n-1}P_n) \qquad \text{and} \qquad 'E^n = -(OP_1 \cdots P_{n-1}'P_n)$$

be the two n-simplexes incident with it. For the orientations given these induce opposite orientations in $(OP_1 \cdots P_{n-1})$. If one now furnishes the faces of E^n and $'E^n$ which lie on $\mathfrak{A}^{n-1}$ with the induced orientations

$$E^{n-1} = +(P_1 \cdots P_{n-1}P_n), \qquad 'E^{n-1} = -(P_1 \cdots P_{n-1}'P_n),$$

then they likewise induce opposite orientations in the common face $(P_1 \cdots P_{n-1})$. It follows from this that a coherent orientation of the n-simplexes of $\mathfrak{St}^n$ implies a coherent orientation of the $(n-1)$-simplexes of $\mathfrak{A}^{n-1}$ and conversely. Obviously the coherently oriented outer boundary is the boundary of the coherently oriented star $\mathfrak{St}^n$.

We have seen that a closed n-chain on a closed orientable pseudomanifold is a multiple of the coherently oriented pseudomanifold. A corresponding theorem is valid for an orientable pseudomanifold with boundary if one considers n-chains upon it whose boundaries lie on the boundary of the pseudomanifold. The proof is the same as in the case of the closed pseudomanifold.

The orientability of a closed pseudomanifold is determined by a condition on the nth Betti number: $p^n = 1$. No corresponding theorem holds for pseudomanifolds with boundary. For example, the homology groups of the (nonorientable) Möbius band are the same as those of the (orientable) annulus. For this reason it is simpler to prove topological invariance of orientability for closed pseudomanifolds than for pseudomanifolds with boundary. For the former, it is an immediate consequence of the invariance of the homology groups (Chapter IV), but in the latter case one requires invariance of the boundary and the deeper methods of Chapter V.

CHAPTER FOUR

SIMPLICIAL APPROXIMATIONS

In this chapter we shall prove the topological invariance of the homology groups derived from the schema of a simplicial complex. To do this we introduce singular k-dimensional simplexes (§25) in the complex $\mathfrak{K}^n$; these are continuous images of rectilinear simplexes of a Euclidean space. We form singular chains (§26) from the singular simplexes and define addition, boundary, and the properties of being closed and null homologous (bounding) for these singular chains. The singular chains are then divided into singular homology classes. The classes form the kth singular homology group of the complex (§27). This group is defined in a topologically invariant manner (§27) and is not dependent upon any particular simplicial decomposition of $\mathfrak{K}^n$. If a simplicial decomposition of $\mathfrak{K}^n$ is now given, it will be shown that each singular chain on $\mathfrak{K}^n$ can be transformed to a simplicial chain of a suitable normal subdivision without changing the essential topological properties of the singular chains. This is achieved by means of the process of simplicial approximation (§30), after a preceding normal subdivision. One obtains, in this way, a proof that the singular and the simplicial homology groups coincide (§28). The fundamental approximation theorem (§28) guarantees the existence of an approximating simplicial chain.

The term "simplicial approximation" generally refers to the transformation from a continuous mapping of a complex to a "simplicial" mapping. The statement that one can transform each continuous mapping to a simplicial mapping, and in fact do so by means of a "deformation," is the content of the deformation theorem (§31). Finally, in this regard, we shall examine the behavior of the homology groups under deformations of mappings.

25. Singular Simplexes

A *singular k-simplex* $\mathfrak{x}^k$ is a point set $\mathfrak{M}$ of an n-dimensional finite or infinite complex* $\mathfrak{K}^n$ which can be regarded in a definite way as a continuous image of a rectilinear simplex $\mathfrak{x}^k$ of a Euclidean space. We shall use German letters to indicate that we are dealing with nonoriented simplexes.

Should it be necessary to regard a given point set as being the image of $\mathfrak{x}^k$ on one occasion and as being the image of another k-simplex $\bar{\mathfrak{x}}^k$ on another occasion, we shall consider the two singular k-simplexes, which both comprise

---

* The fact that $\mathfrak{K}^n$ is a complex is actually unnecessary, since one can define singular simplexes, chains, and homology groups just as well in an arbitrary neighborhood space, for example, in an arbitrary nonempty subset of a complex. However, the definition is of value only for complexes.[16]

the same point set $\mathfrak{M}$ of $\mathfrak{K}^n$ to be *identical* if $\mathfrak{x}^k$ can be mapped linearly onto $\bar{\mathfrak{x}}^k$ so that points associated by the linear mapping correspond to the same point in $\mathfrak{K}^n$.*

The simplexes $\mathfrak{x}^k$ and $\bar{\mathfrak{x}}^k$ are both called *preimages* of the singular simplex $\mathfrak{X}^k$. We shall use lower case letters to denote preimages.

If the preimages of two singular simplexes cannot be mapped linearly in the manner just mentioned then the singular simplexes are not identical, even in the case that they comprise the same point set $\mathfrak{M}$ in $\mathfrak{K}^n$.

The dimension, k of $\mathfrak{X}^k$ can be larger, equal to, or smaller than the dimension n of the complex in which $\mathfrak{X}^k$ lies.

EXAMPLES. A simplex of a simplicial decomposition of $\mathfrak{K}^n$ can be regarded as a singular k-simplex. It is the topological image of a rectilinear preimage simplex of a Euclidean space.

A single point of a complex is a singular k-simplex if one lets all points of a rectilinear k-simplex be mapped into it. A triangle, lying in Euclidean space, which has been arbitrarily crumpled or even in some cases compressed onto a line or a point, is a singular 2-simplex. A Peano curve (Hausdorff [2, p. 202]) in the Euclidean plane is a singular 1-simplex.

Under the mapping of $\mathfrak{x}^k$ into $\mathfrak{K}^n$, an i-dimensional face $\mathfrak{x}^i$ of $\mathfrak{x}^k$ will be sent to a subset $\mathfrak{N}$ of $\mathfrak{M}$. Then $\mathfrak{N}$, regarded as the image of $\mathfrak{x}^i$, is said to be an *i-dimensional face* $\mathfrak{X}^i$ of the singular k-simplex $\mathfrak{X}^k$. It is obviously a singular i-simplex. The singular simplexes $\mathfrak{X}^k$ and $\mathfrak{X}^i$ are said to be *incident*.

A singular k-simplex $\mathfrak{X}^k$ can be *oriented* by orienting the preimage, $\mathfrak{x}^k$. We will use Roman letters to denote oriented simplexes. An oriented singular k-simplex X^k is then a point set $\mathfrak{M}$ of a complex $\mathfrak{K}^n$, which is in a definite way the continuous image of an oriented k-simplex x^k. If one replaces the preimage x^k by another geometric simplex $\bar{x}^k$ and one can map x^k linearly onto $\bar{x}^k$, preserving the orientation, so that associated points are sent to the same point of $\mathfrak{M}$, then the two oriented singular k-simplexes are considered to be identical.

To each oriented singular simplex X^k there corresponds an oppositely oriented simplex, which we denote by $-X^k$. It is obtained when we reverse the orientation of x^k but leave the mapping of x^k onto $\mathfrak{M}$ unchanged. X^k and $-X^k$ determine the same nonoriented singular simplex $\mathfrak{X}^k$.

It may happen that X^k and $-X^k$ are identical, that is, there is an orientation-reversing, linear map of x^k onto itself, such that associated points correspond to the same point of $\mathfrak{X}^k$. In this case, the oriented singular simplex X^k, and also the corresponding nonoriented $\mathfrak{X}^k$, are said to be *degenerate*. A nondegenerate simplex is simultaneously covered by two oppositely oriented simplexes; for a degenerate simplex these two coincide.

EXAMPLE 1. A $(k-1)$-simplex $\mathfrak{E}^{k-1}$ of a simplicial decomposition of $\mathfrak{K}^n$ can be regarded as a singular k-simplex $\mathfrak{X}^k$ if one maps the preimage $\mathfrak{x}^k$, having vertices $p_0, p_1, \ldots, p_k$, linearly onto $\mathfrak{E}^{k-1}$ so that two vertices, for example p_{k-1} and p_k, are mapped to the same vertex of $\mathfrak{E}^{k-1}$. The preimage $\mathfrak{x}^k$ admits an orientation reversing linear mapping onto itself, namely, the mapping

---

* *Editor's note*: This equivalence relation does not appear in modern singular homology theory.

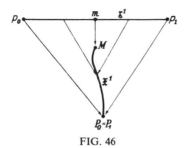

FIG. 46

which transposes the vertices p_k and p_{k-1} and leaves all other vertices fixed. Associated points map to the same point of the point set $\mathfrak{M} = \mathfrak{E}^{k-1}$. The simplex $\mathfrak{X}^k$ is then a degenerate singular simplex on $\mathfrak{R}^n$. In general, a singular k-simplex is degenerate whenever it arises by means of a linear mapping of its preimage $\mathfrak{x}^k$ onto a simplex of lower dimension, $\mathfrak{E}^{k-i}$ ($i > 0$) (§9).

EXAMPLE 2. Let the preimage be a 1-simplex, the segment $(p_0 p_1) = \mathfrak{x}^1$ of the real number line (Fig. 46). A singular simplex $\mathfrak{X}^1$ of the Euclidean plane will arise when one folds together the segment so that points symmetric to the midpoint m fall on one another. It is degenerate, because the mirror image of the segment about the midpoint m is a linear self-mapping of $\mathfrak{x}^1$ which reverses the orientation, taking associated points (that is, mirror image points) to the same point of $\mathfrak{X}^1$. The resulting 1-simplex would be singular but not degenerate if the folding were modified so that p_0 and p_1 again fell together at one end point and m mapped to the other end point, but points which are symmetric with respect to m do not always have the same image.

EXAMPLE 3. If all $k + 1$ vertices of a singular simplex $\mathfrak{X}^k$ are different, then the simplex is not degenerate. Upon an orientation-reversing, linear mapping of the preimage $\mathfrak{x}^k$ onto itself, at least one vertex will be mapped to a different vertex.

On the other hand, a singular k-simplex ($k > 0$) will always be degenerate if it consists of a single point into which all points of the preimage are mapped.

A 0-simplex is never degenerate. This is because we recall that we have also oriented the 0-simplexes (§9).

26. Singular Chains

A *singular k-chain* consists of finitely many, in some cases 0, nondegenerate singular k-simplexes of a complex $\mathfrak{R}^n$, each of which is provided with a definite orientation and a definite positive multiplicity. If the oriented singular simplex X^k appears with multiplicity a in the chain, then we also say that the oppositely oriented simplex $-X^k$ appears with multiplicity $-a$ in the chain, and if a singular simplex does not appear at all in the chain, we say that it appears with multiplicity 0.

Each k-simplex of a simplicial decomposition can be regarded also as a singular k-simplex, for it is the topological image of a geometric simplex. Thus the simplicial chains studied in Chapter III are at the same time singular chains. Simplicial chains are only defined with respect to a particular simplicial decomposition of $\mathfrak{R}^n$ and consist of simplexes of this simplicial decomposition; in contrast, the singular chains are built up out of singular simplexes, independent of any particular simplicial decomposition of $\mathfrak{R}^n$.

The *k-chain* 0, in which no singular simplex appears at all, is included among the singular k-chains. It is also to be counted among the simplicial chains.

Two singular k-chains are *added* by adding the multiplicities with which each oriented singular simplex appears in the two chains.

The singular k-chains of $\Re^n$ form an Abelian group under the operation of chain addition. In general this group has uncountably many generators since these can be taken to be the nondegenerate singular k-simplexes having fixed orientations. The zero element is formed by the k-chain 0; the negative of an element is obtained by reversing the orientation of all singular simplexes appearing in the chain or, what is the same thing, by retaining all of the orientations and multiplying all of the multiplicities by -1. Consequently, a singular chain V^k in which the singular oriented simplexes $X_1^k, X_2^k, \ldots, X_r^k$ (and we require $\mathfrak{X}_1^k \neq \mathfrak{X}_2^k \neq \cdots \neq \mathfrak{X}_r^k$) have multiplicities $v_1, v_2, \ldots, v_r$, respectively, and all other singular simplexes appear with multiplicity 0, can be written as a sum

$$V^k = v_1 X_1^k + v_2 X_2^k + \cdots + v_r X_r^k. \tag{1}$$

From the definition of a singular chain there immediately follows the rule of computation: If $mV^k = 0$ and $m \neq 0$, then $V^k = 0$. This rule would not be valid if we had also allowed degenerate simplexes in the definition of the chains.

The sum (1) also makes sense even if we do not assume that the singular simplexes $\mathfrak{X}_1^k, \ldots, \mathfrak{X}_r^k$ are distinct from one another, that is, we allow identical or oppositely oriented simplexes to appear and these can be added or can cancel. On purely formal grounds perhaps it is even desirable to also allow degenerate singular simplexes to appear; these do not count as terms of the chain, of course, but are equivalent to 0. In the future, when a sum (1) appears, all these possibilities will be allowed unless explicitly stated to the contrary.

The reason why we have not defined a singular chain simply as the continuous image of a particular simplicial chain of preimage simplexes is that we could not define an addition for such chains that would be independent of the choice of the preimage complex. It is important that we define the singular homology groups in such a way that they are topological invariants of the complex $\Re^n$. For this purpose additivity of the singular chains is indispensable.

27. Singular Homology Groups

If x^k $(k > 0)$ is a preimage of the oriented singular simplex X^k, which can be degenerate, and if

$$\Re \partial x^k = \sum_\nu x_\nu^{k-1},$$

we define the *boundary* of X^k to be the singular $(k-1)$-chain $\sum_\nu X_\nu^{k-1}$ and we write

$$\Re \partial X^k = \sum_\nu X_\nu^{k-1}. \tag{1}$$

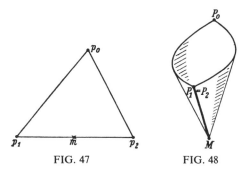

FIG. 47   FIG. 48

Here X_ν^{k-1} is the oriented $(k-1)$-dimensional face of X^k whose preimage is the oriented face x_ν^{k-1} of x^k, provided with the orientation induced by X^k. Degenerate face simplexes X_ν^{k-1} may appear even if X^k is nondegenerate.

An example of this is given by a triangle X^2 (Fig. 47) folded into a horn shape (Fig. 48); the horn has one face, which arises when the preimage face $(p_1 p_2)$ is folded at the midpoint m and the halves of $(p_1 p_2)$ are joined, while no other points of the preimage are identified. As we saw in Section 25, the folded side becomes a degenerate simplex when each pair of points equidistant from m on $(p_1 p_2)$ is brought into coincidence. Since degenerate simplexes are not counted in a singular chain, $\Re \partial X^2$ consists of the two singular 1-simplexes which together form the rim of the horn.

In this example one can assume that the complex into which the singular simplexes are inserted is Euclidean 3-space.

The definition of the boundary of a singular k-simplex X^k is independent of the particular choice of preimage x^k of X^k. For if $\bar{x}^x$ is another preimage of X^k, then according to the definition of equality of singular simplexes there exists a linear mapping T of x^k onto $\bar{x}^k$ with preservation of orientation such that corresponding points have the same image point in X^k when mapped by T. But T also maps the simplexes of $\Re \partial x^k = \sum_\nu x_\nu^{k-1}$ linearly with preservation of orientation onto the simplexes of $\Re \partial \bar{x}^k = \sum_\nu \bar{x}_\nu^{k-1}$. Thus again, according to the definition of equality of singular simplexes, the chains $\sum_\nu x_\nu^{k-1}$ and $\sum_\nu \bar{x}_\nu^{k-1}$ transform to the same singular chain $\Re \partial X^k$.

The boundary of a singular simplex can vanish. This occurs, for example, in the case of a segment x^1 bent to form a circle. Its boundary consists of two 0-simplexes which differ only in orientation.

It is clear that the boundary of the oppositely oriented simplex $-X^k$ is equal to the negative of the boundary of X^k:

$$\Re \partial(-X^k) = -\Re \partial X^k.$$

If, in particular, X^k is degenerate, then X^k coincides with $-X^k$, so one has

$$\Re \partial X^k = \Re \partial(-X^k) = -\Re \partial X^k.$$

But from the rule of computation given in §26, a chain which is equal to its negative is the chain 0. *The boundary of a degenerate singular simplex X^k is therefore the $(k-1)$-chain 0.*

The *boundary of a singular k-chain* is defined to be the sum of the boundaries of its individual k-simplexes:

$$\mathfrak{R}\partial \sum_{\nu} v_{\nu} X_{\nu}^{k} = \sum_{\nu} v_{\nu} \, \mathfrak{R}\partial X_{\nu}^{k}. \tag{2}$$

If equal, oppositely oriented, or degenerate simplexes (which are equivalent to 0) appear in the sum $\sum_{\nu} v_{\nu} X_{\nu}^{k}$, then their boundaries are also equal, oppositely oriented, or 0, respectively; thus the right hand side of (2) is always one and the same singular chain, regardless of how one writes the left-hand side. The boundary of a 0-dimensional chain is always the number 0.

A singular k-chain is said to be *closed* if its boundary vanishes. It is said to be *null homologous* if it is the boundary of a singular $(k + 1)$-chain. It follows that 0-dimensional singular chains are always closed. One can prove that the sum and difference of closed chains are again closed chains and can prove that the sum and difference of null homologous chains are again null homologous, in the same way as shown for the simplicial chains (§16).

The boundary of a singular $(k + 1)$-simplex X^{k+1} and consequently that of an arbitrary singular $(k + 1)$-chain is a closed singular k-chain, since the boundary of the preimage x^{k+1} is closed. In other words: each null homologous singular chain is closed.

Two singular chains, which are not necessarily closed, are said to be *homologous* to one another (on $\mathfrak{R}^{n}$) if their difference is null homologous. In order for two chains to be homologous, their boundaries must coincide (cf. §16). One can define homologies with division for singular chains. A singular chain is said to be division-null homologous (≈ 0) when it has a nonzero integer multiple which is null homologous.

The closed singular k-chains decompose into classes of homologous chains. These classes are the elements of a group, *the kth singular homology group* where the sum of two homology classes is defined to be the homology class containing the sum of two singular chains representing the two homology classes. In contrast, the homology groups which were derived from a simplicial decomposition in Chapter III will be referred to as *simplicial homology groups*, until we have proven that they coincide with the singular homology groups.

Since the proof of this coincidence will establish that the simplicial homology groups are topologically invariant, we should first convince ourselves that the singular homology groups are topologically invariant. We shall, in fact, investigate the more general question of how the singular homology groups behave under a continuous mapping φ of the complex $\mathfrak{R}^{n}$ into a complex $\mathbf{K}^{m}$ (which can also coincide with $\mathfrak{R}^{n}$). Under φ a singular simplex X^{k} of $\mathfrak{R}^{n}$ will transform to a singular simplex, Ξ^{k} of $\mathbf{K}^{m}$. For the preimage x^{k} of X^{k} will first be given a continuous mapping f onto X^{k} and then be given the continuous mapping φ into $\mathbf{K}^{m}$. The product mapping $\varphi \cdot f$

is again a continuous mapping. The reverse oriented simplex $-X^k$ will map to $-\Xi^k$. In the case that X^k and $-X^k$ are equal to one another, Ξ^k and $-\Xi^k$ will also be equal to one another. That is, a degenerate simplex transforms to a degenerate simplex. (On the other hand it is of course possible that the image of a nondegenerate simplex can be degenerate). Consequently, a singular chain

$$V^k = \sum v_\kappa X^k_\kappa$$

will have a completely determined image, $\sum v_\kappa \Xi^k_\kappa$. The image of the sum of two singular chains, $V^k + {}'V^k$ is equal to the sum of their images, that is, to

$$\sum v_\kappa \Xi^k_\kappa + \sum {}'v_\kappa {}'\Xi^k_\kappa.$$

Furthermore, the boundary of a chain transforms under φ to the boundary of the image chain. We can thus state

THEOREM I: *A continuous mapping φ of a complex $\mathfrak{K}^n$ into a complex $\mathbf{K}^m$ carries singular chains to singular chains, and each equation relating the chains of $\mathfrak{K}^n$ to their boundaries is preserved in the passage to $\mathbf{K}^m$.*

In particular, closed chains are transformed to closed chains and null homologous chains are transformed to null homologous chains. Thus each (singular) homology class in $\mathfrak{K}^n$ has a well-defined image homology class in $\mathbf{K}^m$. Since the sum of two homology classes corresponds to the sum of their images, we have proved the important theorem:

THEOREM II: *Under a continuous mapping φ of a complex $\mathfrak{K}^n$ into a complex, $\mathbf{K}^m$, the kth singular homology group of $\mathfrak{K}^n$ will be mapped by a homomorphism ϕ (§83), into the kth singular homology group of $\mathbf{K}^m$. If $\mathbf{K}^n$ and $\mathbf{K}^m$ are homeomorphic and φ is a topological mapping of $\mathfrak{K}^n$ onto $\mathbf{K}^m$, then ϕ is an isomorphism; homeomorphic complexes therefore have the same singular homology groups.*

The last part of the theorem follows from the fact that under a topological mapping the assignment of the singular chains of $\mathfrak{K}^n$ to those of $\mathbf{K}^m$ is one-to-one.

Problems

1. Assuming the approximation theorem, according to which the singular and simplicial homology groups coincide, prove that each homomorphic (automorphic) self-mapping of the first homology group of the torus (§19) can be accomplished by means of a continuous (topological) self-mapping of the torus.

2. Under the same assumption as above, show that an annulus cannot be mapped onto itself topologically so that one boundary circle maps onto itself with preservation of orientation while the other boundary circle maps onto itself with reversal of orientation.

28. The Approximation Theorem, Invariance of Simplicial Homology Groups

Simplicial homology groups can be calculated for a complex which has been provided with a simplicial decomposition. The topological invariance of these groups has not yet been established, for it might happen that they depend upon the choice of the simplicial decomposition. According to their definition the singular homology groups are topologically invariant, but on the other hand we have no method to compute them, that is, to determine their Betti numbers and torsion coefficients. We shall now proceed to prove the coincidence of the singular and simplicial homology groups of a complex and the consequent topological invariance of the simplicial homology groups.

The proof is based upon the following theorem:

APPROXIMATION THEOREM: *If $\Re^n$ is a finite or an infinite complex provided with a fixed simplicial decomposition and A^k is a singular k-chain on $\Re^n$ and furthermore if the boundary* A^{k-1} of A^k is a simplicial chain of the decomposition of $\Re^n$* [in particular it can be the $(k-1)$-chain 0], *then there will exist a simplicial chain $\overline{A}^k$ homologous to A^k.*

From §27 $\overline{A}^k$ will also have A^{k-1} as boundary. The dimension k of the chain A^k can be smaller than, equal to, or even larger than the dimension of the simplicial decomposition of $\Re^n$.

In particular, if one choses A^k to be closed, then $A^{k-1} = 0$ and it follows that:

(I)   Each closed singular k-chain is homologous to a simplicial k-chain.

On the other hand, if one replaces k by $k+1$ in the approximation theorem, there follows:

(II)   If a simplicial k-chain is the boundary of a singular $(k+1)$-chain, then it is also the boundary of a simplicial $(k+1)$-chain.

The coincidence of the simplicial and singular homology groups follows from (I) and (II). The closed k-chains of a simplicial homology class, regarded as singular chains, belong to the same singular homology class. This is true because simplicially homologous chains are also singular homologous. A singular homology class is, then, uniquely associated with each simplicial homology class. Different simplicial homology classes correspond to different singular homology classes, for two simplicial chains which are not simplicially homologous cannot be singular homologous, because of (II). Finally, each singular homology class contains some simplicial homology class, because of (I). The simplicial and singular homology classes are therefore in one-to-one correspondence. The correspondence is also an isomorphism. That is, to the sum $\overline{H}_1 + \overline{H}_2$ of two simplicial homology classes there corresponds to the

---

*For the case $k = 0$ the condition that the boundary A^{k-1} be simplicial is of course, vacuous.

sum $H_1 + H_2$ of the two associated singular homology classes. $H_1 + H_2$ is formed by selecting a representative from the class H_1 and a representative from the class H_2 and adding these; but one can choose closed simplicial chains from $\overline{H}_1$ and $\overline{H}_2$, respectively, as the representatives.

The proof of coincidence of the singular and simplicial homology groups thus reduces to the proof of the approximation theorem. Before proving it (§30) we shall make several observations with regard to prisms in Euclidean spaces.

29. Prisms in Euclidean Spaces

In §9, we saw that a linear self-mapping of a rectilinear simplex x^n preserves orientation if and only if the determinant of the transformation is positive. More generally, if x^n and $'x^n$ are any two oriented rectilinear n-simplexes in n-dimensional Euclidean space $\Re^n$, they are said to have the same orientation if the linear mapping of $\Re^n$ which transforms x^n to $'x^n$ with preservation of orientation has positive transformation determinant. If $''x_n$ is a third n-simplex which has the same orientation as $'x^n$, then x^n obviously has the same orientation as $''x^n$. We say that an *orientation of the whole space* $\Re^n$ is given by an oriented n-simplex and all simplexes with the same orientation as x^n determine the same orientation of $\Re^n$.

Let x_g^k be the g-fold normal subdivision of the rectilinear simplex x^k lying in $\Re^n$. An orientation of the linear subspace $\Re^k$ of $\Re^n$ in which x^k lies is given by means of the orientation of x^k. One can thus orient all of the subsimplexes of x_g^k like x^k. When we speak of the normal subdivision of an oriented simplex x^k we always mean that the subsimplexes are oriented in this way. In the case $k = 0$ the oriented subdivision coincides with the oriented 0-simplex itself.

A *prism* $\mathfrak{z}^{k+1}$ is the point set swept out by a rectilinear simplex $\mathfrak{x}^k$ during a translation in a Euclidean space of dimension at least $k + 1$, which transforms it to another simplex $\mathfrak{y}^k$; we assume that the translation vector does not lie in the k-dimensional linear space which is spanned by $\mathfrak{x}^k$. We call $\mathfrak{x}^k$ and its faces $\mathfrak{x}_\nu^i$ ($0 \le i \le k$) the *floor faces* of the prism $\mathfrak{z}^{k+1}$; we call $\mathfrak{y}^k$ and its faces the *roof faces* of this prism. Obviously each i-dimensional face of $\mathfrak{x}^k$ will describe a prism $\mathfrak{z}^{i+1}$ during the translation, which we call an $(i + 1)$-dimensional *wall face*. The midpoint of $\mathfrak{x}^k$ sweeps out the *axis* of the prism. An arbitrary point of $\mathfrak{x}^k$ sweeps out an *axis parallel interval*. The midpoint of the prism axis is called the *midpoint of the prism*. Each wall face also has a midpoint, since it is itself a prism. In the case $k = 0$ one obtains a segment as a 1-dimensional prism. For $k = 1$ one obtains a parallelogram; for $k = 2$ a triangular prism. In the following discussion we shall always assume $k > 0$.

If $\bar{\mathfrak{z}}^{k+1}$ is another $(k + 1)$-dimensional prism, then there exists a linear mapping of $\mathfrak{z}^{k+1}$ onto $\bar{\mathfrak{z}}^{k+1}$ such that the floor simplex $\mathfrak{x}^k$ transforms to the floor simplex $\bar{\mathfrak{x}}^k$. This linear mapping is uniquely determined by the linear mapping of $\mathfrak{x}^k$ onto $\bar{\mathfrak{x}}^k$, which one can still prescribe arbitrarily; the midpoint of $\mathfrak{z}^{k+1}$ will transform to the midpoint of $\bar{\mathfrak{z}}^{k+1}$.

The prism $\mathfrak{z}^{k+1}$ is convex, due to the convexity of the floor simplex. That is, the prism is a closed bounded point set of a Euclidean space which contains, for each pair of points, the entire line segment connecting them. A ray directed outward from the midpoint of $\mathfrak{z}^{k+1}$ will intersect the boundary of $\mathfrak{z}^{k+1}$, which consists of $\mathfrak{x}^k$, $\mathfrak{y}^k$ and the wall faces $\mathfrak{z}_\nu^k$ ($\nu = 0, 1, 2, \ldots, k$), in exactly one point.

One can confirm all of the above properties of $\mathfrak{z}^{k+1}$ analytically. One selects a vertex of $\mathfrak{x}^k$ as origin of coordinates and chooses the vectors $\mathfrak{v}_1$, $\mathfrak{v}_2, \ldots, \mathfrak{v}^k$ which point to the other vertices of $\mathfrak{x}^k$ as basis vectors, as well as the vector t which is swept out by the origin during the translation generating the prism. The prism will then be formed by the end points of the position vectors:

$$\lambda_1 \mathfrak{v}_1 + \lambda_2 \mathfrak{v}_2 + \cdots + \lambda_k \mathfrak{v}_k + \tau t$$

$$(0 \leq \lambda_i \leq 1, \lambda_1 + \lambda_2 + \cdots + \lambda_k \leq 1, 0 \leq \tau \leq 1).$$

We shall not go into more detail.

We now decompose $\mathfrak{z}^{k+1}$ into simplexes. This can be done in infinitely many ways. Here we are only interested in particular simplicial decompositions which we shall need later. We first form the g-fold normal subdivision of the roof face $\mathfrak{y}^k$. We also allow g to take the value 0; in this case the prism remains undivided. We then divide all wall faces of $\mathfrak{z}^{k+1}$ according to the following procedure: Each 1-dimensional wall face is subdivided into two 1-simplexes through its midpoint. After the i-dimensional wall faces $\mathfrak{z}_\nu^i$ have already been subdivided, one projects the boundary of each $(i + 1)$-dimensional wall face $\mathfrak{z}_\nu^{i+1}$ from its midpoint, and for the case $i = k$ one projects the entire prism boundary from the prism midpoint (cf. §14). The boundary of $\mathfrak{z}_\nu^{i+1}$ consists of the nonsubdivided floor face $\mathfrak{x}_\nu^i$, the g-fold subdivided roof face $\mathfrak{y}_\nu^i$, and certain wall faces $\mathfrak{z}_\mu^i$ which are already subdivided. One obtains the required simplicial decomposition of $\mathfrak{z}^{k+1}$ by dividing all of the wall faces in this way, beginning with the dimension 1 and continuing to the prism $\mathfrak{z}^{k+1}$ itself. Figures 49 and 50 illustrate two simple cases.

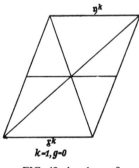

FIG. 49. $k = 1$, $g = 0$.

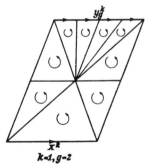

FIG. 50. $k = 1$, $g = 2$.

We now orient the prism $_3^{k+1}$, that is, we orient its individual $(k+1)$-dimensional subsimplexes coherently inside the $(k+1)$-dimensional linear space in which $_3^{k+1}$ lies. The totality of the simplexes oriented in this manner, each taken with multiplicity $+1$, is a chain, which we call z^{k+1}. The floor simplex $_{\mathfrak{x}}^k$ of $_3^{k+1}$ receives the orientation which is induced in it by the incident $(k+1)$-simplex of z^{k+1} and we will subsequently call this oriented simplex x^k. The orientation of the floor face is carried to the subdivided roof face $\mathfrak{y}^k$, by means of the translation. This gives rise to the chain y^k which contains all of the oriented simplexes of the g-fold normal subdivision of $\mathfrak{y}^k$ (cf. Fig. 50). In a corresponding manner, the chains $z_\nu^k, x_\nu^{k-1}, y_\nu^{k-1}$ are defined on the k-dimensional wall faces and on the $(k-1)$-dimensional floor and roof faces of $_3^{k+1}$. One can orient the face simplexes x_ν^{k-1} of the floor simplex x^k so that

$$\mathscr{R}\partial x^k = \sum_\nu x_\nu^{k-1}. \tag{1}$$

Besides (1), there are additional *connection formulas* which hold involving the simplicial chains which were just obtained:*

$$\mathscr{R}\partial y^k = \sum_\nu y_\nu^{k-1} \tag{2}$$

$$\mathscr{R}\partial \sum_\nu z_\nu^k = \sum_\nu x_\nu^{k-1} - \sum_\nu y_\nu^{k-1}, \tag{3}$$

$$\mathscr{R}\partial z^{k+1} = x^k - y^k - \sum_\nu z_\nu^k. \tag{4}$$

Equation (2) expresses as a formula the fact that the boundary of the normal subdivision of the roof face is equal to the normal subdivision of the boundary. Equation (4) states that the boundary of the prism consists of the floor face, roof face, and the k-dimensional wall faces, all subdivided, provided with orientations, and regarded as chains. Equation (3) follows from (1), (2), and (4). Considered as a boundary chain, the right-hand side of (4) is closed and its boundary is the $(k-1)$-chain 0. This fact, taken together with (1) and (2), gives (3) directly.

Formulas (2) and (4) are obviously correct except possibly for signs. A simplex E^k of the boundary of z^{k+1} must always lie on x^k or y^k or on one of the chains z_ν^k, because every other k-simplex of $_3^{k+1}$ is incident with exactly two $(k+1)$-simplexes which induce opposite orientations in that k-simplex because they are coherently oriented.** If E^k is an oriented simplex of the chain z_ν^k, then a particular orientation will be induced in it by the only $(k+1)$-simplex E^{k+1} of z^{k+1} incident with it; this determines the sign with

---

* *Editor's note:* These formulas hold when the orientations of the z_ν^k and the y_ν^{k-1} are chosen properly.

** If a linear self-mapping of the linear $(k+1)$-dimensional space in which $_3^{k+1}$ lies transforms one $(k+1)$-simplex incident with E^k to the other and leaves the points of E^k fixed, then it has a negative transformation determinant.

which E^k appears in the chain $\mathfrak{R}\partial z^{k+1}$. Any other oriented simplex is assigned the same coefficient, because the simplexes of z_ν^k are coherently oriented and the simplexes of z^{k+1} are also coherently oriented. Now the sign of x^k is certainly correct in (4) because the orientation of z^{k+1} was directly determined by the requirement that it induce the given orientation in x^k. Likewise the sign of y^k is correct; for the orientation of y^k was transferred from x^k by means of the translation. Thus we must have the formulas

$$\mathfrak{R}\partial y^k = \sum_\nu \eta_\nu y_\nu^{k-1}, \tag{2'}$$

$$\mathfrak{R}\partial z^{k+1} = x^k - y^k - \sum_\nu \xi_\nu z_\nu^k. \tag{4'}$$

Similarly, we also have the formulas

$$\mathfrak{R}\partial z_\nu^k = x_\nu^{k-1} - y_\nu^{k-1} + \cdots, \tag{3'}$$

where the dots denote certain $(k-1)$-dimensional wall faces z_λ^{k-1}.

When we form the boundary of the right-hand side of (4'), that is, the boundary of the boundary of z^{k+1}, it must vanish. With the aid of (1), (2'), (3') one gets

$$\sum_\nu x_\nu^{k-1} - \sum_\nu \eta_\nu y_\nu^{k-1} - \sum_\nu \xi_\nu \left(x_\nu^{k-1} - y_\nu^{k-1} + \cdots \right) = 0.$$

Comparing coefficients,

$$\eta_\nu = \xi_\nu = 1,$$

completing the proof of formulas (2) through (4).

When one maps a prism $\mathfrak{z}^{k+1}$ continuously into a complex $\mathfrak{R}^n$, and denotes the singular images of the chains x^k, y^k, z^{k+1}, x_ν^{k-1}, y_ν^{k-1}, z_ν^k by X^k, Y^k, Z^{k+1}, X_ν^{k-1}, Y_ν^{k-1}, Z_ν^k, respectively, then from Theorem I of §27 the connection formulas (1) through (4) are also valid when one replaces the symbols in lower case type by the corresponding symbols in capital letters. We shall now map a finite number of prisms, $\mathfrak{z}^{k+1}$, $'\mathfrak{z}^{k+1}$, ..., continuously into $\mathfrak{R}^n$. Using arbitrary integer coefficients, we form a singular chain

$$A^k = aX^k + {}'a'X^k + \cdots$$

from the singular images X^k, $'X^k$, ... of the oriented floor simplexes x^k, $'x^k$, ... of $\mathfrak{z}^{k+1}$, $'\mathfrak{z}^{k+1}$, ... In a corresponding way we form the chains

$$B^k = aY^k + {}'a'Y^k + \cdots,$$

$$C^{k+1} = aZ^{k+1} + {}'a'Z^{k+1} + \cdots,$$

$$A^{k-1} = a\sum X_\nu^{k-1} + {}'a\sum{}'X_\nu^{k-1} + \cdots,$$

$$B^{k-1} = a\sum Y_\nu^{k-1} + {}'a\sum{}'Y_\nu^{k-1} + \cdots,$$

$$C^k = a\sum Z_\nu^k + {}'a\sum{}'Z_\nu^k + \cdots.$$

Based upon the formulas (1) through (4) rewritten in capital letters, we then have the *connection formulas* for these chains:

$$\Re\partial A^k = A^{k-1}, \tag{I}$$

$$\Re\partial B^k = B^{k-1}, \tag{II}$$

$$\Re\partial C^k = A^{k-1} - B^{k-1}, \tag{III}$$

$$\Re\partial C^{k+1} = A^k - B^k - C^k. \tag{IV}$$

30. Proof of the Approximation Theorem

The construction of a chain $\bar{A}^k$ which approximates the singular chain A^k will be accomplished by the process of simplicial approximation. The vertices of the singular k-simplexes of A^k will be replaced by neighboring vertices of the simplicial decomposition of $\Re^n$. This will be done so that the vertices of each simplex of A^k transform to vertices of a simplex (not necessarily k-dimensional) of $\Re^n$. This is possible only when the simplexes of A^k are sufficiently fine. Otherwise, two vertices of a singular simplex of A^k could lie in simplexes of the simplicial decomposition of $\Re^n$ which are far apart, and the vertices neighboring them would not belong to the same simplex of $\Re^n$. One must then, on occasion, subdivide the given singular chain prior to the process of simplicial approximation. Accordingly, the construction proceeds in two steps:

Step 1:   Transformation to a subdivision of A^k.
Step 2:   Simplicial approximation of this subdivision.

Before carrying out the construction, we dispose of the case where the singular k-chain to be approximated, A^k, is 0-dimensional. If

$$A^0 = aX^0 + {}'a'X^0 + \cdots,$$

then one connects the point X^0 linearly with a vertex Y^0 of a simplex of $\Re^n$ to which X^0 belongs (cf. §9). Orient Y^0 with the same sign as X^0. The connecting segment can be regarded as an oriented singular 1-simplex Z^1 having boundary $X^0 - Y^0$. The chain

$$\bar{A}^0 = aY^0 + {}'a'Y^0 + \cdots$$

is then a simplicial 0-chain homologous to A^0.

In what follows we shall now assume that $k > 0$.

We assume further that $\Re^n$ is finite. We may make this assumption without loss of generality because a singular chain A^k in an infinite complex $\Re^n$ is always contained in a finite subcomplex of the simplicial decomposition. This is true, first of all, for an individual singular simplex X^k. For if X^k had points in common with infinitely many simplexes of $\Re^n$, there would exist a sequence of points belonging to X^k which had no accumulation point. One can regard this sequence as the image of an infinite sequence of points of the preimage x^k, which has an accumulation point H. But the mapping of x^k onto X^k could not then be continuous at the point H, contrary to the definition of a singular simplex. Since A^k consists of finitely many singular simplexes, A^k also lies on a finite subcomplex of $\Re^n$. Thus, if the approximation theorem can be proved for finite complexes it will also be valid for infinite complexes.

Step 1: Subdivision of the Singular Chain A^k

(a) Connecting prisms. We first examine an individual oriented singular simplex X^k in $\Re^n$, which may be degenerate. As a preimage of X^k we choose the oriented floor simplex x^k of a prism $\mathfrak{z}^{k+1}$, which we have simplicially decomposed so as to have a g-fold subdivided roof simplex, as described in the previous section. We map the prism $\mathfrak{z}^{k+1}$ continuously into the complex $\Re^n$ so that x^k transforms in the prescribed manner to X^k and each segment parallel to the axis goes to a point. This may be described as parallel projection of the prism into its floor simplex, followed by mapping the image onto X^k. The roof simplex y^k, or rather the chain y^k consisting of the oriented subsimplexes of the g-fold normal subdivision, is thereby mapped to a singular chain Y^k, which is called the *g-fold normal subdivision of X^k*. The chain z^{k+1} is mapped to a singular chain Z^{k+1}, which we call the *connecting chain* of X^k with Y^k.

The g-fold subdivision and the connecting chain are unambiguously determined by X^k and do not depend upon the choice of the prism $\mathfrak{z}^{k+1}$. For if $\bar{\mathfrak{z}}^{k+1}$ is another prism which is both subdivided and oriented like $\mathfrak{z}^{k+1}$, then there exists a linear mapping T of $\mathfrak{z}^{k+1}$ onto $\bar{\mathfrak{z}}^{k+1}$ such that the oriented floor simplex x^k transforms to the oriented floor simplex $\bar{x}^k$. Since corresponding points of x^k and $\bar{x}^k$ have the same image point in X^k and each segment parallel to the axis in $\mathfrak{z}^{k+1}$ and $\bar{\mathfrak{z}}^{k+1}$, respectively, transforms to a point, any two points corresponding to one another by virtue of T, one point lying in $\mathfrak{z}^{k+1}$ and the other in $\bar{\mathfrak{z}}^{k+1}$, will have the same image point in $\Re^n$. Finally, since the subsimplexes of $\mathfrak{z}^{k+1}$ are mapped by T to those of $\bar{\mathfrak{z}}^{k+1}$, preserving orientations, the singular chains Z^{k+1} and $\bar{Z}^{k+1}$ are equal; likewise Y^{k+1} and $\bar{Y}^{k+1}$ are equal according to the definition of equality of singular simplexes.

For an arbitrary singular chain A^k we define the g-fold subdivision B^k to be the sum of the g-fold normal subdivisions of the individual simplexes; we define the connecting chain C^{k+1} to be the sum of the connecting chains of the individual simplexes. Thus if

$$A^k = aX^k + {}'a'X^k + \cdots, \tag{1'}$$

and if we denote the g-fold normal subdivision and connecting chain of ${}^{(\kappa)}X^k$, respectively, by ${}^{(\kappa)}Y^k$ and ${}^{(\kappa)}Z^{k+1}$, then

$$B^k = aY^k + {}'a'Y^k + \cdots, \tag{2'}$$

$$C^{k+1} = aZ^{k+1} + {}'a'Z^{k+1} + \cdots. \tag{3'}$$

If a simplex on the right-hand side of (1') is degenerate, for example X^k, so that $X^k = -X^k$, then it is obviously true that $Y^k = -Y^k$ and $Z^{k+1} = -Z^{k+1}$; hence $Y^k = 0$ and $Z^{k+1} = 0$. The normal subdivision B^k and the connecting chain C^{k+1} are therefore unambiguously determined by A^k and do not depend upon whether one does or does not use degenerate simplexes in (1).

(b) Connection formulas. The mapping of the prism $_3^{k+1}$ onto the singular simplex X^k sends each axis parallel interval of a wall face $_{3\nu}^k$ to a single point. Thus the chains x_ν^{k-1}, y_ν^{k-1}, and z_ν^k which lie on $_{3\nu}^k$ will map, respectively, to the face X_ν^{k-1} of X^k, the g-fold subdivision Y_ν^{k-1} of X_ν^{k-1}, and the chain Z_ν^k connecting X_ν^{k-1} with Y_ν^{k-1}. A corresponding statement is valid for the prisms $'z^{k+1}, \ldots$ belonging to the remaining singular simplexes $'X^k, \ldots$. From formula (1') of this section and formula (1) of the previous section, the boundary of A^k is the chain

$$A^{k-1} = a\sum X_\nu^{k-1} + 'a\sum 'X_\nu^{k-1} + \cdots . \qquad (4')$$

The normal subdivision of A^{k-1} is then the chain

$$B^{k-1} = a\sum Y_\nu^{k-1} + 'a\sum 'Y_\nu^{k-1} + \cdots \qquad (5')$$

and the chain connecting A^{k-1} with B^{k-1} is

$$C^k = a\sum Z_\nu^k + 'a\sum 'Z_\nu^k + \cdots , \qquad (6')$$

so that the connection formulas (I) through (IV) of §29 give

$$\mathfrak{R}\partial A^k = A^{k-1} \qquad (\text{I}')$$

$$\mathfrak{R}\partial B^k = B^{k-1}, \qquad (\text{II}')$$

$$\mathfrak{R}\partial C^k = A^{k-1} - B^{k-1}, \qquad (\text{III}')$$

$$\mathfrak{R}\partial C^{k+1} = A^k - B^k - C^k. \qquad (\text{IV}')$$

Formula (II') states that *the boundary of the g-fold normal subdivision of the singular chain A^k is equal to the g-fold normal subdivision of its boundary.*

One can make the subsimplexes of $y^k, 'y^k, \ldots$ arbitrarily small by making g sufficiently large, and from the theorem of uniform continuity (§7), the singular k-simplexes of the g-fold normal subdivision B^k, of A^k will also become arbitrarily small. More precisely: If one considers $\mathfrak{R}^n$ to be a geometric complex in a Euclidean space of sufficiently high dimension, which is always possible according to Section 11, and takes the distance between two points of $\mathfrak{R}^n$ to be the Euclidean distance in this space, then one can make the diameter of the singular simplexes of B^k arbitrarily small by choosing a sufficiently fine subdivision of A^k.

The subdivision of the singular chain A^k has thereby been achieved and we proceed to step 2.

Step 2: Simplicial Approximation of the Subdivided Chain A^k

(a) Connecting prisms. If B^k is a singular chain and P is a vertex of a singular simplex of B^k, then we will define a *singular simplicial star with center point* P as the totality of singular simplexes of B^k which have P as a vertex. We will now assume that the subdivision B^k of A^k is sufficiently fine that each singular star of B^k will lie in the interior of at least one star of the simplicial decomposition of $\mathfrak{R}^n$. This is possible from Theorem V of §7; one chooses the interior of a simplicial star of $\mathfrak{R}^n$ as a neighborhood $\mathfrak{U}^*(Q/\mathfrak{R}^n)$ of a point Q. There then exists an $\varepsilon > 0$ such that the ε-neighborhood of an arbitrary point of $\mathfrak{R}^n$ lies in the interior of a star of $\mathfrak{R}^n$. One needs only carry the subdivision of A^k far enough so that the diameter of each simplex of B^k is less that $\varepsilon/2$.

In order to emphasize the analogy of the present step to step 1, we shall now denote the singular chains B^{k-1} and B^k, that is, the g-fold subdivisions of the chains A^{k-1} and A^k, respectively, by $\mathbf{A}^{k-1}$ and by

$$\mathbf{A}^k = \alpha \Xi^k + {}'\alpha' \Xi^k + \cdots,$$

where we again leave undecided whether equal, opposite, or even degenerate simplexes appear on the right-hand side. Let $P_1, P_2, \ldots$ be the vertices of the singular simplexes $\Xi^k, {}'\Xi^k, \ldots$, where vertices which might happen to coincide with one another are written down only once. Each vertex P_λ is the center point of a singular star consisting of k-simplexes of $\mathbf{A}^k$. The approximation of $\mathbf{A}^k$ is now made as follows: An *approximating vertex* Q_λ belonging to the simplicial decomposition of $\mathfrak{R}^n$ is assigned to each vertex P_λ. The vertex Q_λ is in fact chosen so that the simplicial star of $\mathfrak{R}^n$ having center point Q_λ will contain the singular star of $\mathbf{A}^k$ having center point P_λ in its interior. The vertex Q_λ which approximates P_λ is in general not uniquely determined by P_λ, since there may be several simplicial stars of $\mathfrak{R}^n$ which contain the singular simplicial star having center point P_λ in their respective interiors. But once the vertex Q_λ has been chosen the remaining part of the construction becomes completely unambiguous. The essential step in the simplicial approximation is the transition from the singular vertices P_λ to the approximating simplicial vertices Q_λ. Figure 51* shows an example of a singular 1-chain (in heavy outline) in a complex $\mathfrak{R}^2$ consisting of rectilinear equilateral triangles.

We shall now first examine an individual singular simplex, for example, Ξ^k. If $P_1, P_2, \ldots, P_\rho$ are its vertices, then the associated vertices $Q_1, Q_2, \ldots, Q_\rho$ are the vertices of a simplex of the simplicial decomposition of $\mathfrak{R}^n$; in contrast to the vertices P_ν, the vertices Q_ν do not all have to be distinct from one another. The stars having center points $Q_1, Q_2, \ldots, Q_\rho$ will all contain

---

Editor's note: Figure 51 contains an error. The condition that each singular simplicial star must lie in the interior of a star of the simplicial decomposition is violated for the singular stars having midpoints P_6 and P_8. One probably needs to redraw the figure anew and replace P_7 by two nearby points.

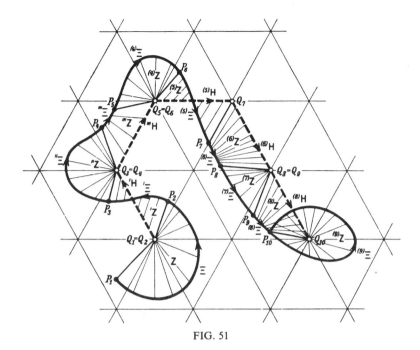

FIG. 51

the singular simplex Ξ^k in their interiors and will therefore have some arbitrary point P of Ξ^k as a common interior point.* The point P belongs to a particular simplex $\mathfrak{G}^j$ of $\mathfrak{R}^n$. A star $\mathfrak{St}$ of $\mathfrak{R}^n$ which contains P in its interior must contain $\mathfrak{G}^j$ and $\mathfrak{G}^j$ cannot lie on the outer boundary of $\mathfrak{St}$; thus the center point of $\mathfrak{St}$ must be a vertex of $\mathfrak{G}^j$. Points $Q_1, Q_2, \ldots, Q_\rho$ are therefore the vertices of an i-dimensional face $\mathfrak{G}^i$ of $\mathfrak{G}^j$ or those of $\mathfrak{G}^j$ itself. Since $\mathfrak{G}^i$ and P both belong to the simplex $\mathfrak{G}^j$ one can connect each point of $\mathfrak{G}^i$ with P and thus with each point of Ξ^k by a straight line segment.**

We now again take the oriented floor simplex x^k of a $(k + 1)$-dimensional prism $\mathfrak{z}^{k+1}$ as the preimage of Ξ^k and we define a continuous mapping of $\mathfrak{z}^{k+1}$ into $\mathfrak{R}^n$ by means of the following conditions:

1. x^k transforms to Ξ^k in the manner previously prescribed.
2. y^k, which is not yet subdivided, is mapped linearly onto $\mathfrak{G}^i$ in the following way: if a vertex P_λ ($\lambda = 0, 1, \ldots, k$) of the floor simplex transforms to a singular vertex P_μ ($\mu = 1, 2, \ldots, \rho$) of Ξ^k, then the corresponding vertex q_λ of the roof simplex y^k transforms to the vertex Q_μ of $\mathfrak{G}^i$, which approximates the vertex P_μ.
3. An axis parallel interval (pq) of $\mathfrak{z}^{k+1}$ is mapped linearly onto the straight line segment connecting the image points P and Q in Ξ^k and $\mathfrak{G}^i$; such

---

* It should be remembered that an interior point of a simplicial star is a point which does not lie on the outer boundary of the star.

** This rectilinearity is to be understood in relation to $\mathfrak{G}^i$ (cf. §9).

a connecting segment exists because, as we have seen, one can connect each point of Ξ^k rectilinearly with each point of $\mathfrak{C}^i$. In Fig. 51 the connecting segments in $\mathfrak{R}^n$ are indicated by straight line segments.

The mapping of $\mathfrak{z}^{k+1}$ into $\mathfrak{R}^n$ has now been determined uniquely. Its continuity follows from the lemma of §14. By means of this mapping the simplex $\mathfrak{C}^i$ becomes an oriented singular simplex with preimage y^k. As such it is denoted by $\mathbf{H}^k$ and from now on it will be called *the* (*simplicial*) *approximation of* Ξ^k.

The chain z^{k+1} lying on the prism $\mathfrak{z}^{k+1}$ transforms, under the continuous mapping into $\mathfrak{R}^n$, to a singular chain $\mathbf{Z}^{k+1}$, which will be called the *connecting chain* of Ξ^k with its approximation $\mathbf{H}^k$.

For the singular chain

$$\mathbf{A}^k = \alpha \Xi^k + {}'\alpha' \Xi^k + \cdots , \tag{1''}$$

we now define the *approximation* $\mathbf{B}^k$ as the sum of the approximations of the individual singular simplexes, that is,

$$\mathbf{B}^k = \alpha \mathbf{H}^k + {}'\alpha' \mathbf{H}^k + \cdots \tag{2''}$$

and we define the connecting chain Γ^{k+1} as the sum of the connecting chains of the individual singular simplexes:

$$\Gamma^{k+1} = \alpha \mathbf{Z}^{k+1} + {}'\alpha' \mathbf{Z}^{k+1} + \cdots . \tag{3''}$$

The fact that the approximating and connecting chains, after initial selection of the approximating vertices, are uniquely determined by $\mathbf{A}^k$, and do not depend upon whether one writes degenerate simplexes on the right-hand side of $\mathbf{A}^k$, follows as in step 1. In Fig. 51 the approximation $\mathbf{B}^1$ is shown in heavy dashes.

$\mathbf{B}^k$ is a simplicial chain, since a singular simplex ${}^{(\lambda)}\mathbf{H}^k$ is either a k-dimensional simplex of the simplicial decomposition of $\mathfrak{R}^n$ or is a degenerate singular simplex (Example 1 in §25). In the latter case it is equivalent to the symbol 0.

(b) Connection formulas. When the prism $\mathfrak{z}^{k+1}$ is mapped according to the conditions 1 through 3 into $\mathfrak{R}^n$, the mapping of each wall face $\mathfrak{z}_\nu^k$ satisfies the corresponding conditions:

1. The floor simplex x_ν^{k-1} of $\mathfrak{z}_\nu^k$ transforms to the face Ξ_ν^{k-1} of Ξ^k.

2. In the linear mapping of y^k onto the simplex $\mathfrak{C}^i$ of the simplicial decomposition of $\mathfrak{R}^n$, the face y_ν^{k-1} of y^k will also be mapped linearly onto a simplex $\mathfrak{C}^h$ that is either $\mathfrak{C}^i$ itself or a face of $\mathfrak{C}^i$. If a vertex p_λ of x_ν^{k-1} now transforms to the vertex P_μ of Ξ_ν^{k-1}, then the corresponding vertex q_λ of y_ν^{k-1} will transform to the approximating vertex Q_μ of P_μ, from the construction of the mapping of the prism $\mathfrak{z}^{k+1}$.

3. Since, under the mapping of $\mathfrak{z}^{k+1}$ each axis parallel interval (pq) will be mapped linearly onto the segment (PQ) connecting the image points, this also holds true in particular for the axis parallel intervals of the wall face $\mathfrak{z}_\nu^k$.

This shows that the chains $x_\nu^{k-1}, y_\nu^{k-1}, z_\nu^k$ lying on $\mathfrak{z}_\nu^k$ map, respectively, to the face Ξ_ν^{k-1} of Ξ^k, to the approximation $\mathbf{H}_\nu^{k-1}$ of Ξ_ν^{k-1} and to the chain $\mathbf{Z}_\nu^k$ connecting Ξ_ν^{k-1} with $\mathbf{H}_\nu^{k-1}$.

A corresponding statement holds for the prisms $'\mathfrak{z}^{k+1}, \ldots$ belonging to the remaining singular simplexes $'\Xi^k, \ldots$. From Eqs. (1″) of this section and (1) of the previous section, the boundary of $\mathbf{A}^k$ is

$$\mathbf{A}^{k-1} = \alpha \sum \Xi_\nu^{k-1} + '\alpha \sum '\Xi_\nu^{k-1} + \cdots . \tag{4″}$$

The approximation of $\mathbf{A}^{k-1}$ is then

$$\mathbf{B}^{k-1} = \alpha \sum \mathbf{H}_\nu^{k-1} + '\alpha \sum '\mathbf{H}_\nu^{k-1} + \cdots , \tag{5″}$$

and the chain connecting $\mathbf{A}^{k-1}$ with $\mathbf{B}^{k-1}$ is

$$\Gamma^k = \alpha \sum \mathbf{Z}_\nu^k + '\alpha \sum '\mathbf{Z}_\nu^k + \cdots , \tag{6″}$$

and the connection formulas (I) through (IV) of §29 hold:

$$\mathcal{R}\,\partial\mathbf{A}^k = \mathbf{A}^{k-1}, \tag{I″}$$

$$\mathcal{R}\,\partial\mathbf{B}^k = \mathbf{B}^{k-1}, \tag{II″}$$

$$\mathcal{R}\,\partial\Gamma^k = \mathbf{A}^{k-1} - \mathbf{B}^{k-1}, \tag{III″}$$

$$\mathcal{R}\,\partial\Gamma^{k+1} = \mathbf{A}^k - \mathbf{B}^k - \Gamma^k. \tag{IV″}$$

Formula (II″) states that *the boundary of the approximation of the singular chain is equal to the approximation of its boundary.*

Final Step

It still remains to bring together the results of steps 1 and 2. We shall add the connecting chains Γ^k and Γ^{k+1} of step 2 to the connecting chains C^k and C^{k+1} of step 1 and obtain new connecting chains

$$V^k = C^k + \Gamma^k,$$
$$V^{k+1} = C^{k+1} + \Gamma^{k+1}.$$

When one remembers that B^k and B^{k-1} were renamed, respectively, $\mathbf{A}^k$ and $\mathbf{A}^{k-1}$, it now follows from the connection formulas (I′) through (IV′) and (I″) through (IV″) that

$$\mathcal{R}\,\partial A^k = A^{k-1}, \tag{I}$$

$$\mathcal{R}\,\partial\mathbf{B}^k = \mathbf{B}^{k-1}, \tag{II}$$

$$\mathcal{R}\,\partial V^k = A^{k-1} - \mathbf{B}^{k-1}, \tag{III}$$

$$\mathcal{R}\,\partial V^{k+1} = A^k - \mathbf{B}^k - V^k. \tag{IV}$$

The previous investigations are valid for an arbitrary singular chain A^k. We shall now, for the moment, make the assumption that A^k is closed, that is,

$A^{k-1} = 0$. In this case, the normal subdivision B^{k-1} and its approximation $\mathbf{B}^{k-1}$, as well as the connecting chains C^k, Γ^k, and therefore also V^k, all vanish. Formula (IV) simplifies to

$$V^{k+1} = A^k - \mathbf{B}^k.$$

In other words, *there exists a simplicial chain homologous to each closed singular chain.*

Now assume further that the boundary A^{k-1} of A^k is simplicial. By a g-fold normal subdivision of A^{k-1} one obtains the chain B^{k-1} and from this one obtains the simplicial approximation $\mathbf{B}^{k-1}$. We claim that $\mathbf{B}^{k-1} = A^{k-1}$, because, in general, *when one simplicially approximates the g-fold subdivision B^i of an arbitrary simplicial chain A^i, one again obtains the chain A^i.*

The claim is correct when $i = 0$. In that case the 0-dimensional chain A^0 coincides with its g-fold normal subdivision and each 0-simplex of A^0 remains fixed during the simplicial approximation. Let us assume that the claim has been proved for $(i-1)$-dimensional chains A^{i-1}, in particular, for the boundary of an i-simplex E^i of A^i. Denote the g-fold normal subdivision of E^i by E^i_g. On simplicially approximating E^i_g one obtains a chain U^i. By the inductive hypothesis, $\mathfrak{R}\partial E^i_g$ transforms to $\mathfrak{R}\partial E^i$. Since the boundary of the approximation is equal to the approximation of the boundary, we have

$$\mathfrak{R}\partial U^i = \mathfrak{R}\partial E^i. \qquad (1)$$

If e^i is a subsimplex of E^i_g, then under the approximation the vertices of e^i will transform to certain vertices of E^i; thus e^i itself will transform either to εE^i ($\varepsilon = \pm 1$) or to a face of E^i. In the latter case the approximation of e^i is degenerate and makes no contribution to the approximation of E^i. The approximation U^i can therefore only be a multiple of E^i and, using (1), one has $U^i = E^i$.

If then,

$$A^i = \gamma E^i + {}'\gamma, {}'E^i + \cdots$$

is an arbitrary simplicial i-chain of $\mathfrak{R}^n$ and if

$$B^i = \gamma E^i_g + {}'\gamma' E^i_g + \cdots$$

is its g-fold normal subdivision, then its simplicial approximation is

$$\mathbf{B}^i = \gamma U^i + {}'\gamma' U^i + \cdots$$

$$= \gamma E^i + {}'\gamma' E^i + \cdots$$

$$= A^i.$$

Thus the claim is proved for the dimension i, and (by induction) in general.

We can therefore set $A^{k-1} = \mathbf{B}^{k-1}$ in the formulas (I) through (IV). Formula (III) becomes

$$\mathfrak{R}\partial V^k = 0,$$

that is, V^k is closed. By construction, moreover, V^k lies* on the subcomplex $\mathfrak{A}^{k-1}$ which is formed by all of the $(k-1)$-simplexes of A^{k-1}. By simplicially approximating V^k in the complex $\mathfrak{A}^{k-1}$ one obtains a homologous simplicial k-chain on $\mathfrak{A}^{k-1}$. But this must be the k-chain 0 since A^{k-1} contains no k-simplexes. Consequently, $V^k \sim 0$. Together with formula (IV), which one can also write as $A^k - \mathbf{B}^k - V^k \sim 0$, this gives $A^k \sim \mathbf{B}^k$.

This completes the proof of the approximation theorem. *The topological invariance of the simplicial homology groups has been demonstrated and at the same time all concepts and theorems which are based upon homology properties have been proved to be topological properties of the complex.*** *For example, the Betti numbers and torsion coefficients are invariant, since they are determined directly by the homology groups. The connectivity numbers are invariant, because they can be expressed in terms of the Betti numbers and the number of even torsion coefficients. The Euler characteristic is invariant and is the same for all decompositions of a complex, since it is the alternating sum of the Betti numbers.* It is easy to show that the properties listed above of the simplicial decompositiom of a complex do not change when we go to new decompositions by means of certain subdivisions and gluings such as we shall encounter in Section 37. We do not have to investigate explicitly those subdivisions of the simplicial decomposition at all, for we have already demonstrated that any two decompositions of one and the same complex are identical as far as properties listed above are concerned, independent of whether or not they have a common subdivision.

The result we have achieved goes much further than this proof of invariance. We could have constructed a bridge from one simplicial decomposition to another using simpler methods.[17] But—and this should be stressed particularly—*we have defined the homology groups in a topologically invariant manner* and have thereby freed ourselves once and for all from the simplicial decomposition by which the complex is constructively presented to us. Thus in the future we shall not have to make step by step approximations. This significant advantage, particularly in regard to mapping theorems for manifolds, repays the effort we expended in obtaining the invariant definition.

31. Deformation and Simplicial Approximation of Mappings

The theorem to which we now turn our attention is also an approximation theorem. However, it deals not with the approximation of singular chains but, rather, with approximations of continuous mappings of a complex $\mathfrak{R}^n$ into a

---

*It is obvious that the chain C^k appearing in the sum $V^k = C^k + \Gamma^k$ lies on $\mathfrak{A}^{k-1}$. That it is also true for Γ^k follows from the previous considerations in this section.

**It should be remembered here that the kth singular homology group and also the kth simplicial homology group (§18) are also defined for $k > n$ (the dimension of the simplicial decomposition) and that the coincidence of the two has been proved.

complex $\mathbf{K}^m$. In order to formulate the theorem we must introduce two concepts which play an important role in topology. These are the concept of the deformation of a mapping and the concept of a simplicial mapping.

For simplicity we shall assume that the complex $\mathfrak{R}^n$ is finite. Let us map it in two different ways into $\mathbf{K}^m$, by means of two continuous mappings g_0 and g_1. We will say that g_0 is homotopically deformable to g_1 if there exists a "continuous family of mappings between g_0 and g_1." By this we mean the following: there exists a family of mappings such that to each value of the parameter t, varying in the interval $0 \leqq t \leqq 1$, there corresponds a mapping g_t; the given mappings g_0 and g_1 correspond, respectively, to the particular values $t = 0$ and $t = 1$, and the image $g_t(P)$ of a point P of $\mathfrak{R}^n$ will depend in a continuous way on the parameter t and the point P.

An equivalent definition of a homotopic deformation is the following: Let $\mathfrak{R}^n \times t$ be the topological product of $\mathfrak{R}^n$ and the unit interval $0 \leqq t \leqq 1$. Then the mapping g_0 will be homotopically deformable to the mapping g_1 if and only if there exists a continuous mapping f taking $\mathfrak{R}^n \times t$ into $\mathbf{K}^m$ in such a way that $f(P \times 0) = g_0(P)$ and $f(P \times 1) = g_1(P)$. For a family of mappings of $\mathfrak{R}^n$ into $\mathbf{K}^m$ is defined by the mapping of $\mathfrak{R}^n \times t$ if one sets

$$g_t(P) = f(P \times t); \tag{1}$$

conversely, * a continuous mapping f od $\mathfrak{R}^n \times t$ is defined by a continuous family of mappings g_t of $\mathfrak{R}^n$ by means of Eq. (1).

The second definition is preferred, because the whole family of mappings g_t is given by a single mapping f of the topological product. $\mathfrak{R}^n \times t$ is called the *deformation complex* of $\mathfrak{R}^n$ and t is called the *deformation parameter*.

It is convenient to embed the deformation complex into a Euclidean space in the following manner: we embed the finite simplicial complex $\mathfrak{R}^n$ in the linear subspace $t = 0$ of the Euclidean space $\mathfrak{R}^{r+1}$ having coordinates $x_1, x_2, \ldots, x_r, t$ preserving the simplicial decomposition of $\mathfrak{R}^n$. According to Section II this is possible whenever $r \geqq 2n + 1$. We then subject $\mathfrak{R}^n$ to the translation which carries the linear subspace $t = 0$ to the linear subspace $t = 1$; $\mathfrak{R}^n$ thereby sweeps out the deformation complex $\mathfrak{R}^n \times t$.

If g_0 is deformable to g_1, then obviously g_1 is deformable to g_0. If g_1 is deformable to still another mapping g_2, then g_0 is also deformable to g_2. Because of this, one can regard the continuous mappings of $\mathfrak{R}^n$ into $\mathbf{K}^m$ as being divided into classes of mappings which are homotopically deformable to one another, the so-called *mapping classes*.

We have only required continuity of the mappings g_t $(0 \leqq t \leqq 1)$. If the mappings are all topological as well, then we speak of an *isotopic deformation* of the mapping g_0 to the mapping g_1. We do not require that the mapping f of the deformation complex into K^m be topological. A congruent displacement

---

* *Editor's note*: The hypothesis that g is *simultaneously* continuous in both t and P is needed here.

of a circular disk, $\Re^2$ in the Euclidean plane $\mathbf{K}^2$ is an example of an isotopic deformation.

It may happen that $\mathbf{K}^m$ coincides with $\Re^n$, so that g_t is a self-mapping of $\Re^n$. In the particular case that g_0 is the identity mapping, the continuous family of self-mappings g_t will be called a *homotopic* (or isotopic, respectively) *deformation of the complex* $\Re^n$ *into itself.* In the treatment to follow we shall deal only with homotopic deformations and we shall, therefore, usually omit the adjective "homotopic."

A *simplicial mapping* of a simplicial complex $\Re^n$ into a simplicial complex $\mathbf{K}^m$ is defined to be a continuous mapping of $\Re^n$ into $\mathbf{K}^m$ such that each simplex $\mathfrak{E}^i$ of the simplicial decomposition of $\Re^n$ ($i = 0, 1, \ldots, n$) is mapped linearly onto a simplex Σ^j ($j \leqq i$) of $\mathbf{K}^m$ (Section 9). If one is given the image vertex in $\mathbf{K}^m$ for each vertex of $\Re^n$, then the simplicial mapping is completely determined from this combinatorial information. The only condition required of the assignment of vertices, is that the vertices of each simplex of $\Re^n$ map to the vertices of some simplex of $\mathbf{K}^m$.

The theorem which we wish to prove states, in its essentials, that each continuous mapping of $\Re^n$ into $\mathbf{K}^m$ is deformable to a simplicial mapping, provided that the simplicial decomposition of $\Re^n$ is sufficiently fine.

Let g_0 be a continuous mapping of $\Re^n$ into $\mathbf{K}^m$. By making a sufficiently fine subdivision of $\Re^n$ we can ensure that the image of each simplicial star of $\Re^n$ lies entirely in the interior of a simplicial star of $\mathbf{K}^m$. To each vertex P_ν of $\Re^n$ we assign a vertex Π_ν of $\mathbf{K}^m$ such that the star with center point Π_ν contains the image of the star with center point P_ν in its interior. If we denote the vertices of an i-simplex $\mathfrak{E}^i$ of $\Re^n$ by $P_0, P_1, \ldots, P_i$, then the stars of $\mathbf{K}^m$ having center points $\Pi_0, \Pi_1, \ldots, \Pi_i$ will, respectively, each contain the image $g_0(\mathfrak{E}^i)$ in its interior; here, the vertices Π_ν need not all be distinct from one another. It follows that $\Pi_0, \Pi_1, \ldots, \Pi_i$ are vertices of a simplex of $\mathbf{K}^m$. For if $g_0(P)$ is the image of an arbitrary point P of $\mathfrak{E}^i$ and Σ is a simplex of $\mathbf{K}^m$, on which the point $g_0(P)$ lies, then each one of the stars with center points $\Pi_0, \Pi_1, \ldots, \Pi_i$ will contain $g_0(P)$ in its interior. Thus $\Pi_0, \Pi_1, \ldots, \Pi_i$ are vertices of Σ and consequently are vertices of a simplex Σ^j of $\mathbf{K}^m$. A simplicial mapping g_1 of $\Re^n$ into $\mathbf{K}^m$ is thus determined by assigning the vertices P_ν of $\Re^n$ to the vertices Π_ν of $\mathbf{K}^m$.

The image point $g_1(P)$ of the previously mentioned point P of $\mathfrak{E}^i$ belongs to Σ^i and thus, like $g_0(P)$, belongs to Σ and can therefore be connected with $g_0(P)$ by a straight line segment. But this also shows that g_0 can be deformed to g_1. One lets the point $g_0(P)$ move along the connecting segment to $g_1(P)$; in some cases this may amount to staying "in place." More precisely: We define a continuous mapping f of the deformation complex $\Re^n \times t$ into $\mathbf{K}^m$ by means of the specifications: $f(P \times 0) = g_0(P)$, $f(P \times 1) = g_1(P)$, and the segment traversed by P on $\Re^n \times t$ when t increases from 0 to 1 maps linearly onto the segment connecting the points $g_0(P)$ and $g_1(P)$ (which may be a single point). Continuity of the mapping f follows from the lemma in §14. Our result is

THEOREM I. (DEFORMATION THEOREM). *Each continuous mapping of a complex $\Re^n$ into a complex $\mathbf{K}^m$ can be deformed (homotopically) to a simplicial mapping, provided that $\Re^n$ has a sufficiently fine simplicial decomposition. During the deformation the image of a point P of $\Re^n$ will leave none of the simplexes of $\mathbf{K}^m$ to which it belonged at the start of the deformation; that is, it will move only on the lowest dimensional simplex of $\mathbf{K}^m$ to which it belonged at the start of the deformation.*

From Theorem II of §27, a continuous mapping of $\Re^n$ into $\mathbf{K}^m$ induces a homomorphism from the kth homology group of $\Re^n$ to the kth homology group of $\mathbf{K}^m$. We shall now examine the influence of a deformation of the continuous mapping upon the homorphic mapping between the homology groups.

THEOREM II. *Let g_0 and g_1 be mappings of $\Re^n$ into $\mathbf{K}^m$. Let A^k be a singular chain on $\Re^n$ having boundary A^{k-1}. Let $\mathbf{A}^k$, having boundary $\mathbf{A}^{k-1}$, and $\mathbf{B}^k$, having boundary $\mathbf{B}^{k-1}$, be the images of A^k and A^{k-1} under g_0 and g_1, respectively. If g_0 is deformable to g_1, then there will exist singular connecting chains Γ^k and Γ^{k+1} which satisfy the connection formulas*:

$$\Re \partial \Gamma^k = \mathbf{A}^{k-1} - \mathbf{B}^{k-1},$$

$$\Re \partial \Gamma^{k+1} = \mathbf{A}^k - \mathbf{B}^k - \Gamma^k.$$

Γ^k *will lie on the point set which the image set of A^{k-1} sweeps out during the deformation; Γ^{k+1} will lie on the point set which the image set of A^k sweeps out during the deformation. In particular, if $A^{k-1} = 0$ (so that $A^{k-1} = 0$ and $\mathbf{B}^{k-1} = 0$), then $\Gamma^k = 0$ and $\mathbf{A}^k \sim \mathbf{B}^k$ (cf. Fig. 52).*

Proof. Embed the deformation complex $\Re^n \times t$ into the Euclidean space $\Re^{r+1}$ as done above in this section. Let the chain A^k be given by the sum

$$A^k = aX^k + {}'a'X^k + \cdots.$$

If x^k is a preimage simplex of X^k and $\mathfrak{z}^{k+1}$ is a prism erected over it having roof face $\mathfrak{y}^k$, then we can map $\mathfrak{z}^{k+1}$ continuously onto the subset of $\Re^n \times t$

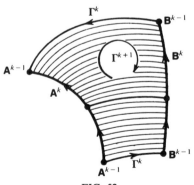

FIG. 52

which is swept out during the translation by the points of the singular simplex X^k: x^k is mapped as before to the singular simplex X^k of $\Re^n \times 0$, while each axis parallel interval of $\mathfrak{z}^{k+1}$ transforms linearly to the segment swept out by the image of its floor point in $\Re^n \times t$ during the translation. We decompose $\mathfrak{z}^{k+1}$ simplicially and orient the subsimplexes as in §29. The oriented simplexes or chains

$$x^k, y^k, z^{k+1}, x_\nu^{k-1}, y_\nu^{k-1}, z_\nu^k$$

will then transform to singular chains

$$X^k, Y^k, Z^{k+1}, X_\nu^{k-1}, Y_\nu^{k-1}, Z_\nu^k$$

on $\Re^n \times t$, under the mapping of $\mathfrak{z}^{k+1}$ into $\Re^n \times t$. We proceed in the corresponding way for the remaining singular simplexes of A^k and adding, we obtain the singular chains

$$B^k = aY^k + {}'a'Y^k + \cdots,$$
$$C^{k+1} = aZ^{k+1} + {}'a'Z^{k+1} + \cdots,$$
$$A^{k-1} = a\sum_\nu X_\nu^{k-1} + {}'a\sum_\nu {}'X_\nu^{k-1} + \cdots,$$
$$B^{k-1} = a\sum_\nu Y_\nu^{k-1} + {}'a\sum_\nu {}'Y_\nu^{k-1} + \cdots,$$
$$C^k = a\sum_\nu Z_\nu^k + {}'a\sum_\nu {}'Z_\nu^k + \cdots,$$

and the connection formulas of §29 hold.

The singular chains C^{k+1} and C^k lie entirely in the subsets of $\Re^n \times t$ which are swept out by the chains A^k and A^{k-1}, respectively, under the translation. This is clear for C^{k+1}. For C^k, we should note that the point set swept out by A^{k-1} is not, in general, the same as the set theoretic union of the point sets which are swept out by the individual singular simplexes $X_\nu^{k-1}, {}'X_\nu^{k-1}, \ldots$. For it may happen that such a simplex may belong formally to the right-hand side of A^{k-1} but will not belong at all to A^{k-1}, either because it is degenerate or because it is cancelled out by other simplexes. If now

$$^{(\kappa)}X_\mu^{k-1} = \pm {}^{(\lambda)}X_\nu^{k-1} \qquad (2)$$

we need only to prove that also

$$^{(\kappa)}Y_\mu^{k-1} = \pm {}^{(\lambda)}Y_\nu^{k-1}$$

and

$$^{(\kappa)}Z_\mu^k = \pm {}^{(\lambda)}Z_\nu^k.$$

But this follows immediately from the construction and from the definition of equality of two singular simplexes.

Under the mapping of $\Re^n \times t$ into $\mathbf{K}^m$ the singular chains A^k, $A^{k-1}, B^k, B^{k-1}, C^k, C^{k+1}$ transform to the chains $\mathbf{A}^k, \mathbf{A}^{k-1}, \mathbf{B}^k, \mathbf{B}^{k-1}, \mathbf{\Gamma}^k, \mathbf{\Gamma}^{k+1}$

of the theorem. From Theorem I in §27, the connection formulas are preserved:

$$\mathfrak{R}\,\partial \mathbf{A}^k = \mathbf{A}^{k-1}, \tag{I}$$

$$\mathfrak{R}\,\partial \mathbf{B}^k = \mathbf{B}^{k-1}, \tag{II}$$

$$\mathfrak{R}\,\partial \mathbf{\Gamma}^k = \mathbf{A}^{k-1} - \mathbf{B}^{k-1}, \tag{III}$$

$$\mathfrak{R}\,\partial \mathbf{\Gamma}^{k+1} = \mathbf{A}^k - \mathbf{B}^k - \mathbf{\Gamma}^k. \tag{IV}$$

Having obtained formulas (III) and (IV), we have proved the connection formulas of the theorem. The chain $\mathbf{\Gamma}^k$ lies on the image of the point set which A^{k-1} sweeps out during the translation on $\mathfrak{R}^n \times t$; the chain $\mathbf{\Gamma}^{k+1}$ lies on the image of the point set which A^k sweeps out during this translation. This completes the proof of Theorem II.

Since, according to Theorem II, the images of a closed singular chain A^k under the mappings g_0 and g_1, which are deformable to one another, are homologous to each other on $\mathbf{K}^m$ we have

THEOREM III: *The homomorphic mapping of the kth homology group of $\mathfrak{R}^n$ to the kth homology group of $\mathbf{K}^m$, which is generated by a continuous mapping of $\mathfrak{R}^n$ into $\mathbf{K}^m$, from Theorem II of §27, will remain unchanged under homotopic deformation of the mapping. It is, consequently, an invariant of the mapping class.*

In particular, this is true when we deal with a deformation of a complex $\mathfrak{R}^n$ into itself. Then $\mathfrak{R}^n$ coincides with $\mathbf{K}^m$ and g^0 is the identity mapping. Since the self-mapping of the homology groups of $\mathfrak{R}^n$ generated by g_0 is the identity mapping, we have

THEOREM IV. *The homomorphic self-mapping of the homology groups of a complex $\mathfrak{R}^n$ generated by a deformation of $\mathfrak{R}^n$ into itself is the identity mapping.*

Theorem III gives us a necessary though by no means sufficient condition that two mappings of $\mathfrak{R}^n$ into $\mathbf{K}^m$ belong to the same mapping class. For example, let $\mathfrak{R}^n = \mathbf{K}^m$ be the n-sphere $\mathfrak{S}^n$, that is, the boundary of an $(n+1)$-simplex $\mathfrak{E}^{n+1} = (P_0 P_1 \cdots P_n P_{n+1})$, and let the self-mapping g_1 of $\mathfrak{S}^n$ be defined by means of the linear mapping of $\mathfrak{E}^{n+1}$ which fixes the vertices $P_2, P_3, \ldots, P_{n+1}$ individually while interchanging P_0 and P_1. We ask: Does g_1 belong to the mapping class of the identity? That is, can one deform g_1 homotopically to the identity mapping? The chain* $\mathfrak{R}\,\partial E^{n+1}$ on $\mathfrak{S}^n$ forms a homology basis for the nth homology group of $\mathfrak{S}^n$, which is the free cyclic group. Under the mapping, the oriented simplex $E^{n+1} = +(P_0 P_1 \cdots P_{n+1})$ transforms to $-E^{n+1} = +(P_1 P_0 \cdots P_{n+1})$; thus $\mathfrak{R}\,\partial E^{n+1}$ transforms to $-\mathfrak{R}\,\partial E^{n+1}$. Thus the homorphism of the nth homology group, induced by g_1

---

* As always, E^{n+1} denotes the simplex $\mathfrak{E}^{n+1}$ provided with a definite orientation.

is not the identity. Consequently, from Theorem IV, g_1 cannot belong to the mapping class of the identity.

Problem

Consider the mapping which interchanges diametrically opposite points of the unit n-sphere in Euclidean $(n + 1)$-space. For which dimensions will it belong to the mapping class of the identity? (Express the diametral point interchange as a product of mirror reflections.)

As a further application of Theorem II, we shall determine the homology groups of projective n-space $\mathfrak{P}^n$.

We first determine the dimensions for which $\mathfrak{P}^n$ is orientable. For this purpose, we remember that $\mathfrak{P}^n$ arises from the n-sphere $\mathfrak{S}^n$ by means of identification of diametrically opposite points. We coherently orient a centrally symmetric simplicial decomposition of the n-sphere, for example, the octahedral decomposition (§14). Since the diametral point interchange is the product of $n + 1$ mirror reflections and the orientation of $\mathfrak{S}^n$ is reversed by each reflection, the orientation of $\mathfrak{S}^n$ will be preserved or reversed according to whether n is odd or even. For odd n, two oriented n-simplexes of $\mathfrak{S}^n$ map to the same oriented n-simplex of $\mathfrak{P}^n$ and one obtains in this way a coherent orientation of $\mathfrak{P}^n$. For even n, on the other hand, it is impossible to orient $\mathfrak{P}^n$, because a coherent orientation could be "pushed back" to give a coherent orientation of $\mathfrak{S}^n$, whereas in the transformation of $\mathfrak{S}^n$ to $\mathfrak{P}^n$, paired coherently oriented n-simplexes of $\mathfrak{S}^n$ map with opposite orientations to an n-simplex of $\mathfrak{P}^n$. (One should verify the relationships for dimensions $n = 2$ and $n = 3$.) Thus *$\mathfrak{P}^n$ is orientable for odd n and nonorientable for even n.*

We now introduce projective coordinates $x_1, x_2, \ldots, x_{n+1}$ into $\mathfrak{P}^n$, as in §14, and denote by $\mathfrak{P}^k$ the projective subspace described by the equations

$$x_{k+2} = 0, \qquad \ldots, \qquad x_{n+1} = 0. \qquad (\mathfrak{P}^k)$$

In the sequence $\mathfrak{P}^n, \mathfrak{P}^{n-1}, \ldots, \mathfrak{P}^1, \mathfrak{P}^0$ each of the projective spaces is contained in its predecessors.

In order to determine a homology basis for dimension k $(0 < k < n)$ we begin with a closed singular k-chain U^k. If U^k happens to pass through the point $(0, 0, \ldots, 0, 1)$, then we push it away from this point by means of a simplicial approximation in a simplicial decomposition of $\mathfrak{P}^n$ which has this point as the midpoint of an n-simplex. We then remove from $\mathfrak{P}^n$ the interior points of a small ball about the point $(0, 0, \ldots, 0, 1)$ which does not intersect U^n to form a bounded pseudomanifold and we transform the latter to the projective space $\mathfrak{P}^{n-1}$ by means of a family of transformations

$$x_1' = x_1, \qquad x_2' = x_2, \qquad \ldots, \qquad x_n' = x_n, \qquad x_{n+1}' = tx_{n+1},$$

where t runs from 1 to 0. This transforms U^k to a homologous chain $'U^k$ (Theorem II). If $k < n - 1$, one repeats the same procedure in $\mathfrak{P}^{n-1}$ and pushes $'U^k$ into $\mathfrak{P}^{n-2}$. In this way one finally arrives at a chain $V^k \sim U^k$ which lies in $\mathfrak{P}^k$. If k is even, so that $\mathfrak{P}^k$ is nonorientable, then all closed

k-chains of $\mathfrak{P}^k$ are homologous to 0 on $\mathfrak{P}^k$; that is, for even k the kth homology group of $\mathfrak{P}^n$ consists of the null element alone. If k is odd, then there exists a closed k-chain P^k on $\mathfrak{P}^k$ which arises by means of the coherent orientation of $\mathfrak{P}^k$ such that any other closed k-chain V^k on $\mathfrak{P}^k$ is homologous to a multiple of P^k. For odd k, the kth homology group of $\mathfrak{P}^n$ is cyclic and P^k forms a homology basis. More precisely, it has order 2. For $\mathfrak{P}^{k+1}$ is a nonorientable pseudomanifold and consequently has exactly one k-dimensional torsion coefficient, which has the value 2 (§24). Thus $2P^k \sim 0$ in $\mathfrak{P}^{k+1}$. It is certain that P^k itself is not already null homologous in $\mathfrak{P}^{k+1}$. It is also not ~ 0 in $\mathfrak{P}^n$. For if P^k were the boundary of a chain U^{k+1} of $\mathfrak{P}^n$, then one could push U^{k+1} into the subspace $\mathfrak{P}^{k+1}$ by means of the procedure already applied to $\mathfrak{P}^n$, while leaving P^k pointwise fixed. P^k would then be the boundary of a chain V^{k+1} of $\mathfrak{P}^{k+1}$, which is not the case. Our result is

The odd dimensional homology groups of projective n-space $\mathfrak{P}^n$ have order 2, and those of even dimension consist of the null element alone; only the 0th homology group is an exception for each value of n and the nth for odd n; these are free cyclic groups.

One can use the coherently oriented projective subspaces $P^1, P^3, P^5, \ldots$ as homology bases of the odd dimensions.[18]

CHAPTER FIVE

LOCAL PROPERTIES

The invariance proof given in the preceding chapter has given us the homology groups as our first topological invariants of a complex. But the homology groups give only a coarse classification of complexes. Complexes which have the same homology groups are not necessarily homeomorphic. This is demonstrated by the example of an interval (1-simplex) and a disk (2-simplex). These complexes have the same homology groups; the 0th homology group is the free group having one generator, while all other groups consist of the null element alone (§18 and §19). Nevertheless, the two complexes cannot be mapped topologically one onto the other. This was demonstrated in §1.

Homology groups were invariants "in the large," and cannot be calculated until the whole complex is known. On the other hand, the difference between a line segment and a disk is already apparent in arbitrarily small neighborhoods of points belonging to those respective complexes. Even the neighborhoods cannot be mapped topologically one onto the other. In this chapter we shall produce "local" invariants, which we shall call properties of a complex at a point, and which are determined when one studies an arbitrarily small neighborhood of the point in question. The most important of these local invariants are the "homology groups at a point." With their aid we shall prove the invariance of dimension, which up to now has been defined only for simplicial decompositions, and we shall prove the invariance of boundary, of the property of being a pseudomanifold, of orientability, and other properties.[19]

Our results will be valid for infinite complexes as well as for finite complexes.

32. Homology Groups of a Complex at a Point

Let $\Re^n$ be a connected simplicial complex of dimension $n > 0$ and let P be a point of $\Re^n$. Draw a straight ray outward from P in the i-simplex to which P belongs, in the simplicial decomposition of $\Re^n$. This ray will leave the simplex at a point Q. The totality of all intervals (PQ) which one can extend outward from P exactly fills the subcomplex Ω of all simplexes of $\Re^n$ to which P belongs. We call Ω the *simplicial neighborhood* of P in $\Re^n$; Ω is a neighborhood of P in $\Re^n$ because of condition $(k4)$ of §10. On the other hand, the set of endpoints Q of all intervals (PQ) forms the so-called *neighborhood complex* $\mathfrak{A}$ of P in $\Re^n$; $\mathfrak{A}$ is not a neighborhood of P, but on the contrary consists of the totality of simplexes of Ω which do not contain P. In particular, if P is a vertex of the simplicial decomposition, then Ω is the simplicial star with center point P and outer boundary $\mathfrak{A}$. If P is not a vertex, then one can always find another simplicial decomposition in which P is a vertex, simply by projecting from P to the neighborhood complex $\mathfrak{A}$ of P.

We now define the *kth homology group of $\mathfrak{R}^n$ at the point P* to be the kth homology group of the neighborhood complex $\mathfrak{A}$ of P. The homology groups of $\mathfrak{R}^n$ at P are, then, defined with respect to a particular simplicial decomposition of $\mathfrak{R}^n$. We will show, however, that they are in reality independent of the simplicial decomposition and are therefore topological invariants of $\mathfrak{R}^n$.[20]

EXAMPLE 1. Let P be an inner point of an n-simplex $\mathfrak{E}^n$ of $\mathfrak{R}^n$ ($n > 1$). The neighborhood complex $\mathfrak{A}$ will be formed by the boundary of $\mathfrak{E}^n$ and is therefore an $(n-1)$-sphere. The homology groups of dimensions 1 through $n-2$ of $\mathfrak{R}^n$ at P will then consist of the null element, those of dimensions 0 and $n-1$ are free cyclic groups (§19).

EXAMPLE 2. Let P be an inner point of an $(n-1)$-simplex $\mathfrak{E}^{n-1}$ of $\mathfrak{R}^n$ such that $\mathfrak{E}^{n-1}$ is incident with n-simplexes $\mathfrak{E}_1^n, \mathfrak{E}_2^n, \ldots, \mathfrak{E}_k^n$ (Fig. 53). If $n > 1$ and $B_1, B_2, \ldots, B_k$ are the vertices of $\mathfrak{E}_1^n, \mathfrak{E}_2^n, \ldots, \mathfrak{E}_k^n$ which lie opposite to the common face, then the neighborhood complex $\mathfrak{A}^{n-1}$ of P consists of k simplicial stars $\mathfrak{St}_1^{n-1}, \mathfrak{St}_2^{n-1}, \ldots, \mathfrak{St}_k^{n-1}$ having center points $B_1, B_2, \ldots, B_k$ and whose common outer boundary is $\mathfrak{S}^{n-2}$, the boundary of $\mathfrak{E}^{n-1}$. We wish to compute the $(n-1)$th homology group at P. For this purpose we orient the $(n-1)$-simplexes of $\mathfrak{St}_\mu^{n-1}$ coherently and such that the resulting $(n-1)$-chain St_μ^{m-1} together with the simplex E^{n-1} having fixed orientation forms a closed chain.* This is possible because St_μ^{n-1} and E^{n-1} together form the boundary of the simplex E_μ^n. We then have

$$\Re \partial St_\mu^{n-1} = -\Re \partial E^{n-1}. \tag{1}$$

The $k-1$ closed $(n-1)$-chains

$$St_1^{n-1} - St_k^{n-1}, St_2^{n-1} - St_k^{n-1}, \ldots, St_{k-1}^{n-1} - St_k^{n-1} \tag{2}$$

are homologously independent on $\mathfrak{A}^{n-1}$. For it follows from

$$\sum_{\nu=1}^{k-1} a_\nu \left(St_\nu^{n-1} - St_k^{n-1} \right) \sim 0$$

that

$$\sum_{\nu=1}^{k-1} a_\nu St_\nu^{n-1} - \left(\sum_{\nu=1}^{k-1} a_\nu \right) St_k^{n-1} \sim 0.$$

We can replace the homology symbol by the equality sign since no n-simplexes appear in the complex $\mathfrak{A}^{n-1}$ on which this homology holds (§18). Thus $a_1 = a_2 = \cdots = a_{k-1} = 0$, since no two chains St_μ and St_ν have a common $(k-1)$-simplex for $\mu \neq \nu$.

---

*Recall that we indicate the transformation from nonoriented to oriented objects by using Roman instead of German letters.

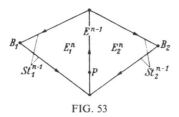

FIG. 53

On the other hand, each closed $(n-1)$-chain U^{n-1} on $\mathfrak{A}^{n-1}$ is a linear combination of the chains St_μ^{n-1}. For if an $(n-1)$-simplex of St_μ^{n-1} appears in U^{n-1} with multiplicity b_μ, then, because U^{n-1} is closed, all $(n-1)$-simplexes of St_μ^{n-1} must appear in it with multiplicity b_μ. Consequently,

$$U^{n-1} = b_1 St_1^{n-1} + b_2 St_2^{n-1} + \cdots + b_k St_k^{n-1}.$$

Because of (1), $\mathfrak{R}\partial U^{n-1}$ is equal to zero only if $\sum_{\mu=1}^{k} b_\mu$ vanishes. In that case, however, one can bring U^{n-1} to the form

$$U^{n-1} = \sum_{\nu=1}^{k-1} b_\nu \left(St_\nu^{n-1} - St_k^{n-1} \right).$$

Thus the $(n-1)$-chains (2) describe a homology basis for the dimension $n-1$. *For $n > 1$, the $(n-1)$th homology group of $\mathfrak{A}^{n-1}$ is the free Abelian group having $k-1$ generators.*

In the case $n = 1$, P is the common vertex of the 1-simplexes so that $\mathfrak{A}^0$ consists of k points, and the $(n-1)$th homology group of $\mathfrak{A}^0$ is the free Abelian group having k generators.

In order to prove that the homology groups at P are independent of the particular simplicial decomposition of $\mathfrak{R}^n$, we shall consider two simplicial stars on $\mathfrak{R}^n$, Ω and Ω', having the same center point P. We only require of these simplicial stars that they are neighborhoods of P in $\mathfrak{R}^n$; we do not require that there exist a simplicial decomposition of $\mathfrak{R}^n$ such that Ω or Ω' is a simplicial neighborhood of P. It is not possible to prove that the outer boundary $\mathfrak{A}$ of Ω is homeomorphic to the outer boundary $\mathfrak{A}'$ of Ω'. For the existence of such a homeomorphism of $\mathfrak{A}$ onto $\mathfrak{A}'$ would imply that there exists a continuous mapping φ of $\mathfrak{A}$ into $\mathfrak{A}'$ and a continuous mapping ψ of $\mathfrak{A}'$ into $\mathfrak{A}$ such that the mapping $\psi\varphi$ (first φ, then ψ) of $\mathfrak{A}$ into itself and, likewise, the mapping $\varphi\psi$ of $\mathfrak{A}'$ into itself are each the identity mapping. Rather, we shall be able to prove

THEOREM I. *There exists a continuous mapping φ from $\mathfrak{A}$ into $\mathfrak{A}'$ and a* · *continuous mapping ψ from $\mathfrak{A}'$ into $\mathfrak{A}$ such that the mappings $\psi\varphi$ of $\mathfrak{A}$ into itself and $\varphi\psi$ of $\mathfrak{A}'$ into itself are both deformable to the identity.*

The proof appears later in this section.
From Theorem I there follows

THEOREM II. *$\mathfrak{A}$ and $\mathfrak{A}'$ have the same homology groups.*

Proof. From §27, under the mapping φ of $\mathfrak{A}$ into $\mathfrak{A}'$ the kth homology group $\mathfrak{H}^k$ of $\mathfrak{A}$ will be mapped by a homomorphism ϕ into the kth homology group $'\mathfrak{H}^k$ of $\mathfrak{A}'$. Likewise, a homomorphic mapping, ψ, will be carried out under the mapping ψ. Since the self-mapping $\psi\varphi$ of $\mathfrak{A}$ is deformable to the identity mapping, the associated homomorphic self-mapping $\psi\phi$ of the group $\mathfrak{H}^k$ is the identity mapping of the group (§31, Theorem IV). By the same reasoning, the homomorphic self-mapping $\phi\psi$ is the identity. This implies, however, that the homomorphic mappings ϕ and ψ are one-to-one correspondences and that ϕ is the reciprocal mapping of ψ. Groups $\mathfrak{H}^k$ and $'\mathfrak{H}^k$ are therefore isomorphic.

If, in particular, $\mathfrak{A}$ and $\mathfrak{A}'$ are neighborhood complexes of P in two different simplicial decompositions of $\mathfrak{R}^n$, then from Theorem II follows

THEOREM III. *The homology groups at P are independent of the simplicial decomposition used.*

One gets the result, at the same time, that the determination of the homology groups at P does not require a simplicial decomposition of all of $\mathfrak{R}^n$. It suffices to know a simplicial star Ω which is a neighborhood of P and whose center point coincides with P. The homology groups at P are then the homology groups of the outer boundary $\mathfrak{A}$ of Ω and are therefore already determined by an arbitrarily small neighborhood of P.

Proof of Theorem I. First, we make a few introductory comments. The simplicial star Ω consists of the totality of straight line segments (PQ) connecting P to the points Q of $\mathfrak{A}$. Starting from P, when one marks off a fraction of each of the segments (PQ), say $1/k$, then one obtains from the totality of segments $(PR) = (1/k)(PQ)$ a new simplicial star Ω_1 having outer boundary $\mathfrak{A}_1$. The star Ω_1 arises from Ω by means of a "proportional shrinking," and it is therefore homeomorphic to Ω. Each arbitrary neighborhood $\mathfrak{U}(P/\mathfrak{R}^n)$ of P contains such proportional diminutions of Ω. For the intersection of $\mathfrak{U}(P/\mathfrak{R}^n)$ with Ω is a neighborhood $\mathfrak{U}(P/\Omega)$ of P, from §5, and in it lies a proportional diminution of Ω. On the other hand, each proportional diminution Ω_1 of Ω is a neighborhood of P in Ω and therefore, from Theorem VII of §7, is also a neighborhood of P in $\mathfrak{R}^n$. The totality of segments (RQ) forms a "zone," which we will denote by $(\mathfrak{A}, \mathfrak{A}_1)$ (Fig. 54). To "project" this zone onto $\mathfrak{A}$ (or A_1, respectively) will mean that we map each segment (RQ) into the endpoint Q (or R, respectively). In like manner we can define a proportional diminution, a zone, and a projection of the zone for the simplicial star Ω', where it is to be noted that a straight line segment in Ω' is not in general a straight line segment in Ω.

We now look at a nested sequence of six simplicial stars

$$\Omega_1, \Omega_1', \Omega_2, \Omega_2', \Omega_3, \Omega_3', \tag{3}$$

where $\Omega_1, \Omega_2, \Omega_3$ are proportional diminutions of Ω, and $\Omega_1', \Omega_2', \Omega_3'$ are proportional diminutions of Ω' and each star of the sequence lies entirely in

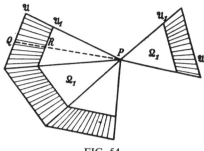

FIG. 54

the interior of its predecessor. One can construct these stars sequentially by setting $\Omega = \Omega_1$ and then choosing Ω_1' as a proportional diminution of Ω' such that it lies entirely in the interior of Ω_1. One constructs Ω_2 in the same way, and so forth. We shall denote the outer boundaries of the stars (3) by

$$\mathfrak{A}_1, \mathfrak{A}_1', \mathfrak{A}_2, \mathfrak{A}_2', \mathfrak{A}_3, \mathfrak{A}_3'. \tag{4}$$

They give rise to four zones:

$$(\mathfrak{A}_1, \mathfrak{A}_2), (\mathfrak{A}_1', \mathfrak{A}_2'), (\mathfrak{A}_2, \mathfrak{A}_3), (\mathfrak{A}_2', \mathfrak{A}_3'),$$

in which the outer boundaries $\mathfrak{A}_1', \mathfrak{A}_2, \mathfrak{A}_2', \mathfrak{A}_3$, respectively, lie. For example, $\mathfrak{A}_1'$ lies in the zone $(\mathfrak{A}_1, \mathfrak{A}_2)$ because $\mathfrak{A}_1'$ belongs to the point set $\Omega_1 - \Omega_2$, for by assumption $\mathfrak{A}_1'$ belongs to Ω_1 and, on the other hand, Ω_2 lies in the interior of Ω_1' and thus can have no point in common with $\mathfrak{A}_1'$.

Since $\mathfrak{A}_1, \mathfrak{A}_2, \mathfrak{A}_3$ are homeomorphic to $\mathfrak{A}$, while $\mathfrak{A}_1', \mathfrak{A}_2', \mathfrak{A}_3'$ are homeomorphic to $\mathfrak{A}'$, it will suffice to introduce mappings φ and ψ of the theorem as mappings of $\mathfrak{A}_2$ into $\mathfrak{A}_2'$ and $\mathfrak{A}_2'$ into $\mathfrak{A}_2$, respectively. We define mappings φ and ψ as follows: Project the zone $(\mathfrak{A}_1', \mathfrak{A}_2')$ onto $\mathfrak{A}_2'$. The outer boundary $\mathfrak{A}_2$, which belongs to this zone, is thereby mapped by a continuous mapping φ into $\mathfrak{A}_2'$. On the other hand, when we project the zone $(\mathfrak{A}_2, \mathfrak{A}_3)$ onto $\mathfrak{A}_2$, $\mathfrak{A}_2'$ is mapped by a continuous mapping, ψ into $\mathfrak{A}_2$ (Fig. 55).

If g_0 denotes the identity self-mapping of $\mathfrak{A}_2$ and g_1 denotes the self-mapping $\psi\varphi$, then the claim that g_0 is deformable to g_1 is equivalent to

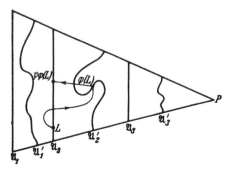

FIG. 55

the following claim: We can map the topological product $\mathfrak{A}_2 \times \mathfrak{t}$ of $\mathfrak{A}_2$ with the unit interval $(0 \leqq t \leqq 1)$ into $\mathfrak{A}_2$ by a continuous mapping f in a way that, for each point L of $\mathfrak{A}_2$, $f(L \times 0) = g_0(L) = L$ and $f(L \times 1) = g_1(L)$. To construct f we decompose $\mathfrak{A}_2 \times \mathfrak{t}$ into two parts, $\mathfrak{A}_2 \times \mathfrak{r}$ and $\mathfrak{A}_2 \times \mathfrak{s}$, where $\mathfrak{r}$ is the half-interval $0 \leqq t \leqq \frac{1}{2}$ and $\mathfrak{s}$ is the half-interval $\frac{1}{2} \leqq t \leqq 1$. We map these two parts in the following way: For each point L of $\mathfrak{A}_2$ the half-interval $L \times \mathfrak{r}$ is mapped linearly onto the straight line segment connecting $L = g_0(L)$ with $\varphi(L)$, that is, onto the projection ray which L describes during the projection of the zone $(\mathfrak{A}_1', \mathfrak{A}_2')$ onto $\mathfrak{A}_2'$. The half-interval $L \times \mathfrak{s}$ is mapped linearly onto the straight line segment connecting $\varphi(L)$ with $\psi\varphi(L) = g_1(L)$, that is, onto the projection ray which $\varphi(L)$ describes during projection of the zone $(\mathfrak{A}_2, \mathfrak{A}_3)$ onto $\mathfrak{A}_2$. Given the two partial mappings, one obtains a mapping $\bar{f}$ of the whole of $\mathfrak{A}_2 \times \mathfrak{t}$; this mapping is continuous, from §14. The image of $\mathfrak{A}_2 \times \mathfrak{r}$ belongs to the zone $(\mathfrak{A}_1', \mathfrak{A}_2')$, the image of $\mathfrak{A}_2 \times \mathfrak{s}$ belongs to $(\mathfrak{A}_2, \mathfrak{A}_3)$; thus the image $\bar{f}(\mathfrak{A}_2 \times \mathfrak{t})$ belongs to $(\mathfrak{A}_1, \mathfrak{A}_3)$. If we project the zones $(\mathfrak{A}_1, \mathfrak{A}_2)$ and $(\mathfrak{A}_2, \mathfrak{A}_3)$ simultaneously onto $\mathfrak{A}_2$, then the point set $\bar{f}(\mathfrak{A}_2 \times \mathfrak{t})$ will likewise be pushed onto $\mathfrak{A}_2$. Since the points $f(L \times 0) = g_0(L) = L$ and $\bar{f}(L \times 1) = \psi\varphi(L) = g_1(L)$ already lie on $\mathfrak{A}_2$, they remain unchanged under this projection. From Theorem IV of §6, the mapping $\bar{f}$ followed by this projection is a continuous mapping, $f(\mathfrak{A}_2 \times \mathfrak{t})$, which is what we are looking for. Thus the deformability of $\psi\varphi$ to the identity mapping has been proved. The deformability of $\varphi\psi$ to the identify mapping follows in a corresponding manner.

Another way of expressing the invariance of the homology groups at a point is the following

THEOREM IV. *If two complexes $\mathfrak{R}$ and $\mathfrak{R}'$ are homeomorphic in neighborhoods of the points P of $\mathfrak{R}$ and P' of $\mathfrak{R}'$, that is, there exist neighborhoods ω of P and ω' of P' which can be mapped topologically onto one another so that P and P' correspond to one another, then the homology groups of $\mathfrak{R}$ at P coincide with the homology groups of $\mathfrak{R}'$ at P'.*

Proof. Choose a simplicial star Ω on ω having center point P so that Ω is a neighborhood of P. Under the topological mapping of ω onto ω', Ω will transform to a simplicial star Ω' having center point P'. Since neighborhoods transform to neighborhoods under topological mappings, Ω' is a neighborhood of P'. The homology groups at P and at P' are the homology groups of the homeomorphic outer boundaries of Ω and Ω', respectively, and are therefore isomorphic.

Problems

1. What are the homology groups at a point of Euclidean n-space?

2. Show that it is impossible to map an open subset of a Euclidean space topologically onto an open subset of a Euclidean space of higher dimension.

3. If $\mathfrak{E}^k$ $(k < n)$ is a topological simplex in $\mathfrak{R}^n$, show that there exist points which do not belong to $\mathfrak{E}^k$ in each neighborhood of an arbitrary point, P of $\mathfrak{E}^k$.

33. Invariance of Dimension

Several of the concepts introduced previously were defined in terms of a particular simplicial decomposition of a complex, for example, dimension, pseudomanifold, and orientability. They might, then, not really describe properties of the complex but, instead, describe properties of the simplicial decomposition. Given two distinct decompositions of one and the same complex, it is conceivable that one decomposition might be 3-dimensional and the other 4-dimensional, one might be orientable and the other nonorientable, one might be bounded and the other unbounded. We now possess the means to show that something of this nature cannot, in fact, occur. All possible simplicial decompositions of a complex will, for example, be simultaneously n-dimensional or orientable or bounded. We shall free these concepts from any simplicial decomposition and, in doing so, shall ensure their invariance with respect to topological mappings. For example, if the complexes $\Re_1$ and $\Re_2$ are homeomorphic and $\Re_1$ is n-dimensional, then $\Re_2$ is also n-dimensional. That is, an n-dimensional simplicial decomposition of $\Re_1$ will transform to an n-dimensional simplicial decomposition of $\Re_2$ under the homeomorphism from $\Re_1$ to $\Re_2$.

We shall prove the invariance of the following concepts: dimension, pure complex, boundary, closed pseudomanifold, orientability, bounded pseudo-manifold. The proofs will be carried out in a way that the concept in question is characterized independently of any simplicial decomposition. In this context, the homology groups at a point, which we know to be topologically invariant, play a decisive role.

We begin with the invariance of dimension. We have already characterized the 0-dimensional complexes in an invariant way, without making reference to a simplicial decomposition: The 0-dimensional complexes are those complexes which consist only of isolated points.

In an n-dimensional simplicial complex $\Re^n$ ($n > 0$) there exists at least one n-simplex $\mathfrak{E}^n$, but no simplexes of higher dimension. At the midpoint of $\mathfrak{E}^n$, the $(n-1)$th homology group differs from the null element; it is the free Abelian group having one or, for $n = 1$, two generators, respectively (Example 1, §32). At each point of $\Re^n$ the homology groups for each and every dimension greater than $n - 1$ consist of the null element alone, since the neighborhood complex of the point only contains simplexes having dimension $n - 1$ or less. This gives the following invariant characterization of dimension:

34. Invariance of the Purity of a Complex

Let $\Re^n$ be an arbitrary complex, not necessarily pure, having a particular simplicial decomposition. Let $\mathfrak{E}^i$ be the set of all i-simplexes of $\Re^n$ which are not faces of $(i + 1)$-simplexes; $\mathfrak{E}^i$ can also be empty. The set $\mathfrak{E}^i$ can be characterized independently of a simplicial decomposition, as follows: $\mathfrak{E}^0$

consists of all isolated points of $\mathfrak{R}^n$; for $i > 0$, $\mathfrak{C}^i$ is the closed hull (i.e., the closure) $'\mathfrak{C}^i$ of those points P of $\mathfrak{R}^n$ which possess the following two properties:

1. The $(i - 1)$th homology group at P is the free Abelian group on two generators if $i = 1$, and the free cyclic group when $i > 1$.
2. There exists a neighborhood of P whose points also have property 1.

It is clear that $\mathfrak{C}^i$ is contained in $'\mathfrak{C}^i$. For all inner points of the i-simplexes of $\mathfrak{C}^i$ possess properties 1 and 2 and therefore belong to $'\mathfrak{C}^i$. Since $'\mathfrak{C}^i$ is closed, the boundary points of the i-simplexes of $\mathfrak{C}^i$ will also belong to $'\mathfrak{C}^i$. Conversely, if a point P possesses properties 1 and 2, then let $\mathfrak{C}^j$ be a simplex of highest dimension on which P lies. Because of property 1, j cannot be smaller than i. If it were true that $j > i$, then in each neighborhood of P there would exist inner points of $\mathfrak{C}^j$ at which the $(i - 1)$th homology group would be the free cyclic group for $i = 1$ and the null element for $i > 1$, in contradiction to property 2. The simplex of highest dimension on which P lies, $\mathfrak{C}^j$, is therefore an i-simplex and belongs to $\mathfrak{C}^i$.

THEOREM I. *The dimension, n of a complex $\mathfrak{R}$ which does not consist exclusively of isolated points is the smallest integer having the property that the homology groups of dimension n, $n + 1$, ... at each point of $\mathfrak{R}$ consist only of the null element.*

In particular, homogeneous complexes of different dimension cannot be homeomorphic. This follows immediately from Theorem IV of §32. We then get

THEOREM II. *In a homogeneous complex $\mathfrak{R}^n$, the homology groups at each point are those of the $(n - 1)$-sphere.*

For each point P of $\mathfrak{R}^n$ has as a neighborhood a simplicial star whose outer boundary is homeomorpbic to the $(n - 1)$-sphere.

One is relieved of the temptation to view the theorem of the topological invariance of dimension as an intuitively obvious fact when one observes that the invariance holds only for topological mappings. It does not hold for mappings that are one-to-one but not continuous, nor does it hold for mappings which are continuous but not one-to-one. A line segment, for example, can be mapped one-to-one onto the surface of a triangle, because both point sets have the same cardinality; it can also be mapped continuously onto the triangle, as demonstrated by the Peano curve which passes through each point of the triangle but has multiple points.[21]

Since $\mathfrak{C}^i$ is closed, the closed hull of all the points P, that is, $'\mathfrak{C}^i$, belongs to $\mathfrak{C}^i$. Thus $'\mathfrak{C}^i$ is also contained in $\mathfrak{C}^i$, so that $'\mathfrak{C}^i = \mathfrak{C}^i$.

If, therefore, $\mathfrak{R}_1^n$ and $\mathfrak{R}_2^n$ are distinct simplicial decompositions of the same complex $\mathfrak{R}^n$, and $\mathfrak{C}_1^i$ and $\mathfrak{C}_2^i$ are the subcomplexes formed in $\mathfrak{R}_1^n$ and $\mathfrak{R}_2^n$, respectively, by all i-simplexes which are not faces of $(i - 1)$-simplexes, then $\mathfrak{C}_1^i$ and $\mathfrak{C}_2^i$ are both the same point set, $'\mathfrak{C}^i$.

A pure n-dimensional complex is a complex such that the subsets $'\mathfrak{C}^0$ through $'\mathfrak{C}^{n-1}$ are empty. It has thus been characterized in an invariant manner.

35. Invariance of Boundary

Let $\mathfrak{R}^n$ ($n > 1$) be a pure complex having a particular simplicial decomposition and let $\mathfrak{Q}_\nu^{n-1}$ ($\nu = 1, 3, 4, 5, \ldots$) be the set of those $(n-1)$-simplexes with which exactly ν n-simplexes are incident.

We shall characterize $\mathfrak{Q}_\nu^{n-1}$ topologically. At an inner point of an $(n-1)$-simplex of $\mathfrak{Q}_\nu^{n-1}$ the $(n-1)$th homology group is the free Abelian group having $\nu - 1$ generators (§32, Example 2). When we denote by $\mathfrak{Q}_\nu'$ the closed hull of all points P at which the $(n-1)$th homology group is the free Abelian group having $\nu - 1$ generators, then accordingly $\mathfrak{Q}_\nu^{n-1}$ belongs to $\mathfrak{Q}_\nu'$. In general, $\mathfrak{Q}_\nu^{n-1}$ and $\mathfrak{Q}_\nu'$ will not coincide, but we can show that $\mathfrak{Q}_\nu^{n-1}$ is determined by $\mathfrak{Q}_\nu'$ in a topologically invariant manner.*

$\mathfrak{Q}_\nu'$ *is a subcomplex of the simplicial decomposition of $\mathfrak{R}^n$ (or is empty).* For if the free Abelian group having $(\nu - 1)$ generators is the $(n-1)$th homology group at P, then this group is also the $(n-1)$th homology group at all inner points of the simplex $\mathfrak{C}$ of lowest dimension on which P lies. All of these points have the same neighborhood complex and, therefore, the same homology groups. Thus all inner points of $\mathfrak{C}$ belong to $\mathfrak{Q}_\nu'$. Since $\mathfrak{Q}_\nu'$ is closed, all boundary points of $\mathfrak{C}$ will also belong to $\mathfrak{Q}_\nu'$. Therefore, along with P, the simplex of lowest dimension on which P lies will also belong to $\mathfrak{Q}_\nu'$. But $\mathfrak{Q}_\nu'$ is *at most* $(n-1)$-dimensional. For at the inner points of an n-simplex of $\mathfrak{R}^n$ the $(n-1)$th homology group is the free Abelian group of one and not $\nu - 1$ generators, since the value $\nu = 2$ was excluded.

If in $\mathfrak{Q}_\nu'$ there exists an $(n-1)$-simplex $\mathfrak{C}^{n-1}$, then there are exactly ν n-simplexes of $\mathfrak{R}^n$ incident with this simplex. For only in this case is the $(n-1)$th homology group at an inner point of $\mathfrak{C}^{n-1}$ the free Abelian group having $\nu - 1$ generators (Example 2, §32). Thus $\mathfrak{C}^{n-1}$ belongs to $\mathfrak{Q}_\nu^{n-1}$ and consequently $\mathfrak{Q}_\nu^{n-1}$ consists of the totality of $(n-1)$-simplexes of $\mathfrak{Q}_\nu'$. From §34, however, this subcomplex of $\mathfrak{Q}_\nu'$ is determined in a topologically invariant manner by $\mathfrak{Q}_\nu'$. Since $\mathfrak{Q}_\nu'$ in turn is determined in a topologically invariant manner by the given complex, $\mathfrak{Q}_\nu^{n-1}$ is associated in a topologically invariant manner with $\mathfrak{R}^n$. In the case $n = 1$, $\mathfrak{Q}_\nu^{n-1}$ consists of the totality of points of $\mathfrak{R}^1$ at which the 0th homology group is the free Abelian group having ν generators.

Previously, we defined the boundary of a pure n-dimensional complex to be the totality of $(n-1)$-simplexes which are incident with an odd number of n-simplexes. From now on we define the boundary of a pure complex, in a topologically invariant manner, as the set theoretic union of the subcomplexes

---

*The following example shows that, in general, $\mathfrak{Q}_\nu'$ is larger than $\mathfrak{Q}_\nu^{n-1}$: let $\mathfrak{R}^n$ consist of two tetrahedrons which have one common vertex; $\mathfrak{Q}_3'$ is formed by the common vertex; $\mathfrak{Q}_3^1$ is empty.

$\mathfrak{L}_1^{n-1}, \mathfrak{L}_3^{n-1}, \mathfrak{L}_5^{n-1}, \ldots$ With this definition the topological invariance of the boundary of a simplex, of a closed n-ball, and of a convex region are assured.

Problems

Prove that the annulus and the Möbius band are not homeomorphic.

36. Invariance of Pseudomanifolds and of Orientability

We have already shown that property (PM1) of a closed pseudomanifold, that it is a pure finite complex, is topologically invariant (§12 and §35). We can formulate property (PM2) in a topologically invariant manner, as follows: the invariant subcomplexes $\mathfrak{L}_\nu^{n-1}$ of those $(n-1)$-simplexes which are incident with exactly $\nu \neq 2$ n-simplexes are all empty. It was shown in §24 that the connectability condition (PM3) is equivalent to the requirement that the nth connectivity number $q^n = 1$; the topological invariance of the connectivity numbers was proved in §30.

The orientability of a closed pseudomanifold can be expressed by the fact that its nth Betti number is $p^n = 1$. The invariance of this was shown in §30.

If an orientable closed pseudomanifold $\mathfrak{K}^n$ has been simplicially decomposed in two distinct ways and the decompositions have been oriented coherently, then we can regard the two oriented decompositions as two singular closed k-chains B_1^n and B_2^n. Since the nth homology group is the free cyclic group and since the homology classes of B_1^n and B_2^n both generate the nth homology group, $B_1^n \sim \pm B_2^n$. If the $+$ sign holds, then we say that the two simplicial decompositions *have the same orientation*.

We can define the orientation of $\mathfrak{K}^n$ without making reference to a particular simplicial decomposition by letting B_1^n be an arbitrary singular closed n-chain which has the property that each closed n-chain is homologous to a multiple of B_1^n. We will call such an n-chain an *orienting n-chain* on $\mathfrak{K}^n$. If B_2^n is another orienting n-chain, then B_1^n and B_2^n will determine the same or opposite orientation on $\mathfrak{K}^n$ according, respectively, to whether $B_1^n \sim B_2^n$ or $B_1^n \sim -B_2^n$.

If an orientable pseudomanifold $\mathfrak{K}^n$, which has been oriented by a chain B^n, is mapped topologically onto another pseudomanifold $\mathbf{K}^n$ (which can also coincide with $\mathfrak{K}^n$), which in turn has been oriented by the chain $\mathbf{B}^n$, then B^n will transform to an oriented chain $'B^n$ of $\mathbf{K}^n$. In this case, $'B^n \sim \pm \mathbf{B}^n$. We speak of a mapping with *preservation* or *reversal of orientation* according to whether the positive or the negative sign holds. The "mirror reflection" of the n-sphere, considered in §31, is an example of a mapping with reversal of orientation.

The pseudomanifolds with boundary can easily be related to the closed pseudomanifolds with the help of the concept of *doubling*. We define doubling for a bounded pure complex $\mathfrak{K}^n$ $(n > 0)$ as follows: We take a homeomorphic copy $'\mathfrak{K}^n$ of $\mathfrak{K}^n$, and we identify points of the boundary of $\mathfrak{K}^n$ with points of the boundary of $'\mathfrak{K}^n$ which correspond to them by virtue of the homeomorphism between $\mathfrak{K}^n$ and $'\mathfrak{K}^n$. For example, we double a disk by attaching a second disk to it, joining boundary to boundary; a 2-sphere then results.

We can now characterize a pseudomanifold with boundary in a topologically invariant manner, as follows: *A pseudomanifold with boundary is a bounded pure complex whose doubling is a closed pseudomanifold. The pseudomanifold with boundary is orientable if and only if its doubling is orientable.*

We leave the proof, which is purely combinatorial, to the reader.

CHAPTER SIX

SURFACE TOPOLOGY

The principal problem of topology, the homeomorphism problem, can be solved in dimension 2, using methods which cannot be generalized to higher dimensions. For this reason we shall develop surface topology independently of our previous results, which were valid for arbitrary dimension. We shall start with polygons, instead of with a simplicial complex, and we shall construct closed polyhedral surfaces by identifying sides of polygons.

37. Closed Surfaces

A topological *polygon* is a closed disk in the Euclidean plane whose circumference is divided by r points ($r \geqq 2$) into r segments. The points are called the *vertices* of the polygon and the segments are called the *sides* of the polygon. Each topological image of this disk is also a polygon. The designations of vertices and sides in the preimage carries over to the image.

For $r \geqq 3$ each polygon can be represented as a bounded convex closed 2-ball subset of the Euclidean plane having straight sides.

Let α^2 ($\geqq 1$) disjoint polygons be given in the Euclidean plane. Let certain sides of the polygons be mapped topologically onto one another, so that the boundary points of a given side necessarily map to the boundary points of the image (Example 1, §6). We will call the set $\overline{\mathfrak{M}}$ of polygons whose sides are mapped onto one another in this manner a *system of polygons.*

We first devote our attention only to systems of polygons for which the total number of sides is even and each side is paired with exactly one other side by the topological mapping.

In the system of polygons $\overline{\mathfrak{M}}$, points which map into one another are to be considered equivalent in the sense that points in a neighborhood space are equivalent (§8). We can then recognize the following classes of equivalent points in $\overline{\mathfrak{M}}$: an inner point of a polygon is equivalent only to itself; an inner point of a polygonal side is equivalent to exactly one other point; a vertex may be equivalent to one, several, or even no other points.

The pairing of polygonal sides can be arbitrary, except that it must satisfy the condition that the polygons cannot be partitioned into two classes so that the sides of the polygons in each class are paired only among themselves (connectedness condition).

When equivalent points of $\overline{\mathfrak{M}}$ are identified, a point set $\mathfrak{M}$ will be defined. Like $\overline{\mathfrak{M}}$, this point set is a neighborhood space. We call $\mathfrak{M}$ a *closed surface* or a closed 2-dimensional manifold. Here the adjective "closed" implies, first, that the surface can be built of finitely many polygons, that is to say, it has no infinitely distant points and, second, it possesses no free edges, that is, it has no boundary.

The identification of polygonal sides will determine those polygonal vertices which belong to a single equivalence class and form a single point of the surface. Assume that there are α^0 different equivalence classes of vertices in the system of polygons. That is, let the polygonal vertices map to α^0 distinct points on the surface. We will call these α^0 points *vertices of the polyhedron*. They are to be regarded as distinct from the polygonal vertices from which they arise by identification. We will call the images of polygonal sides (*polyhedral*) *edges*; we call the images of polygons (*polyhedral*) *surface elements*. Let them be, respectively, α^1 and α^2 in number. The polyhedral edges and surface elements are continuous images of the polygonal sides and polygons but, in general, are not topological images. For example, the two boundary points of a polygonal side may become equivalent as a consequence of the association of sides; that is, the points map into the same polyhedral vertex. The polyhedral edge will then have a self-intersection and will be a topological circle on the surface. Surface elements can also exhibit self-intersections. These occur along one or more polyhedral edges or possibly only at vertices; they always appear, for example, whenever a single polygon is used in the construction of the closed surface. An example of this is the system of polygons consisting of a single quadrilateral polygon which covers the torus (§1).

In the future, when we deal with a surface whose points are divided in a particular way into surface elements, edges, and vertices, we shall speak of a *polyhedral surface* or of a polygonally decomposed surface. On the other hand, when we are not concerned with such a particular classification and wish to study the surface as a neighborhood space, we shall speak simply of a surface. The concepts of surface and polyhedral surface are related to one another in the same way that the concepts of complex and simplicial complex are related (§10). As an example, the cubic surface and the dodecahedron are distinct polyhedral surfaces which are homeomorphic to the same surface, the 2-sphere.

In all we have three concepts, which must be kept distinct; the system of individual disjoint polygons $\overline{\mathfrak{M}}$, for which certain points have been declared to be equivalent; the polyhedral surface $\mathfrak{M}$ which results from $\overline{\mathfrak{M}}$ after equivalent points have been identified, but for which the assignment of

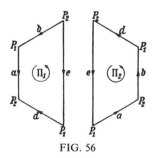

FIG. 56

surface points to individual vertices, edges, and surface elements is still essential; and the closed surface determined by the system of polygons, in which such a polygonal division has been dispensed with.

We now give a less obvious example of a polyhedral surface. The polygonal decomposition of this surface, unlike that of the cubic surface or dodecahedron, cannot be embedded in Euclidean 3-space with planar surface elements. The system of polygons consists of two quadrilaterals, Π_1 and Π_2 (Fig. 56); equivalent sides are indicated with the same letter, and equivalent vertices are likewise indicated by the same letter. Arrows indicate the sense of association of equivalent polygon sides, i.e., arrow tips should always fall on arrow tips. We have $\alpha^0 = 2$, $\alpha^1 = 4$, $\alpha^2 = 2$. It will turn out that the surface determined by this system of polygons is Klein's bottle (§2 and §39).

In the treatment which follows, our interest shall be not in the particular system of polygons but, rather, in the surface which it determines. We now set ourselves the task of *discovering when two systems of polygons determine the same surface*, that is, finding when the polyhedral surfaces produced by the identification of equivalent points are different polygonal decompositions of the same surface.

Two systems of polygons will certainly generate the same surface if they can be topologically mapped onto one another so that equivalent points in one system of polygons transform to equivalent points in the other system. We will not regard such systems of polygons as being distinct.

This implies that to define the surface we need only to know how the boundary points A' and B' of a polygonal side a' map to the boundary points A'' and B'' of the associated polygon side a'', not the detailed definition of the topological mapping of a' onto a''. *There exist only two essentially different mappings of two associated polygon sides from one onto the other.* One mapping pairs A' with A''; the other pairs A' with B''.

Proof. If two topological mappings T and T^* of a' onto a'' both send A' to A'' and B' to B'', then the self-mapping S of a', which arises when one first carries out T and then $(T^*)^{-1}$, is topological and leaves the boundary points A' and B' of a' fixed.

If a' is a side of the polygon Π, then there exists a topological mapping of Π onto itself such that a' is mapped by S to itself while all other sides of Π

remain fixed pointwise (Problem 2, §6). One gets the same system of polygons when one replaces Π by its image. This image is again the polygon Π. However, the points of a' and a'' are now equivalent as a consequence of the mapping T^*. Thus, one can replace T by any other topological mapping T^* which maps boundary points in the same manner as T and obtain the same surface by identification of equivalent points.

If, for example, we represent all polygons by straight sided polygons of the Euclidean plane, then we may assume that associated sides are mapped onto one another by a linear mapping.

We now orient the polygonal sides. That is, we choose one of the two boundary points of each side as initial point and the other as endpoint of the side; associated sides coincide, so that initial point maps to initial point and endpoint maps to endpoint. We also orient the polygons. To orient a polygon, we orient all of the polygonal sides coherently so that each vertex occurs once as initial point and once as endpoint of the adjacent sides. The orientation of a side in the coherent orientation of the polygonal boundary is called the orientation induced by the polygonal orientation. We can orient a polygon, like a side, in two different ways. In the figure we shall denote the orientation of a side by means of an arrow along the side, that of a polygon by means of a circular arrow inscribed in the polygon.

The orientation of the polygon will determine a *sense of traversal* of the boundary, that is, a cyclic ordering of the sides. We think of the polygon and its sides as having fixed orientations and designate associated sides by the same letter. We can then describe the system of polygons by a purely combinatorial schema. We write down the sides of each polygon in the order in which they follow in a traversal of the polygonal boundary, all on one row, and we give them the exponents $+1$ (which we shall omit in the future) or -1, according to whether their given orientation does or does not coincide, respectively, with that induced by the polygon. This schema will completely determine the system of polygons and therefore, also the surface. In the row of ordered sides of a polygon, the labels assigned to the sides indicate the pairwise identification of sides, and the exponents determine which of the two essentially distinct ways the associated sides are to be topologically mapped onto one another.

We can make the following changes in the schema without changing the system of polygons: we can cyclically permute the entries in the row; we can reverse the orientation of any edge, which we indicate by changing its exponent; we can reverse the orientation of the polygon, which amounts to making an anticyclic interchange of the letters of the corresponding row, together with a simultaneous change of all exponents of this row.

As an example we give the schema for the previously introduced system of polygons:

Polygon Π_1: $bad^{-1}e^{-1}$,
Polygon Π_2: $abde$.

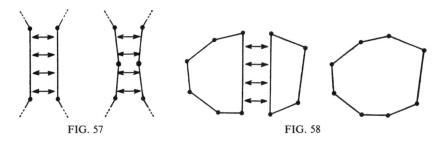

FIG. 57   FIG. 58

We now examine the question: When do two systems of polygons generate the same surface? From one system of polygons we may generate other systems of polygons which determine the same surface; we do this by means of the following *elementary transformations*.

A *cutting* or a *subdivision of dimension* 1 involves dividing each side of an associated pair of sides into two sides by adding a pair of associated vertices, while keeping fixed the original mapping of corresponding points. The passage from the subdivided system back to the original is called a *composition* or a *gluing of dimension* 1 (Fig. 57). A *composition* or *gluing of dimension* 2 involves uniting two different polygons along a pair of equivalent sides to form a single polygon by identifying these sides. The reverse process, cutting a polygon into two polygons, is called a *subdivision* or a *cutting of dimension* 2 (Fig. 58).

Two systems of polygons which transform one to the other by means of finitely many such elementary transformations are said to be *elementarily related*. Since neither an elementary transformation of dimension 1 nor an elementary transformation of dimension 2 will change the surface determined by the system of polygons, and since from §8 one may perform the identifications step by step, it follows that *elementarily related systems of polygons determine homeomorphic surfaces*.

Elementarily related polyhedral surfaces have two important properties in common, Euler characteristic and orientability. These concepts were defined in §23 and §24 for complexes and pseudomanifolds, respectively, from a simplicial decomposition. Since we have not yet defined surfaces as simplicial complexes, we now define Euler characteristic and orientability anew for a polyhedral surface and derive these properties from the schema of its system of polygons. We shall explain the connection between our new and old definitions in §39.

We define the *Euler characteristic* of a polyhedral surface to be the number

$$N = -\alpha^0 + \alpha^1 - \alpha^2,$$

where α^0, α^1, α^2 denote the number of polyhedral vertices, edges, and surface elements, respectively. As an example, for the quadrilateral which covers the torus it is

$$N = -1 + 2 - 1 = 0;$$

for the cube surface it is

$$N = -8 + 12 - 6 = -2.$$

Elementarily related polyhedral surfaces have the same Euler characteristic. Under an elementary transformation of dimension 1, α^0 and α^1 each change by 1 while α^2 is unchanged. Under an elementary transformation of dimension 2 only α^1 and α^2 change, each by 1.

We say that a polyhedral surface is *orientable* when the polygons of its system of polygons can be oriented so that opposite orientations are induced in the two members of a pair of associated sides. This is expressed in the schema of the system of polygons by the fact that upon appropriately orienting the polygons, each side will appear once with the exponent $+1$ and once with the exponent -1. It can be checked easily that the property of orientability remains unchanged under subdivision and under gluing. The polyhedral surface in the example earlier in this section is nonorientable.

We therefore obtain this result: *A necessary condition for two polyhedral surfaces to be elementarily related is that they have the same Euler characteristic and that either both are orientable or both are nonorientable.*

These conditions are also sufficient. This will be proved by showing that all polyhedral surfaces which have the same Euler characteristic and orientability can be brought to a common normal form by means of subdivision and gluing.

38. Transformation to Normal Form

The reduction to normal form is made in six steps.

Step 1

We start with a system of α^2 polygons ($\alpha^2 > 1$). By means of $\alpha^2 - 1$ gluings of dimension 2 we obtain *a single polygon*. The sides of this polygon are identified pairwise with one another. The schema of this system of polygons, consisting of but a single polygon, can be written as a single row in which each letter appears twice. If a letter appears once with exponent $+1$ and once with exponent -1, then the sides designated by that letter are said to be of the first type; otherwise, they are said to be of the second type. We have defined the polyhedral surface to be orientable if and only if all sides are identified so as to be of the first type.

As an example, if one deals with a surface homeomorphic to the 2-sphere, that is, a so-called Eulerian surface, one can obtain the single polygon by erasing from the polyhedral surface the planar network contained in it. This is a network such as one constructs to form a model of the polyhedral surface by folding together a piece of cardboard. One then needs only to erase the inner edges of the network.

Step 2: Side Cancellation

If a pair of sides appear in the sequence aa^{-1} and other sides are also present, so there are at least four sides, then the adjacent sides a and a^{-1} may be identified or, stated in another way, *cancelled*. This results in a polygon whose schema can be obtained from the schema of the original polygon by crossing out the sequence of edges aa^{-1}.

Cancellation can be reduced to subdivision and gluing as follows. If

$$\sim\!\sim\!\sim\!\sim\!\sim aa^{-1} \sim\!\sim\!\sim\!\sim\!\sim$$

is the schema of the polygon (Fig. 59), where wavy lines denote sequences of sides which do not need detailed description, we can assume that there is at least one side to the left and one to the right of aa^{-1}. A subdivision of dimension 2 gives two polygons

$$\sim\!\sim\!\sim\!\sim\!\sim ax \quad \text{and} \quad x^{-1}a^{-1}\sim\!\sim\!\sim\!\sim\!\sim,$$

where the newly created edge is designated by x (Fig. 60). By means of a gluing, the sides a and x can be merged into new side, y; similarly, x^{-1} and a^{-1} are merged into y^{-1}. The result is two polygons (Fig. 61),

$$\sim\!\sim\!\sim\!\sim\!\sim y \quad \text{and} \quad y^{-1}\sim\!\sim\!\sim\!\sim\!\sim$$

These two polygons are again united to form a single polygon by means of a gluing of dimension 2.

We proceed in this manner until we either arrive at a 2-gon or a polygon of at least four sides in which no sequence of sides aa^{-1} appears. The two possible schemata of a 2-gon are

$$aa^{-1} \quad \text{and} \quad aa. \tag{0}$$

Both have the desired normal form. From now on, we may assume that we are dealing with a polygon of at least four sides in which no sequence of sides aa^{-1} appears.

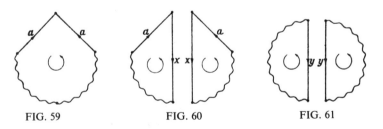

FIG. 59   FIG. 60   FIG. 61

Step 3: Transformation to a Polyhedral Surface Having a Single Vertex

Let the polygon obtained in step 2 be an r-gon ($r \geqq 4$). Denote equivalent vertices by the same letter. Either all vertices are equivalent or else, besides a class P of vertices, there exists still another class. In the latter case we can

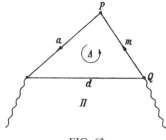

FIG. 62

transform the polygon into a polygon in which the number of vertices in P has been diminished by 1. For on the boundary of the polygon there exists a Q-vertex Q which together with a P-vertex bounds an edge m. A sequence QmP or PmQ will then occur on the boundary. Let us designate the second polygonal side adjacent to the P-vertex by a. Its second boundary point is either a Q-point Q, a P-point, or a point of another equivalence class. We connect this second boundary point by means of a diagonal d to the first polygon vertex Q and by a subdivision of dimension 2 we decompose the r-gon into a triangle Δ and an $(r - 1)$-gon Π (Fig. 62). The side a' associated with a belongs to Π. Otherwise, a would have to coincide with either m or m^{-1}. In the former case an edge sequence mm would appear in Δ and we would have $Q = P$, contrary to our assumption. In the second case m and m^{-1} could be canceled, contrary to our assumption concerning the nature of the r-gon. We now identify a and a' by means of a gluing of dimension 2 and thus obtain a new r-gon which contains one less occurrence of P and one more of Q.

Either the new polygon has sides which can be canceled or the new polygon will have the same properties as the original one. In the latter case we repeat the procedure just given and again reduce the number of P-vertices by 1. We proceed further until we can again cancel sides, which will occur at the latest when only one P-vertex remains on the boundary. Cancellation then yields either a 2-gon, which is a desired normal form, or a polygon having all its vertices equivalent, or a polygon having fewer than r vertices, not all of which are equivalent, in which case we repeat the whole process.

Step 4: Cross-Cap Normalization

After having brought the normalization to this stage, if the polyhedral surface is nonorientable then at least two associated sides will be of the second type, for example the two sides c of Fig. 63. The polygon schema is then

$$\text{〜〜〜〜〜} c \text{〜〜〜〜〜} c. \tag{1}$$

We decompose this polygon into two polygons by means of a diagonal a

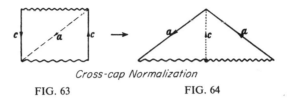

Cross-cap Normalization

FIG. 63   FIG. 64

which joins the endpoints of the 2 sides c. The schema of the transformed system of polygons is

$$\text{\~\~\~\~\~} ca, \qquad\qquad (2)$$
$$a^{-1} \text{\~\~\~\~\~} c.$$

We now glue the two polygons together along c. This gluing gives the polygon

$$\text{\~\~\~\~\~} aa \text{\~\~\~\~\~} ,$$

having the sequence of sides aa, which we call a *cross-cap* (cf. §2). The process of transformation of the schema is illustrated in Fig. 63 and 64. If a pair of sides of the second type appears in the new polygon, then we can unite these sides to a cross-cap without destroying the cross-cap already present. We proceed until we have united all pairs of sides of the second type to cross-caps. If this exhausts all of the sides, then we have arrived at the normal form (cross-cap form) of the nonorientable surface:

$$a_1 a_1 a_2 a_2 \text{\~\~\~\~\~} a_k a_k. \qquad\qquad (k)$$

For $k = 1$ we get the 2-gon obtained previously.

Step 5: Handle Normalization

If, on the other hand, the surface is orientable or side pairs of the first type still appear after the cross-cap normalization, then there must exist two pairs of sides of the first type which alternate on the boundary, i.e., with appropriate orientation these sides form a sequence (Fig. 65)

$$\text{\~\~\~\~\~} c \text{\~\~\~\~\~} d^{-1} \text{\~\~\~\~\~} c^{-1} \text{\~\~\~\~\~} d \text{\~\~\~\~\~}$$

on the polygon boundary.

For if the pair of sides c did not alternate with another pair of sides of the first type then all sides of the sequence $c \text{\~\~\~\~\~} c^{-1}$ would be paired among themselves, since we have assumed that the cross-cap normalization has been completed; the vertices of the sides lying between c and c^{-1} would be equivalent to one another and to the end points of the two sides c. By virtue of the connectedness condition for the association of sides the initial point of a side c could never be equivalent to these vertices. This however contradicts the result of step 3, that the polygon has been normalized to have only equivalent vertices.

The two crossed pairs of sides can be transformed by means of elementary transformations to a sequence of sides $aba^{-1}b^{-1}$, as illustrated in Figs. 65–67. Such a sequence is called a *handle*, based on the description given in §2. The

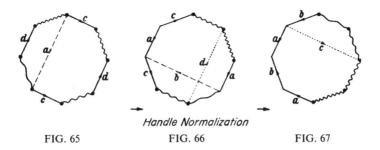

Handle Normalization

FIG. 65   FIG. 66   FIG. 67

uniting of crossed pairs of sides will not destroy previously obtained cross-caps or handles for these belong to sequences of sides included in the wavy lines, and are not disturbed by subsequent modifications. If the given surface is orientable, then we can bring it to the normal form (handle form)

$$a_1b_1a_1^{-1}b_1^{-1}a_2b_2a_2^{-1}b_2^{-1} \cdots a_hb_ha_h^{-1}b_n^{-1}. \tag{h}$$

Step 6: Transformation of the Handles into Cross-Caps

The only case remaining is where both handles and cross-caps appear. In that case we can replace each handle by two cross-caps by means of elementary subdivisions and gluings of dimension 2. The transformation process consists of cutting apart the given polygon (Fig. 68) and gluing it together again so that the six sides in question become of the second type (Fig. 69) and subsequently applying the cross-cap normalization of step 4 (Figs. 70–72).

We have thus reduced each system of polygons to one of the following normal forms consisting of a single polygon, the *fundamental polygon*:*

$$aa^{-1}, \tag{0}$$

$$a_1b_1a_1^{-1}b_1^{-1}a_2b_2a_2^{-1}b_2^{-1} \cdots a_hb_ha_h^{-1}b_h^{-1}, \tag{h}$$

$$a_1a_1a_2a_2 \cdots a_ka_k. \tag{k}$$

The fundamental polygons (0) and (*h*) on the one hand and (*k*) on the other differ in orientability and they differ from each other in the value of the Euler characteristic, which can be calculated from the fundamental polygon to be

$$N = -2 + 1 - 1 = -2, \tag{0'}$$

$$N = -1 + 2h - 1 = 2(h - 1), \tag{h'}$$

$$N = -1 + k - 1 = k - 2. \tag{k'}$$

These formulas also give the relationship between the number of handles *h*, or

---

*Illustrations of the fundamental polygon are found in: Fig. 11 for (0); Fig. 5 for $h = 1$ and Fig. 10 for $h = 2$; Fig. 18 for $k = 1$ and Fig. 19 for $k = 2$.

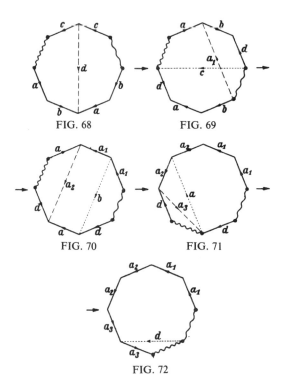

FIG. 68   FIG. 69

FIG. 70   FIG. 71

FIG. 72

the number of cross-caps k, and the Euler characteristic of a surface. Orientable surfaces will always have an even Euler characteristic. We have seen that the characteristic N and the orientability do not change under elementary transformations, so we have the following result:

THEOREM. *Two polyhedral surfaces are elementarily-related if and only if they have the same Euler characteristics and orientability.*[22]

With this result we have made major progress towards our goal of classifying all closed surfaces. We now know that there are no closed surfaces other than these given in the rows (0), (h), and (k). However, it is still conceivable that two fundamental polygons determine the same surface even though they are not elementarily related. In the next section we shall show with the help of the invariance theorems established in the previous chapter that this is not the case.

Problems

1. Show that a closed surface of Euler characteristic N can be covered by a single polygon having the following ordering of sides (symmetric normal form):

$$c_1 c_2 \cdots c_{N+2} c_1^{-1} c_2^{-1} \cdots c_{N+2}^{\pm 1}.$$

Here the negative sign occurs for the last exponent only if the surface is orientable, that is, N is even. What characteristic will the surface have in the case of odd N, if one retains the minus sign in the last exponent?

2. Bring the system of polygons of the example in §37 to normal form.

39. Types of Normal Form: The Principal Theorem

By means of repeated subdivision we can derive from a given system of polygons $\mathfrak{M}$ an elementarily related system of polygons which consists entirely of triangles and which becomes a simplicial complex when equivalent points are identified. This complex obviously possesses the properties (PM1) through (PM3) of a closed pseudomanifold (§24).

Neither the Euler characteristic nor the orientability changes during the cutting. The definitions in §37 of the Euler characteristic and orientability coincide with the definitions given in §23 and §24 in the particular case that the polyhedral surface is a simplicial complex. Since we have shown, in §36 and §30 that the orientability of a pseudomanifold and the Euler characteristic of a simplicial complex are topologically invariant properties, it follows that two closed polyhedral surfaces can be homeomorphic only if they have the same Euler characteristic and orientability. In particular, the surfaces given by the fundamental polygons (0), (h), and (k) are all distinct. If two polyhedral surfaces coincide in characteristic and orientability however, they are elementarily related and, from the theorem of §38, are therefore homeomorphic. We summarize our result in the

PRINCIPAL THEOREM OF SURFACE TOPOLOGY. *Two closed surfaces are homeomorphic if and only if they have the same Euler characteristic and orientability. The most general orientable closed surface is the 2-sphere with h handles attached ($h \geq 0$). The most general nonorientable closed surface is the 2-sphere with k cross-caps attached ($k \geq 1$).*

The number of handles h of an orientable surface is called the *genus* of the orientable surface. The number of cross-caps k of a nonorientable surface is called the genus of the nonorientable surface. Genus is related to the Euler characteristic N by the appropriate formula (0′), (h′), or (k′) of §38.

Previously, we defined a closed surface as a neighborhood space which can be derived from a system of polygons. An equivalent definition, which takes note of the fact that a closed surface is also a complex, is the following: *A closed surface is a finite, connected homogeneous 2-dimensional complex.*

Proof. Since a closed surface is a pseudomanifold, it is a finite connected 2-dimensional complex. The homogeneity condition, that is, the existence at each point of a neighborhood homeomorphic to the interior of a disk, is obviously fulfilled for interior points of the fundamental polygon. It is fulfilled for interior points of a polygonal side because there is exactly one side equivalent to it. One obtains such a disk neighborhood for a polygonal vertex by cutting a small triangle from the fundamental polygon at each polygonal vertex. By identifying sides of the triangles one forms a cycle homeomorphic to a disk (or in the trivial exceptional case of two cycles, to the 2-sphere).

Conversely, if $\mathfrak{R}^2$ is an arbitrary finite connected homogeneous complex, then from Theorem II of §33 the homology groups at each point P of $\mathfrak{R}^2$ are those of the circle and it follows from Example 2 of §32 that exactly two triangles are incident with each edge of a simplicial decomposition of $\mathfrak{R}^2$ and that the neighborhood complex of a vertex of $\mathfrak{R}^2$ is a circle. One may then regard $\mathfrak{R}^2$ as arising from its triangles by means of a stepwise identification of sides. The connectedness condition (§37) is fulfilled because we assumed that $\mathfrak{R}^2$ was connected (§12). Thus $\mathfrak{R}^2$ is a closed surface in the sense of the definition of §37.

Problems

1. Show that each closed 2-dimensional pseudomanifold $\mathfrak{R}^2$ of Euler characteristic $N = -2$ is homeomorphic to the 2-sphere.

2. In $\mathfrak{R}^3$ we are given a finite 1-dimensional connected complex $\mathfrak{R}^1$ of Euler characteristic N, which is constructed of straight line segments. A small 2-sphere of fixed radius is drawn about each point of $\mathfrak{R}^1$. Show that the envelope of the spheres is a closed surface of Euler characteristic $2N$.

40. Surfaces with Boundary

A *surface with boundary* is a neighborhood space which arises from a system of polygons when equivalent points are identified by gluing polygonal sides. Points of two polygonal sides are defined to be equivalent when they are equivalent under a given topological mapping of the sides. Whereas for closed surfaces there is exactly one other side equivalent to each side, for surfaces with boundary certain polygonal sides remain free. At least one such must exist. The free sides are called *boundary sides*. We require that the connectedness condition of §37 must hold for surfaces with boundary, as well as for the closed surfaces. Just as for a closed surface, the sides and polygons of a surface with boundary can be oriented and the surface with boundary can be described by a combinatorial schema.

If α^0, α^1, α^2 are the number of nonequivalent vertices, sides, and polygons, respectively, of the system of polygons, then we also define the number

$$N = -\alpha^0 + \alpha^1 - \alpha^2$$

to be the *Euler characteristic* of the surface with boundary.

A surface with boundary is *orientable* if its polygons can be oriented so as to induce opposite orientations in each pair of associated sides.

Euler characteristic and orientability remain unchanged under elementary subdivisions and gluings, just as for the case of surfaces which have no boundaries. By means of a sequence of subdivisions one can also transform a surface with boundary to a simplicial complex which is a pseudomanifold with boundary; thus the characteristic and orientability are topological invariants of the surface with boundary.

The method of classifying surfaces with boundary is the same as that used to classify closed surfaces. We present certain normal forms, to which each surface with boundary can be brought

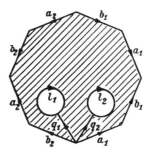

FIG. 73. $h = 2$, $r = 2$.

by elementary transformations, and we show that nonhomeomorphic surfaces have different normal forms. Each normal form consists of a single polygon and this polygon has the schema

$$q_1 l_1 q_1^{-1} \cdot q_2 l_2 q_2^{-1} \cdots q_r l_r q_r^{-1} \cdot a_1 b_1 a_1^{-1} b_1^{-1} \cdots a_h b_h a_h^{-1} b_h^{-1} \qquad (r > 0, h \geqq 0)$$

for the orientable case, and:

$$q_1 l_1 q_1^{-1} \cdot q_2 l_2 q_2^{-1} \cdots q_r l_r q_r^{-1} \cdot a_1 a_1 \cdots a_k a_k \qquad (r > 0, k > 0)$$

for the nonorientable case. The schema arises from the normal form of a closed surface by inserting a sequence of sides of the form $q_1 l_1 q_1^{-1} \cdots q_r l_r q_r^{-1}$. Sides $l_1, l_2, \ldots, l_r$ are the *boundary sides*. When we identify each pair of sides q_i and q_i^{-1} of the normal polygon, we get the fundamental polygon of a closed surface into which r holes have been cut (Fig. 73).

To bring an arbitrary surface with boundary given by a system of polygons to normal form, we start with a simplicial decomposition of the surface with boundary. As stated previously, the surface with boundary is a pseudomanifold with boundary whose boundary is a 1-dimensional subcomplex, $\Re$.

We assume in advance that each triangle of the simplicial decomposition either has one side or one vertex lying on the boundary or that it does not intersect the boundary at all. We are allowed to make this assumption; if a given simplicial decomposition does not have this property, we need only transform to the normal subdivision.

If Δ is a triangle which meets the boundary, that is, Δ has either one vertex or one side in common with $\Re$, then there are exactly two sides which protrude from the boundary and are incident with Δ. By a side protruding from $\Re$ we mean a side which has one vertex lying on $\Re$ but does not itself belong to $\Re$, that is, the other vertex does not lie on $\Re$. Conversely, corresponding to each side which protrudes from the boundary there are exactly two triangles which protrude from the boundary and are incident with that side. The set of all triangles and sides which protrude from $\Re$ will, then, necessarily decompose into finitely many cyclic sequences of alternating triangles and sides, such that each member of such a sequence is incident with both of its neighboring members.

If

$$\Delta_1 a_1 \Delta_2 a_2 \cdots \Delta_s a_s$$

is such a cyclic sequence, then we form a single polygon Π from the triangles $\Delta_1, \Delta_2, \ldots, \Delta_s$ by first adjoining Δ_1 and Δ_2 along a_1, then Δ_2 and Δ_3 along a_2, and so forth until we finally adjoin Δ_{s-1} to Δ_3 along a_{s-1}. The edge a_s will then appear twice in Π, once as a side of Δ_1, in which case we designate it by a_s', and once as a side of Δ_s, which we designate as a_s. We proceed in like manner for the remaining sequences. It is clear that an arbitrarily selected edge of $\Re$ must lie on the boundary of one of these polygons, let us say on Π. But the totality of edges of $\Re$ which appear on Π will then form a single connected sequence of edges. For if P_i is the vertex of a_i which lies on $\Re$, then for $i > 1$ the totality of points of Δ_i which lie on $\Re$ will consist of the edge

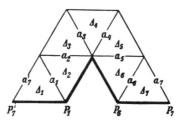

FIG. 74. Polygon $\Pi, s = 7$.

$P_{i-1}P_i$ if the point P_{i-1} is distinct from P_i and will consist of the point $P_{i-1} = P_i$ when points P_{i-1} and P_i coincide. When $i = 1$, P_{i-1} is to be replaced by the vertex P_s', that is, the vertex of a_s' which lies on $\Re$. Consequently the edges of the sequence $P_s'P_1P_2 \cdots P_s$ in Π, and only these edges, belong to $\Re$. Certain of the vertex points can coincide; for example, in Fig. 74, $P_2 = P_3 = P_4 = P_5$. If we now join the edges of the sequence of edges to form a single side l of Π by means of gluings, and we omit the prime mark from the side a_s' so as to return to our previous side designation, then there results a sequence

$$\sim\!\sim\!\sim\!\sim a_s l a_s^{-1} \sim\!\sim\!\sim\!\sim$$

on the boundary of Π. The polygons obtained in this manner are then composed with the remaining triangles of the simplicial decomposition which do not protrude from $\Re$ by means of gluings to form a single polygon. This polygon will have the schema

$$\sim\!\sim\!\sim b_1 l_1 b_1^{-1} \sim\!\sim\!\sim b_2 l_2 b_2^{-1} \sim\!\sim\!\sim b_r l_r b_r^{-1} \sim\!\sim\!\sim.$$

The l_i denote the only boundary sides and the wavy lines denote sides which are associated in pairs among themselves and do not require detailed specification (Fig. 75).

The sequences of sides $b_i l_i b_i^{-1}$ can be brought to a single unbroken sequence by cutting and reassembling the polygon. Figure 76 shows, for example, how one unites the first pair of sequences. Here, neither the other sequences of sides of the form $b_i l_i b_i^{-1}$ nor sequences of sides denoted by wavy lines are ripped apart. By repeating this unification procedure we arrive at a polygon which has only pairwise associated sides, apart from a sequence $q_1 l_1 q_1^{-1} \cdots q_r l_r q_r^{-1}$. The latter sequence has the same surface point as initial point and endpoint; for the initial point of q_1 is equivalent to the endpoint of q_1^{-1} because of the association of sides, and is therefore equivalent to the initial point of q_2 and so forth and is, finally, equivalent to the endpoint of q_r^{-1}. We can, then, cut off the sequence by means of a diagonal $\bar{l}$ having two equivalent boundary points, both of which we denote by O. This gives rise to a polygon which has a single boundary side $\bar{l}$. We can carry over the normalization procedure for closed surfaces to this polygon, and decrease the number of vertices not equivalent to O, replacing them by O vertices. All vertices of

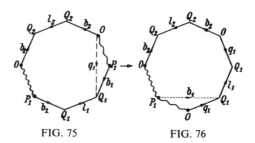

FIG. 75   FIG. 76

the polygon will finally become O vertices. After completing the normalization procedure, we rejoin the cut-off polygon along $\bar{l}$. The desired normal form of the surface with boundary will then be obtained. The normal forms differ in the number of boundary sides, Euler characteristic, and orientability, three properties which we know to be topological invariants. Thus different normal forms describe nonhomeomorphic surfaces.

Problem. Show that the double (§36) of an orientable surface of genus h having r holes is a closed surface of genus $\bar{h} = 2h + r - 1$. Show that the double of a nonorientable surface of genus k having r roles is a closed surface of genus $\bar{k} = 2k + 2r - 2$.

41. Homology Groups of Surfaces

Since the homology groups of a complex are determined by its numerical invariants (Betti numbers and torsion coefficients), it is easy to calculate them for closed surfaces. From the generalized Euler polyhedron formula (§23), the Euler characteristic is

$$N = -p^0 + p^1 - p^2.$$

Since the surface is a connected complex, from §18, $p^0 = 1$. Since it is either an orientable or a nonorientable pseudomanifold, $p^2 = 1$ in the case of orientability and 0 in the case of orientability. From §24 we have

$$N = p^1 - 2 \qquad \text{in the orientable case,}$$

$$N = p^1 - 1 \qquad \text{in the nonorientable case.}$$

Thus we have obtained the first Betti number in terms of the Euler characteristic. From §38 it can also be given in terms of the genus h or k, respectively, of the surface:

$$p^1 = N + 2 = 2h \qquad \text{in the orientable case,}$$

$$p^1 = N + 1 = k - 1 \qquad \text{in the nonorientable case.}$$

There are no torsion coefficients for the orientable case; for the nonorientable case there is a single 1-dimensional torsion coefficient, which has value 2 (§24).

The 1-dimensional homology group of the orientable closed surface of genus h is therefore the free Abelian group with $2h$ generators, and that of the nonorientable closed surface of genus k is the direct sum of a free Abelian group having $k - 1$ generators and a group of order 2.

The homology groups of dimensions 0 and 2 are not of special interest, since that for dimension 0 is the free cyclic group for each connected complex (§18) and that for dimension 2 is already determined by the general property that a closed surface is a closed orientable or nonorientable pseudomanifold;

from §24 it is the free cyclic group in the orientable case and consists of the null element alone in the nonorientable case.

To find a 1-dimensional homology basis we use the

LEMMA. *The vertices, edges and surface elements of a polyhedral surface form a block system.*

A block system (§22) is always defined with respect to a simplicial decomposition of a complex. We shall use any simplicial decomposition $\mathfrak{R}^2$ which can be obtained from the given polyhedral surface by subdivisions. Instead of referring to vertices, edges, and surface elements we must, more precisely, refer to the chains which are formed from oriented vertices or the chains which result from subdivided and coherently oriented edges or surface elements, for blocks are simplicial chains.

Proof. It is simple to check that conditions (Bl1) and (Bl2) are satisfied for each dimension, that (Bl3) is satisfied for dimensions 0 and 2, and (Bl4) is satisfied for dimension 1; we omit the proof here (cf. §22). We shall simultaneously prove (Bl3) for dimension $k = 1$ and (Bl4) for dimension $k = 0$ by showing that if U^1 is a simplicial 1-chain on $\mathfrak{R}^2$ and if the boundary of U^1 is an 0-block chain, that is, $\mathfrak{R}\partial U^1$ is formed from polyhedron vertices, then there exists a 1-block chain V^1 which is homologous to U^1. (Bl4) is implied by this statement for $k = 0$, and likewise (Bl3) for $k = 1$, if in particular one chooses U^1 to be a closed chain. V^1 is constructed by pushing U^1 across surface pieces to the polyhedron edges. Let us look at the subchain U_m^1 of U^1, which lies over the inner 1-simplexes of a particular surface element Π of the polyhedral surface; inner 1-simplexes of $\mathfrak{R}^2$ are those which do not belong to a polyhedral edge of $\mathfrak{R}^2$. The chain U_m^1 determines a chain $'U_m^1$ on the polygon $'\Pi$ of the system of polygons which generates the polyhedral surface. Now $\mathfrak{R}\partial' U_m^1$ is a 0-chain on the boundary of $'\Pi$ which has a sum of coefficients equal to 0. Since a 0-chain which has a sum of coefficients equal to 0 is always ~ 0 on a connected complex (§18), there exists a 1-chain $'U_r^1$ on the boundary of $'\Pi$ which has the same boundary as $'U_m^1$. Then $'U_m^1 - 'U_r^1$ is closed and is therefore null homologous in Π, as is each closed 1-chain on the disk. The chains U_m^1 and U_r^1 of $\mathfrak{R}^2$, which correspond to the chains $'U_m^1$ and $'U_r^1$, are likewise homologous to one another. If we replace U_m^1 by U_r^1 and in the same way we push the chain U_1 out of the remaining surface elements, then we obtain a chain V^1 homologous to U^1, and all of its simplexes belong to the polyhedral edges. Since, by assumption, $\mathfrak{R}\partial V^1 = \mathfrak{R}\partial U^1$ is a block chain, all 1-simplexes of a coherently oriented polyhedral edge appear in V^1 equally often. That is, V^1 is a block chain, as desired.

The lemma allows us to compute the homology groups of a polyhedral surface generated by a system of polygons in a new way. It also allows us to find a homology basis. We need only to construct the block incidence matrices (§22) and bring them simultaneously to normal form. We shall perform the computation only for the fundamental polygons (0), (h), and (k) of §38 (cf. Figs. 11, 10, and 19). The matrices are

$\mathbf{E}^0$	a
P	1
O	-1

$\mathbf{E}^1$	Π
a	0

(0)

$\mathbf{E}^0$	a_1	b_1	$\cdots$	a_h	b_h
O	0	0	$\cdots$	0	0

$\mathbf{E}^1$	Π
a_1	0
b_1	0
$\vdots$	$\vdots$
a_h	0
b_h	0

(h)

$\mathbf{E}^0$	a_1	a_2	$\cdots$	a_k
O	0	0	$\cdots$	0

$\mathbf{E}^1$	Π
a_1	2
a_2	2
$\vdots$	$\vdots$
a_k	2

(k)

To bring the incidence matrix $\mathbf{E}^0$ of (0) to normal form we add the first row to the second row. The normal form $\mathbf{H}^0$ is then

$\mathbf{H}^0$	a
$P - O$	1
O	0

and a 0-dimensional homology basis is formed by the point O. The incidence matrix $\mathbf{E}^1$ of (0) is already in normal form, as are both of the incidence matrices of (h). Consequently, for the closed surface of genus h the $2h$ 1-chains $a_1, b_1, \ldots, a_h, b_h$ form a 1-dimensional homology basis, which is at the same time a Betti basis.

To bring the incidence matrices (k) to normal form we subtract the first row in $\mathbf{E}^1$ from the remaining rows, obtaining

$\mathbf{H}^0$	$\sum_{i=1}^{k}$	a_2	$\cdots$	a_k
O	0	0	$\cdots$	0

$\mathbf{H}^1$	Π
$\sum_{i=1}^{k} a_i$	2
a_2	0
$\vdots$	$\vdots$
a_k	0

Here the 1-chains $\sum_{i=1}^{k} a_i, a_2, \ldots, a_k$ form a homology basis. The 1-chains $a_2, \ldots, a_k$ form a Betti basis, in agreement with the fact that the first Betti number p^1 is equal to $k - 1$.

While the first Betti number p^1 differs for orientable and nonorientable surfaces, the connectivity number for all surfaces is $q^1 = N + 2$. This is because $N = -q^0 + q^1 - q^2$ (§23), and $q^0 = q^2 = 1$ (§23).

We obtain a connectivity basis when we construct the incidence matrices mod 2 $\check{\mathbf{E}}^0$ and $\check{\mathbf{E}}^1$. We do this by replacing all even integers in the incidence matrices $\mathbf{E}^0$ and $\mathbf{F}^1$ by the residue class $\check{0}$ and replacing all odd integers by the residue class $\check{1}$. The matrices $\check{\mathbf{E}}^0$ and $\check{\mathbf{E}}^1$ are already in normal form for the surfaces (h) and (k). Thus the edges of the fundamental polygon form a 1-dimensional connectivity basis, when considered as chains mod 2. The 2-sphere $(h = 0)$ is the trivial exceptional case. The geometrical meaning of the number q^1 is indicated by the following theorem.

THEOREM. *The connectivity number for dimension* 1, $q^1 = N + 2$, *is equal to the maximum number of essential cuts which do not disconnect the surface.*

By an essential cut* we mean a cut sequence of edges forming a closed path free of double points on a simplicial decomposition of the surface. That is, an essential cut is a 1-dimensional subcomplex (§23) in which each vertex is incident with exactly two edges. A system of r essential cuts, where we assume that no two have an edge in common, "disconnects" the surface if there exist two triangles of the simplicial decomposition which cannot be connected by a sequence of alternating incident triangles and edges (that is, edges not belonging to the essential cuts). The fact that there are at least $N + 2$ essential cuts which do not disconnect the surface (upon making an appropriate choice of simplicial decomposition) is shown by inspection of the edge complex of the fundamental polygon (§38). But if the system of essential cuts consists of $r > N + 2$ essential cuts $\mathfrak{f}_1, \mathfrak{f}_2, \ldots, \mathfrak{f}_r$, then as chains mod 2 they are homologously dependent because the connectivity number $N + 2$ is the maximum number of homologously independent chains mod 2 (§23). There thus exists a 2-chain mod 2, that is, a 2-dimensional subcomplex $\mathfrak{U}^2$, whose boundary is formed by a linear combination of these essential cuts

$$\mathfrak{R} \, \partial \mathfrak{U}^2 = \check{\varepsilon}_1 \mathfrak{f}_1 + \check{\varepsilon}_2 \mathfrak{f}_2 + \cdots + \check{\varepsilon}_r \mathfrak{f}_r,$$

where not all $\check{\varepsilon}$ vanish. Since we assume that no two essential cuts $\mathfrak{f}_i$ and $\mathfrak{f}_j$ ever have an edge in common, tbe right-hand side of the above equation is $\neq 0$ and thus $\mathfrak{U}^2$ is neither the chain 0 nor the whole surface. If, now, $\mathfrak{B}^2$ is the complementary complex of $\mathfrak{U}^2$, that is, those triangles of the simplicial decomposition of the surface which do not belong to $\mathfrak{U}^2$, then we cannot connect a 2-simplex of $\mathfrak{B}^2$ with a 2-simplex of $\mathfrak{U}^2$ without passing across an essential cut. The surface will therefore be disconnected by any $q^1 + 1$ essential cuts.[23]

Remark. A narrow strip which surrounds an essential cut on a nonorientable surface will be either an annulus or a Möbius band. Accordingly one says that the essential cut is respectively 2-sided or 1-sided.

---

* *Translator's note*: The term *regressive cut* also appears in the literature.

(We can obtain such a strip, for example, by selecting all triangles of the 2-fold normal subdivision of the given simplicial complex which have vertices or edges incident with the essential cut.) If a nonorientable surface becomes orientable by cutting it along the essential cut, then the essential cut is said to be orientability producing.

Problem

Using the results of Problem 1 of §38, show that there exists an orientability producing essential cut on each nonorientable surface. Each orientability producing essential cut is respectively 2-sided or 1-sided according to whether the genus k is even or odd. (The Euler characteristic remains unchanged on cutting. Use the fact that the Euler characteristic of an orientable closed surface is always even.)

CHAPTER SEVEN

THE FUNDAMENTAL GROUP

It is only in 2 dimensions that the homology groups are sufficient to completely classify all manifolds. In higher dimensions, the most important topological invariant which allows one to distinguish different complexes and manifolds is the fundamental group. This invariant will often distinguish between two manifolds or complexes in cases where their homology groups coincide. There exists only one fundamental group of a complex. It provides information with regard to the behavior of 1-dimensional paths of the complex. In general it is a non-Abelian group. Efforts to define fundamental groups for each dimension of a complex, in analogy to the homology groups, have not yet led to important results.*

Like the homology groups, the fundamental group of a complex is a global invariant. But also, like the homology groups, it allows a local invariant to be defined, the fundamental group at a point.

42. The Fundamental Group

If an oriented interval $\bar{w}$ has been mapped into a finite or infinite complex $\mathfrak{R}^n$, then we say that a *path* w has been determined in $\mathfrak{R}^n$. The image points of $\bar{w}$ are called the points of the path. If $\bar{P}$ and $\bar{Q}$ are, respectively, the initial point and endpoint of $\bar{w}$, and P and Q are their respective images, then P is called the *initial point* and Q is called the *endpoint* of w, and we say that w "leads" from P to Q. If $\bar{w}$ is traversed continuously from $\bar{P}$ to $\bar{Q}$, then the image point is said to *traverse* the path w continuously from P to Q. The image point may remain in fixed position in the complex $\mathfrak{R}^n$, for the definition does not exclude the possibility that $\bar{w}$ is mapped to a single point. In this case w consists of a single point.

If there is a topological mapping between the preimage $\bar{w}$ and another oriented interval $\bar{w}'$ under which initial and endpoints correspond, then $\bar{w}'$ is also mapped uniquely and continuously into $\mathfrak{R}^n$, via a detour through $\bar{w}$, and it therefore determines a path w'. We do not regard w and w' as different paths.[24] Obviously the points of w coincide with those of w'.

---

* *Editor's note*: A rich and extensive theory of higher homotopy groups has developed since 1933.

If we choose the unit interval $0 \leqq s \leqq 1$ of the real number line as $\bar{w}$, with $s = 0$ as initial point and $s = 1$ as endpoint, then each s-value in the interval $0 \leqq s \leqq 1$ determines a unique point of w. The points of the path w can be related in infinitely many ways to a parameter which runs from 0 through 1. If s' is another such parameter, then s and s' are related by a topological transformation, that is, a monotone coordinate transformation. We can interpret s as the time at which a point of the path is traversed.

If u is a path from P to Q and v is a path from Q to R, then we can join these paths to form a single path: we compose the preimages $\bar{u}$ and $\bar{v}$ to give a single preimage $\bar{w}$, and we map them as before. The image of $\bar{w}$ is a path w and we denote this path by

$$w = uv.$$

The path w arises, then, by first traversing u and then v.

The *inverse* path w^{-1} of w is obtained by reversing the orientation of the preimage $\bar{w}$.

A path whose initial point coincides with its endpoint is said to be *closed*.

A sequence of finitely many directed edges of a simplicial decomposition of $\Re^n$ such that the endpoint of each edge coincides with the initial point of the succeeding edge forms an *edge path* which leads from P to Q, where P is the initial point of the first edge and Q is the endpoint of the last edge. It is to be regarded as the product of the individual edges. As a 1-simplex, each oriented edge is a continuous (in fact topological) image of an oriented geometric 1-simplex, that is, of an oriented interval, and is therefore a path in the sense of our definition.

A path will be (homotopically) *deformed* if the mapping of its preimage $\bar{w}$ into $\Re^n$ is homotopically deformed holding the initial point P and the endpoint Q of the path fixed. Let t be a parameter varying from 0 to 1. If a mapping g_t of $\bar{w}$ into $\Re^n$ is defined for each value of t, in particular the mappings g_0 of $\bar{w}$ onto w_0 and g_1 of $\bar{w}$ onto w_1, and if the image point $g_t(\bar{R})$ of a point $\bar{R}$ of $\bar{w}$ is simultaneously continuous in $\bar{R}$ and t and, furthermore, if the image points of the initial point $\bar{P}$ and final point $\bar{Q}$ of $\bar{w}$ are the previously named points P and Q, respectively, then the path w_0 is said to be homotopically deformable to w_1 and the paths w_0 and w_1 are said to be *homotopic to one another in the complex* $\Re^n$. Obviously w_1 is also deformable to w_0. We shall assume from now on that homotopic paths always have a common initial point and a common endpoint.

As in the case of a deformation of a mapping, we can now describe all of the mappings g_t of the path $\bar{w}$ by means of a single mapping f of the deformation complex $\bar{w} \times t$, where $\bar{w} \times t$ denotes the topological product of $\bar{w}$ and the unit interval t $(0 \leqq t \leqq 1)$. In particular, we may choose $\bar{w} \times t$ to be a rectangle of the Euclidean plane, the *deformation rectangle*. We can now give the following interpretation to the definition of the deformation: w_0 is deformable to w_1 if one can map the deformation rectangle $\bar{w} \times t$ into $\Re^n$ by a continuous mapping so that the sides $\bar{w} \times 0 = \bar{w}_0$ and $\bar{w} \times 1 = \bar{w}_1$ transform

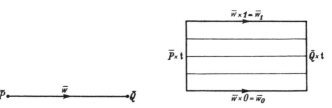

FIG. 77. The preimage interval.   FIG. 78. The deformation rectangle.

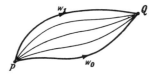

FIG. 79. The mapping into $\mathfrak{R}^n$.

to the paths w_0 and w_1, respectively, while the other two sides $\overline{P} \times t$ and $\overline{Q} \times t$ transform to the points P and Q, respectively. Here $\overline{w} \times 0$ denotes the topological product of the preimage interval $\overline{w}$ and the point 0 of the interval t.

The relation between f and the mappings g_t is given by the equations

$$f(\overline{R} \times t) = g_t(\overline{R}),$$

as in the case of deformation of mappings. The difference between the present case and our earlier case lies only in the additional requirement that f must transform the sides $\overline{P} \times t$ and $\overline{Q} \times t$ to the respective points P and Q (Figs. 77–79).

The deformation of a string or rubber band with fixed endpoints in Euclidean 3-space is a special case of the deformations considered here. The string has no double points and it cannot cut itself during the deformation, whereas, for example, knots can be unknotted by moving through themselves during a homotopic deformation. In addition, deformations of a string are realizable by isotopic deformations of the surrounding space into itself, and this is not in general true for arbitrary homotopic deformations of a path in a complex.

If w_0 is deformable to w_1 and w_1 is deformable to w_2, then w_0 is also deformable to w_2. We need only to identify the sides which map to w_1 of the deformation rectangles mapping w_0 to w_1 and w_1 to w_2, and do so in a manner that points to be identified have the same image point in $\mathfrak{R}^n$. This is always possible since both of the sides are preimages of the same path w_1. We then obtain a new rectangle which is mapped continuously into $\mathfrak{R}^n$ so that two parallel sides transform to w_0 and w_2, respectively, and the other two sides transform to P and Q, respectively. That is, w_0 is deformable to w_2. The relation of deformability is, then, transitive. Based on this fact, one can divide the paths connecting two points into classes of mutually deformable paths.

If

$$w_0 = a \cdots bc_0 d \cdots e$$

is a path in $\Re^n$ which is the product of the paths $a, \ldots, b, c_0, d, \ldots, e$ and if c_0 is deformable to c_1, then w_0 is deformable* to

$$w_1 = a \cdots b c_1 d \cdots e.$$

We show this as follows. By our assumptions there exists a continuous mapping of the "center" rectangle of Fig. 80 such that $\bar{c}_0$ maps to c_0, $\bar{c}_1$ to c_1, and the two vertical sides to the common initial and endpoints, respectively, of c_0 and c_1 (Fig. 81). The mapping of the center rectangle is then extended to a mapping of the entire rectangle under which $\bar{a} \times 0, \ldots, \bar{b} \times 0, \bar{d} \times 0, \ldots, \bar{e} \times 0$ map to the paths $a, \ldots, b, d, \ldots, e$ and each vertical interval of the subrectangles other than those in the center rectangle maps to a single point.

If in a product path ab the first factor consists of a point path, then ab is deformable to b; if the second factor consists of a point path, then ab is deformable to a.

Proof. In the rectangle of Fig. 82 the lower side has been divided into two parts, $\bar{a}_0$ and $\bar{b}_0$. Each point of $\bar{b}_0$ has been connected by a straight line segment to a point on the opposite side $\bar{b}_1$ which corresponds to the point on $\bar{b}_0$ by virtue of a linear mapping. Likewise, all points of $\bar{a}_0$ have been connected by straight line segments to the initial point of $\bar{b}_1$. We can map the rectangle continuously onto $\bar{b}_1$ so that each connecting segment transforms to its endpoint lying on $\bar{b}_1$. If we map $\bar{b}_1$ onto the path b, we then have a map of the rectangle into $\Re^n$ such that $\bar{a}_0 \bar{b}_0$ transforms to ab and the two vertical sides of the rectangle transform to the initial point and endpoint, respectively, of b. But this says that ab is deformable to b.

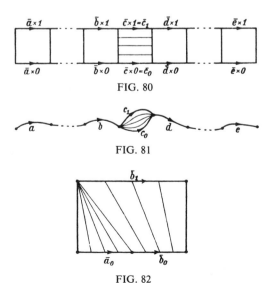

FIG. 80

FIG. 81

FIG. 82

*The path c_1 can also be the first or last factor in the product.

When a path w is deformable to $\bar{w}_0$, where w_0 consists of a single point, then w is said to be deformable to a point, or *null homotopic*. A closed path w is therefore null homotopic in $\mathfrak{K}^n$ if one can map a rectangle continuously into $\mathfrak{K}^n$ so that one side $\bar{w}$ transforms to w and the remaining three sides transform to the initial point of w.

EXAMPLE. Let the complex $\mathfrak{K}^n$ be Euclidean n-space $\mathfrak{R}^n$. Each closed path in this space is null homotopic. To show this, let each point of w run with a constant speed to the initial point P of w in one second; then construct the corresponding mapping of the deformation rectangle.

The same conclusion holds when $\mathfrak{K}^n$ is an n-simplex, $\mathfrak{E}^n$, for example, a rectilinear simplex lying in $\mathfrak{R}^n$. In particular, in the case $n = 1$, a path consisting of an interval a which runs out and back from its initial point is null homotopic on a. It follows from this that any path w which runs out and back, that is, a path of the form $w = uu^{-1}$ in an arbitrary complex, is null homotopic. We need only to map the oriented interval a continuously onto the path u to obtain a continuous mapping of the deformation rectangle into the complex such that one side maps to the path w and the remaining sides to the initial point of w.

The null homotopic paths can also be characterized in the following way: *A closed path w is null homotopic if and only if it can be spanned by a singular disk*. A singular disk is the continuous image of a closed disk in $\mathfrak{R}^n$.

For instead of mapping the deformation rectangle directly into $\mathfrak{K}^n$ by a continuous mapping χ which sends $\bar{w}$ to w and the other sides to the initial point, one can perform the mapping in two steps. First map the rectangle by a continuous mapping φ onto a closed circular disk $\mathfrak{K}^2$ so that the side $\bar{w}_1$ maps to the circumference of the circle, the other three sides to a point O (Fig. 83), and the vertical intervals of the rectangle map to the corresponding chords of the circle; then map the circular disk into the complex $\mathfrak{K}^n$ by means of a mapping ψ such that $\chi = \psi\varphi$, which is obviously possible. From Theorem IV of §8, ψ is continuous. Thus if w is null homotopic, it spans a singular disk. On the other hand, if we are given the continuous mapping ψ of the closed circular disk, then the mapping $\psi\varphi = \chi$ of the rectangle is also determined.

We now make the additional assumption that the complex $\mathfrak{K}^n$ is connected. In that case any two points can be connected by a path. We shall examine all possible closed paths in $\mathfrak{K}^n$ which emanate from some arbitrarily chosen fixed initial point O. These paths will be decomposed into classes of mutually

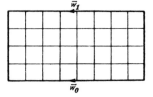

FIG. 83

deformable paths, so-called *path classes*. For the moment we shall denote the path class in which the path w lies by

$$\{w\}.$$

The path classes become elements of a group, the so-called *fundamental group*[25] or *path group* $\mathfrak{F}$ of the connected complex $\mathfrak{K}^n$ when we define the product of two classes, $\{w_1\}$ and $\{w_2\}$ is the path class $\{w_1w_2\}$ of the product path w_1w_2. This specification is independent of the particular choice of representative paths w_1 and w_2 selected from their path classes. For if we replace w_1 and w_2 by two other paths w_1' and w_2' which are deformable, respectively, to w_1 and w_2, then (from earlier in this section) $w_1'w_2'$ is deformable to the path w_1w_2.

The product operation satisfies the group axioms:

1. The associative law is obviously satisfied.
2. There exists a unit element, namely, the class of null homotopic paths. For if one places a null homotopic path in front of or behind an arbitrary path then, from earlier in this section, the latter does not change its path class.
3. To each element there corresponds a reciprocal element. This is the path class of the inverse path. For a path running out from and then back to its initial point is null homotopic.

In general, the commutative law does not hold, and it is not necessarily true that $\{w_1\}\{w_2\} = \{w_2\}\{w_1\}$. In contrast to the homology groups, the fundamental group is not, in general, Abelian. We shall discuss the relationship between the fundamental group and the homology groups in §48.

The path class $\{w_1\}\{w_2\}$ contains not only the product path w_1w_2 but also each path arising from it by means of a deformation in which the initial and endpoints (which coincide with O) are kept fixed. This is true in particular for a path w which arises by pulling the middle of w_1w_2 away from the point O (Fig. 84). Paths w_1 and w_2 do not have to belong to different path classes; they can in fact coincide.

Considered as an abstract group, the fundamental group is independent of the choice of initial point O and is in fact completely determined by the connected complex $\mathfrak{K}^n$. For if we choose a point O' as initial point of the closed paths, instead of O, we can traverse a path v from O to O' and we can associate a path class

$$\{w\} = \{vw'v^{-1}\}$$

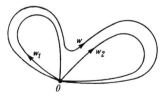

FIG. 84. $\{w\} = \{w_1\}\{w_2\}$.

with the class $\{w'\}$ of paths departing from O'. This association does not depend upon the choice of representative. Conversely, we can associate the class $\{v^{-1}wv\}$ of paths departing from O' with the class $\{w\}$ of paths departing from O. This assignment is just the inverse of the previous one. For by our first prescription we go from $\{w'\}$ to $\{w\} = \{vw'v^{-1}\}$, while by our second prescription we go from $\{w\}$ to

$$\{v^{-1}wv\} = \{v^{-1}(vw'v^{-1})v\} = \{w'\}.$$

The association of path classes belonging to O and O' is therefore a one-to-one correspondence. It is also an isomorphism. To the product

$$\{w_1'\}\{w_2'\} = \{w_1'w_2'\}$$

there corresponds the path class

$$\{vw_1'w_2'v^{-1}\} = \{vw_1'v^{-1} \cdot vw_2'v^{-1}\} = \{vw_1'v^{-1}\}\{vw_2'v^{-1}\},$$

which is the product of the associated path classes. In the above transformations we have made use of the facts that the path vv^{-1} is null homotopic and the path class is not altered by the addition of null homotopic paths.

The isomorphism between the fundamental groups $\mathfrak{F}(O)$ (with initial point O) and $\mathfrak{F}(O')$ (with initial point O') is dependent upon the particular choice of auxiliary path v. If we choose an auxiliary path u in place of v, then instead of the association

$$\{w\} \rightarrow \{v^{-1}wv\} \tag{1}$$

one has the association

$$\{w\} \rightarrow \{u^{-1}wu\} = \{u^{-1}vv^{-1}wvv^{-1}u\} = \{(u^{-1}v)v^{-1}wv(u^{-1}v)^{-1}\}. \tag{2}$$

We must, then, transform* all elements of $\mathfrak{F}(O')$ by the fixed element $\{u^{-1}v\}$ in order to obtain the isomorphism (2) from the isomorphism (1). That is, *the isomorphism between $\mathfrak{F}(O)$ and $\mathfrak{F}(O')$ is only determined up to an inner automorphism of $\mathfrak{F}(O')$.*

We saw in §27 (Theorem II) that when a complex $\mathfrak{K}^n$ is mapped into a complex $\mathbf{K}^m$ by a continuous mapping φ, the kth homology group of $\mathfrak{K}^n$ is mapped by a homomorphism into the kth homology group of $\mathbf{K}^m$. We shall now derive a corresponding theorem for the fundamental group.

The mapping φ will transform an arbitrary path w on $\mathfrak{K}^n$ to a path w' on $\mathfrak{K}^m$. For the preimage $\overline{w}$ of w will first be mapped by a continuous mapping f onto w and will then be mapped continuously by φ. The product mapping φf is again continuous and therefore determines a path w' in $\mathbf{K}^m$. If w_0 and w_1 are

---

*That is, we must multiply on the left by $\{u^{-1}v\}$ and on the right by $\{u^{-1}v\}^{-1}$. Such a transformation produces an inner automorphism of the fundamental group (see, for example, Speiser [1, p. 121]).

two paths in $\Re^n$ which are deformable one to the other, then the image paths w_0' and w_1' are also deformable one to the other. The deformation rectangle $0 \leq s \leq 1$, $0 \leq t \leq 1$ can be mapped continuously into $\Re^n$ so that the parallel sides $t = 0$ and $t = 1$ transform, respectively, to w_0 and w_1 and the other parallel sides transform, respectively, to the common initial point A' and the common endpoint B' of w_0' and w_1'. That is, w_0' and w_1' are homotopic in $\mathbf{K}^m$. Finally, the product of two paths obviously transforms to the product of their images.

If O is the initial point of the closed paths of the fundamental group $\mathfrak{F}$ of $\Re^n$, then we select the image point O' as the initial point of the closed paths of the fundamental group $\mathbf{F}$ of $\mathbf{K}^m$. Since homotopic paths map to homotopic paths, to each path class of $\mathfrak{F}$ there will correspond a particular image path class of $\mathfrak{F}$. Since the image of the product of two classes is the product of the images, then this mapping of $\mathfrak{F}$ into $\mathbf{F}$ is a homomorphism. Even if one does not choose O' as initial point, but instead chooses another point Ω of $\mathbf{K}^m$, we can still speak of a homomorphic mapping of the fundamental group $\mathfrak{F}$. For there holds an isomorphism between the groups $\mathbf{F}(O')$ and $\mathbf{F}(\Omega)$ which is determined by means of an auxiliary path joining O' to Ω. The homomorphic mapping of $\mathfrak{F}$ into $\mathbf{F}(O)$ followed by the isomorphic mapping onto $\mathbf{F}(\Omega)$ is a homomorphic mapping of $\mathfrak{F}$ into $\mathbf{F}(\Omega)$. We shall call this homomorphism ϕ. If we change the auxiliary path, then ϕ will be changed by an inner automorphism $\mathbf{F}(\Omega)$ (from earlier in this section). We then have the following theorem:

THEOREM . *When a complex $\Re^n$ is mapped into a complex $\mathbf{K}^m$ by a continuous mapping φ, then the fundamental group $\mathfrak{F}$ of $\Re^n$ will be mapped into the fundamental group $\mathbf{F}$ of $\mathbf{K}^m$ by a homomorphism ϕ. The mapping ϕ is determined only up to an inner automorphism of $\mathbf{F}$. If $\Re^n$ and $\mathbf{K}^m$ are homeomorphic and φ is a topological mapping of $\Re^n$ onto $\mathbf{K}^m$, then ϕ is an isomorphism. That is, homeomorphic complexes have isomorphic fundamental groups.*

The last part of the theorem follows immediately from the fact that under a topological mapping there is a one-to-one correspondence between the paths $\Re^n$ and the paths of $\mathbf{K}^m$.

43. Examples

EXAMPLE 1. *The fundamental group of the topological product of two complexes $\Re_1$ and $\Re_2$ is equal to the direct product of the fundamental groups of the two factors.*

Proof is as follows. From §14, each point P of the topological product $\Re_1 \times \Re_2$ can be uniquely represented in the form

$$P = P_1 \times P_2. \qquad (P)$$

If $\bar{\Re}$ is an arbitrary complex and g is a continuous mapping of $\bar{\Re}$ into $\Re_1 \times \Re_2$ which sends the point $\bar{P}$ of $\bar{\Re}$ to the point P of $\Re_1 \times \Re_2$, that is,

$$\bar{P} \to P, \qquad (g)$$

then because of equation (P) there is determined a continuous mapping g_1

$$\bar{P} \to P_1 \tag{g_1}$$

of $\bar{\Re}$ into $\Re_1$ and a continuous mapping g_2

$$\bar{P} \to P_2 \tag{g_2}$$

of $\bar{\Re}$ into $\Re_2$. Conversely, if the continuous mappings g_1 and g_2 are given, then a continuous mapping g is determined by virtue of equation (P). In particular, if $\bar{\Re}$ is an oriented interval, then paths w, w_1, and w_2 in $\Re_1 \times \Re_2$, $\Re_1$, and $\Re_2$, respectively, are determined by the mappings g, g_1, and g_2. *To each path w in $\Re_1 \times \Re_2$ there corresponds a pair of paths w_1 and w_2, and conversely.*

If, on the other hand, $\bar{\Re}$ is a rectangle which is mapped by g into $\Re_1 \times \Re_2$ so that three sides transform to a point $O = O_1 \times O_2$ while the fourth side transforms to a path w, then under the mapping g_i $(i = 1, 2)$ the same three sides will transform to the point O_i and the fourth side will transform to the path w_i. Thus w_1, w_2 is the pair of paths which is associated with w. If w is then null homotopic in $\Re_1 \times \Re_2$ it follows that w_1 will be null homotopic in $\Re_1$ and w_2 will be null homotopic in $\Re_2$. The converse is also true: If w_1 and w_2 are both null homotopic, then there exists a mapping g_i of the rectangle $\bar{\Re}$ into $\Re_i$ such that three sides transform to O_i and the fourth side transforms to w_i. The mapping g determined by g_1 and g_2 transforms these three sides of $\bar{\Re}$ to $O_1 \times O_2$ and the fourth side to w.

If we choose a point O_i in $\Re_i$ as initial point of the closed paths and likewise choose $O_1 \times O_2$ as initial point in $\Re_1 \times \Re_2$, then the path classes of $\Re_1 \times \Re_2$ will be in one-to-one correspondence with the pairs of path classes of $\Re_1$ and $\Re_2$. Since the product of two path classes of $\Re_1 \times \Re_2$ is obviously formed by forming the products of the corresponding path classes of $\Re_1$ and $\Re_2$, respectively, the fundamental group of $\Re_1 \times \Re_2$ is the direct product of the fundamental groups of $\Re_1$ and $\Re_2$.

EXAMPLE 2. The fundamental group of the n-simplex and also that of Euclidean n-space $\Re^n$ consists of the unit element alone, because each closed path w is null homotopic.

EXAMPLE 3. For $n > 1$, the fundamental group of the n-sphere $\mathfrak{S}^n$ consists of the unit element alone. For if a closed path w of $\mathfrak{S}^n$ does not pass through the point O' diametrically opposite to the base point O of w, we may deform w to O by moving each point of w to O linearly in one second along its great circle through O and O'. If, however, w passes through O', then we may approximate the mapping of the preimage $\bar{w}$, as was done previously in §31, in a simplicial decomposition of $\mathfrak{S}^n$ such that the diametral point O' becomes an inner point of an n-simplex and the point O is a vertex. This deforms the path w so that it no longer passes through O', although O remains as initial point, whence the present case reduces to the previous case. We cannot avoid the use of a simplicial approximation in the proof, since it can happen that a path passes through every point of the n-sphere.

EXAMPLE 4. If the fundamental group of a complex consists of the unit element alone, the complex is said to be *simply connected*. Such a complex is characterized by the fact that one can deform each closed path in the complex to a point or, what amounts to the same thing, each closed path can be spanned by a closed disk. Of course this cannot always be done topologically. On the contrary, in some circumstances one may have to allow singularities (branch points, fold lines, self-intersections, and the like). The question of to what extent a complex is determined by the condition that it be simply connected is related to important unsolved problems of topology, even when one restricts oneself to homogeneous complexes. The question is easily answered in 1 dimension. There exists only one finite 1-dimensional connected homogeneous complex, the circle (1-sphere); there exists only one infinite 1-dimensional connected homogeneous complex, the real number line. The real number line is simply connected. In the case of 2 dimensions we shall show in §47, by determining the fundamental groups of all homogeneous finite complexes (that is, of all closed surfaces), that the 2-sphere is the only simply connected finite homogeneous complex. The Euclidean plane is the only simply connected infinite homogeneous 2-dimensional complex.[26] In 3 dimensions it appears likely that the 3-sphere is the only simply connected finite homogeneous complex (Poincaré's conjecture). In 4 dimensions the topological product of two 2-spheres is an example of a homogeneous finite complex which is simply connected, from Example 1, but is not homeomorphic to the 4-sphere, because the Betti number for the dimension 2 is $p^2 = 2$, while $p^2 = 0$ for the 4-sphere (from §19).*

44. The Edge Path Group of a Simplicial Complex

The fundamental group has been defined for a connected complex in a topologically invariant manner. However, in order to calculate this topological invariant for a connected complex we must proceed as we did in the case of the homology groups and go to a particular simplicial decomposition of the complex. With its aid we define a new group, the *edge path group*, which is not defined as a topological invariant, but it is computable. We shall subsequently prove, by means of a simplicial approximation, that the edge path group and fundamental group coincide. The fundamental group $\mathfrak{F}$ of a connected complex has already been defined in a topologically invariant way. The topological invariance of the edge path group will then follow.

We first consider a finite or infinite complex $\mathfrak{K}^n$ having a fixed simplicial decomposition. Let its edges be oriented in some particular way and let them be denoted by $a_1, a_2, \dots$. We shall first deal only with edge paths on $\mathfrak{K}^n$, that is, with products

$$w = a_l^{\varepsilon_l} a_m^{\varepsilon_m} \cdots a_z^{\varepsilon_z} \qquad (\varepsilon_l = \pm 1, \varepsilon_m = \pm 1, \dots, \varepsilon_z = \pm 1),$$

where the endpoint of $a_l^{\varepsilon_l}$ will always coincide with the initial point of $a_m^{\varepsilon_m}$ and so forth. We shall also include paths having no edges, that is, paths consisting of an isolated vertex as an edge path.

To avoid confusion we shall on occasion refer to the paths and deformations considered in §42 as continuous paths and continuous deformations, in contrast to the edge paths and the combinatorial deformations, which we shall now define.

When we traverse the boundary of a 2-simplex $\mathfrak{E}^2$ of $\mathfrak{K}^n$ in a particular sense, starting from a particular vertex, we obtain a closed edge path, for example

$$a_1^{\varepsilon_1} a_2^{\varepsilon_2} a_3^{\varepsilon_3};$$

we call this path a *boundary path* of the triangle. Each triangle has six different boundary paths, since we can depart from each of the vertices in two opposite directions.

---

* We can show that the Betti number p^2 of the topological product of two 2-spheres $\mathfrak{K}_1$ and $\mathfrak{K}_2$ is not equal to zero, as follows: Under the mapping g_1 of Example 1 the direct product $\mathfrak{K}_1 \times P_2$ (P_2 is a point of $\mathfrak{K}_2$) will be mapped topologically onto the 2-sphere $\mathfrak{K}_1$. After simplicially decomposing and coherently orienting $\mathfrak{K}_1 \times P_2$, we shall find a 2-chain U^2 which is not null homologous in $\mathfrak{K}_1 \times \mathfrak{K}_2$. For if it were null homologous, then its image $'U^2$ under the mapping g_1 would also be null homologous, from Theorem I of §27. But $'U^2$ is the coherently oriented 2-sphere $\mathfrak{K}_1$ and therefore is not null homologous on $\mathfrak{K}_1$.

We can generate new edge paths from an edge path by means of the following elementary combinatorial deformations, (α) and (β):

(α) Insert or remove an edge which runs back and forth. The edge path

$$a_l^{\varepsilon_l} \cdots a_p^{\varepsilon_p} a_q^{\varepsilon_q} \cdots a_z^{\varepsilon_z}$$

is replaced by

$$a_l^{\varepsilon_l} \cdots a_p^{\varepsilon_p} a_\rho^{\varepsilon_\rho} a_\rho^{-\varepsilon_\rho} a_q^{\varepsilon_q} \cdots a_z^{\varepsilon_z};$$

here the initial point of $a_\rho^{\varepsilon_\rho}$ must coincide with the endpoint of $a_p^{\varepsilon_p}$ and thus must also coincide with the initial point of $a_q^{\varepsilon_q}$. In the other direction, a term $a_\rho^{\varepsilon_\rho} a_\rho^{-\varepsilon_\rho}$ may be deleted from the expression for an edge path.*

By a repeated application of (α) we can insert an arbitrary edge path u which runs back and forth into an edge path $w = w_1 w_2$ so that the initial point of u coincides with the endpoint of w_1. We thereby obtain the edge path $w' = w_1 u u^{-1} w_2$ from w.

(β) Insert or remove the boundary path of a 2-simplex. One traverses a given path w up to a particular vertex, then traverses the boundary path of a 2-simplex beginning and ending at this vertex and, finally, traverses the remainder of path w.

We now introduce a third elementary combinatorial deformation which is derived from the previous deformations.

(γ) In an edge path w we replace an edge $a_1^{\varepsilon_1}$ which belongs to the boundary path $a_1^{\varepsilon_1} a_2^{\varepsilon_2} a_3^{\varepsilon_3}$ of a 2-simplex by $(a_2^{\varepsilon_2} a_3^{\varepsilon_3})^{-1}$ or, inversely, we replace a sequence $a_3^{-\varepsilon_3} a_2^{-\varepsilon_2}$ by $a_1^{\varepsilon_1}$. That is, we can let the path "jump over" a triangle. The operation (γ) can be produced by carrying out the deformations (β) and (α) (see Fig. 85, for which $\varepsilon_1 = 1$, $\varepsilon_2 = -1$, $\varepsilon_3 = -1$).

Two edge paths w and w' which can be transformed one to the other by finitely many elementary combinatorial deformations are said to be combinatorially deformable to one another or *combinatorially homotopic*. Combinatorially homotopic edge paths always have the same initial point and end point.

Let us now assume that the complex $\mathfrak{R}^n$ is connected. The closed edge paths departing from a given fixed vertex O are decomposable into classes of combinatorially equivalent paths, the so-called edge path classes. The edge path classes are elements of a group $\mathfrak{F}_k$, the *edge path group* of the given simplicial decomposition of $\mathfrak{R}^n$, provided one takes the product of two edge

---

* If w is closed, however, and the first edge is the inverse of the last edge, that is,

$$w = a_l^{\varepsilon_l} \cdots a_l^{-\varepsilon_l},$$

then $a_l^{\varepsilon_l}$ cannot be canceled out against $a_l^{-\varepsilon_l}$. That is, the initial and end points of an edge path remain fixed under combinatorial deformations.

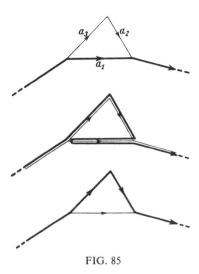

FIG. 85

path classes $[w_1]$ and $[w_2]$ to be the edge path class $[w_1w_2]$. This determination is clearly independent of the particular selection of the representative paths w_1 and w_2 from the classes $[w_1]$ and $[w_2]$. We easily check that the group axioms are satisfied, as for the definition of the fundamental group, and that the group does not change when we select another vertex O' in place of O to be the initial point of the edge path.

The edge path group $\mathfrak{F}_k$ of the complex $\mathfrak{K}^n$ coincides with its fundamental group. To show this we first notice that each combinatorial deformation can be realized by a continuous deformation, for the boundary of a triangle and a path running back and forth are both continuously null homotopic and each may be eliminated after contraction to a point (§42), assuming of course that some path remains after this. If we then choose the same vertex O of the simplicial decomposition of $\mathfrak{K}^n$ as the initial point of the continuous paths of $\mathfrak{F}$ and as the initial point of the edge paths of $\mathfrak{F}_k$, then each edge path class $[w]$ lies within the continuous path class $\{w\}$. The subsequent proof that $\mathfrak{F}_k$ coincides with $\mathfrak{F}$ proceeds in much the same way as the proof of invariance of the homology groups. We must establish two properties:

(I)   Each closed continuous path having O as initial point is continuously deformable to an edge path. That is, at least one $[w]$-class lies within each $\{w\}$-class.

(II)   If an edge path having initial point O is continuously null homotopic in $\mathfrak{K}^n$, then it can also be transformed to O by means of combinatorial deformations.[27] That is, at most one $[w]$-class lies in each $\{w\}$-class. The elements of $\mathfrak{F}$ and $\mathfrak{F}_k$ are therefore in one-to-one correspondence. Since we can use edge paths in particular to form products of elements of $\mathfrak{F}$, we have an isomorphism as well.

We shall prove both claim (I) and claim (II) with the help of the deformation theorem of §31.

(I) If w is an arbitrary closed path departing from O and having the preimage $\overline{w}$, then we can deform the mapping g_0 of $\overline{w}$ onto w to a simplicial mapping g_1 of a sufficiently fine subdivision of $\overline{w}$. The initial point and end point of $\overline{w}$ will both map to the point O during the deformation, since the deformation does not move any image point from the simplex of $\mathfrak{R}^n$ to which it initially belonged. The path $g_1(\overline{w})$, which arises from $w = g_0(\overline{w})$ by deformation is therefore either the point O or consists only of edges of $\mathfrak{R}^n$, between which some point paths may be inserted. We may eliminate the latter, from §42, so that w is deformed to an edge path.

(II) If w is a closed edge path which can be contracted to the initial point O by means of a continuous deformation, then a deformation rectangle $\overline{PP_0Q_0Q}$ can be mapped continuously into $\mathfrak{R}^n$ so that the side $\overline{PQ}$ transforms to w and the remaining three sides transform to O (Fig. 86). We mark the points on $\overline{w}$ which correspond to the vertices of the subintervals of w, construct vertical lines through these points of division, and complete the decomposition of the rectangle, which was started in this way, by means of additional vertical and horizontal lines to form a grid which is sufficiently fine so that the rectangles meeting in each vertex of the division will map into the interior of some simplicial star of $\mathfrak{R}^n$. This is always possible, from the uniform continuity theorem (§7). We subdivide the grid by means of diagonals to form a simplicial complex $\overline{\mathfrak{R}}^2$. From the deformation theorem, the mapping g_0 of $\overline{\mathfrak{R}}^2$ into $\mathfrak{R}^n$ can be deformed to a simplicial mapping g_1. If $\overline{w}$ denotes the subdivided rectangle edge $\overline{w}$, then the image $g_0(\overline{w})$ is the subdivided edge path w. Under the deformation of g_0 to g_1 the image of a subinterval of the path $\overline{PP_0Q_0Q}$ will always be the point O, since during the deformation of the mapping no image point leaves a simplex of $\mathfrak{R}^n$ to which it belonged at the start of the deformation. The sides $\overline{PP_0}$, $\overline{P_0Q_0}$, $\overline{Q_0Q}$, however, all map under g_0 to the 0-simplex O. For the same reason, a dividing point of $g_0(\overline{w})$ will continually remain fixed under the deformation if it is a vertex of $\mathfrak{R}^n$; if, however, the dividing point is an inner point of an edge a of w, then it will transform to a vertex of a. The path $g_1(\overline{w})$, which arises from the deformation of $g_0(\overline{w})$, will then consist of the same edges as w (aside from certain interpolated point paths, which we eliminate) but will also contain, in

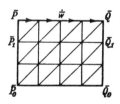

FIG. 86

general, certain edges which run back and forth. In any case, we can arrive at $g_1(\bar{w})$ from w by means of a series of combinatorial deformations of the type (α).

We now transform $\ddot{w}$ on $\overline{\Re}^2$ to the path $\overline{P}\overline{P}_0\overline{Q}_0\overline{Q}$ by a series of combinatorial deformations. This can be done, for example, by first replacing each subinterval of $\ddot{w}$ by the two other sides of the subtriangle adjoining it. We thereby obtain a sequence of segments leading from $\overline{P}$ to $\overline{Q}$ which consists alternately of vertical and diagonal segments. We replace each diagonal segment by the other two sides of the triangle lying underneath it and then eliminate the resulting segments which run back and forth. The new interval path is $\overline{P}\overline{P}_1\overline{Q}_1\overline{Q}$. We apply the same procedure to $\overline{P}_1\overline{Q}_1$ as was previously applied to $\overline{PQ}$ and we proceed in this manner until we finally arrive at $\overline{P}\overline{P}_0\overline{Q}_0\overline{Q}$. Each combinatorial deformation applied to $\ddot{w}$ will determine a combinatorial deformation of $g_1(\bar{w})$, as a consequence of the simplicial mapping g_1, and this may, of course, also be the identity mapping. An operation (β) applied to a triangle $\mathfrak{E}^2$ of $\overline{\Re}^2$ will correspond in $\Re^n$ either to an operation (α), an operation (β), or to the identity, depending upon whether $\mathfrak{E}^2$ transforms to a triangle, an edge, or a vertex, respectively, of $\Re^n$. An operation (α) applied to a segment $\mathfrak{E}^1$ of $\overline{\Re}^2$ will correspond to either an operation (α) or to the identity, depending upon whether $\mathfrak{E}^1$ transforms to an edge or to a vertex of $\Re^n$.

We can then transform w to $g_1(\ddot{w})$ and then into the image of the path $\overline{P}\overline{P}_0\overline{Q}_0\overline{Q}$ by means of combinatorial deformations. But this image is the point O. Thus w is combinatorially null homotopic.

This proves the coincidence of the invariantly defined fundamental group $\mathfrak{F}$ with the combinatorially defined edge path group $\mathfrak{F}_k$. We shall, from now on, denote both groups with the same symbol.

It also follows, because these groups are the same, that *the fundamental group of a complex $\Re^n$ is not changed when one eliminates the inner points of all simplexes of dimension higher than 2*. For the edge path group depends only upon the 0-, 1-, and 2-simplexes of the decomposition, that is, upon the 2-dimensional "skeleton complex" of the simplicial decomposition of $\Re^n$.

45. The Edge Path Group of a Surface Complex

The result of §44 has reduced the problem of representing the fundamental group of a given complex $\Re^n$ by means of generators and relations to the simpler problem of finding the edge path group of a 2-dimensional simplicial complex. The generators and relations can then be found, in principle, by using a procedure which we shall learn in §46. The relations which one obtains with the help of a simplicial complex are quite numerous and difficult to understand, and if one wishes to use them one must simplify them by eliminating redundant generators. We can arrive at considerably simpler relations if we apply the computational procedure not to a simplicial complex but to a surface complex built up of polygons. We shall now define the surface complex.

Although previous investigations were carried out for arbitrary complexes, we shall restrict ourselves here to finite complexes, for the sake of simplicity.

An *edge complex* consists of finitely many intervals with certain boundary points identified. We do not exclude the case where the end points of a single interval are identified. We shall call the image of an interval in the edge complex an edge. Examples are the edge complex of the normal forms of closed surfaces or the edge complex of the cube.

An edge complex can be transformed to a 1-dimensional simplicial complex by subdividing its edges. It wlll then be a complex in the sense of §10.

We orient an edge by orienting its preimage. We shall denote the arbitrarily oriented edges, which we can regard as particular paths on the edge complex, by a_ν and the reverse oriented edges by a_ν^{-1}.

We form "edge paths" out of the paths a_ν and their inverse paths; these are paths of the form

$$a_l^{\varepsilon_l} a_m^{\varepsilon_m} \cdots a_z^{\varepsilon_z} \qquad (\varepsilon_\nu = \pm 1),$$

where the endpoint of an edge will always coincide with the initial point of the next edge.

A *surface complex* will arise from the edge complex by "spanning" certain closed edge paths by disks. Let

$$w = a_l^{\varepsilon_l} a_m^{\varepsilon_m} \cdots a_z^{\varepsilon_z}$$

be a closed path which consists of at least one edge. Take a circular disk of radius 1 and parametrize its boundary by arc length s: $0 \leqq s < 2\pi$. Similarly, describe the closed path w by the same parameter s. Points of the disk boundary are then to be identified with path points having the same value of the parameter s. We describe the resulting neighborhood space by saying it arises from the edge complex by *spanning* a *disk* over the closed path w, namely, the image of the circular disk for which w is now the boundary. Paths w, w^{-1}, and the paths arising from them by cyclically permuting their edges are then the *boundary paths* of the disk. It should be noted that w does not have to be free of double points; for example, it is possible that $w = aa^{-1}$.

When one repeats this process of spanning disks finitely often, a surface complex will be obtained from the edge complex. We shall also speak of an edge complex itself as a surface complex (with no spanning disks).

We can define an edge path group in a surface complex $\Re$ as in a simplicial complex. For this purpose we choose a fixed vertex O as the initial point of the edge paths and we divide the closed edge paths departing from O into edge path classes. Two such closed edge paths are put into the same class if they can be combinatorially deformed into one another. The combinatorial deformation is defined as in the case of a simplicial complex (§44), except that we must modify the operation (β) to deal with polygons instead of triangles:

(α)   Insert or remove an edge which runs back and forth.
(β)   Insert or remove the boundary path of a disk.

An edge a which is the boundary path of a disk can be eliminated from any path in which it appears (Fig. 87).

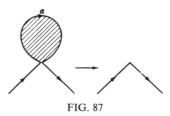

FIG. 87

The edge path classes are the elements of the edge path group $\mathfrak{F}_k$ of the complex $\mathfrak{K}$, when multiplication of classes is defined as earlier.

We shall now show that the edge path group $\mathfrak{F}_k$ is the same as the fundamental group $\mathfrak{F}$ of $\mathfrak{K}$. In the special case that $\mathfrak{K}$ has been simplicially decomposed this has already been proved in §44, since the surface complex $\mathfrak{K}$ is then either a simplicial complex $\mathfrak{K}^2$ or $\mathfrak{K}^1$. We shall reduce the general case of an arbitrary surface complex $\mathfrak{K}$ to this known case by repeatedly subdividing to arrive at a simplicial complex and showing that the edge path group does not change during each subdivision.

We distinguish two different types of subdivision:

(U1)   Subdivision of an edge: an inner point on an edge of the surface complex $\mathfrak{K}$ is declared to be a new vertex, whereupon the edge is decomposed into two subedges.

(U2)   Subdivision of a surface element: the preimage of the surface element, which is a Euclidean disk, is decomposed by a chord into two parts. The endpoints of the chord transform under the mapping of the boundary to vertices of the edge complex. The chord transforms to a new edge, and the two parts of the disk transform to two new surface elements.

Upon subdividing, a surface complex is transformed to another surface complex. By repeatedly subdividing, we can arrive at a simplicial complex, starting from an arbitrary surface complex $\mathfrak{K}$. We first divide all surface elements into triangles and then form the 2-fold normal subdivision of the resulting triangle complex.

If we denote by $\dot{\mathfrak{K}}$ the surface complex obtained from $\mathfrak{K}$ by a single subdivision (U1) or (U2), then the claim that $\mathfrak{K}$ and $\dot{\mathfrak{K}}$ have the same fundamental group is equivalent to the following two claims:

1.   A closed edge path $\dot{w}$ of $\dot{\Re}$ emanating from O is combinatorially deformable in $\dot{\Re}$ to an edge path of $\Re$.

2.   A closed edge path w of $\Re$ emanating from O which is combinatorially null homotopic in $\dot{\Re}$ is also combinatorially null homotopic in $\Re$.

Proof of 1. Suppose $\dot{\Re}$ arises from $\Re$ by subdivision of an edge a at a point T. Then edge a is divided into two subedges b and c which may be regarded as paths in such a way that $bc = a$ (Fig. 88). The endpoint Q of a does not necessarily have to differ from the initial point P. If one of the sequences bb^{-1}, cc^{-1}, $b^{-1}b$ or $c^{-1}c$ appears in an edge path $\dot{w}$ of $\dot{\Re}$, then we can remove it by a combinatorial deformation (α). We repeat this until $b^{\pm 1}$ and $c^{\pm 1}$ no longer appear in w or occur only in the sequences bc $(= a)$ or $c^{-1}b^{-1}$ $(= a^{-1})$. In this way we transform $\dot{w}$ combinatorially to a path of $\Re$.

If $\dot{\Re}$ arises from $\Re$ by subdivision of a surface element Π, then Π decomposes into two surface elements $\mathbf{P}$ and $\mathbf{\Sigma}$ which are separated from one another by the new edge d (Fig. 89). Let the boundary paths of $\mathbf{P}$ and $\mathbf{\Sigma}$ be du^{-1} and dv^{-1}. Should the edges $d^{\pm 1}$ appear in an edge path $\dot{w}$ of $\dot{\Re}$, we can replace them by $v^{\pm 1}$ by means of a combinatorial deformation (β) followed by (α). This will deform $\dot{w}$ combinatorially to a path w of $\Re$.

Proof of 2. If the edge path w of $\Re$ is transformed by a sequence of single combinatorial deformations inside $\dot{\Re}$ to the sequence of edge paths

$$\dot{w}_1, \dot{w}_2, \ldots, \dot{w}_r = O,$$

then these edge paths of $\dot{\Re}$ will not in general be edge paths of $\Re$. We shall now give a rule which assigns an edge path w_i of $\Re$ to each path $\dot{w}_i$ so that each path in the sequence of paths

$$w = w_0, w_1, w_2, \ldots, w_r$$

arises from its predecessor by means of one or more combinatorial deformations inside $\Re$ and so that $w_r = O$. This will show that w is combinatorially deformable to O inside $\Re$.

If $\dot{\Re}$ arises from $\Re$ by subdividing edge a, the rule of assignment is: to each edge of $\dot{\Re}$ which differs from b or c we assign the self-same edge of $\Re$. The edge $a^{\pm 1}$ is assigned to the edge $b^{\pm 1}$ and no edge at all is assigned to the edge c. The edge path w_i which is associated with the edge path $\dot{w}_i$ is then obviously a closed edge path in $\Re$ emanating from O.

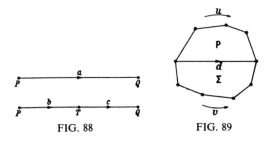

FIG. 88   FIG. 89

If we now complete the transformation from $\dot{w}_i$ to $\dot{w}_{i+1}$ by means of a combinatorial deformation (α) applied to an edge different than b or c, or by means of a deformation (β), then the transformation from w_i to w_{i+1} is accomplished by the same deformation. But if $\dot{w}_{i+1}$ arises from $\dot{w}_i$ by means of a deformation (α) applied to b or c, then w_{i+1} arises from w_i by a deformation (α) applied to a (or else $w_i = w_{i+1}$).

If, on the other hand, $\dot{\Re}$ arises from $\Re$ by subdivision of a surface element, then we choose the following rule of assignment: each of the edges of $\dot{\Re}$ different than d will be assigned the self-same edge of $\Re$, and the edge d will be assigned the path v. The addition or removal of dd^{-1} in $\dot{\Re}$ will then correspond to the addition or removal of vv^{-1} in $\Re$, which can be achieved by means of deformations (α). To the deformations (β) applied to the polygons **P** and Σ there will correspond combinatorial deformations in $\Re$ which can be reduced to deformations (β) and (α). In this way we have proved the invariance of the edge path group $\mathfrak{F}_k$ under subdivision and have thus proved that $\mathfrak{F}_k$ coincides with the fundamental group $\mathfrak{F}$.

46. Generators and Relations

We again denote the oriented edges of the surface complex $\Re$ by $a_1, a_2, \ldots, a_{\alpha_1}$ and denote the initial point of the closed edge paths by O. In order to arrive at a system of generators we select for each vertex of $\Re$ a fixed path, the so-called *auxiliary path*, which leads from O to that vertex. The auxiliary path of O will consist of the point O alone. To each oriented edge a_i there will then correspond a unique closed path A_i departing from O, which is obtained by traversing the auxiliary path from O to the initial point of a_i, traversing a_i to its endpoint and, finally, returning to O along the inverse auxiliary path of the endpoint. A_i is a particular edge path

$$A_i = \varphi_i(a_\nu) \qquad (A)$$

and, as such, belongs to a particular edge path class, which we shall henceforth also denote by A_i, omitting the rectangular brackets. Not all of the classes $A_1, A_2, \ldots, A_{\alpha_1}$ need be distinct.

These edge path classes can serve as generators of the edge path group. We can combinatorially deform each closed edge path emanating from O to a product of edge paths A_j. If $w = F(a_i)$, for example, $w = a_l^{\varepsilon_l} a_m^{\varepsilon_l} \cdots a_z^{\varepsilon_z}$ is such a path, then it is in fact combinatorially deformable to the edge path

$$W = F(A_i) = A_l^{\varepsilon_l} A_m^{\varepsilon_m} \cdots A_z^{\varepsilon_z}$$

by general application of the operation (α). The path W arises from w by introducing an auxiliary path running back and forth between each pair of successive edges.

Now for the relations! The existence of a relation

$$R(A_i) = 1$$

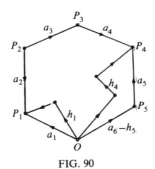

FIG. 90

means that the edge path class $R(A_i)$ is the identity, that is, the edge path $P(A_i)$ is combinatorially deformable to the vertex O.

It is possible to present a finite number of relations from which all other relations follow. These relations are of two types:

(I)   If, in equation (A) we replace the edge a_ν by the closed edge path A_ν, we get

$$A_i = \varphi_i(A_\nu). \tag{I}$$

This is a correct relation, for the left-hand side comes from the right-hand side by removing auxiliary paths which run back and forth. There are as many relations of the type (I) as there are edges in the simplicial decomposition of our (finite) complex.

Figure 90 shows an example. Certain edges are shown which may be considered to lie on an arbitrary complex. Choose the path $h_2 = a_1 a_2^{-1}$ as the auxiliary path to P_2; choose $h_3 = a_6 a_5 a_4^{-1}$ as the auxiliary path to P_3. Let h_0 denote the auxiliary path from O which consists of a single point. The auxiliary paths to P_1, P_4, and P_5 are indicated in the figure. We then have, for example, the path

$$A_3 = h_2 a_3 h_3^{-1} = a_1 a_2^{-1} \cdot a_3 \cdot a_4 a_5^{-1} a_6^{-1} = \varphi_3(a_\nu)$$

and the path

$$\varphi_3(A_\nu) = \left(h_0 a_1 h_1^{-1}\right)\left(h_1 a_2^{-1} h_2^{-1}\right) \cdot h_2 a_3 h_3^{-1} \cdot \left(h_3 a_4 h_4^{-1}\right)\left(h_4 a_5^{-1} h_5^{-1}\right)\left(h_5 a_6^{-1} h_0\right),$$

where we still must regard the paths h_ν as being composed of their individual edges. The above paths are combinatorially deformable into one-another usimg only opearations (α). The path classes A_3 and $\varphi_3(A_\nu)$ then do, in fact, coincide.*

(II)   If $a_l^{\varepsilon_l} a_m^{\varepsilon_m} \cdots a_z^{\varepsilon_z}$ is a boundary path of a surface element Π, then the path $A_l^{\varepsilon_l} A_m^{\varepsilon_m} \cdots A_z^{\varepsilon_z}$ is combinatorially deformable to O and therefore

$$A_l^{\varepsilon_l} A_m^{\varepsilon_m} \cdots A_z^{\varepsilon_z} = 1 \tag{II}$$

is a correct relation. For the path is, first of all, deformable to that path which leads from O along the auxiliary path to the initial point of $a_l^{\varepsilon_l}$ on the surface element Π, runs around Π, and then returns to O along the same auxiliary

---

* *Editor's note*: The intended relation in this example is $A_3 = A_1 A_2^{-1} A_3 A_4 A_5^{-1} A_6^{-1}$. The expression $\varphi_i(A_\nu)$ means "substitute the symbol A_ν for a_ν in the word $\varphi_i(a_\nu)$."

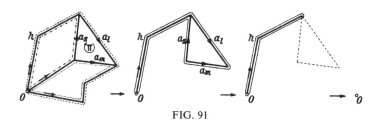

FIG. 91

path (or rather, its inverse path). This deformation is made by removing the auxiliary paths running back and forth to the other vertices of Π, that is, by repeated application of the operation (α). By applying the operation (β) to the surface element Π we obtain the path hh^{-1} which can be deformed combinatorially to O by repeated application of (α).

This deformation process is illustrated in Fig. 91 for a triangle Π. Here we choose $\varepsilon_l = \varepsilon_z = 1$, $\varepsilon_m = -1$, and we deform the corresponding path $A_l A_m^{-1} A_z$ to the point O.

There are as many relations of the type (II) as there are polygons in $\mathfrak{R}$.

The relations (I) *and* (II) *form a system of defining relations, that is, the left-hand side of an arbitrary relation* $R(A_i) = 1$ *can be reduced to 1 by applying the relations* (I) *and* (II) *and the trivial relations* $A_i A_i^{-1} = A_i^{-1} A_i = 1.$

Proof. The relation $R(A_i) = 1$ states that the edge path $R(\varphi_i(a_\nu))$, which we shall denote by $w_0(a_\nu)$, is combinatorially null homotopic. Therefore there exists a sequence of paths

$$w_0(a_\nu), w_1(a_\nu), \ldots, w_m(a_\nu)$$

such that $w_j(a_\nu)$ $(j = 1, 2, \ldots, m)$ arises from $w_{j-1}(a_\nu)$ by means of an operation (α) or (β) and such that $w_m(a_\nu)$ contains no factors at all. By applying the relations (I) the left-hand side $R(A_i)$ of the given relation will transform to $R(\varphi_i(A_\nu)) = w_0(A_\nu)$. We then form the sequence of products

$$w_0(A_\nu), w_1(A_\nu), \ldots, w_m(A_\nu) = 1.$$

If $w_j(a_\nu)$ follows from $w_{j-1}(a_\nu)$ by applying the operation (α), then $w_j(A_\nu)$ follows from $w_{j-1}(A_\nu)$ by applying the trivial relation $A_\nu A_\nu^{-1} = 1$ or $A_\nu^{-1} A_\nu = 1$. But if $w_j(a_\nu)$ arises from $w_{j-1}(a_\nu)$ by applying the operation (β) to a polygon Π of $\mathfrak{R}$, then we get $w_j(A_\nu)$ from $w_{j-1}(A_\nu)$ by applying the relation (II) associated with the polygon Π. We can therefore transform $R(A_i)$ by repeated application of relations (I) and (II) and the trivial relations to the successive terms of the sequence $w_0(A_\nu), w_1(A_\nu), \ldots, w_m(A_\nu)$. The last of these is the empty product.

As an example we choose the 1-dimensional simplicial complex which consists of the sides of a traingle (Fig. 92), that is, a simplicial decomposition of the circle or 1-sphere. The initial point O is a vertex of the traingle. The auxiliary paths are $h_1 = a_1$ to the vertex P_1 and $h_2 = a_3^{-1}$ to the vertex P_2; the path h_0 consists only of the point O.

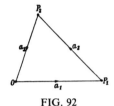

FIG. 92

The edge paths are

$$A_1 = h_0 a_1 h_1^{-1} = a_1 a_1^{-1} = \varphi_1(a_\nu),$$

$$A_2 = h_1 a_2 h_2^{-1} = a_1 a_2 a_3 = \varphi_2(a_\nu),$$

$$A_3 = h_2 a_3 h_0^{-1} = a_3^{-1} a_3 = \varphi_3(a_\nu).$$

The relations are

$$\left.\begin{array}{l} A_1 = A_1 A_1^{-1} = 1 \\ A_2 = A_1 A_2 A_3 \\ A_3 = A_3^{-1} A_3 = 1 \end{array}\right\} \quad \text{(I); no relations (II)}.$$

There then remains one generator, A_2, and all relations become trivial. *The fundamental group of the circle is therefore the free cyclic group.* The path classes are obviously represented by closed paths which traverse the circle $0, 1, 2, \ldots$ times in one or the other sense.

Likewise, the fundamental group of the annulus, of the solid torus, and, more generally, of the topological product of the circle and the closed n-ball is the free cyclic group. For it is the direct product of the fundamental group of the circle and the fundamental group of the closed n-ball (§43, Example 1) but the latter consists of the unit element (§43, Example 2).

If we span a surface element over the triangle (Fig. 92), then we obtain an additional relation

$$A_1 A_2 A_3 = 1$$

so that now $A_2 = 1$, in agreement with the fact that the fundamental group of the closed n-ball consists of the unit element.

Problem

If r is a rotationally symmetric torus lying in $\mathfrak{R}^3$, show that it is impossible to map $\mathfrak{R}^3$ topologically onto itself so that r is transformed to itself and meridian and longitude circles are interchanged. (Consider the meridians and longitudes as elements of the fundamental group of the solid torus bounded by r.)

47. Edge Complexes and Closed Surfaces

As a further application of the methods developed in the previous section we will compute the fundamental groups of edge complexes and closed surfaces.

Let $a_1, a_2, \ldots, a_{\alpha^1}$ be the edges of an edge complex $\mathfrak{R}^1$. We first consider the simple case where $\mathfrak{R}^1$ has only a single vertex, O. In that case all a_i are closed paths, the auxiliary paths disappear, and the closed paths A_i, which represent the generating path classes of the fundamental group, become equal to the a_i. The relations of type (I) become $A_i = A_i$ and are therefore trivial.

There are no relations of type (II). Thus the fundamental group is the free group having

$$\alpha^1 = N + 1 = -\alpha^0 + \alpha^1 + 1$$

generators; here $\alpha^0 = 1$ and α^1 are, respectively, the number of vertices and edges of the edge complex and $N = -\alpha^0 + \alpha^1$ denotes the Euler characteristic. The result is obvious immediately. For each path is a product of closed edge paths $a_i = A_i$, so that the path classes A_i are generators of the fundamental group, and a product $\prod_i A_i^{\varepsilon_i}$ can be the unit element of the fundamental group only if the path $\prod_i A_i^{\varepsilon_i}$ can be deformed to O by applying the operation (α), that is group theoretically, if $\prod_i A_i^{\varepsilon_i}$ can be reduced to 1 by applying the trivial relations $A_i A_i^{-1} 1$ and $A_i^{-1} A_i = 1$.

For an edge complex $\Re^1$ having more than one vertex, we can reduce the computation of the fundamental group to this simple case. There exists an edge, let us say a_1, for which the initial point P and the endpoint Q are different. If h is the auxiliary path to P, we choose ha_1 as the auxiliary path to Q. The closed path A_1 which belongs to a_1 is then the path $h \cdot a_1 \cdot (ha_1)^{-1}$, that is, the relation $A_1 = 1$ belonging to a_1.

If we now have

$$A_i = \varphi_i(a_\nu) \qquad (i = 2, 3, \ldots, \alpha^1; \nu = 1, 2, \ldots, \alpha^1),$$

the associated realtion is

$$A_i = \varphi_i(A\nu).$$

Using the relation $A_1 = 1$ we get from this the system of defining relations

$$A_i = \psi_i(A_\mu) \qquad (i, \mu = 2, 3, \ldots, \alpha^1). \tag{I}$$

Here $\psi_i(A_\mu)$ is that product path which arises from $\varphi_i(a_\nu)$ by omitting all factors a_i. But the above relations are exactly the relations for the complex $\Re_1^1$ which one gets from $\Re^1$ when one omits the edge a_1 and identifies the vertices P and Q, that is, one contracts the edge a_1 to a point. For upon cancellation of a_1 the system of auxiliary paths of $\Re^1$ transforms to a system of auxiliary paths in $\Re_1^1$ and the closed paths A_i' belonging to the edges of $\Re_1^1$ are just the same paths $A_i' = \psi_i(a_\mu)$, so that we get

$$A_i' = \psi_i(A_\mu')$$

as relations of the fundamental group of $\Re_1^1$, that is, the relations (I). Thus $\Re^1$ and $\Re_1^1$ have the same fundamental group.

If there is more than one vertex in $\Re_1^1$, we apply the above procedure once again and do so repeatedly until we arrive at a complex $\Re_s^1$ having a single vertex. The Euler characteristic $N = -\alpha^0 + \alpha^1$ is obviously the same for all of the complexes $\Re^1, \Re_1^1, \ldots, \Re_s^1$. But, as we have seen, the fundamental group of $\Re_s^1$ is the free group having $N + 1$ generators. We have achieved the following result: *The fundamental group of an edge complex $\Re^1$ is the free group* *having $N + 1$ generators, where N is the Euler characteristic, $-\alpha^0 + \alpha^1$, of the edge complex.*

After equivalent sides have been identified, the fundamental polygons of closed surfaces (§38) will belong to the surface complexes. For example, the fundamental polygon of the torus consists of one vertex O and two edges a and b (meridian and longitude circles), and one surface element that is spanned over the edge path $aba^{-1}b^{-1}$. The generators of the fundamental group correspond to the edges a and b. Let them be denoted by A and B. No auxiliary paths appear, since only one vertex is present. The relations of type (I) become identities and there is only one relation of type (II), which corresponds to the one surface element, and which states that the boundary path of the surface element is null homotopic in the surface complex. It is

$$ABA^{-1}B^{-1} = 1.$$

This relation shows that the generators A and B commute. *The fundamental group of the torus is, then, the free Abelian group having two generators.*

Exactly the same considerations can be applied to other fundamental polygons, and lead to the result:

THEOREM *The fundamental group of the orientable closed surface of genus h (sphere with h handles attached, $h = 0, 1, 2 \ldots$) can be generated by $N + 2 = 2h$ elements which are connected by the one relation*

$$A_1 B_1 A_1^{-1} B_1^{-1} \cdots A_h B_h A_h^{-1} B_h^{-1} = 1.$$

The fundamental group of the nonorientable surface of genus k (sphere with k cross-caps attached $k + 1, 2 \ldots$) has $N + 2$ generators and the one relation

$$A_1^2 A_2^2 \cdots A_k^2 = 1.$$

Except for the fundamental group of the 2-sphere ($h = 0$), which consists of the unit element alone, and that of the projective plane ($k = 1$), which has order 2, all of the fundamental groups of closed surfaces are of infinite order. For they have a Betti number of value $N + 2 = 2h > 0$ in the orientable case and a Betti number of value $N + 1 = k - 1 > 0$ in the nonorientable case (§86).

Problems

1. Determine the fundamental groups of the surfaces with boundary.
2. Using Problem 1 of §38, show that the group having the one relation

$$A_1 B_1 A_1^{-1} B_1^{-1} \cdots A_h B_h A_h^{-1} B_h^{-1} = 1$$

is isomorphic to the group having the one relation

$$C_1 C_2 \cdots C_{2h} C_1^{-1} C_2^{-1} \cdots C_{2h}^{-1} = 1.$$

48. The Fundamental and Homology Groups

The preimage $\overline{w}$ of a path w in a complex $\Re^n$ is a geometric 1-simplex which is mapped continuously onto w. The path w then represents an oriented singular simplex, and consequently, a singular 1-chain. If we choose a different preimage, $\overline{w}'$ of w instead of $\overline{w}$, then we obtain in the same way a singular simplex w' which is in general different than w, even though w and w' coincide as paths. This is due to the fact that we have defined equality of paths so as to allow their preimages to map to one another topologically, while our definition of equality of singular simplexes requires that the preimages map to one another linearly. It is possible to show, however, that the singular simplexes w and w', which may on occasion be different, describe homologous chains. For there exists a topological mapping of $\overline{w}$ onto $\overline{w}'$ with preservation of orientation such that corresponding points have the same image point in the complex $\Re^n$. The preimage $\overline{w}$ thereby transforms to an oriented singular simplex $\overline{w}_1$ on $\overline{w}'$ which has the same boundary as $\overline{w}'$. Thus $\overline{w}' - \overline{w}_1 \sim 0$, as is the case for each closed 1-chain on an interval. This homology is preserved, however, under a mapping of the interval $\overline{w}'$ onto w (§27), and since $\overline{w}_1$ transforms to w and $\overline{w}'$ transforms to w', it follows that $w \sim w'$. Accordingly, to each path w there will correspond a 1-chain with the same designation and which will in fact depend upon the choice of preimage $\overline{w}$ of w but whose homology class will be uniquely determined. We can therefore speak of paths which are homologous to one another. If a path is closed, so also is the singular chain belonging to it.

THEOREM I. *To the product* $w = w_1 w_2$ *of two paths there corresponds the sum of the associated chains, that is, the chain* $w \sim w_1 + w_2$.

Proof. The preimage of $w_1 w_2$ is an interval $\overline{w}$ which is composed of two subintervals, $\overline{w}_1$ and $\overline{w}_2$. On that interval $w_1 + w_2 \sim \overline{w}$ and this homology is preserved in their images in $\Re^n$.

THEOREM II. *If two paths* w_0 *and* w_1 *are homotopic, then they are also homologous. (The converse is not necessarily true.)*

Proof. If $\overline{w}_0$, $\overline{w}_1$, $\overline{u}$, $\overline{v}$ are the sides of the deformation rectangle belonging to the deformation of w_0 to w_1 (Fig. 93), then on this rectangle there holds the homology

$$\overline{w}_1 - \overline{w}_0 + \overline{u} - \overline{v} \sim 0.$$

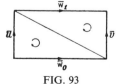

FIG. 93

One then has for the images

$$w_1 - w_0 + u - v \sim 0.$$

However, since $\bar{u}$ and $\bar{v}$ map to points, u and v are degenerate singular simplexes (§25), so that $w_1 - w_0 \sim 0$.

As an example to show that homologous paths do not have to be homotopic, we have the waist section path l on the double torus (later in this section), which is null homologous but is not null homotopic.

On a connected complex, $\mathfrak{R}^n$, we can uniquely associate a homology class to each path class and since the sum of the corresponding homology classes is associated to the product of two path classes, we have a homorphic mapping χ of the fundamental group $\mathfrak{F}$ into the homology group $\mathfrak{H}^1$. To study this homorphism more closely it will be sufficient to look at edge paths and simplicial chains on a simplicial decomposition of $\mathfrak{R}^n$. We choose a vertex O as the initial point of the closed paths. Since we can convert an arbitrary closed simplicial 1-chain to a closed edge path departing from O, by means of adding back and forth running edge paths, it follows that each homology class is the image of at least one path class. That is, χ is a homorphism of $\mathfrak{F}$ onto $\mathfrak{H}^1$. It is therefore characterized by means of the normal subgroup $\mathfrak{N}$ of $\mathfrak{F}$ which corresponds to the null element of $\mathfrak{H}^1$. We claim that $\mathfrak{N}$ is the commutator group of $\mathfrak{F}$.

Proof. Each commutator $F_1 F_2 F_1^{-1} F_2^{-1}$ of $\mathfrak{F}$ will transform to the null element of $\mathfrak{H}^1$ because $\mathfrak{H}^1$ is Abelian. We need only to show the converse, that each element of $\mathfrak{N}$ belongs to the commutator group; that is, the path class of an arbitrary null homologous path w belongs to the commutator group. If

$$U^2 = \sum_{i=1}^{\alpha} \lambda_i E_i^2$$

is a 2-chain with the boundary w, then for each triangle E_i^2 one chooses an edge path r_i which departs from O along a path u_i to a vertex of E_i^2, runs around E_i^2, and returns to O along u_i^{-1}. The traversal around E_i^2 is to be made in the sense required for the chain r_i to be the boundary of E_i^2. Since each path r_i is null homotopic, the edge path w is deformable to the edge path

$$w' = w\left(r_1^{\lambda_1} \cdots r_\alpha^{\lambda_\alpha}\right)^{-1}.$$

Also, since

$$r_1^{\lambda_1} \cdots r_\alpha^{\lambda_\alpha}$$

is, like w, the boundary of U^2, both paths are equal when considered as chains. Thus w is the chain 0 and it traverses each 1-simplex equally often in both directions. We can now transform

$$w' = f(a_j)$$

by means of combinatorial deformations to the edge path

$$W' = f(A_j),$$

where A_j denotes the closed edge path belonging to a_j(§46). In W' each path A_j appears equally often with the exponents $+1$ and -1, so that upon making the fundamental group Abelian W' transforms to the null element. That is, the path class of W', which is the path class of w' and w, belongs to the commutator group.

This proves

THEOREM III. *The first homology group $\mathfrak{H}^1$ of a connected complex $\mathfrak{R}^n$ is the Abelianized fundamental group (factor group of the fundamental group $\mathfrak{F}$ relative to its commutator group).*

We can easily verify this theorem in the special case of closed surfaces by comparison of the results of §41 and §47.

We have characterized a null homotopic path as a path which may be spanned by a singular disk. We now give a similar characterization for null homologous closed paths.

THEOREM IV. *A closed path w is null homologous if and only if it is the boundary of an oriented surface which lies in the complex, possibly with singularities.*

Proof. When considered as an element of the fundamental group, w belongs to the commutator subgroup, from Theorem III, and is therefore deformable to a product of commutators

$$u_1 v_1 u_1^{-1} v_1^{-1} \cdots u_h v_h u_h^{-1} v_h^{-1}.$$

The closed path

$$w^{-1} u_1 v_1 u_1^{-1} v_1^{-1} \cdots u_h v_h u_h^{-1} v_h^{-1}$$

is therefore null homotopic and can be spanned by a disk which has as its preimage a $(4h + 1)$-gon having boundary

$$\overline{w}^{-1} \overline{u}_1 \overline{v}_1 \overline{u}_1^{-1} \overline{v}_1^{-1} \cdots \overline{u}_h \overline{v}_h \overline{u}_h^{-1} \overline{v}_h^{-1}.$$

Upon being mapped into $\mathfrak{R}^n$, this polygon closes to a surface with boundary of genus h which lies in the complex (possibly with singularities) and whose boundary is the path w. (The case $h = 1$ is illustrated in Figs. 6–8.) Conversely, if a path w can be spanned by such a surface, that is, one can map a surface with boundary continuously into the complex $\mathfrak{R}^n$ so that the boundary of the single hole transforms to w, then w is null homologous. For it follows from the orientability of the spanned surface that there will exist a singular 2-chain on it whose boundary is precisely the path w, regarded as a 1-chain.

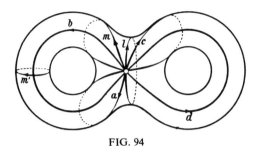

FIG. 94

An example of a path which may be spanned by a surface of genus 1 but not by a disk is the waist section l of the double torus. This path separates the two rings of the double torus (Fig. 94). It is the boundary of a torus with boundary, as can be seen by cutting the double torus along l. This confirms the fact that the element of the fundamental group which is represented by l, namely $L = (ABA^{-1}B^{-1})^{-1}$, is a commutator. On the other hand, l cannot be spanned by a disk because this element is not the unit element of the fundamental group.

We can see this as follows. The fundamental group $\mathfrak{F}$ has a single defining relation

$$ABA^{-1}B^{-1}CDC^{-1}D^{-1} = 1. \tag{1}$$

Upon introducing additional relations

$$A = D, \qquad B = C$$

we get a factor group $\overline{\mathfrak{F}}$ of $\mathfrak{F}$ (§83). Elimination of C and D shows that $\overline{\mathfrak{F}}$ is the free group having two generators, A and B. The relation $ABA^{-1}B^{-1} = 1$ therefore does not hold in $\overline{\mathfrak{F}}$ and thus can not hold in $\mathfrak{F}$.

The waist section is thus null homologous but not null homotopic on the double torus.

49. Free Deformation of Closed Paths

Until now *we have always kept the initial point fixed* when we deformed closed paths. On occasion we must also deal with deformations under which the initial point moves. We will call these *free deformations*, in contrast to the *constrained deformations* previously considered. When we refer simply to "a deformation" without supplying a qualifying adjective, we shall always mean a constrained deformation.

The exact definition of a free deformation is:

Two closed paths w_0 and w_1 on a complex $\mathfrak{K}^n$ are said to be freely deformable to one another or *freely homotopic* if it is possible to map a deformation rectangle $\overline{w} \times \tau$ continuously into $\mathfrak{K}^n$ so that the side $\overline{w} \times 0 = \overline{w}_0$ transforms to w_0 and the side $\overline{w} \times 1 = \overline{w}_1$ transforms to w_1, while each pair of corresponding points $\overline{P} \times t$ and $\overline{Q} \times t$ of the other two sides map to the same point of $\mathfrak{K}^n$.

The oriented sides $\overline{P} \times \tau$ and $\overline{Q} \times \tau$ then transform to the same path v which connects the initial point P_0 of w_0 with the initial point P_1 of w_1. If the side $\overline{w} \times 0$ is displaced parallel to itself across the rectangle to the opposite side, then the image path will experience the free deformation and the initial point P_0 of w_0 will traverse the path v.

The relation between the free and constrained deformations is provided by the following lemma.

LEMMA . *If, during a free deformation of a closed path w_0 to a path w_1, the initial point P_0 of w_0 describes a path v, then w_0 can be deformed to $v w_1 v^{-1}$ by a constrained deformation. Conversely, if a path w_0 can be deformed to a path $v w_1 v^{-1}$ by means of a constrained deformation, then w_0 and w_1 are freely homotopic.*

Proof. We consider two rectangles, $\mathfrak{R}$ and $\mathfrak{T}$ whose boundaries are divided and designated as in Fig. 95. We may consider $\mathfrak{T}$ to arise from $\mathfrak{R}$ by letting the two sides $\bar{u}'$ and $\bar{u}''$ of $\mathfrak{R}$ shrink to points, that is, by identifying all of the points of each individual side. This can be done by means of a continuous mapping φ of $\mathfrak{R}$ onto $\mathfrak{T}$ (indicated in the figure) which transforms the paths $\bar{w}_0$, $\bar{w}_1$, $\bar{v}'$, and $\bar{v}''$ to the paths of $\mathfrak{T}$ having like designation. If w_0 is freely deformable to w_1 and the initial point of w_0 describes the path v, then there exists a continuous mapping ψ of $\mathfrak{T}$ into $\mathfrak{R}^n$ which transforms $\bar{w}_0$ to w_0 and $\bar{w}_1$ to w_1 and transforms each of the other two sides $\bar{v}'$ and $\bar{v}''$ to v. The mapping $\psi\varphi = \chi$ will then transform the sides $\bar{u}'$ and $\bar{u}''$ of $\mathfrak{R}$ to points and will transform the other two sides to the paths w_0 and $v w_1 v^{-1}$, which is equivalent to the existence of a constrained deformation between these two paths.

Conversely, when there is a constrained deformation from w_0 to $v w_1 v^{-1}$, the mapping χ of $\mathfrak{R}$ into $\mathfrak{R}^n$ is specified and a continuous mapping ψ of $\mathfrak{T}$ is defined by means of the equation $\chi = \psi\varphi$, from §8, Theorem IV. The mapping ψ transforms $\bar{w}_0$ to w_0 and $\bar{w}_1$ to w_1, but transforms the other two sides to v.

When there is a constrained homotopy between two paths with common initial point, they are also freely homotopic. We can see that the converse is not generally true from the paths m and a of Fig. 94. Path m is freely deformable to a via the position m'. But m and a are not homotopic. For m is homotopic to bab^{-1} and the path classes of m and a, namely, BAB^{-1} and A, are different because

$$A(BAB^{-1})^{-1} = ABA^{-1}B^{-1} \neq 1,$$

which was proved at the end of §48.

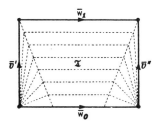 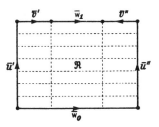

FIG. 95

THEOREM. *Two closed paths, w_0 and w_1 which start from a common initial point O are freely homotopic if and only if they represent conjugate elements W_0 and W_1 of the fundamental group, that is, if there exists an element V such that $W_0 = VW_1V^{-1}$.*

All paths which belong to a single conjugacy class of the fundamental group will then be freely homotopic to one another. [28]

The proof is an immediate consequence of the lemma.

We review our three possible ways of dividing into classes, the closed paths emanating from O:

1. The narrowest division into classes is division into path classes. Two paths belong to the same path class if they are deformable to one another, keeping O fixed. The path classes form the elements of the fundamental group.

2. All paths in a single conjugacy class in the fundamental group coincide with the set of all paths freely deformable to one another. In general these classes do not form a group.

3. All paths of residue class of the fundamental group relative to its commutator group form a class of homologous paths. These homology classes form the homology group $\mathfrak{H}^1$.[29]

50. Fundamental Group and Deformation of Mappings

A continuous mapping φ of a connected complex $\mathfrak{K}^m$ into a connected complex $\mathbf{K}^m$ induces a homomorphism $\mathbf{\Phi}$ of the fundamental group $\mathfrak{F}$ of $\mathfrak{K}^n$ into the fundamental group $\mathbf{F}$ of $\mathbf{K}^m$. The homomorphism $\mathbf{\Phi}$ is only determined up to an inner automorphism of $\mathbf{F}$ (§42). We shall now prove that *this homomorphism does not change under deformation of the mapping φ.*

If the initial point O of $\mathfrak{F}$ transforms under φ to the initial point Ω of $\mathbf{F}$ and if Ω describes a path v with end point Ω_1 under the deformation of φ to a mapping φ_1 and, finally, if w' and w_1' are the images of a closed path w departing from O under the mappings φ and φ_1, then w' will be freely deformed to w_1' during the deformation of φ to φ_1. From the lemma of §49 the paths w' and $vw_1'v^{-1}$ are then homotopic. Path w is then mapped by φ and φ_1 into the same path class of $\mathbf{F}$, up to an inner automorphism.

In particular, under a deformation of a complex $\mathfrak{K}^n$ into itself, the fundamental group of the complex will be mapped to itself by an inner automorphism.

51. The Fundamental Group at a Point

Earlier, in Theorem III of §32, we saw that the homology groups of the neighborhood complex of a point P are the same for all simplicial decompositions of a complex $\mathfrak{K}^n$. This theorem is also valid for the fundamental group.

THEOREM. *If $\mathfrak{A}_1$ and $\mathfrak{A}_2$ are the outer boundaries of two simplicial stars with center point P on a complex $\mathfrak{K}^n$ then $\mathfrak{A}_1$ and $\mathfrak{A}_2$ have the same fundamental group, provided that the simplicial stars are neighborhoods of P in $\mathfrak{K}^n$ and that $\mathfrak{A}_1$ (and consequently also $\mathfrak{A}_2$) is a connected complex.*

This fundamental group is called the fundamental group of the point P.

Proof. From Theorem I of §32 there exists a continuous mapping φ of $\mathfrak{A}_1$ into $\mathfrak{A}_2$ and a continuous mapping ψ of $\mathfrak{A}_2$ into $\mathfrak{A}_1$ such that the self-mappings $\psi\varphi$ and $\varphi\psi$ of $\mathfrak{A}_1$ and $\mathfrak{A}_2$, respectively, are deformable to the identity. The homomorphisms ϕ and ψ induced by φ and ψ on the fundamental group $\mathfrak{F}_1$ of $\mathfrak{A}_1$ and the fundamental group $\mathfrak{F}_2$ of $\mathfrak{A}_2$ then have the property that the homomorphisms induced by $\psi\phi$ of $\mathfrak{F}_1$ and $\phi\psi$ of $\mathfrak{F}_2$ are inner automorphisms, from §50. But this implies that ϕ is a one-to-one mapping of $\mathfrak{F}_1$ onto $\mathfrak{F}_2$, that is, an isomorphism. For if $\phi(F_1') = \phi(F_1'')$ were to hold for distinct elements F_1' and F_1'', then we would also have $\psi\phi(F_1') = \psi\phi(F_1'')$, which is impossible if $\psi\phi$ is an inner automorphism of $\mathfrak{F}_1$. Furthermore, if F_2 is that element of $\mathfrak{F}_2$ which transforms to a given element F_2' under the automorphism $\phi\psi$ of $\mathfrak{F}_2$, then the element $\psi(F_2)$ of $\mathfrak{F}_1$ transforms to F_2' under the homomorphism ϕ. That is, each element of $\mathfrak{F}_2$ is the image of an element of $\mathfrak{F}_1$. This proves that ϕ is one-to-one.

52. The Fundamental Group of a Composite Complex

Frequently, the determination of the fundamental group of a complex $\mathfrak{K}$ can be simplified by decomposing $\mathfrak{K}$ into two subcomplexes having known fundamental groups. Let $\mathfrak{K}'$ and $\mathfrak{K}''$ be two connected subcomplexes of a connected n-dimensional simplicial complex $\mathfrak{K}$. We require that each simplex of $\mathfrak{K}$ belongs to at least one of the two subcomplexes. The intersection $\mathfrak{D}$ of $\mathfrak{K}'$ and $\mathfrak{K}''$ is not empty, because $\mathfrak{K}$ is assumed to be connected. We require that $\mathfrak{D}$ also be connected.

Let $\mathfrak{F}$, $\mathfrak{F}'$, $\mathfrak{F}''$, and $\mathfrak{F}_{\mathfrak{D}}$ be the fundamental groups of $\mathfrak{K}$, $\mathfrak{K}'$, $\mathfrak{K}''$, and $\mathfrak{D}$. Choose a point O of $\mathfrak{D}$ as the initial point of the closed paths. Each closed path of $\mathfrak{D}$ will then also be a closed path of $\mathfrak{K}'$ and of $\mathfrak{K}''$. Consequently, to each element of $\mathfrak{F}_{\mathfrak{D}}$ there will correspond an element of $\mathfrak{F}'$ and an element of $\mathfrak{F}''$. We then have:

THEOREM I. *$\mathfrak{F}$ is a factor group of the free product $\mathfrak{F}' \circ \mathfrak{F}''$. We obtain $\mathfrak{F}$ from the free product, when one identifies any pair of elements of $\mathfrak{F}_1'$ and $\mathfrak{F}_1''$ which correspond to the same element of $\mathfrak{F}_{\mathfrak{D}}$. That is, we introduce relations between the generators of $\mathfrak{F}'$ and the generators of $\mathfrak{F}''$ by setting those elements equal to one another.*

Proof. According to the general prescription for the computation of the fundamental group, each vertex of $\mathfrak{K}$ is to be connected to O by an auxiliary path. If the vertex belongs to $\mathfrak{D}$, then we may assume that the auxiliary path lies entirely in $\mathfrak{D}$, since we have required that $\mathfrak{D}$ be connected. Likewise, the auxiliary path is to lie entirely in $\mathfrak{K}'$ when the vertex lies in $\mathfrak{K}^1$ and is to lie entirely in $\mathfrak{K}''$ when the vertex lies in $\mathfrak{K}''$.

A simplex of arbitrary dimension of $\mathfrak{K}$ will belong either to $\mathfrak{K}'$ and not to $\mathfrak{K}''$, or to $\mathfrak{K}''$ and not to $\mathfrak{K}'$, or to both $\mathfrak{K}'$ and $\mathfrak{K}''$, that is, to $\mathfrak{D}$. This determimes a partition of the simplexes of $\mathfrak{K}$ into three disjoint subsets: $\overline{\mathfrak{K}}'$, $\overline{\mathfrak{K}}''$, and $\mathfrak{D}$.

The generators A_ν of $\mathfrak{F}$ may be placed in one-to-one correspondence with the edges a_ν of $\mathfrak{K}$ (§46). According to whether a_ν belongs to $\overline{\mathfrak{K}}'$, $\overline{\mathfrak{K}}''$, or $\mathfrak{D}$ we shall rename the corresponding generator A_ν of $\mathfrak{F}$, respectively,

$$\overline{K}_j', \qquad \overline{K}_j'', \qquad \text{or} \qquad D_j.$$

The relations of type (I) of $\mathfrak{F}$ are in one-to-one correspondence with the edges of $\mathfrak{K}$; the relations of type (II) of $\mathfrak{F}$ are in one-to-one correspondence with the triangles of $\mathfrak{K}$. We shall partition the relations into three classes, according to whether the edge (or triangle) belonging to the relation lies in $\overline{\mathfrak{K}}'$, $\overline{\mathfrak{K}}''$, or $\mathfrak{D}$. The relations for these three classes are then

$$R_i'\left(D_j, \overline{K}_j'\right) = 1 \qquad (i = 1, 2, \ldots, \kappa'), \qquad\qquad (\overline{\mathfrak{K}}')$$

$$R_i''\left(D_j, \overline{K}_j''\right) = 1 \qquad (i = 1, 2, \ldots, \kappa''), \qquad\qquad (\overline{\mathfrak{K}}'')$$

$$R_i^{(\mathfrak{D})}(D_j) = 1 \qquad (i = 1, 2, \ldots, \delta). \qquad\qquad (\mathfrak{D})$$

Because of the particular choice of auxiliary paths, the $\overline{\mathfrak{K}}_j''$ do not appear in the R_i', the $\overline{K}_j'$ do not appear in the R_i'' and neither appear in the $R_i^{(\mathfrak{D})}$.

The relations $(\mathfrak{D})$ are obviously the defining relations of the fundamental group $\mathfrak{F}_{\mathfrak{D}}$ of $\mathfrak{D}$. The relations $(\mathfrak{D}) + (\overline{\mathfrak{K}}')$ and $(\mathfrak{D}) + (\overline{\mathfrak{K}}'')$, respectively, define the fundamental groups $\mathfrak{F}'$ of $\mathfrak{K}'$ and $\mathfrak{F}''$ of $\mathfrak{K}''$. Finally, since $(\mathfrak{D}) + (\overline{\mathfrak{K}}') + (\overline{\mathfrak{K}}'')$ are the defining relations of $\mathfrak{K}$, they can obviously be replaced by the following system of relations:

$$\left.\begin{array}{l} R_i'\left(D_j, \overline{K}_j'\right) = 1 \\[2mm] R_i^{(\mathfrak{D})}(D_j') = 1 \end{array}\right\} \qquad\qquad (\mathfrak{F}')$$

$$\left.\begin{array}{l} R_i''\left(D_j'', \overline{K}_j''\right) = 1 \\[2mm] R_i^{(\mathfrak{D})}(D_j'') = 1 \end{array}\right\} \qquad\qquad (\mathfrak{F}'')$$

$$D_j' = D_j'' \qquad \text{for all } j. \qquad\qquad (\mathfrak{d})$$

The relations $(\mathfrak{F}')$ together with the relations $(\mathfrak{F}'')$ are the relations of the free product of $\mathfrak{F}'$ and $\mathfrak{F}''$ (§85). The relations $(\mathfrak{d})$ indicate that two elements D_j' of $\mathfrak{F}'$ and D_j'' or $\mathfrak{F}''$ which correspond to the same element, D_j of $\mathfrak{D}$ are to be identified.

The theorem becomes particularly simple to state in the case that the fundamental group of $\mathfrak{D}$ reduces to the unit element. In that case, *the fundamental group of $\mathfrak{R}$ is the free product of the fundamental groups of $\mathfrak{R}'$ and $\mathfrak{R}''$.*

If we Abelianize all of the fundamental groups in Theorem I, then we get, taking account of Theorem III of §48:

THEOREM II. *The homology group $\mathfrak{H}^1$ of $\mathfrak{R}$ is a factor group of the direct product of the homology groups $'\mathfrak{H}^1$ of $\mathfrak{R}'$ and $''\mathfrak{H}^1$ of $\mathfrak{R}''$. It is obtained by setting each element of $'\mathfrak{H}^1$ equal to that element of $''\mathfrak{H}^1$ which corresponds to the same element of the homology group $\mathfrak{H}^1_{\mathfrak{D}}$ of $\mathfrak{D}$.*

We shall not derive the corresponding theorem for the homology groups of higher dimension. Its statement is not as simple.[30]

As an application of Theorem I, we shall determine the fundamental group of a *torus knot*. To define the torus knot we draw m axis parallel lines on the surface of a cylinder of finite length located in Euclidean 3-space $\mathfrak{R}^3$ the lines being uniformly spaced at an angular interval of $2\pi/m$ radians. We shall bend the cylinder into a torus. However, before bending the cylinder, we twist the floor and roof faces by an angle of $2\pi n/m$ radians. This is shown in Fig. 96 for the case $m = 3$ and $n = 5$. We then bend the cylinder to form the torus so that points which lay over one another before the bending are identified. In the case when m and n are relatively prime integers, the m axis parallel lines will close on the torus to a single closed curve, which we shall call the torus knot belonging to the integers m, n.* We now bore the knot out of $\mathfrak{R}^3$. That is, we let a small ball, having its center point on the knot, glide over the knot and we remove from $\mathfrak{R}^3$ those points which were swept over by the interior points of the ball. We then complete $\mathfrak{R}^3$ to form the 3-sphere $\mathfrak{S}^3$ by adding a single improper point (§14). The part of $\mathfrak{S}^3$ remaining after the knot has been bored out is a 3-dimensional complex $\mathfrak{R}$ which has a thin tube (topologically a torus) lying around the knot as its boundary. This can easily be rigorously formulated, as can the process described intuitively above. The complex $\mathfrak{R}$ is called the *exterior space of the knot*. The fundamental group of $\mathfrak{R}$ is called the *knot group*. To determine the knot group we decompose $\mathfrak{S}^3$

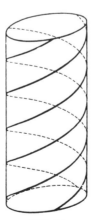

FIG. 96

* Next to the circle, the torus knot $2, 3$ is the simplest of all knots, the trefoil knot, which is shown in Fig. 2.

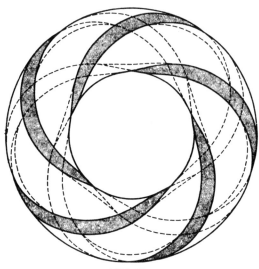

FIG. 97

into two subsets whose common boundary is the torus on which the knot lies. The tube will then be split along its length to form two half-tubes. The complex $\Re$ will be decomposed to form two solid tori, from Problem 4 of §14, such that a winding groove of semicircular profile has been milled out of each torus. We shall designate the "finite" solid torus as the subcomplex $\Re'$ and we shall designate the other solid torus, which contains the point at infinity, as the subcomplex $\Re''$. The intersection $\mathfrak{D}$ of the two subcomplexes is a twisted annulus which covers the ungrooved portion of the torus on which the knot lies (Fig. 97). From §46, the fundamental group $\mathfrak{F}'$ of $\Re'$ is the free group on one generator A (renamed $\overline{K}_1'$ in the proof of Theorem I). The fundamental group of $\Re''$ is the free group on one generator B. We represent A by the core of the solid torus $\Re'$ and we bend this curve so that it passes through a point O of the median line of the annulus $\mathfrak{D}$. We choose this point O as the initial point. We represent B by the core of the other solid torus $\Re''$; for example, it can be the axis of rotation of the torus on which the knot lies. We likewise deform B so that it passes through the point O. We take the median line D of the annulus as the generator of the fundamental group of $\mathfrak{D}$, which is also the free group on one generator (§46). The group $\mathfrak{F}' \circ \mathfrak{F}''$ is the free group on two generators A and B. Considered as an element of $\mathfrak{F}'$, D is equal to A^m, but considered as an element of $\mathfrak{F}''$, D is equal to B^n, when the paths A and B are appropriately oriented. *We then obtain the knot group of the torus knot m, n by introducing the relation*

$$A^m = B^n$$

between the two generators.

It follows from this that a torus knot m, n such that $m > 1$ and $n > 1$ cannot be transformed to a circle by means of an isotopic deformation of Euclidean 3-space, because its knot group is not infinite cyclic.[31]

Problems

1. Assume that the space $\Re^4$ has been completed by adding a point at infinity to form the 4-sphere $\mathfrak{S}^4$. A closed sequence of edges consisting of straight line segments, free of double points, has been bored from $\Re^4$. Show that the fundamental group of the exterior space consists of the unit element. (Apply Theorem I to $\mathfrak{S}^4$, which is decomposed into two subcomplexes, the exterior space $\mathfrak{A}$ and the bore $\mathfrak{B}$. The bore is the topological product of the circle and the 3-ball.

The intersection $\mathfrak{D}$ of $\mathfrak{A}$ and $\mathfrak{B}$ is the topological product of the circle and the 2-sphere. The fundamental group of $\mathfrak{A}$ is unchanged when $\mathfrak{A}$ is completed with $\mathfrak{B}$. Why does the argument fail in three dimensions?)

2. Find a 4-dimensional homogeneous complex whose fundamental group is the free group on r generators. (Bore out a small 4-ball from each of r topological products of the circle and 3-sphere, and by using a 4-sphere with r holes construct a complex having the desired properties.)

3. Find a 4-dimensional homogeneous complex having an arbitrarily prescribed fundamental group with finitely many generators. (Bore out a path w which is free of double points, from a homogeneous complex $\mathfrak{K}^4$. Complete the resulting complex with boundary $\bar{\mathfrak{K}}^4$ with a topological product of the 2-sphere and the disk. This gives a homogeneous complex $'\mathfrak{K}^4$ whose fundamental group is obtained from that of $\mathfrak{K}^4$ by adding the relation $w = 1$.)

CHAPTER EIGHT

COVERING COMPLEXES

The fundamental group is closely related to the coverings of a complex. We illustrate this by an example. When we let a cylindrical ring of radius 1 roll on a plane, it will roll over an infinite strip on the plane (Fig. 98). This strip can be wound infinitely often around the ring, so that it covers the ring infinitely many times. The initial point O of the closed paths on the ring will repeat periodically on the strip at distances of 2π. A path on the strip which leads from one of these periodically repeated points to another such point can be projected through the strip to a path on the ring which wraps around the ring several times. Each element of the fundamental group of the ring can be represented by a path which loops the ring. On the other hand, the path connecting two of the periodically repeated points on the strip is determined up to a homotopy by the initial point and the endpoint of the path. That is, the initial point of the path can be transformed to the endpoint by making a displacement of the strip. This displacement determines the path, up to a homotopy. We can, therefore, relate the fundamental group of the ring to the group of covering movements of the strip, also called the group of "covering transformations," which displaces the strip by multiples of 2π into itself.

In like manner we can regard the fundamental group of each complex as being the group of covering transformations of a particular covering complex, the universal covering complex. The rest of the unbranched coverings of the complex, and we shall only deal with unbranched coverings, corresponds to the subgroups of the fundamental group. Thus a knowledge of the fundamental group will provide us with a complete picture of the possible covering complexes.

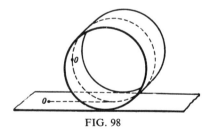

FIG. 98

53. Unbranched Covering Complexes

Let $\mathfrak{R}$ and $\tilde{\mathfrak{R}}$ be finite or infinite connected complexes. We say that $\tilde{\mathfrak{R}}$ *covers* $\mathfrak{R}$ or $\tilde{\mathfrak{R}}$ is a *covering* of $\mathfrak{R}$ if there exists a continuous mapping G of $\tilde{\mathfrak{R}}$ onto $\mathfrak{R}$ which satisfies the following conditions:

(U11)   At least one point $\tilde{P}$ of $\tilde{\Re}$ is mapped to each point, P of $\Re$. We say that $\tilde{P}$ *lies above P*, and that P is the *ground point* or *base point* of $\tilde{P}$.

(U12)   If $\tilde{P}_1, \tilde{P}_2, \ldots$ are all the points lying above P, then there exist "distinguished neighborhoods" $\mathfrak{U}(P), \mathfrak{U}(\tilde{P}_1), \mathfrak{U}(\tilde{P}_2), \ldots$ such that G maps $\mathfrak{U}(\tilde{P}_1), \mathfrak{U}(\tilde{P}_2), \ldots$ topologically onto $\mathfrak{U}(P)$. (This is the nonbranching condition.)

(U13)   Each point of $\tilde{\Re}$ which lies over a point of $\mathfrak{U}(P)$ belongs to at least one of the distinguished neighborhoods $\mathfrak{U}(\tilde{P}_1), \mathfrak{U}(\tilde{P}_2), \ldots$. (This is the nonbounding condition.)

The condition (U12) characterizes the mapping G as being locally topological and thereby excludes such objects as fold lines, or the branch points which occur when Riemann surfaces are regarded as covering spaces of the 2-sphere. We shall deal only with unbranched covering complexes.

The condition (U13) insures that each path f of the base complex lifts to the covering complex, as we shall show in §54. The following example shows that (U13) is not a consequence of (U11) and (U12). Let the base complex be an annulus. Let the covering complex be a half-open rectangular strip which includes the two length sides but not the two width sides, a and b of Fig. 99. We map the covering complex onto the annulus so that the open ends overlap. In this case conditions (U11) and (U12) are satisfied but no (U13). That is, given a point P lying under either line a or line b, there exist points lying above an arbitrary neighborhood of P which do not belong to any distinguished neighborhood. A path which winds twice about the inner circle of the annulus cannot be lifted to the covering complex in such a way that the covering path is projected to the base complex by the mapping G.

We should note that the base complex is obtained from the covering complex by identifying points which have the same image. Theorem II of §8 is applicable, because an arbitrary neighborhood of a point $\tilde{P}$ maps to a neighborhood of the base point P under the mapping G.

The torus is a simple example of a covering complex. We shall map a torus, $\tilde{\Re}$, shown cut open to a rectangle in Fig. 101, onto another torus $\Re$ so that each of the four squares of $\tilde{\Re}$ transforms congruently to the single square of $\Re$. We thereby produce a fourfold covering of the torus $\Re$ by $\tilde{\Re}$. We can regard the covering torus as arising from the base torus by taking four congruent copies of $\Re$ (sheets) lying above $\Re$, cutting each of them along a meridian circle (the vertical sides of Fig. 100) and joining them cyclically to one another (Fig.101). If we think of the torus as lying in ordinary 3-space as a ring surface generated by a rotation, the meridian circle will become a closed curve along which $\tilde{\Re}$ penetrates itself.* But when we regard $\tilde{\Re}$ as a two-dimensional

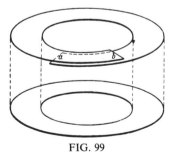

FIG. 99

* *Editor's note*: The intended meaning is evidently that the torus is to be regarded as a torus of revolution. The covering torus K is to be thought of as a cylinder which is embedded in 3-space so that the four points which project to a given point P lie directly above P; also the two boundary circles of the cylinder are to be identified.

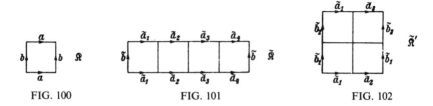

FIG. 100   FIG. 101   FIG. 102

manifold, existing independently of an embedding into 3-space, then the points of this curve are distinguished in no way from other points of $\tilde{\Re}$.

Two covering complexes are regarded as being equivalent if there is a topological mapping between them such that corresponding points lie above the same ground point. It is possible, however, that homeomorphic complexes can be defined to be covering complexes of one and the same base complex in equivalent ways.

Examples: We can find a fourfold covering of a torus $\Re$ by a torus $\tilde{\Re}'$ such that $\tilde{\Re}'$ and the torus $\tilde{\Re}$ used previously are inequivalent covering complexes. To construct $\tilde{\Re}'$ cut open the four sheets of $\tilde{\Re}$ along meridian and longitude circles and reidentify with a cross-over. Figure 102 shows $\tilde{\Re}'$ cut open to form a square, and subdivided into four squares so that points located congruently with respect to each of the four subdivision squares project to the same point of $\Re$.

We now give two additional examples of covering complexes, which will be used later. Let the base complex $\Re$ be the double torus, which we embed in ordinary 3-space as a sphere with two handles. We consider three congruent copies to be laid over it, with their handles cut open. In the first example, we sew the first and second copies together with a cross-over at the left handle cut (Fig. 103) and we rejoin the third copy to itself there so that it again runs smoothly; at the right handle cut we sew copies 2 and 3 together with a cross-over, while the first copy is rejoined to itself and again runs smoothly. In the second example, we rejoin each of the three copies to itself at the left handle cut, as if the left handle cut had never been made; at the right handle cut the three copies are sewn cyclically to one another. The meridian circles at the handle cuts become closed curves along which the covering surface has self-intersections, in so far as the surface does not run smoothly over the cuts, and either two or all three of the copies cross through one another there. The self-intersections are indicated schematically in Fig. 103.

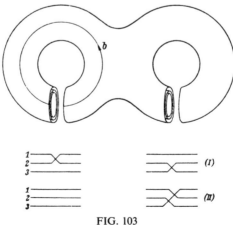

FIG. 103

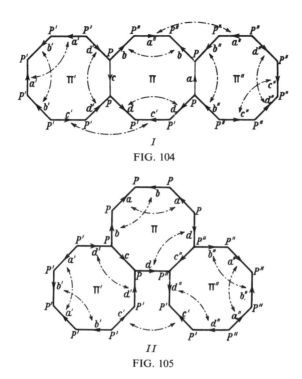

I

FIG. 104

II

FIG. 105

We obtain another model of the covering surface when we cut open the base double torus to form its fundamental polygon, an octagon, and join two congruent copies to this polygon in an appropriate manner. The covering surface then appears as a single polygon having pairwise equivalent sides. In Fig. 104 and 105 equivalent vertices and equivalent edges have been given the same designations; those vertices, edges, and surface elements which lie above the same base element are distinguished by means of accent marks. In each of the two examples, the covering surface is orientable and its Euler characteristic is three times that of the base surface: $\tilde{N} = 3N = 6$. For each vertex, edge and surface element in the base space is covered by exactly three vertices, edges, and surface elements respectively. The covering surface has genus $h = 4$ (§39).

54. Base Path and Covering Path

If w is the preimage of a path $\tilde{W}$ in $\tilde{\Re}$ and $\tilde{T}$ is the mapping of w onto $\tilde{W}$, then $T = G\tilde{T}$ is a continuous mapping of w into $\Re$ and this mapping determines a path W in $\Re$. We call W the *ground path* or *base path* belonging to $\tilde{W}$ and we say that $\tilde{W}$ *covers* the path W. We will now prove not only that each path $\tilde{W}$ of $\tilde{\Re}$ projects to a particular ground path W of $\Re$ but also the converse. Each path of the base complex can be lifted to the covering complex in as many ways as there are points lying above the initial point A of W.

THEOREM I. *If W is a path in $\Re$ leading from A to B and $\tilde{A}$ is a point of $\tilde{\Re}$ lying above A, then there exists exactly one path $\tilde{W}$ which covers W and has initial point $\tilde{A}$.*

Proof. We choose the oriented unit interval w ($0 \leqq s \leqq 1$) as the preimage of W. Let T be the mapping of w onto W. If $\tilde{T}'$ is then an arbitrary continuous mapping of w into $\tilde{\Re}$, the resulting path $\tilde{W}$ will be a covering path of W if and only if $G\tilde{T}'$ maps w to W; that is, there exists a topological self-mapping S of w such that corresponding points s and $S(s)$ have the same image in $\Re$ under T and $G\tilde{T}'$ respectively: $T(s) = G\tilde{T}'S(s)$. From the definition of equality of two paths, the same path will be defined by the mapping $\tilde{T}'$ as by $\tilde{T}'S = \tilde{T}$. We thus obtain the most general covering path $\tilde{W}$ by mapping w into $\tilde{\Re}$ by a continuous mapping $\tilde{T}$ such that $G\tilde{T} = T$. Expressed in another way: For each value of s the point $\tilde{T}(s)$ lies above s. The proof of Theorem I is thereby reduced to the proof of the following lemma.

LEMMA. *Let the unit interval $0 \leqq s \leqq 1$ be mapped continuously into $\tilde{\Re}$ by a mapping T. If $\tilde{A}$ is a point of $\tilde{\Re}$ lying above $A = T(0)$, then there exists exactly one continuous mapping, $\tilde{T}$ of the unit interval into $\tilde{\Re}$ such that $\tilde{T}(0) = \tilde{A}$ and $\tilde{T}(s)$ lies above $T(s)$ for each value of s.*

Proof. (a)   Existence of $\tilde{T}$. Assign an arbitrary distinguished neighborhood $\mathfrak{U}(P)$ to each point P of $\Re$. We divide the unit interval w into n equal parts $r_1, r_2, \ldots, r_n$, making n sufficiently large so that the image $T(r_i)$ of each subinterval belongs entirely to a distinguished neighborhood $\mathfrak{U}_i$ (of a point whose identity is of no special interest). The fact that this is possible follows from the uniform continuity theorem (§7, Theorem IV) when one chooses a distinguished neighborhood $\mathfrak{U}(P)$ of each point P as the neighborhood $\mathfrak{U}^*(P/\mathfrak{B})$ of the theorem. We now select a distinguished neighborhood $\tilde{\mathfrak{U}}_1$ which lies above $\mathfrak{U}_1$ and contains the point $\tilde{A}$. Such a neighborhood exists because of (U13). The mapping G maps $\tilde{\mathfrak{U}}_1$ topologically onto $\mathfrak{U}_1$; thus we can map the image $T(r_1)$, which lies in $\mathfrak{U}_1$, topologically into $\tilde{\mathfrak{U}}_1$ and thereby obtain a mapping $\tilde{T}_1$ of the first subinterval, r_1, into $\tilde{\mathfrak{U}}_1$ such that the initial point $s = 0$ of r_1 transforms to $\tilde{A}$. The second subinterval, r_2, is mapped in the same way. We select a neighborhood $\tilde{\mathfrak{U}}_2$ lying above the distinguished neighborhood $\mathfrak{U}_2$ in which $T(r_2)$ lies such that $\mathfrak{U}_2$ contains the image of the endpoint of r_1, that is, the point $\tilde{T}_1(1/n)$, and maps $\tilde{T}(r_2)$ topologically into $\mathfrak{U}_2$. We obtain in this way a continuous mapping $\tilde{T}_2$ of r_2 into $\tilde{\mathfrak{U}}_2$ such that $\tilde{T}_1(1/n) = \tilde{T}_2(1/n)$. We proceed in this manner and obtain in sequence the mappings $\tilde{T}_1, \tilde{T}_2, \ldots, \tilde{T}_n$ of the subintervals $r_1, r_2, \ldots, r_n$ such that $\tilde{T}_i(i/n) = \tilde{T}_{i+1}(i/n)$. These partial mappings taken together define a mapping $\tilde{T}$ of the unit interval into $\tilde{\Re}$, of the required nature.

(b)   Uniqueness of $\tilde{T}$. Let $\tilde{T}'$ be another continuous mapping of the unit interval, which satisfies the conditions of the lemma. Then $\tilde{T}$ and $\tilde{T}'$ will coincide with each other up to a certain value of the parameter s^* ($\geqq 0$). They will also coincide at s^* itself, because the mappings are continuous. There will then exist a distinguished neighborhood $\tilde{\mathfrak{U}}^*$ of the point $\tilde{T}(s^*) = \tilde{T}'(s^*)$ which is mapped topologically by G onto a distinguished neighborhood $\mathfrak{U}^*$ of the point lying underneath, $T(s^*)$. Furthermore, because $\tilde{T}$ and $\tilde{T}'$ are continuous, there exists an $\varepsilon > 0$ such that all points $\tilde{T}(s)$ and $\tilde{T}'(s)$ for which $\varepsilon \geqq |s - s^*|$ lie in $\tilde{\mathfrak{U}}^*$. Since $\tilde{T}(s)$ and $\tilde{T}'(s)$ both lie above the same base point $T(s)$ of $\mathfrak{U}^*$, $\tilde{T}(s) = \tilde{T}'(s)$ because the points of $\mathfrak{U}^*$ are in one-to-one correspondence with the points of $\tilde{\mathfrak{U}}^*$. Thus $\tilde{T}$ and $\tilde{T}'$ also coincide up to the parameter value $s + \varepsilon$, and therefore coincide everywhere.

We now apply Theorem I to define the number of sheets of the covering complex.

Let P and Q be two arbitrary points of $\mathfrak{R}$. It is possible to show that the points lying above P are in one-to-one correspondence with the points lying above Q. To do this we construct a path W from P to Q and we assign to each point $\tilde{P}_i$ lying above P the endpoint of the path lying above W which has $\tilde{P}_i$ as initial point. In the other direction, to each point $\tilde{Q}_k$ lying above Q we assign the endpoint of the path which lies above W^{-1} and has $\tilde{Q}_k$ as initial point. From Theorem I these paths exist and are unique. The latter assignment is clearly the inverse of the former assignment, so that there exists a one-to-one correspondence between the points lying above P and the points lying above Q. Consequently, *the same number of points of $\tilde{\mathfrak{R}}$ lie above each point of $\mathfrak{R}$, let us say g for example. This number is called the multiplicity of the covering or the number of sheets.* The number g can be either finite or infinite.

Theorem I allows us to compare paths in the base and the covering complexes. The following theorem does the same for the path classes. It states that one can project a deformation of a path in the covering complex to the base complex and one can lift a deformation of a path in the base complex to the covering complex.

THEOREM II. *Let $\tilde{W}_0$ and $\tilde{W}_1$ be two paths in $\tilde{\mathfrak{R}}$ from $\tilde{A}$ to $\tilde{B}$ and let W_0 and W_1 be the corresponding ground paths leading from A to B. If $\tilde{W}_0$ is deformable to $\tilde{W}_1$, then W_0 is deformable to W_1. Conversely, if W_0 and W_1 are two paths in $\mathfrak{R}$ from A to B which are deformable one to the other and if $\tilde{W}_0$ and $\tilde{W}_1$ are the two covering paths starting out from point $\tilde{A}$ lying above A, then these covering paths lead to the same endpoint $\tilde{B}$ above B and they are also deformable one to the other.*

The proof of the first part follows from the continuity of the mapping G (§42).

The converse statement appears to be just as apparent but it is, in fact, far less simple to prove because we must map the deformation rectangle from the base complex into the covering complex, although no globally continuous mapping is available; rather, the mapping is topological only in a (distinguished) neighborhood.

The deformability of W_0 to W_1 signifies that one can map the deformation rectangle $\mathfrak{R}$, having coordinates $0 \leqq s \leqq 1$, $0 \leqq t \leqq 1$ (Fig. 106), continuously into $\mathfrak{R}$ by means of a mapping T so that the oriented sides $t = 0$, $t = 1$

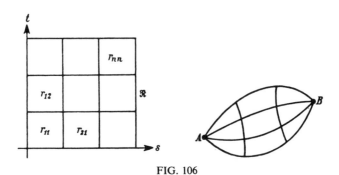

FIG. 106

transform to the paths W_0, W_1 and the sides $s = 0$, $s = 1$ transform to the points A and B. We shall construct a continuous mapping $\tilde{T}$ of $\Re$ into $\tilde{\Re}$ such that, for each point (s, t) of $\Re$, $\tilde{T}(s, t)$ lies above $T(s, t)$ and $T(0, 0) = \tilde{A}$. This will map the unit interval $t = 0$ continuously so that the point $\tilde{T}(s, 0)$ lies above the point $T(s, 0)$ for each value of s and $T(0, 0) = \tilde{A}$. From our previous lemma, this interval will then transform to the path $\tilde{W}_0$ lying above W_0 and starting at $\tilde{A}$. In like manner, the unit intervals $s = 0$, $s = 1$ transform under $\tilde{T}$ to $\tilde{A}$ and the endpoint $\tilde{B}$, respectively. Finally, the side $t = 1$ transforms to the covering path $\tilde{W}_1$ which, consequently, must terminate at $\tilde{B}$. This will prove the deformability of $\tilde{W}_0$ to $\tilde{W}_1$.

To construct $\tilde{T}$ we subdivide the deformation rectangle $\Re$ by n equidistant horizontal and n equidistant vertical lines into n^2 rectangles r_{ik},

$$(i - 1)/n \leqq s \leqq i/n, \qquad (k - 1)/n \leqq t \leqq k/n$$

and make n sufficiently large so that the image $T(r_{ik})$ of each subdivision rectangle belongs completely to a distinguished neighborhood $\mathfrak{U}_{ik}$ (of an appropriate point). This is always possible, by the uniform continuity theorem. We now construct $\tilde{T}$ stepwise by mapping the individual subdivision rectangles into $\tilde{\Re}$. We begin with r_{11} and select a distinguished neighborhood $\tilde{\mathfrak{U}}_{11}$ lying above $\mathfrak{U}_{11}$ and containing the point $\tilde{A}$. Such a neighborhood exists, by (U13). The image $T(r_{11})$, which lies wholly in $\mathfrak{U}_{11}$, will be mapped topologically into $\tilde{\mathfrak{U}}_{11}$. We thereby have found a topological mapping $\tilde{T}_{11}$ of r_{11} into $\tilde{\mathfrak{U}}_{11}$ under which $\tilde{T}_{11}(0, 0) = \tilde{A}$ and $\tilde{T}_{11}(s, t)$ lies above $T(s, t)$. We next select a distinguished neighborhood lying above $\mathfrak{U}_{21}$ which contains the point $\tilde{T}_{11}(1/n, 0)$, that is, the image of the lower right-hand vertex of r_{11}, and we map $T(r_{21})$ topologically into $\tilde{\mathfrak{U}}_{21}$ to give a continuous mapping $\tilde{T}_{21}$ of r_{21} into $\tilde{\mathfrak{U}}_{21}$. The mappings $\tilde{T}_{11}$ and $\tilde{T}_{21}$ coincide on the common side of r_{11} and r_{21}. For the initial point $(1/n, 0)$ of this segment will be mapped to the same point by both mappings and, since $\tilde{T}_{11}(1/n, t)$ and $\tilde{T}_{21}(1/n, t)$ both lie over the same point $T(1/n, t)$ it follows from the lemma that $\tilde{T}_{11}(1/n, t) = \tilde{T}_{21}(1/n, t)$. Repeating the same procedure we now map the rectangle r_{31} continuously by $\tilde{T}_{31}$ into $\tilde{\Re}$ so that $\tilde{T}_{21}(2/n, 0) = \tilde{T}_{31}(2/n, 0)$ and we prove in like manner that the common side of r_{21} and r_{31} is mapped in the same way by $\tilde{T}_{21}$ and $\tilde{T}_{31}$.

After n steps we arrive at a continuous mapping $\tilde{T}_1$ of the whole strip r_{11}, $r_{21}, \ldots, r_{n1}$ and such that $\tilde{T}_1(s,t)$ lies above $T(s,t)$ and the side $s = 0$, $0 \le t \le 1/n$ transforms to $\tilde{A}$. In the same way we obtain mappings T_i of the remaining strips $r_{1i}, r_{2i}, \ldots, r_{ni} (i = 1, 2, 3, \ldots, n)$ such that $\tilde{T}_i(s,t)$ always lies above $T(s,t)$ and $\tilde{T}_i(0,t) = \tilde{A}$. Since the common unit interval of the ith and the $(i + 1)$th strips are mapped in the same way by $\tilde{T}_i$ and $\tilde{T}_{i+1}$ (again by the lemma), we have obtained a continuous mapping $\tilde{T}$ of the whole deformation rectangle $\Re$ such that $\tilde{T}(s,t)$ lies above $T(s,t)$ and $\tilde{T}(0,0) = \tilde{A}$.

55. Coverings and Subgroups of the Fundamental Group

Let us select a point O of the base complex $\Re$ and a point $\tilde{O}_1$ lying above O in the covering complex $\tilde{\Re}$ as the initial points of the closed paths of $\Re$ and $\tilde{\Re}$, respectively. From the theorem of §42, the continuous projection mapping G induces a mapping which sends each path class of the fundamental group $\tilde{\mathfrak{F}}$ of $\tilde{\Re}$ to a particular path class of the fundamental group $\mathfrak{F}$ of $\Re$. This induced mapping is a homomorphic mapping of $\tilde{\mathfrak{F}}$ onto a particular subgroup $\mathfrak{H}_1$, of $\mathfrak{F}$.

The second part of Theorem II of §54 states that two nonhomotopic paths of $\tilde{\mathfrak{F}}$ will project to two nonhomotopic paths of $\mathfrak{F}$. That is, the homomorophic mapping of $\tilde{\mathfrak{F}}$ onto $\mathfrak{H}_1$ is also an isomorphism.

Consequently, the fundamental group $\tilde{\mathfrak{F}}$ of the covering complex is isomorphic to a particular subgroup $\mathfrak{H}_1$ of the fundamental group of the base complex. The subgroup $\mathfrak{H}_1$ is obtained by projecting $\tilde{\mathfrak{F}}$ into the base complex.

One should note that the subgroup $\mathfrak{H}_1$ is not determined by the covering $\tilde{\Re}$ alone but also depends, as we shall see, upon the choice of initial point $\tilde{O}_1$. Now let

$$\mathfrak{F} = \mathfrak{H}_1 + \mathfrak{H}_1\{F_{12}\} + \mathfrak{H}_1\{F_{13}\} + \cdots$$

be the decomposition of $\mathfrak{F}$ into residue classes modulo $\mathfrak{H}_1$. Here as in §42, $\{F_{1i}\}$ denotes the path class of the path F_{1i}. If H_1 is an arbitrary path in $\mathfrak{H}_1$ and $\tilde{H}_1$ is the covering path starting at $\tilde{O}_1$, then $\tilde{H}_1$ is closed. It follows from this that when we lift the paths of a particular residue class $\mathfrak{H}_1\{F_{1i}\}$ from $\Re$ to $\tilde{\Re}$ all of the paths starting out from O_1 lead to the same endpoint O_i, which is the endpoint of the path $\tilde{F}_{1i}$ lying above F_{1i} and starting at $\tilde{O}_1$. Consequently, each residue class $\mathfrak{H}_1\{F_{1i}\}$ determines a point $\tilde{O}_i$ lying above O. Different residue classes $\mathfrak{H}_1\{F_{1i}\}$ and $\mathfrak{H}_1\{F_{1j}\}$ determine different points $\tilde{O}_i \ne O_j$. Otherwise $\tilde{F}_{1i}\tilde{F}_{1j}^{-1}$ would be closed so that $\{F_{1i}F_{1j}^{-1}\}$ would be an element of $\mathfrak{H}_1$, whereas $\{\tilde{F}_{1i}\}$ and $\{F_{1j}\}$ belong to different residue classes. Since the endpoints of the paths $\tilde{F}_{11}, \tilde{F}_{12}, \ldots$, include all points of $\tilde{\Re}$ lying above O, a one-to-one correspondence has been produced between the right residue classes of $\mathfrak{H}_1$ in $\mathfrak{F}$ and the points lying above O. It follows that *the number of sheets of the covering is equal to the index of $\mathfrak{H}_1$ in $\mathfrak{F}$.*

Each closed path $\tilde{H}_i$ starting out from $\tilde{O}_i$ can be deformed to a path having the form $\tilde{F}_{1i}^{-1}\tilde{H}_1\tilde{F}_{1i}$, for example to the path $\tilde{F}_{1i}^{-1} \cdot (\tilde{F}_{1i}H_i\tilde{F}_{1i}^{-1}) \cdot F_{1i}$. Here $\tilde{H}_1$ is a closed path starting out from $\tilde{O}_1$. The path $\tilde{F}_{1i}^{-1}\tilde{H}_1\tilde{F}_{1i}$ projects to the path $F_{1i}^{-1}H_1F_{1i}$. The latter is a path of the subgroup $\mathfrak{H}_i = \{F_{1i}\}^{-1}\mathfrak{H}_1\{F_{1i}\}$, which is conjugate to $\mathfrak{H}_1$. Conversely, each path of $\mathfrak{H}_i$ lifts to a closed path in $\tilde{\mathfrak{R}}$ based at $\tilde{O}_i$. For each path having the particular form $F_{1i}^{-1}H_1F_{1i}$ will be lifted into $\tilde{\mathfrak{R}}$ to a path $\tilde{F}_{1i}^{-1}\tilde{H}_1\tilde{F}_{1i}$ starting out from $\tilde{O}_i$; this path is closed because $\tilde{F}_{1i}^{-1}$ leads from $\tilde{O}_i$ to O_1, $\tilde{H}_1$ leads from O_1 back to O_1, and $\tilde{F}_{1i}$ leads from O_1 again back to $\tilde{O}_i$. The same is true for an arbitrary path of $\mathfrak{H}_i$ because one can deform it to a path having the form $F_{1i}^{-1}H_1F_{1i}$ and then lift the deformed path into $\tilde{\mathfrak{R}}$; this does not change the endpoint of the path. This means: If we make $\tilde{O}_i$ instead of $\tilde{O}_1$ the initial point of the closed paths of $\tilde{\mathfrak{R}}$, then the fundamental group $\tilde{\mathfrak{F}}$ of $\tilde{\mathfrak{R}}$ projects to the subgroup $\{F_{1i}\}^{-1}\mathfrak{H}_1\{F_{1i}\}$ conjugate to $\mathfrak{H}_1$. Moreover, by choosing $\tilde{O}_i$ appropriately, we obtain each subgroup conjugate to $\mathfrak{H}_1$. Thus *the covering of $\mathfrak{R}$ by $\tilde{\mathfrak{R}}$ determines* an entire class of conjugate subgroups of $\mathfrak{F}$.

We now prove the converse, that each class of conjugate subgroups determines a covering. The problem of determining all coverings having a given number of sheets g is thereby reduced to the group theoretic question of constructing all classes of conjugate subgroups of index g of the fundamental group of the base complex. We shall carry through this construction in §58 for finite complexes and a finite number of sheets.

Existence Proof

I.   Construction of $\tilde{\mathfrak{R}}$. Let $\mathfrak{H}$ be an arbitrary subgroup of $\mathfrak{F}$.* We wish to find a covering complex $\tilde{\mathfrak{R}}$ whose fundamental group $\tilde{\mathfrak{F}}$ projects to $\mathfrak{H}$ when the initial point $\tilde{O}_1$ above O is chosen appropriately. We shall assume that O is a vertex of a simplicial decomposition of $\mathfrak{R}$; this can be accomplished if need be by subdividing $\mathfrak{R}$. We shall construct the covering complex $\tilde{\mathfrak{R}}$ so that it has a simplicial decomposition such that when one maps $\tilde{\mathfrak{R}}$ onto $\mathfrak{R}$ each simplex of $\tilde{\mathfrak{R}}$ will transform linearly to a simplex of $\mathfrak{R}$. We shall call this decomposition *the simplicial decomposition which is lifted from $\mathfrak{R}$ to $\tilde{\mathfrak{R}}$.*

We construct $\tilde{\mathfrak{R}}$ by presenting its schema (§11). The elements of the schema will be certain path classes which we now define.

Let A be a vertex of $\mathfrak{R}$. We divide all of the paths leading from O to A into classes, so-called $\mathfrak{H}$-classes. We consider two paths U and U' to be in the same $\mathfrak{H}$-class if and only if the closed path UU'^{-1} belongs to the subgroup $\mathfrak{H}$. Then to each vertex A of $\mathfrak{R}$ there belong certain $\mathfrak{H}$-classes; we denote these by

$$\overline{A}_1, \overline{A}_2, \ldots .$$

In particular, the paths of the subgroup $\mathfrak{H}$ themselves form such an $\mathfrak{H}$-class, the unit $\mathfrak{H}$-class. It belongs to the point O and will be denoted by $\overline{O}_1$.

---

* For simplicity we shall now write $\mathfrak{H}$ instead of $\mathfrak{H}_1$.

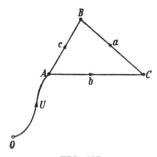

FIG. 107

Two $\mathfrak{H}$-classes $\overline{A}_i$ and $\overline{B}_k$ are said to be *neighboring* if their endpoints A and B are the vertices of a 1-simplex c of $\mathfrak{R}$ and, also, the closed path UcV^{-1} which one can form from a path U of $\overline{A}_i$ and a path V of $\overline{B}_k$ belongs to $\mathfrak{H}$. Here it is obviously unimportant how one chooses the paths U and V from $\overline{A}_i$ and $\overline{B}_k$.

If we are given the $\mathfrak{H}$-class $\overline{A}_i$, then there exists exactly one neighboring $\mathfrak{H}$-class, $\overline{B}_k$ belonging to the vertex B. The $\mathfrak{H}$-class $\overline{B}_k$ consists of all of the paths V for which UcV^{-1} is a path of $\mathfrak{H}$ for fixed U. Its existence is insured by giving a representative path, for example $V = Uc$. We now have the following lemma.

LEMMA. *If $\mathfrak{E}^i$ is a simplex of the base complex $\mathfrak{R}$ and has vertices $A, B, \ldots, C$ and one has chosen an $\mathfrak{H}$-class $\overline{A}_i$ of path connecting to A, there exists exactly one $\mathfrak{H}$-class $\overline{B}_k, \ldots, C_l$ for each of the remaining vertices $B, \ldots, C$ such that all of these $\mathfrak{H}$-classes are pairwise neighboring.*

For $\overline{B}_k, \ldots, \overline{C}_l$ are uniquely determined by the requirement that they are neighboring to $\overline{A}_i$. We shall show that the $\mathfrak{H}$-classes are pairwise neighboring using $\overline{B}_k$ and $\overline{C}_l$ as an example. If U is a path belonging to $\overline{A}_i$ (Fig. 107) and a, b, c are the oriented simplexes CB, AC, AB, then Ub and Uc are representative classes from $\overline{C}_l$ and $\overline{B}_k$. The condition that $\overline{B}_k$ and $\overline{C}_l$ be neighboring classes can be stated: $Ub \cdot a \cdot (Uc)^{-1}$ must belong to $\mathfrak{H}$. But this is true since this path is null homotopic and thus belongs to the unit path class of the fundamental group.

It now follows that *the $\mathfrak{H}$-classes form the schema of a simplicial complex, $\tilde{\mathfrak{R}}$,* when we choose as characteristic subsets of the schema (§11) each system of pairwise neighboring $\mathfrak{H}$-classes whose paths lead to the vertices of a simplex of $\mathfrak{R}$. We can easily confirm that the conditions required for a schema, (Sch1), (Sch2), and (Sch3) of §11, are in fact fulfilled.

We can also describe the simplicial complex $\tilde{\mathfrak{R}}$ in slightly different words: Each vertex $\tilde{A}_i$ of $\tilde{\mathfrak{R}}$ is in one-to-one correspondence with an $\mathfrak{H}$-class $\overline{A}_i$ of $\mathfrak{R}$, which in turn is in one-to-one correspondence with a particular vertex of $\mathfrak{R}$, namely, the endpoint A of the paths of the $\mathfrak{H}$-class $\overline{A}_i$. A collection of vertices $\tilde{A}_i, \tilde{B}_k, \ldots, \tilde{C}_l$ will form the vertices of a simplex if and only if the corresponding vertices $A, B, \ldots, C$ of $\Re$ are the vertices of a simplex and, moreover, the $\mathfrak{H}$-classes $\overline{A}_i, \overline{B}_k, \ldots, \overline{C}_l$ are neighboring.

The complex $\widetilde{\Re}$ is a connected complex. To prove this it will suffice to show that each $\mathfrak{H}$-class $\overline{A}_i$ can be connected with the unit $\mathfrak{H}$-class $\overline{O}_1$ by a sequence of $\mathfrak{H}$-classes in which each two successive classes are neighboring. In $\overline{A}_i$ choose an edge path U from O to A. Such a path exists because one can deform each path from O to A to an edge path (§31). The partial paths of U which lead from O to the vertices traversed by U, or rather the $\mathfrak{H}$-classes corresponding to them, form a sequence having the desired properties.

II.   We shall now prove that $\widetilde{\Re}$ is a covering complex of $\Re$ when we assign the vertex A_i as base point to each vertex $\widetilde{A}_i$ of $\widetilde{\Re}$ (which corresponds to the $\mathfrak{H}$-class $\overline{A}_i$) and map each simplex of $\widetilde{\Re}$ linearly onto that simplex of $\Re$ which is spanned by the base points of its vertices. The three conditions for a covering are satisfied. There exists a system of neighboring $\mathfrak{H}$-classes for each simplex of $\Re$ and therefore there exists at least one simplex lying above that simplex. Consequently, (U11) holds.

Let P be an arbitrary point of $\Re$ and let $\mathfrak{E}^i$ be the uniquely determined simplex having P as an inner point. The totality of simplexes which have $\mathfrak{E}^i$ as a face form a neighborhood $\mathfrak{U}(P)$ of P in $\Re$. We claim that $\mathfrak{U}(P)$ is a distinguished neighborhood. A point $\widetilde{P}$ lying above P is an inner point of a simplex $\widetilde{\mathfrak{E}}^i_\lambda$ lying above $\mathfrak{E}^i$, where $\widetilde{\mathfrak{E}}^i_\lambda$ is determined by a system of $i+1$ neighboring $\mathfrak{H}$-classes which lead to the vertices of $\mathfrak{E}^i$. If $\mathfrak{E}^{i+k}$ is a simplex incident with $\mathfrak{E}^i$, then by the lemma there exists exactly one system of neighboring $\mathfrak{H}$-classes leading to the vertices of $\mathfrak{E}^{i+k}$ which contains the given system of neighboring $\mathfrak{H}$-classes as a subsystem. The simplexes having $\mathfrak{E}^i$ as a face are consequently in one-to-one correspondence with the simplexes having $\widetilde{\mathfrak{E}}^i_\lambda$ as a face. That is, the neighborhood $\mathfrak{U}(\widetilde{P})$ of $\widetilde{P}$ in $\widetilde{\Re}$ consisting of the totality of simplexes incident with $\widetilde{\mathfrak{E}}^i_\lambda$ maps topologically onto the neighborhood $\mathfrak{U}(\widetilde{P})$. Thus (U12) is satisfied.

Let $\widetilde{Q}_\mu$ be a point of $\widetilde{\Re}$ which lies above a point Q of $\mathfrak{U}(P)$. Point Q belongs to a simplex $\mathfrak{E}^{i+k}_\mu$ of $\mathfrak{U}(P)$ incident with $\mathfrak{E}^i$ and thus $\widetilde{Q}_\mu$ belongs to a simplex $\widetilde{\mathfrak{E}}^{i+k}_\mu$ lying above $\mathfrak{E}^{i+k}_\mu$ and containing a point $\widetilde{P}_\mu$ lying above P. Thus $\widetilde{Q}_\mu$ belongs to the neighborhood $\mathfrak{U}(\widetilde{P}_\mu)$ and (U13) is satisfied.

III.   The covering $\widetilde{\Re}$ belongs to the subgroup $\mathfrak{H}$ of the fundamental group $\mathfrak{F}$ of $\Re$.

We choose the vertex $\widetilde{O}_1$ lying above O as the initial point of the closed paths in $\widetilde{\Re}$, where $\widetilde{O}_1$ belongs to the unit $\mathfrak{H}$-class $\overline{O}_1$. Let U be an edge path in $\Re$ running from O to a vertex A and let U_τ be a subpath of U consisting of the first τ edges of U. The $\mathfrak{H}$-classes of the paths $U_0, U_1, \ldots, U_l = U$ then form a sequence of successively neighboring $\mathfrak{H}$-classes. The vertices of $\widetilde{\Re}$ which are determined by each two successive $\mathfrak{H}$-classes are then also neighboring and as a consequence the vertices of the sequence of vertices corresponding to the $\mathfrak{H}$-classes are connected by an edge path $\widetilde{U}$ lying above

U. The initial point of this path belongs to the $\mathfrak{H}$-class of U_0 and is then the vertex $\tilde{O}_1$.

If U is in particular a closed edge path belonging to $\mathfrak{H}$, that is, U lies in the unit $\mathfrak{H}$-class $\overline{O}_1$, then the covering path $\tilde{U}$ starting at $\tilde{O}_1$ must return to $\tilde{O}_1$. But if U does not belong to $\mathfrak{H}$, that is, U lies in an $\mathfrak{H}$-class $\overline{O}_i \neq \overline{O}_1$, then the lifted edge path $\tilde{U}_i$ leads from $\tilde{O}_1$ to $\tilde{O}_i \neq \tilde{O}_1$ and thus is not closed. Since each closed path starting out from O can be deformed to an edge path, we have shown that a closed path U starting out from O will be covered by a closed covering path $\tilde{U}$ having initial point $\tilde{O}_1$ if and only if U belongs to $\mathfrak{H}$. But this means that the fundamental group $\tilde{\mathfrak{F}}$ of $\tilde{\mathfrak{R}}$ projects to just the subgroup $\mathfrak{H}$ of $\mathfrak{F}$ when one makes $\tilde{O}_1$ the initial point of the closed paths in $\tilde{\mathfrak{R}}$. Thus $\tilde{\mathfrak{R}}$ is a covering complex belonging to $\mathfrak{H}$.

Uniqueness of $\tilde{\mathfrak{R}}$

Let $\tilde{\mathfrak{R}}'$ be another covering complex* which determines the same class of conjugate subgroups of the fundamental group $\mathfrak{F}$ of $\mathfrak{R}$ as $\tilde{\mathfrak{R}}$; this is the class to which the subgroup $\mathfrak{H}$ belongs. There then exists a point $\tilde{O}_1'$ of $\tilde{\mathfrak{R}}'$ lying above O such that the fundamental group of $\tilde{\mathfrak{R}}'$ projects to the subgroup $\mathfrak{H}$ when $\tilde{O}_1'$ is chosen as the initial point of the closed paths. (It is possible that several points of $\tilde{\mathfrak{R}}'$ lying above O might have this property.) We now construct a mapping of $\tilde{\mathfrak{R}}$ onto $\tilde{\mathfrak{R}}'$ by which $\tilde{O}_1$ transforms to $\tilde{O}_1'$ and corresponding points lie above the same point of $\mathfrak{R}$.

If $\tilde{P}$ is an arbitrary point of $\tilde{\mathfrak{R}}$, then we draw a path $\tilde{W}$ from $\tilde{O}_1$ to $\tilde{P}$. The corresponding ground path is the path W starting from O. Let $\tilde{W}'$ be the covering path above W starting from $\tilde{O}_1'$. Call its endpoint $\tilde{P}'$. It will be assigned as the image point of P. We must show that this assignment is independent of the choice of the connecting path $\tilde{W}$. Let $\tilde{V}$ be another connecting path of $\tilde{O}_1$ with $\tilde{P}$. Then $\tilde{W}\tilde{V}^{-1}$ is closed and WV^{-1} belongs to $\mathfrak{H}$. But if this is the case, the covering path above WV^{-1} starting from $\tilde{O}_1'$ is also closed, since the fundamental group of $\tilde{\mathfrak{R}}'$ is to project to $\mathfrak{H}$ when one selects $\tilde{O}_1'$ as the initial point of the closed paths of $\tilde{\mathfrak{R}}'$. Thus $\tilde{W}'$ and $\tilde{V}'$ lead to the same endpoint $\tilde{P}'$. The assignment $\tilde{P} \rightarrow \tilde{P}'$ is a one-to-one correspondence since one can transform from $\tilde{P}'$ uniquely back to $\tilde{P}$ by the reverse construction.

We need to prove continuity of the mapping in only one direction, for example in the direction $\tilde{P} \rightarrow \tilde{P}'$, because $\tilde{\mathfrak{R}}$ and $\tilde{\mathfrak{R}}'$ are alike in their properties. For a given neighborhood $\tilde{\mathfrak{U}}'(\tilde{P}')$ we must find a neighborhood $\tilde{\mathfrak{U}}(\tilde{P})$ whose image lies entirely in $\tilde{\mathfrak{U}}'(\tilde{P}')$. Let $\tilde{\mathfrak{U}}_1'$ be a distinguished neighborhood of $\tilde{P}'$ with $\tilde{\mathfrak{U}}'$. We can, for example, choose this to be the

---

*We shall not assume that there exists a simplicial decomposition of $\tilde{\mathfrak{R}}'$ which projects to a simplicial decomposition of $\mathfrak{R}$. In fact, we now proceed without using any simplicial decomposition.

intersection of an arbitrary distinguished neighborhood of $\tilde{P}'$ with $\tilde{\mathfrak{U}}'$. Let $\mathfrak{U}_1$ be the distinguished neighborhood of P onto which $\tilde{\mathfrak{U}}_1'$ is mapped topologically. In addition, let $\tilde{\mathfrak{U}}$ be a neighborhood of $\tilde{P}$ which is sufficiently small that it projects to a subset $\mathfrak{U}$ of $\mathfrak{U}_1$. Furthermore, we require that each point of $\tilde{\mathfrak{U}}$ can be connected to $\tilde{P}$ by means of a path lying entirely in $\tilde{\mathfrak{U}}$. Since $\tilde{\mathfrak{R}}$ is a complex, we can certainly find such a neighborhood $\tilde{\mathfrak{U}}$. For example, a suitable simplicial star with center point $\tilde{P}$ will satisfy all requirements. We claim that $\tilde{\mathfrak{U}}$ will be mapped into $\tilde{\mathfrak{U}}'$ under the mapping of $\tilde{\mathfrak{R}}$ onto $\tilde{\mathfrak{R}}'$. In fact, if $\tilde{W}$ is a path from $\tilde{O}_1$ to $\tilde{P}$, choose $\tilde{W}$ extended by a path $\Delta\tilde{W}$ from $\tilde{P}$ to $\tilde{Q}$ and lying in $\tilde{\mathfrak{U}}$ as the path from $\tilde{O}_1$ to a point $\tilde{Q}$ of $\tilde{\mathfrak{U}}$. The base path $W \cdot \Delta W$ runs from O to P and then runs subsequently in $\mathfrak{U}_1$. The path lifted to $\tilde{\mathfrak{R}}'$, that is $\tilde{W}' \cdot \Delta\tilde{W}'$, runs from $\tilde{O}_1'$ to $\tilde{P}'$ and subsequently lies in $\tilde{\mathfrak{U}}_1'$ and thus also in $\tilde{\mathfrak{U}}'$. The image point of Q, which is the endpoint of the path $\tilde{W}' \cdot \Delta\tilde{W}'$, therefore lies in $\tilde{\mathfrak{U}}'$.

We summarize our results:

THEOREM. *The coverings of a complex are in one-to-one correspondence with the classes of conjugate subgroups. More precisely: If $\tilde{\mathfrak{R}}$ is a covering complex of a complex $\mathfrak{R}$ and one chooses as initial point of the closed paths of $\tilde{\mathfrak{R}}$ a point $\tilde{O}$ lying over the initial point O of the closed paths of $\mathfrak{R}$, then the fundamental group $\tilde{\mathfrak{F}}$ of $\tilde{\mathfrak{R}}$ will project to a subgroup $\mathfrak{H}$ of the fundamental group $\mathfrak{F}$ of $\mathfrak{R}$ in a way such that $\mathfrak{H}$ is isomorphic with $\tilde{\mathfrak{F}}$. If one chooses a different point as initial point $\tilde{O}$ lying above O, then $\tilde{\mathfrak{F}}$ will project to a subgroup of $\mathfrak{F}$ which is conjugate to $\mathfrak{H}$.*

Problems

1. What is the relationship between the Euler characteristic of a finite complex and that of a g-sheeted covering?

2. If $\mathfrak{F}$ is the free group on a generators and $\mathfrak{H}$ is a subgroup of index i, show that $\mathfrak{H}$ is the free group on $i(a - 1) + 1$ generators. [Note that $\mathfrak{F}$ is the fundamental group of an edge complex having Euler characteristic $N = a - 1$ (§47).]

3. Show that a simply connected complex has a exactly one covering complex, namely, the complex itself.

56. Universal Coverings

Among the covering complexes of a complex $\mathfrak{R}$ one of particular importance is the covering which belongs to the subgroup $\mathfrak{H} = 1$. It is called the *universal covering complex of* $\mathfrak{R}$. It is characterized by the fact that its fundamental group $\tilde{\mathfrak{F}} = \mathfrak{H}$ consists of the unit element alone, so that it is simply connected. As a corollary to the theorem of §55 we have

THEOREM I. *Each connected complex $\mathfrak{R}$ possesses a unique simply connected covering complex, the universal covering complex $\hat{\mathfrak{R}}$.*

One can speak of the universal covering complex as being the strongest, for one has

THEOREM II. *The universal covering complex $\hat{\mathfrak{R}}$ is characterized by the fact that it covers every covering complex $\tilde{\mathfrak{R}}$ of $\mathfrak{R}$.*

The proof rests on the following lemma.

LEMMA. *If $\mathfrak{R}$ is covered by $\tilde{\mathfrak{R}}$ and $\tilde{\mathfrak{R}}$ is covered by $\tilde{\tilde{\mathfrak{R}}}$, then $\mathfrak{R}$ is also covered by $\tilde{\tilde{\mathfrak{R}}}$.*

Condition (U11), that at least one point $\tilde{\tilde{P}}$ of $\tilde{\tilde{\mathfrak{R}}}$ lies above each point P of $\mathfrak{R}$, is obviously satisfied. To prove (U12) and (U13) for $\tilde{\tilde{\mathfrak{R}}}$ we lift a simplicial decomposition of $\mathfrak{R}$ to $\tilde{\mathfrak{R}}$ and from there to $\tilde{\tilde{\mathfrak{R}}}$ and we choose as the distinguished neighborhood of a point P the totality of simplexes containing this point. We select distinguished neighborhoods of the points $\tilde{\tilde{P}}_1, \tilde{\tilde{P}}_2 \ldots$, lying above P, in the same manner as we did in the existence proof of the covering (§55). These simplicial stars all map topologically one onto the other, whereby (U12), (U13), and the lemma follow.

Now let $\hat{\mathfrak{R}}$ be the universal covering complex of $\tilde{\mathfrak{R}}$ and let $\tilde{\mathfrak{R}}$ be a covering complex of $\mathfrak{R}$. By the lemma, $\hat{\mathfrak{R}}$ covers $\mathfrak{R}$ and, since it is simply connected, it is the universal covering complex of $\mathfrak{R}$. Thus the universal covering complex of $\mathfrak{R}$ also covers $\tilde{\mathfrak{R}}$. Conversely, if a covering complex of $\mathfrak{R}$ covers each covering complex, then it also covers the universal complex $\hat{\mathfrak{R}}$. Since the latter is simply connected, it has only itself as a covering complex (§55, Problem 3).

We have already met several examples of universal covering complexes. The Euclidean plane covers the torus universally by means of the continuous mapping produced by identification (§8). The real number line is a universal covering complex of the circle. More generally, Euclidean n-space is a universal covering complex of the topological product of n circles. The n-sphere is the universal covering complex of projective n-space.

Problems

1. Let $\mathfrak{R}^1$ be the edge complex of two triangles having a vertex in common. Find the universal covering complex $\hat{\mathfrak{R}}^1$ of $\mathfrak{R}^1$.

2. On the boundary of the circular disk of unit radius identify each pair of points which lie at a distance $2\pi/p$ measured along the boundary (so that p points in all are to be identified with any given point). Show that the fundamental group of the resulting complex $\mathfrak{R}^2$ is cyclic of order p and the universal covering complex $\hat{\mathfrak{R}}^2$ consists of p circular disks having a common boundary.

57. Regular Coverings

DEFINITION 1. The fundamental group $\tilde{\mathfrak{F}}$ of $\tilde{\mathfrak{R}}$ will project to a subgroup of $\mathfrak{F}$ which depends, in general, upon one's choice of the initial point of the closed paths of $\tilde{\mathfrak{R}}$. That is, when one selects $\tilde{O}_1$ or $\tilde{O}_2$ or $\ldots$ as the initial

point lying above O one will accordingly obtain the subgroup $\mathfrak{H}_1$ or $\mathfrak{H}_2$ or
.... All of these subgroups are conjugate in $\mathfrak{F}$ and form a complete system
of conjugate subgroups. But these subgroups are not necessarily different
from one another. *In the extreme case that all of these subgroups are the same,
the subgroup to which the fundamental group $\tilde{\mathfrak{F}}$ projects is a normal subgroup of
$\mathfrak{F}$ and we say that the covering is regular.*

The following are examples of regular coverings: all coverings of the torus, because the
fundamental group of the torus is Abelian; each universal covering, because it belongs to the
normal subgroup $\mathfrak{H} = 1$; each two-sheeted covering, because a subgroup of index 2 is always a
normal subgroup.

DEFINITION 2. *A regular covering can also be characterized by the fact that
the paths lying above a closed path in the base space are either all closed or all
nonclosed.*

For if $\tilde{W}_i$ is the covering path starting out from $\tilde{O}_i$ which lies above a
closed path W, then $\tilde{W}_i$ is closed or not closed according to whether W does
or does not, respectively, belong to $\mathfrak{H}_i$. If all of the conjugate subgroups
coincide, and for example $= \mathfrak{H}$, then the path W lies in all $\mathfrak{H}_i$, in which case
all paths $\tilde{W}_i$ are closed, or W lies in none of the groups $\mathfrak{H}_i$, in which case all
of the paths $\tilde{W}_i$ are not closed. If, on the other hand, the $\mathfrak{H}_i$ are not all the
same, for example $\mathfrak{H}_1 \neq \mathfrak{H}_2$, then there exists a path which belongs to $\mathfrak{H}_1$ but
not to $\mathfrak{H}_2$. Then $\tilde{W}_1$ is closed but not $\tilde{W}_2$. The fact that we have chosen the
point O as the initial point of W is not an essential restriction, since we can
always make an arbitrary closed path in $\mathfrak{R}$ into a closed path starting out
from O by means of an auxiliary path running back and forth from O.

The first of the triple coverings of the double torus [Fig. 103 (I)] is not
regular. The path in the covering surface lying above the longitude circle b is
either closed or not closed according to whether, when one lifts b to the
covering surface, one starts the covering path on sheet 3 or on sheet 2,
respectively.

DEFINITION 3. We can characterize a regular covering in still a third way,
making use of the concept of a *covering transformation*. A covering
transformation of the covering complex $\tilde{\mathfrak{R}}$ is a topological mapping of $\tilde{\mathfrak{R}}$ onto
itself such that each point remains above its base point and only points lying
above one and the same base point are interchanged with one another. There
exists at most one covering transformation which transforms $\tilde{O}_1$ to another
given point lying above O, for example $\tilde{O}_2$. To show this let $\tilde{P}$ be an arbitrary
point of $\tilde{\mathfrak{R}}$. Draw a path $\tilde{W}_1$ from $\tilde{O}_1$ to $\tilde{P}$. Let W be the corresponding
ground path. Under the covering transformation $\tilde{W}_1$ will transform to the
path $\tilde{W}_2$ lying above W and starting out from $\tilde{O}_2$. The image point of $\tilde{P}$ is the
uniquely determined end point of $\tilde{W}_2$. All of the covering transformations of
$\tilde{\mathfrak{R}}$ obviously form a group, the group $\mathfrak{D}$ of covering transformations of the
covering. This group may possibly consist only of the identity mapping; at
most, its order is equal to the number of sheets of the covering, and this

occurs in the case when one can transform $\tilde{O}_1$ to all of the points $\tilde{O}_2, \tilde{O}_3, \ldots$ or, what is the same, if $\mathfrak{D}$ acts transitively over the points lying above O.

The third characterization of a regular covering is: $\tilde{\mathfrak{R}}$ *is a regular covering of* $\mathfrak{R}$ *if and only if the group of covering transformations acts transitively on the set of points lying above each point of* $\mathfrak{R}$.

If the covering is not regular then, as we know, there exists a path W of $\mathfrak{R}$ which is covered at the same time by both closed and nonclosed paths. In such a case it is obvious that no transitive group of covering transformations exists, since a closed path will again transform to a closed path under a covering transformation.We know from §55 that two covering complexes $\tilde{\mathfrak{R}}$ and $\tilde{\mathfrak{R}}'$ which determine the same class of conjugate subgroups (in our case the same normal subgroup $\mathfrak{H}$) can be mapped one onto the other topologically so that corresponding points have the same base point in $\mathfrak{R}$. It follows that one can transform a point $\tilde{O}_1$ lying above O to an arbitrary point $\tilde{O}_1'$ of $\tilde{\mathfrak{R}}'$ only if the following condition is satisfied: on using $\tilde{O}_1'$ as the initial point of the closed paths of $\tilde{\mathfrak{R}}'$ and $\tilde{O}_1$ as the initial point of the closed paths of $\tilde{\mathfrak{R}}$, the fundamental groups $\tilde{\mathfrak{F}}'$ of $\tilde{\mathfrak{R}}'$ and $\tilde{\mathfrak{F}}$ of $\tilde{\mathfrak{R}}$ both project to the same subgroup of $\mathfrak{F}$. If $\mathfrak{H}$ is a normal subgroup of $\mathfrak{F}$, as in the present case, then this condition is always satisfied independently of the choice of point $\tilde{O}_1'$, above O. If we let $\tilde{\mathfrak{R}}$ and $\tilde{\mathfrak{R}}'$ coincide, we obtain a covering transformation of $\tilde{\mathfrak{R}}$ such that $\tilde{O}_1$ transforms to an arbitrary point lying above O.

The two coverings of the double torus given in §53 serve to illustrate this. The first covering is nonregular, as we saw, and thus does not possess a transitive group of covering transformations. The only possible covering transformation is the identity covering transformation; this should be self-evident since sheet 2 is cut through on both handles while sheets 1 and 3 are each cut on only one handle. In contrast, in the second covering we can cyclically permute the three sheets and the covering is regular.

In a regular covering the group of covering transformations $\mathfrak{D}$ *is isomorphic to the factor group* $\mathfrak{F}/\mathfrak{H}$.

The residue classes of the decomposition relative to the subgroup $\mathfrak{H}_1 = \mathfrak{H}$

$$\mathfrak{F} = \mathfrak{H}\{F_{11}\} + \mathfrak{H}\{F_{12}\} + \cdots,$$

are in one-to-one correspondence with the points $\tilde{O}_1, \tilde{O}_2, \ldots$ lying above O. On the other hand, to each point $\tilde{O}_i$ there corresponds a covering transformation D_{1i} which transforms $\tilde{O}_1$ to $\tilde{O}_i$. We thus obtain a one-to-one correspondence between the residue classes $\mathfrak{H}\{F_{1i}\}$ and the covering transformations:

$$\mathfrak{H}\{F_{1i}\} \leftrightarrow D_{1i}.$$

To show that this correspondence establishes an isomorphism between the factor group $\mathfrak{F}/\mathfrak{H}$ and the group of covering transformations, we shall determine the covering transformation corresponding to the product of two residue classes $\mathfrak{H}\{F_{1i}\} \cdot \mathfrak{H}\{F_{1j}\} = \mathfrak{H}\{F_{1i}\}\{F_{1j}\} = \mathfrak{H}\{F_{1k}\}$. One can arrive at

FIG. 108

the point $\tilde{O}_k$ by proceeding from $\tilde{O}_1$ to $\tilde{O}_i$ along a path $\tilde{F}_{1i}$ and then proceeding further to $\tilde{O}_k$ along a path $\tilde{F}_{ik}$ lying above F_{1j} (Fig. 108). Since $\tilde{F}_{ik}$ and $\tilde{F}_{1j}$ lie above the same ground path F_{1j}, $\tilde{F}_{ik}$ is the image of $\tilde{F}_{1j}$ under the covering transformation D_{1i}. Thus D_{1i} transforms the point $\tilde{O}_j$ to $\tilde{O}_k$ hence the covering transformation $D_{1i}D_{1j}$ (first D_{1j}, then D_{1i}!) transforms the point $\tilde{O}_1$ to $\tilde{O}_k$. Thus $D_{1i}D_{1j} = D_{1k}$. This equation states that to the product $\mathfrak{H}\{F_{1i}\} \cdot \mathfrak{H}\{F_{1j}\}$ there corresponds the product $D_{1i}D_{1j}$ of the corresponding covering transformations. We should remember here that one obtains the product of two paths by first traversing the left path and then traversing the right one (§42), but obtains the product of two transformations by first carrying out the right transformation and then the left (§6).

If, in particular, $\mathfrak{H} = 1$, so that $\tilde{\mathfrak{K}}$ is the universal covering complex, then $\mathfrak{F}/\mathfrak{H} = \mathfrak{F}$ and the group of covering transformations is isomorphic to the fundamental group of the base complex. This proves that *the fundamental group of a complex $\mathfrak{K}$ is the group of covering transformations of its universal covering complex.*

As an example, the fundamental group of the projective plane is of order 2, in agreement with the fact that the group of covering transformations of its universal covering surface, which is the 2-sphere, contains two covering transformations: the identity transformation and the interchange of diametrically opposite points.

The fundamental group of the torus has the relation $ABA^{-1}B^{-1} = 1$; it is the free Abelian group having two generators (§47). The group of covering transformations is formed by translations of the Euclidean plane which transform a grid of rectangles into itself (§8).

The group $\mathfrak{D}$ of covering transformations can also be regarded as a factor group in the case of a nonregular covering. If $\mathfrak{H}_i$ denotes, as previously, the subgroup of $\mathfrak{F}$ to which the fundamental group of $\tilde{\mathfrak{K}}$ projects, using initial point $\tilde{O}_i$, and $\mathfrak{Z}_i$ is the normalizer of $\mathfrak{H}_i$ in $\mathfrak{F}$ (that is the totality of elements $\{W\}$ of $\mathfrak{F}$ for which $\{W\}^{-1}\mathfrak{H}_i\{W\} = \mathfrak{H}_i$ is a subgroup of $\mathfrak{F}$), then $\mathfrak{D}$ is isomorphic to $\mathfrak{Z}_i/\mathfrak{H}_i$.

Problems

1. Prove the above assertion.
2. If an n-dimensional complex $\mathfrak{K}^n$ has the n-sphere $\mathfrak{S}^n$ as its universal covering complex, show

that the order of the fundamental group of $\mathfrak{K}^n$ is either 1 or 2 in the case of even n. (The Euler characteristic of $\mathfrak{K}^n$ is a divisor of the Euler characteristic of $\mathfrak{S}^n$.

3. Prove that the fundamental group of the orientable surface of genus $h = 4$ is contained in the fundamental group of the orientable surface of genus $h = 2$ both as a normal subgroup, and also as one of three conjugate subgroups. (cf. the examples in §53 of coverings of the double torus).

4. The nonoriented line elements of the projective plane form a 3-dimensional manifold, the so-called *quaternion space*. Its fundamental group is the quaternion group. Given the subgroups of the quaternion group (see, for example, Speiser [1, p. 767]) determine all 2-, 4-, and 8-sheeted coverings of this manifold. Appropriately assign to them the space of oriented line elements of the projective plane and the spaces of nonoriented and oriented line elements, respectively, of the 2-sphere. (See Problem 3, §14, and Comment 12.)

58. The Monodromy Group

We shall now solve the problem of determining all possible finite sheeted coverings of a finite complex $\mathfrak{K}$. Assume that g is the number of sheets; we wish to find the number of coverings of $\mathfrak{K}$ which have g sheets.

If W is a closed path in $\mathfrak{K}$ having initial point O, then there exist g covering paths $\tilde{W}_1, \tilde{W}_2, \ldots, \tilde{W}_g$ having distinct initial points $\tilde{O}_1, \tilde{O}_2, \ldots, \tilde{O}_g$, respectively. Their endpoints are also distinct. Let us call them $\tilde{O}_{k_1}, \tilde{O}_{k_2}, \ldots, \tilde{O}_{kg}$. We shall assign the permutation

$$\begin{pmatrix} 1 & 2 & \cdots & g \\ k_1 & k_2 & \cdots & k_g \end{pmatrix}$$

to the path W. It is evident that homotopic paths will be assigned the same permutation, and the product of the permutations assigned to two paths W_1 and W_2 will be assigned to the product $W_1 W_2$ of the paths. We thus have found a homomorphism which maps the fundamental group $\mathfrak{F}$ onto a particular group $\mathfrak{M}$ of permutations of g integers. We call $\mathfrak{M}$ the *monodromy group* of the covering complex.

It is easy to find those elements of $\mathfrak{F}$ which map to the unit element of $\mathfrak{M}$; they are the path classes of closed paths W in $\mathfrak{K}$ which lift to closed paths in $\tilde{\mathfrak{K}}$. This property is independent of which point $\tilde{O}_1, \tilde{O}_2, \ldots, \tilde{O}_g$ is chosen as initial point. As before, let us denote the conjugate subgroup into which the fundamental group of $\tilde{\mathfrak{K}}$ projects, when one selects $\tilde{O}_1, \tilde{O}_2, \ldots, \tilde{O}_g$, respectively, as the initial point, by $\mathfrak{H}_1, \mathfrak{H}_2, \ldots, \mathfrak{H}_g$, respectively. The identity permutation will then be assigned to W if and only if W belongs to the intersection $\mathfrak{T}$ of the subgroups $\mathfrak{H}_1, \mathfrak{H}_2, \ldots, \mathfrak{H}_g$. This gives the following theorem:

THEOREM. *The monodromy group $\mathfrak{M}$ of a covering $\tilde{\mathfrak{K}}$ is isomorphic with the factor group $\mathfrak{F}/\mathfrak{T}$ of the fundamental group $\mathfrak{F}$ of the base complex, relative to the intersection $\mathfrak{T}$ of the conjugate subgroups $\mathfrak{H}_1, \mathfrak{H}_2, \ldots, \mathfrak{H}_g$ to which the covering $\tilde{\mathfrak{K}}$ belongs. The order of $\mathfrak{M}$ is thus $\geqq g$ (the number of sheets). The equality sign is valid only if $\mathfrak{H}_1, \mathfrak{H}_2, \ldots, \mathfrak{H}_g$ coincide, that is, when $\tilde{\mathfrak{K}}$ is a* *regular covering. For a regular covering, the monodromy group is isomorphic with this group of covering transformations.*

In general, one speaks of a *representation* of a group $\mathfrak{F}$ when a permutation of g integers is assigned to correspond with each of the elements of the group and the product of the assigned permutations corresponds to the product of two elements. Two representations of the same group are considered to be the same if one representation can be obtained from the other by rearranging the order of the integers $1, 2, \ldots, g$. A covering complex $\tilde{\mathfrak{K}}$ will induce a representation of the fundamental group $\mathfrak{F}$ of the base complex $\mathfrak{K}$. The fact that $\tilde{\mathfrak{K}}$ is a connected complex is expressed by the fact that the representation is transitive: if i and j are any integers appearing in the permutations, then there exists an element $\{W\}$ of $\mathfrak{F}$ which transforms i to j. We need only to select W as the ground path of a path $\tilde{W}$ of $\tilde{\mathfrak{K}}$ which connects the points $\tilde{O}_i$ and $\tilde{O}_j$ with one another.

We shall now show the converse: exactly one covering complex $\tilde{\mathfrak{K}}$ belongs to each transitive representation $\mathfrak{P}$ of the fundamental group $\mathfrak{F}$. Let $\mathfrak{H}_1$ be the subgroup of those elements of $\mathfrak{F}$ which fix the integer 1. Thus $\mathfrak{H}_1$ consists of those paths which are covered by closed paths starting from $\tilde{O}_1$ in the desired covering complex $\tilde{\mathfrak{K}}$. This means that $\tilde{\mathfrak{K}}$ belongs to the group $\mathfrak{H}_1$. Thus when there is any covering complex whatsoever which induces the given representation $\mathfrak{P}$ of $\mathfrak{F}$, it must be the covering complex $\tilde{\mathfrak{K}}$ belonging to the subgroup $\mathfrak{H}_1$ of $\mathfrak{F}$. A representation $\mathfrak{P}'$ of $\mathfrak{F}$, induced by $\tilde{\mathfrak{K}}$, will assign permutations leaving the integer 1 fixed to paths belonging to $\mathfrak{H}_1$ and only to these paths. This implies that $\mathfrak{P}$ and $\mathfrak{P}'$ must coincide, as a consequence of the following group theoretic lemma:

LEMMA. *A transitive representation $\mathfrak{P}$ of a group $\mathfrak{F}$ is determined by the subgroup $\mathfrak{H}_1$ of all elements which fix the integer 1.*

Proof. The elements of one of the residue classes of the decomposition*

$$\mathfrak{F} = \mathfrak{H}_1\{F_{11}\} + \mathfrak{H}_1\{F_{12}\} + \cdots$$

will all transform the integer 1 to the integer 1, while elements of a different residue class transform the integer 1 to a different integer. We may assume that 1 is transformed to i by the residue class $\mathfrak{H}_1\{F_{1i}\}$. Only the integers 1 through g appear in the permutations of the representation $\mathfrak{P}$ and since the integer 1 can be transformed to any of the integers $1, 2, \ldots, g$ by use of appropriate elements of $\mathfrak{F}$—for $\mathfrak{P}$ is to be transitive—then there must exist exactly g residue classes in the decomposition of $\mathfrak{F}$ relative to $\mathfrak{H}_1$:

$$\mathfrak{F} = \mathfrak{H}_1\{F_{11}\} + \mathfrak{H}_1\{F_{12}\} + \cdots + \mathfrak{H}_1\{F_{1g}\}.$$

---

*Even though the lemma deals with abstract group theory and the elements of $\mathfrak{F}$ do not have to be path classes, we shall, for the sake of consistency, retain the curly brackets in denoting the group elements.

If $\{W\}$ is an arbitrary element of $\mathfrak{F}$ and one wishes to know how the integer i will be permuted by $\{W\}$, then one should take into consideration the fact that $\{F_{1i}\}\{W\}$ is a member of a well-determined residue class, let us say $\mathfrak{H}_1\{F_{1k}\}$. The class $\{F_{1i}\}\{W\}$ transforms 1 to k and thus $\{W\}$ transforms i to k.

The problem of determining all g-sheeted coverings of a finite complex $\mathfrak{R}$ has now been reduced to the problem of finding all representations of the fundamental group $\mathfrak{F}$ of $\mathfrak{R}$ by transitive permutation groups on g symbols. [32] To treat the latter problem we begin by determining the generators

$$A_1, A_2, \ldots, A_s$$

and relations

$$R_1(A_i) = 1, \qquad R_2(A_i) = 1, \qquad \ldots, \qquad R_r(A_i) = 1$$

of the fundamental group $\mathfrak{F}$, using the procedure of §46. A representation $\mathfrak{P}$ of $\mathfrak{F}$ is obviously determined by the permutations assigned to the generators. We should, then, assign arbitrary permutations $P_1, P_2, \ldots, P_s$ of the integers 1 through g to the generators $A_1, A_2, \ldots, A_s$ and examine whether the following two conditions are satisfied:

(M1)   The permutation $R_j(P_i)$ corresponding to the left-hand side $R_j(A_i)$ of a defining relation of $\mathfrak{F}$ is the identity permutation.

(M2)   The group of permutations generated by the permutations $P_1, P_2, \ldots, P_s$ is transitive.

Condition (M1) is equivalent to the statement that two products $\prod_i(A_i)$ and $\prod_i'(A_i)$ which represent the same group element of $\mathfrak{F}$ will correspond to the same permutation, so that a permutation is uniquely assigned to each element of $\mathfrak{F}$. The requirement that the product of the associated permutations corresponds to the product of two group elements is then satisfied automatically. Since there are only finitely many possible ways to assign permutations of g integers to the finitely many generators $A_1, A_2, \ldots, A_s$, we can in principle find all of the representations of $\mathfrak{F}$ by trial. When we have found a representation, we can find the subgroup $\mathfrak{H}_1$ of elements which leave the integer 1 fixed and we can construct the covering complex using the method of §55.

Since each covering transformation produces a specific permutation of the points $\tilde{O}_1, \tilde{O}_2, \ldots, \tilde{O}_g$ lying above O, we can also represent the group of covering transformations $\mathfrak{D}$ of a regular covering complex by a regular group of permutations* on the integers $1, 2, \ldots, g$. However, these permutations are the same as the permutations of the monodromy group $\mathfrak{M}$ only when the group of covering transformation is Abelian. For it is known from group theory that there are two representations of the group $\mathfrak{F}/\mathfrak{H}$ by a regular group of permutations; these are the

* A group of permutations is said to be regular if the number of integers permuted is equal to the order of the group.

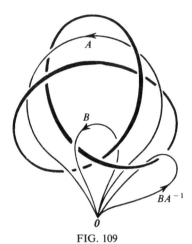

FIG. 109

two representations which arise when one assigns the permutation

$$\begin{pmatrix} X_1 & X_2 & \cdots & X_g \\ X_1 Q & X_2 Q & \cdots & X_g Q \end{pmatrix}$$

to the element Q of $\mathfrak{F}/\mathfrak{H}$ in one case, and assigns the permutation

$$\begin{pmatrix} X_1 & X_2 & \cdots & X_g \\ Q X_1 & Q X_2 & \cdots & Q x_g \end{pmatrix}$$

in the other case. Here $X_1, X_2, \ldots, X_g$ are the elements of $\mathfrak{F}/\mathfrak{H}$. The first group of permutations is the monodromy group. The second is the group of covering transformations.*

In these permutations we need only to replace the elements of $\mathfrak{F}/\mathfrak{H}$ by the points $\tilde{O}_1, \tilde{O}_2, \ldots, \tilde{O}_g$ (which are in one-to-one correspondence with the residue classes of $\mathfrak{F}/\mathfrak{H}$) to obtain the monodromy group and the group of covering transformations, respectively, as permutation groups of the points $\tilde{O}_1, \tilde{O}_2, \ldots, \tilde{O}_g$. The permutations of these two groups coincide only if $X_i Q = Q X_i$ for each Q, that is, if $\mathfrak{F}/\mathfrak{H}$ is Abelian.

As an application of the theory of coverings we will determine the 3- and 4-sheeted coverings of the exterior space of the trefoil knot.** This knot is shown in projection in Fig. 109 as the darker line. From §52 the fundamental group has the single relation

$$A^2 = B^3.$$

Here A is homotopic to the core of the properly located solid torus; B is homotopic to the core of the complementary solid torus in the 3-sphere.

To find the 3-sheeted coverings one must assign a permutation of three integers to the element B such that the third power of this permutation is the square of another permutation of three integers. It is obvious that there are only two ways to do this. One can assign either the identity permutation to a cyclic permutation to B. The identity permutation is not allowable because the permutation group is required to be transitive; thus we assign[†]

$$B \to (1\ 2\ 3).$$

---

* *Editor's note*: The mapping which is defined here from the group of covering transformations to the group of permutations on the g symbols $(X_1, \ldots, X_g)$ appears to be an antihomomorphism, hence not a representation.

** This instructive example was communicated to us in an exchange of correspondence with Mr. H. Kneser.

[†] See Speiser [1, 106], for the cyclic notation of permutations.

Two possibilities exist for A, either the identity permutation

$$A = (1)(2)(3) \tag{I}$$

or

$$A = (1\ 2)(3). \tag{II}$$

In case (I) the monodromy group is a cyclic group of order 3 and the covering is regular. In case (II) the monodromy group is the dihedral group of order 6 (or the double ratio group); since the order 6 is larger than the number of sheets 3, the covering is not regular. The cyclic permutation in case (I) corresponds to the "meridian circle" BA^{-1} in the covering space one first arrives back at the initial point only after traversing the knot three times. Case (II) is different; here $BA^{-1} = (1)(2\ 3)$. The covering path of the meridian circle, beginning at $\tilde{O}_1$ is closed; the covering path beginning at $\tilde{O}_2$ first closes after a twofold traversal of the knot. This completes the description of the 3-sheeted coverings.

If one wishes to determine the 4-sheeted covering spaces, the cyclic permutation $(1\ 2\ 3\ 4)$ and the permutation $(1\ 2)(3)(4)$ are not allowable for B since B^3 will not be the square of a permutation in that case. The identity permutation fails because of the requirement that the monodromy group be transitive. There remain only two possibilities,

$$B \to (1\ 2\ 3)(4), \tag{I}$$

$$B \to (1\ 2)(3\ 4). \tag{II}$$

In case (I), $B^3 = (1)(2)(3)(4)$ and we must have $A^2 = (1)(2)(3)(4)$; thus for A we can allow only an interchange of two elements or a pair of interchanges of two elements. Taking into account the transitivity of the monodromy group, we have only two possibilities in case (I),

$$A \to (1\ 4)(2)(3), \tag{I$_1$}$$

$$A \to (1\ 4)(2\ 3), \tag{I$_2$}$$

and one possibility in case (II),

$$A \to (1\ 2\ 3\ 4). \tag{II}$$

There are then three 4-sheeted coverings of the trefoil knot. Only the last covering is regular, having a cyclic group of covering transformations. In the case (I_1) the monodromy group is the symmetric group on four symbols (the octahedral group); in case (I_2) it is the alternating group (the tetrahedral group). In the three cases the sheet permutations

$$(1\ 2\ 3\ 4), \tag{I$_1$}$$

$$(1\ 3\ 4)(2), \tag{I$_2$}$$

$$(1\ 2\ 3\ 4), \tag{II}$$

correspond to the meridian circle BA^{-1}.

We call the g-fold covering of a knot *locally cyclic* when a small loop around the knot must be traversed g times so that the covering path is closed. Accordingly, (I_1) and (II) are locally cyclic. A covering is defined to be *globally cyclic* if it is a regular covering having a cyclic group of covering transformations. The covering corresponding to (II), but not that corresponding to (I_1), is also globally cyclic. In §77 we shall learn another definition of a globally cyclic covering. It will turn out that a covering which is globally cyclic is also locally cyclic. When we speak simply of a cyclic covering we shall always mean a covering which is globally cyclic.

The trefoil knot possesses a cyclic covering for an arbitrary number of sheets g; this covering is determined by the permutations

$$B \to (1\ 2\ \cdots\ g)^2, \quad A \to (1\ 2\ \cdots\ g)^3.$$

The superscripts 2 and 3 denote here exponents of the permutation. We shall see in §77 that an arbitrary knot has a unique cyclic covering with a given number of sheets.

Problems

    1. Determine the 5-sheeted coverings of the trefoil knot (there are two of them).

    2. Show that there are fifteen different 2-sheeted coverings of the double torus.

We have previously used group theory to survey the coverings of a complex. In the other direction, it is possible to use the theory of covering complexes to examine group theoretic questions, with regard to finding the generators and relations of a subgroup $\mathfrak{H}$ when the generators and relations of the whole group $\mathfrak{F}$ are given. If $\mathfrak{F}$ has the generators $A_1, \ldots, A_a$ and the relations $R_1(A_\nu) = 1, \ldots, R_r(A_\nu) = 1$, we construct a surface complex $\mathfrak{K}$ having a single vertex O, a edges $A_1, \ldots, A_a$ which after orienting become the generators of the fundamental group of $\mathfrak{K}$, and having r surface elements which span the closed paths $R_1(A_\nu), \ldots, R_r(A_\nu)$. From §46, $\mathfrak{F}$ is the fundamental group of $\mathfrak{K}$. The subgroup $\mathfrak{H}$ can now be regarded as the fnndamental group of a covering complex $\tilde{\mathfrak{K}}$ belonging to $\mathfrak{H}$. For simplicity we assume that the number of sheets g, which is equal to the index of $\mathfrak{H}$ in $\mathfrak{F}$, is finite. Then $\tilde{\mathfrak{K}}$ will be a surface complex consisting of g vertices, ag edges, and rg surface elements. The generators and relations can now be found by the procedure of §46.

One can make this method, which is due to Reidemeister [1, 7] independent of geometry and express it in a purely group theoretic form.

CHAPTER NINE

3-DIMENSIONAL MANIFOLDS

59. General Properties

A 3-dimensional closed manifold, $\mathfrak{M}^3$, is a 3-dimensional connected finite homogeneous complex.* The adjective "closed" implies both finiteness and empty boundary, as in the case of surfaces. Because we shall at first deal only with closed manifolds we shall normally omit this adjective. Homogeneity implies that each point of $\mathfrak{M}^3$ possesses a neighborhood which can be mapped topologically onto the interior of the unit 3-dimensional ball. From theorem II of §33, the homology groups at a point of $\mathfrak{M}^3$ are the same as those of a 2-sphere; the Betti numbers of the neighborhood complex are $p^0 = p^2 = 1$, $p^1 = 0$.

We shall interpret the homogeneity condition combinatorially, and for this purpose we shall look at a particular simplicial decomposition of $\mathfrak{M}^3$. Exactly two 3-simplexes must be incident with each 2-simplex $\mathfrak{E}^2$ of the decomposition, so that $p^2 = 1$ for the neighborhood complex of the midpoint of $\mathfrak{E}^2$, as shown in Example 2 of §32. It follows that the 2- and 3-simplexes which lie around an edge $\mathfrak{E}^1$ of $\mathfrak{M}^3$ must form one or more cycles, let us say l cycles, of alternately incident simplexes of dimensions 2 and 3. But then $l = 1$, because the neighborhood complex of the midpoint of $\mathfrak{E}^1$ consists of l 2-spheres which have only the boundary points of $\mathfrak{E}^1$ in common. Consequently, the second Betti number of the neighborhood complex is equal to l; but this should be equal to 1. Finally, if one considers the simplicial star formed by the simplexes of $\mathfrak{M}^3$ lying about one of its vertices $\mathfrak{E}^0$, then the outer boundary of this simplicial star is a 2-dimensional complex in which exactly two 2-simplexes are incident with each 1-simplex, while the simplexes which lie about a vertex form exactly one cycle (for after projection from $\mathfrak{E}^0$

---

* *Editor's Note*: Modern usage does not require a "manifold" to be triangulable, however for $n \leqslant 3$ this is not a restriction. For $n = 1$ this is an elementary result; for $n = 2$ it was established by Rado [1] in 1925; for $n = 3$ it was established by Moise [1]. For $n > 4$ the restriction is genuine (Kirby and Siebenmann [1]). The case $n = 4$ is open.

the corresponding properties must hold for the 2- and 3-dimensional simplexes). It follows that this outer boundary consists of one or more closed surfaces. Since $p^0 = 1$ for the outer boundary, it must consist of only one closed surface (§18). Only the 2-sphere or projective plane are possibilities for this surface, because $p^1 = 0$. The latter possibility is eliminated because $p^2 = 1$ (§41), thus the neighborhood complex of $\mathfrak{C}^0$ is the 2-sphere.

Conversely, if the neighborhood complex of each vertex of a simplicial complex is a 2-sphere, then the complex is a 3-dimensional manifold. For it follows from our assumption that exactly two 3-simplexes are incident with each 2-simplex and that the simplexes lying about an edge form exactly one cycle, implying homogeneity. Our result is:

THEOREM I. *A 3-dimensional connected finite simplicial complex is a closed manifold if and only if the neighborhood complex of each vertex is a 2-sphere.**

From this follows:

THEOREM II. *Each 3-dimensional closed manifold is also a closed pseudomanifold.*

Only the connectability of each pair of 3-simplexes remains to be proven. The totality of 3-simplexes which can be connected to a given 3-simplex is a subcomplex. If it did not exhaust $\mathfrak{M}^3$ then it would have only edges or vertices but no 2-simplexes in common with the subcomplex which remained. But in that case the neighborhood complex of such a vertex or of the midpoint of such an edge would certainly not be a 2-sphere; on the contrary, it would consist of two subcomplexes which had at most common vertices.

As a consequence of Theorem II one can distinguish between orientable and non-orientable 3-dimensional manifolds.

We will later prove a duality theorem for n-dimensional manifolds. This theorem states that relations $q^i = q^{n-i}$ hold for the connectivity numbers of a closed n-dimensional manifold (§69). We have, for the case $n = 3$,

$$q^0 = q^3 \qquad \text{and} \qquad q^1 = q^2 \tag{1}$$

and it follows from this that the Euler characteristic is

$$N = -q^0 + q^1 - q^2 + q^3 = 0. \tag{2}$$

THEOREM III. *The Euler characteristic of a closed 3-dimensional manifold is 0.*

From §23, the Euler characteristic is also equal to the alternating sum of the Betti numbers,

$$N = -p^0 + p^1 - p^2 + p^3.$$

Editor's Note: It is interesting that the authors anticipated here the concept of a "combinational manifold" (see Siebenmann [1]).

Since $p^0 = 1$ for any connected complex and $p^3 = 1$ or 0, respectively according to whether $\mathfrak{M}^3$ is or is not orientable (§24), we have

$$p^2 = p^1 \qquad \textit{for orientable manifolds,}$$

$$p^2 = p^1 - 1 \qquad \textit{for nonorientable manifolds.}$$

We also know from §24 that there are torsion coefficients in dimension 2 only in the nonorientability case and, in fact, exactly one torsion coefficient exists of value 2. The numerical invariants of a closed 3-dimensional manifold whose orientability character is known are then determined by the numerical invariants for dimension 1. The latter can be obtained from the fundamental group, from §48. It will turn out, in fact, that the fundamental group is the most important feature used for showing the distinctness of 3-dimensional manifolds.

Since, by the last equation: $p^1 = p^2 + 1$ for a nonorientable manifold, $p^1 > 0$. This implies

THEOREM IV. *The homology group for the dimension 1 and hence also the fundamental group of a nonorientable 3-dimensional manifold are infinite groups.*

This theorem is true only for 3-dimensional manifolds. A counterexample for higher dimensions is the topological product of the projective plane and the $(n - 2)$-sphere (see Example 1 of §43).

60. Representation by a Polyhedron

Just as one can represent a surface by an appropriate polygon having pairwise associated sides, one can represent a 3-dimensional manifold $\mathfrak{M}^3$ by a 3-dimensional full (solid) polyhedron having pairwise associated surface faces. In fact, $\mathfrak{M}^3$ is constructed by selecting finitely many 3-simplexes and identifying certain of their faces (§10). We can carry through the construction step by step by starting with one simplex and attaching a second simplex along a surface face, then attaching a third simplex, and so forth. We then get a simplicial complex which is homeomorphic to the closed 3-ball. The boundary of the closed 3-ball is simplicially decomposed and we match all of its 2-simplexes in pairs so that both triangles of a pair correspond to the same triangle in $\mathfrak{M}^3$. That is, $\mathfrak{M}^3$ is constructed from the closed 3-ball by defining pairs of equivalent 2-simplexes on the boundary of the ball and identifying the two 2-simplexes of each pair.

We can free ourselves from the restriction that the associated faces are required to be triangles. For this purpose we shall define a 3-dimensional full (solid) polyhedron to be a closed 3-ball (or a topological image of a closed 3-ball) whose boundary $\mathfrak{R}$ has been divided into polygons so that the following conditions are satisfied:

each polygon is at least a 2-gon;
each point of $\Re$ belongs to at least one polygon;
two polygons are either disjoint or have certain common edges or vertices.*

We call the vertices, edges, and polygons of the boundary $\Re$ *the vertices, edges, and faces (surface faces) of the full polyhedron.*

As an example, a solid dodecahedron is a full polyhedron with 20 vertices, 30 edges, and 12 faces. As another example, a closed 3-ball whose boundary has been decomposed into two hemispheres by a great circle also becomes a full polyhedron when one subdivides the great circle by two or more vertices.

We can obtain a simplicial complex by normally subdividing a full polyhedron. The normal subdivision is performed in the following manner. First we decompose each polyhedral edge into two subsimplexes by means of an inner point; more precisely, we topologically map a normally subdivided 1-simplex, which is a simplicial star $\mathfrak{S}t^1$ onto the polyhedral edge. The boundary of a polyhedron face $\mathfrak{a}^2$ will thereby become a simplicial complex homeomorphic to the circle. We now choose a simplicial star $\mathfrak{S}t^2$ whose outer boundary has been simplicially decomposed like the boundary of $\mathfrak{a}^2$ and map this simplicial star topologically onto $\mathfrak{a}^2$ so that the simplicial decomposition of the outer boundary of $\mathfrak{S}t^2$ comes into cover with that of the boundary of $\mathfrak{a}^2$. In this way $\mathfrak{a}^2$ itself becomes a simplicial star. By proceeding in this way with all of the other faces of the polyhedron, we make the boundary of the full polyhedron $\mathfrak{a}^3$ into a simplicial complex. We now choose a simplicial star $\mathfrak{S}t^3$ whose outer boundary has been decomposed like the boundary of the polyhedron and map it topologically onto $\mathfrak{a}^3$ so that the simplicial decomposition of the outer boundary of $\mathfrak{S}t^3$ comes into cover with the simplicial decomposition of the boundary of $\mathfrak{a}^3$. The simplicial decomposition of the polyhedron which we have obtained in this way is called the (first) normal subdivision.

In the full polyhedron we now let the faces be associated pairwise and, in fact, we require that the associations be topological mappings which transform vertices to vertices and edges to edges. This requires that associated faces must have the same number of vertices. We also require that the subsimplexes of the respective normal subdivisions each two associated faces be mapped linearly one onto the other by the topological mapping. Due to the association of faces certain polyhedral edges will become equivalent to one another and certain polyhedral vertices will become equivalent to one another.

We shall require, once and for all, that the associations be such that no oriented edge ever becomes equivalent to itself with opposite orientation:

---

*The boundary of the full polyhedron is a special surface, in the sense of §37, which covers the 2-sphere. It is special because the polyhedral surface elements are now themselves polygons, so that self-intersections are excluded.

hence an edge will never have inner points which are equivalent to one another. This requirement is not essential, because an oriented edge which is equivalent to the same edge having opposite orientation can always be transformed to two edges which no longer have equivalent inner points, by means of a subdivision. On the other hand, both boundary points of an edge can be equivalent.

We denote vertices which become equivalent under the association of faces, by $'a_\nu^0, ''a_\nu^0, \ldots$; after identification of equivalent points these vertices collapse to one and the same point, a_ν^0; the index ν runs from 1 to α^0. In like manner, a set of equivalent edges will be denoted by $'a_\nu^1, ''a_\nu^1, \ldots$; let the number of equivalent edges be α^1 so that ν runs here from 1 to α^1.[*] Two equivalent faces will be denoted by $'a_\nu^2$ and $''a_\nu^2$; let there be α^2 such pairs of equivalent faces. Finally, we shall call the full polyhedron $'a^3$. It is clear that a complex is formed from $'a^3$ by means of identification of equivalent points. But the normal subdivision of $'a^3$ will not yet transform to a simplicial complex under the identification, because certain nonequivalent subsimplexes of $'a^3$ have equivalent vertices, so that in making the identification one will have two different simplexes with the same vertices. Thus condition $(k3)$ of §10 will be violated for the simplicially decomposed point sets arising from the identification. We can insure that a simplicial complex $\mathfrak{R}^3$ will be formed by once again normally subdividing all simplexes of the normally divided full polyhedron before making the identification (see the problem in §13).

Another result of this construction is that $\mathfrak{R}^3$ is uniquely determined topologically, that is, up to a homeomorphic mapping, by the schema of the polyhedron, that is, by a directory listing all vertices, edges, and faces together with their incidences and associations. One will easily recognize that the schema of the first and consequently the second normal subdivision will be determined with all of its associations from the schema of the polyhedron, and the schema of $\mathfrak{R}^3$ will thereby be determined.

In general, of course, $\mathfrak{R}^3$ will not be a manifold when one makes an arbitrary association of faces of a polyhedron. We now ask the question: What condition must the pairwise association satisfy in order that $\mathfrak{R}^3$ be a manifold $\mathfrak{M}^3$?

The homogeneity condition can be violated only at points

$$a_1^0, a_2^0, \ldots, a_{\alpha^0}^0,$$

which correspond to the vertices of the polyhedron. For if P is an inner point of $'a^3$, then there obviously exists a neighborhood of P which is homeomorphic to the interior of a solid ball. And if P' is an inner point of a face $'a_\nu^2$ and P'' is the corresponding point on the associated face $''a_\nu^2$ then we may choose "half-ball" neighborhoods of P' and P'' which join together to form a ball neighborhood when the identification of $'a_\nu^2$ and $''a_\nu^2$ is made.

---

[*] Here we are dealing with the vertices, edges, and faces of the polyhedron itself and not with those of the normal subdivision.

Finally, if $'a_\nu^1, ''a_\nu^1, \ldots, {}^{(r)}a_\nu^1$ is a set of equivalent polyhedral edges and if $P', P'', \ldots, P^{(r)}$ are equivalent inner points of these edges, then we may choose "spherical sector" neighborhoods of these points such that the neighborhoods join together to form a ball neighborhood when associated polyhedron faces are identified.

If, on the other hand, P is one of the points $a_1^0, a_2^0, \ldots, a_{\alpha^0}^0$, then it is possible for the neighborhood complex to be an arbitrary closed surface. Examples can be constructed easily.

In order that $\Re^3$ be homogeneous at these points, i.e. that it be a manifold, it is necessary that its Euler characteristic be zero (§59). This condition is also sufficient. We have the following theorem:

THEOREM I. *A complex, $\Re^3$ which is formed by identifying the faces of a polyhedron will be a manifold if and only if its Euler characteristic $N = 0$.*

Since it is easy to compute N directly by counting, this theorem gives a useful criterion for determining the homogeneity of such a complex.

Proof. The condition that $\Re^3$ arises from a polyhedron by pairwise identification of faces implies that $\Re^3$ can be inhomogeneous only at finitely many points $a_1^0, a_2^0, \ldots, a_{\alpha^0}^0$ and that the neighborhood complexes of these points are closed surfaces in any simplicial decomposition of $\Re^3$. We choose the simplicial decomposition so that

$$a_1^0, a_2^0, \ldots, a_{\alpha^0}^0$$

become vertices and that the simplicial stars

$$\mathfrak{S}t_1^3, \mathfrak{S}t_2^3, \ldots, \mathfrak{S}t_{\alpha^0}^3$$

about these vertices are disjoint. The outer boundary $\mathfrak{A}_\nu$ of $\mathfrak{S}t_\nu^3$ is a closed surface of Euler characteristic N_ν and we must infer from $N = 0$ that $N_\nu = -2$, so that the neighborhood complex is a 2-sphere. Let $\overline{\Re}^3$ be the complex which is formed from $\Re^3$ when one removes all of the stars $\mathfrak{S}t_1^3, \mathfrak{S}t_2^3, \ldots, \mathfrak{S}t_{\alpha^0}^3$ but leaves their outer boundaries in place. This is a 3-dimensional complex having a boundary which is formed by the surfaces $\mathfrak{A}_\nu$. The Euler characteristic of this complex is

$$N + \sum_{\nu=1}^{\alpha^0} (N_\nu + 1).$$

For when one removes all of the simplexes of $\mathfrak{S}t_\nu^3$ (including those of its outer boundary) the Euler characteristic of $\Re^3$ is increased by 1, because the Euler characteristic of a simplicial star is always -1 (§23, Problem 1). If one then replaces the simplexes of $\mathfrak{A}_\nu$, the characteristic is increased by N_ν. We now form a doubling $\overline{\overline{\Re}}_2^3$ of $\overline{\Re}^3$. This is done by identifying corresponding boundary points in two copies of $\overline{\Re}^3$. Its Euler characteristic is then twice as large as that of $\overline{\Re}^3$ minus the sum of the characteristics of the boundary surfaces and is thus equal to

$$2N + 2 \sum_{\nu=1}^{\alpha^0} (N_\nu + 1) - \sum_{\nu=1}^{\alpha^0} N_\nu = 2N + \sum_{\nu=1}^{\alpha^0} (N_\nu + 2).$$

As one can easily check, $\overline{\mathfrak{R}}_2^3$ is homogeneous at all points and is therefore a manifold with Euler characteristic 0. Since we also assumed that $N = 0$,

$$\sum_{\nu=1}^{\alpha^0} (N_\nu + 2) = 0.$$

But since N_ν is the Euler characteristic of a surface, $N_\nu \geqq -2$ and it follows that $N_\nu = -2$, which was to be shown.

The Euler characteristic of a complex $\mathfrak{R}^3$ is

$$N = -\overline{\alpha}^0 + \overline{\alpha}^1 - \overline{\alpha}^2 + \overline{\alpha}^3,$$

where $\overline{\alpha}^i$ denotes the number of i-simplexes of a simplicial decomposition of $\mathfrak{R}^3$ or, more generally, the number of i-blocks of a block system (§23). If $\mathfrak{R}^3$ is given by means of a full polyhedron $'\mathfrak{a}^3$ having pairwise associated faces, then such a block system will be formed by the oriented vertices, edges, faces, and the oriented full polyhedron, where, of course, equivalent vertices, edges, and faces will be counted as nonequivalent blocks. We shall show this in the next section. We need now only to set the numbers of nonequivalent vertices, edges, and faces α^0, α^1, α^2 of $'\mathfrak{a}^3$ for the values of $\overline{\alpha}^0$, $\overline{\alpha}^1$, $\overline{\alpha}^2$ and set $\overline{\alpha}^3 = 1$ to obtain the correct value of the Euler characteristic N.

We now give some examples of polyhedra which will be used later.

EXAMPLE 1. *Lens spaces.* If one measures the complexity of a polyhedron by the number of pairs of associated faces, then the lenses are the simplest polyhedra. A lens is a region of 3-space bounded by two spherical caps which meet in an equatorial circle (the sharp edge of the lens). Divide the equatorial circle into p equal circular arcs. Its two caps then become p-gons $'\mathfrak{a}^2$ and $''\mathfrak{a}^2$. We can accomplish the association of $'\mathfrak{a}^2$ and $''\mathfrak{a}^2$ in several ways. We can, for example, reflect the lower cap $'\mathfrak{a}^2$ in a plane passing through the equatorial circle to bring it into coincidence with the upper cap $''\mathfrak{a}^2$. More generally, we can let the reflection follow a rigid rotation of one of the caps onto itself, let us say $'\mathfrak{a}^2$, by the angle $2\pi q/p$ radians. We can describe this more briefly by saying that the association of $'\mathfrak{a}^2$ with $''\mathfrak{a}^2$ occurs via a screw rotation through the angle $2\pi q/p$. The assignment of points to the upper lens surface is completely determined by the screw angle $2\pi q/p$. It is therefore no restriction to assume that the integers q and p are relatively prime. We can also assume that $0 \leqq q \leqq p/2$, since it is obviously unimportant for the resulting manifold whether the screw rotation is clockwise or counterclockwise. The p vertices and edges into which the equatorial circle decomposes become equivalent, so that $\alpha^0 = \alpha^1 = \alpha^2 = \alpha^3 = 1$ and thus $N = 0$. The complex which is formed by means of the identifications described above is then in fact homogeneous. This is easily checked directly. The complex is called the *lens space* (p, q). In Fig. 110, $p = 3$, $q = 1$. In the case $p = 2$ the association becomes an "interchange of diametrically opposite points" when one takes the solid ball as the lens and a great circle on its surface as the equatorial circle. The resulting lens space is the projective space $\mathfrak{P}^3$, from §14. For $p = 1$ no polyhedron results, because the equatorial circle would be divided by only one vertex so that as a polyhedron edge it would exhibit an unallowable self-intersection. If one inserts an additional point of division, then a polyhedron will arise and it closes to form the 3-sphere $\mathfrak{S}^3$, since the rotation is eliminated here (§14).

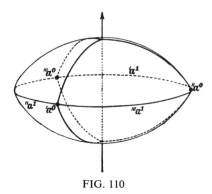

FIG. 110

It is indicative of the difficulty of the topology of three dimensions that the homeomorphism problem has not even been solved for these simple lens spaces. That is, it has not been determined in general when two lens spaces (p, q) and (p', q') are homeomorphic.* Additional discussion is given in §62 and §77.

EXAMPLE 2. *The topological product of three circles*. Just as we can form the topological product of two circles, the torus, from a square whose opposite sides are identified by means of translations, we can form the topological product of three circles from a cube whose opposite sides are identified by means of translations. There exist three nonequivalent edges and three faces, while all vertices become equivalent to one another; therefore $N = 0$.

61. Homology Groups

In the previous section we derived the condition under which the complex arising from a polyhedron by identification of faces becomes a manifold. In the present discussion we do not assume that this condition is satisfied by the full polyhedron under consideration; that is, the complex $\Re^3$ may be inhomogeneous at the polyhedral vertices.

The full polyhedron $'\mathfrak{a}^3$, its faces $'\mathfrak{a}^2_\nu$ and $''\mathfrak{a}^2_\nu$, its edges $'\mathfrak{a}^1_\nu, ''\mathfrak{a}^1_\nu, '''\mathfrak{a}^1_\nu, \ldots$, and its vertices $'\mathfrak{a}^0_\nu, ''\mathfrak{a}^0_\nu, '''\mathfrak{a}^0_\nu, \ldots$ are respectively 3-, 2-, 1-, and 0-dimensional elements (closed n-balls). That is, they are orientable pseudomanifolds with boundary. It is therefore possible to coherently orient an arbitrary simplicial decomposition of these elements. Among the simplicial decompositions we shall consider only the twofold normal subdivision of $'\mathfrak{a}^3$. We orient the vertices with the + sign. We orient the edges and the faces so that associated edges and associated faces are given the same orientation, so that orientations are preserved under the identification mapping of equivalent elements. In our illustrations we do not show the simplicial decomposition; we denote the orientation of an edge by an arrowhead set upon it and denote the orientation of a face by a circular arrow set into it. If $'\mathfrak{a}^k_\nu, ''\mathfrak{a}^k_\nu, \ldots$ is a set of equivalent

---

Editor's Note: Two lens spaces (p, q), (p', q') are homeomorphic if and only if $p = p'$ and $q' \equiv \pm q \pmod p$ or $qq' \equiv \pm 1 \pmod p$. See Brody [1] for a self-contained proof. Earlier proofs are due to combined results of Reidemeister and Moise. The homotopy classification of lens spaces is also known (e.g., see Hilton and Wylie [1, pp. 223–225]).

k-dimensional elements of the polyhedron, then we denote the k-chain formed from the oriented elements by the corresponding Roman letters $'a_\nu^k, ''a_\nu^k, \ldots$. The symbol a_ν^k will denote the k-chain in the simplicial complex $\Re^3$ which results from the chains $'a_\nu^k, ''a_\nu^k, \ldots$ when equivalent points are identified. We then have the following chains on $\Re^3$, having dimensions 0 through 3:

$$a_1^0, \quad a_2^0, \quad \ldots, \quad a_{\alpha^0}^0,$$
$$a_1^1, \quad a_2^1, \quad \ldots, \quad a_{\alpha^1}^1,$$
$$a_1^2, \quad a_2^2, \quad \ldots, \quad a_{\alpha^2}^2,$$
$$a^3.$$

We claim that these chains form a system of blocks on $\Re^3$ (§22). The chains are linearly independent because the simplexes are disjoint. The boundary of each a_χ^k is composed of chains a_i^{k-1}. To show that (Bl3) and (Bl4) are also satisfied let us consider a simplicial chain U^k on $\Re^3$ such that the boundary of U^k is a linear combination of the chains a_i^{k-1} and can be the $(k-1)$-chain 0 under certain conditions. We attempt to construct a homologous chain $V^k \sim U^k$ which is composed entirely out of chains a_χ^k. It will be useful to consider a chain $'U^k$ on the polyhedron $'a^3$ as a "preimage" of U^k, where $'U^k$ arises from U^k by associating with each k-simplex of U^k one of the corresponding simplexes of $'a^3$ and providing the latter simplex with the same orientation and multiplicity as the former simplex. In the case $k = 3$, each of the coherently oriented 3-simplexes of $'a^3$ must appear with the same multiplicity in $'U^k$. Otherwise $\Re \partial U^k$ would not be a linear combination of the chains a_ν^2. Thus $'U^k$ is a multiple of $'a^3$ and our goal has been achieved for the case $k = 3$. In the case $k = 2$ we consider chains $'U_m^2$ which are formed from those simplexes of $'U^2$ which do not lie on the boundary of $'a^3$. The boundary of $'U_m^2$ lies on the boundary of $'a^3$ and is itself, when considered as a closed 1-chain on a 2-sphere, the boundary of a chain $'U_r^2$ on the boundary of $'a^3$. Then $'U_m^2 - 'U_r^2$ is a closed chain and is null homologous on $'a^3$. It is then also true on $\Re^3$ that $U_m^2 \sim U_r^2$, so when we replace U_m^2 by U_r^2 we go from the chain U^2 to a homologous chain V^2. A preimage $'V^2$, will contain no inner simplexes of $'a^3$ and we can standardize its selection so that, of the two possible preimages of a 2-simplex of V^2, we choose that one which belongs to $'a_\nu^2$ (and not to $''a_\nu^2$). Then $'V^2$ will be a linear combination of the chains $'a_\nu^2$. For, since the boundary of V^2 is a linear combination of the a_ν^1, each of the coherently oriented 2-simplexes of $'a_\nu^2$ must appear with the same multiplicity in $'V^2$. Thus V^2 is a chain such as we are seeking. This push out procedure also leads to our goal in the case $k = 1$. We first push $'U^1$ onto the boundary of $'a^3$ and subsequently push it in the same way onto the boundary of each individual $'a_\nu^2$. The chain V^1 which one obtains from U^1 is itself a linear combination of the chains a_ν^1.

We can consequently compute the homology groups of $\Re^3$ from the block system of the a_ν^k using the procedure of §22. In particular, the simple computation of the Euler characteristic given in §60 is thereby justified.

Obviously $\mathfrak{R}^3$ is orientable if and only if $\mathfrak{R}\partial a^3 = 0$. That is, in the equation

$$\mathfrak{R}\partial' a^3 = \sum_{\nu=1}^{\alpha^2} \left({}'\varepsilon_\nu\, 'a_\nu^2 + {}''\varepsilon_\nu\, ''a_\nu^2 \right)$$

we have ${}'\varepsilon_\nu + {}''\varepsilon_\nu = 0$. This implies that ${}'a_\nu^2$ and ${}''a_\nu^2$ must be oppositely oriented on the boundary of the polyhedron, where orientation is assigned relative to a coherent orientation of the whole polyhedron. One say that for an observer standing outside the polyhedron the orientation arrows of any two associated faces must have opposite senses of traversal. Two oriented faces ${}'a_\nu^2$ and ${}''a_\nu^2$, for which this is the case are said to *form an association of type one* and in the other case *form an association of type two*. The complex $\mathfrak{R}^3$ is then orientable if and only if all associations of faces are of type one. The lens spaces and the topological product of three circles are orientable.

EXAMPLE 1. *Lens spaces.* The block incidence matrices of the lens space (p, q) are

E^0	a^1		E^1	a^2		E^2	a^3
a^0	0		a^1	p		a^2	0

These block incidence matrices already stand in normal form. We list the number of elements α^k, the ranks γ^k, and the Betti numbers p^k calculated from the formula $p^k = \alpha^k - \gamma^k - \gamma^{k-1}$ (§21) together in a table:

$k =$	0	1	2	3
α^k	1	1	1	1
γ^k	0	1	0	–
p^k	1	0	0	1

There is a single torsion coefficient, a 1-dimensional torsion coefficient having the value p. The integer q does not appear at all in the homology groups.

The significance of a 1-dimensional torsion coefficient becomes clear from this example. The existence of such a torsion coefficient having the value p imples that there exists a closed 1-chain which bounds a 2-chain only when first traversed p times. The equatorial circle of the lens, which consists of p equivalent edges ${}'a^1, {}''a^1, \ldots, {}^{(p)}a^1$, bounds a surface element, the lens cap. In the case $p = 2$ the lens space is the projective space $\mathfrak{P}^3$ and there exist two equivalent edges which become homotopic, after identification, to a projective line. We already know that a projective line will bound a surface element when traversed twice; this is a projective plane which has been cut open by the line (§17). The nomenclature "torsion coefficient" now becomes justified as well. The full polyhedron becomes closed to form a lens space after one has twisted it about the lens axis, that is, one associates the lower lens cap with the upper lens cap by means of a screw rotation.

EXAMPLE 2. *Octahedron space* is obtained when one twists opposite lying triangles of a full octahedron by $\pi/3$ radians relative to one another, associates them in pairs with one another, and then identifies them. The edge network of the polyhedron is shown in Fig. 111, which is in fact the stereographic projection of the octahedral surface into the plane, which we think of as being closed by a point at infinity to form the 2-sphere. The point at infinity is a vertex of the edge network. For ease of recognition, associated elements in the figure are designated with the same symbol and not with ${}^{(\lambda)}a_\nu^k$ as in the general theory.

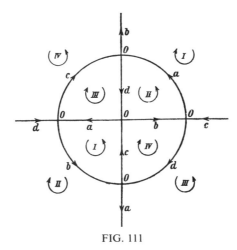

FIG. 111

By use of the procedure of §87, the incidence matrix $\mathbf{E}^1$ can be brought to normal form:

$$\begin{bmatrix} 3 & 0 & 0 & 0 \\ 0 & 1 & 0 & 0 \\ 0 & 0 & 1 & 0 \\ 0 & 0 & 0 & 1 \end{bmatrix}$$

The number of systems of equivalent vertices, edges, and faces, as well as the ranks of the incidence matrices and the Betti numbers p^k, are summarized in a table as in the previous example:

$k =$	0	1	2	3
α^k	1	4	4	1
γ^k	0	4	0	–
p^k	1	0	0	1

There exists a single 1-dimensional torsion coefficient, having the value 3. The homology group for dimension 1 is therefore of order 3.

Problem

In octahedral space show that the four edges a, b, c, d are homologous to one another and each edge becomes null homologous only after it has been traversed three times.

62. The Fundamental Group

Just as we determined the homology groups in the previous section, we shall now determine the fundamental group of the 3-dimensional manifold $\Re^3$ from its polyhedron $'\mathfrak{a}^3$. We achieve this by means of the following theorem.

THEOREM I. *The fundamental group of $\Re^3$ is the same as the fundamental group of the surface complex $\Re^2$ which arises from the boundary of the polyhedron $'\mathfrak{a}^3$ by means of the identification of equivalent points.*

Proof. We must demonstrate two facts. (I) Each closed path w whose initial point is a fixed point O of $\Re^2$ is deformable to a path w_1 of $\Re^2$. (II) If a path w_1 of $\Re^2$ is null homotopic in $\Re^3$, then it is also null homotopic in $\Re^2$.

To prove (I) we examine the point set w' of all points of the polyhedron $'\mathfrak{a}^3$ which map to point of w. If there exists an interior point P of $'\mathfrak{a}^3$ which does not belong to w', then we deform w' to lie upon the boundary of $'\mathfrak{a}^3$ by "projecting outward from P," that is, we let the points of w' run at constant speed to the boundary of $'\mathfrak{a}^3$ along straight line rays drawn from P. Corresponding to this projection there exists a deformation in $\Re^3$ of w to a path w_1 of $\Re^2$. If w' is to exhaust all of the interior points of $'\mathfrak{a}^3$, one must first free such a point P. We can do this, for example, by means of a simplicial approximation of w in $\Re^3$ (which, from §31 is, in fact, a deformation of w).

One can prove (II) in the same way. If $\mathfrak{A}$ denotes a "singular deformation rectangle" which sweeps over w_1 during the deformation to the point O and if $'\mathfrak{A}$ denotes the set of all points of $'\mathfrak{a}^3$ which map to $\mathfrak{A}$, then project $'\mathfrak{A}$ outward to lie upon the boundary of $'\mathfrak{a}^3$ from an interior point P of $'\mathfrak{a}^3$ such that P does not belong to $'\mathfrak{A}$. This will correspond to a pushing out of $\mathfrak{A}$ from $\Re^3$ into the subcomplex $\Re^2$. If no such point P exists, then approximate $\mathfrak{A}$ simplicially, as was done previously. That is, the continuous mapping of the sufficiently finely divided simplicially decomposed deformation rectangle $\overline{\mathfrak{A}}$ onto $\mathfrak{A}$ will be simplicially approximated in $\Re^3$.

The relations belonging to the fundamental group can be determined by the procedure of §46. This is particularly easy when all vertices of the polyhedron are equivalent. The vertices then correspond to one and the same point O in $\Re^3$, which will be taken to be the initial point of the closed paths. Since the auxiliary paths do not appear, the oriented edges of the polyhedron or, rather, the closed paths in $\Re^3$ corresponding to these edges are the generators of the fundamental group. The group relations are obtained by running around the polyhedron faces.

This simple case occurs for example with the lens spaces (p, q). A single generator a exists (we now omit the indices of the previous section). We obtain the single relation $a^p = 1$ by running around one of the two equivalent polygons. *The fundamental group of the lens space (p, q) is therefore the cyclic group of order p and is independent of q.*

Two cases are now conceivable. Either two lens spaces (p, q) and (p, q') are homeomorphic when $q \neq q'$ or else our previously known invariants, the homology groups and the fundamental group, are insufficient to prove that the two lens spaces are distinct. It turns out that both cases occur in actuality, depending upon the nature of the integers q and q'. With regard to the homeomorphism we can easily prove the following theorem:

THEOREM II. *The lens spaces (p, q) and (p, q') are homeomorphic if q and q' satisfy the congruence*

$$qq' \equiv \pm 1 \pmod p. \tag{1}$$

Since this congruence has exactly one solution satisfying the condition $0 \leq q' \leq p/2$ for a given q, q and q' each determine the other uniquely.

Proof. We decompose the lens $\mathfrak{L}$ having screw angle $2\pi q/p$ into p equal pie-shaped tetrahedra $\mathfrak{T}_1, \mathfrak{T}_2, \ldots, \mathfrak{T}_p$ by means of p half-planes passing through the lens axis (Fig. 110 shows the case

$p = 3$). We construct a new lens $\mathfrak{L}'$ having screw angle $2\pi q'/p$ from these pieces. The tetrahedron $\mathfrak{T}_i$ has a triangle $\underline{\Delta}_i$ in common with the lower lens cap and has a triangle $\overline{\Delta}_i$ in common with the upper lens cap. The lens axis b is common to all $\mathfrak{T}_i$. The lens edges lying opposite to b are all equivalent to one another and form one and the same edge a in the lens space. The triangles are equivalent in pairs to one another, as a consequence of the association of the lens caps. In particular, $\underline{\Delta}_i$ is equivalent to $\overline{\Delta}_{i+q}$, where $i + q$ is to be reduced mod p if necessary. We now construct the new lens $\mathfrak{L}'$ from the p tetrahedra. To do this we first join tetrahedron $\mathfrak{T}_{1+q}$ to tetrahedron $\mathfrak{T}_1$ by identifying the equivalent triangles $\overline{\Delta}_{1+q}$ and $\underline{\Delta}_1$. We then join $\mathfrak{T}_{1+2q}$ to $\mathfrak{T}_{1+q}$ by identifying triangles $\overline{\Delta}_{1+2q}$ and $\underline{\Delta}_{1+q}$, and so forth. We ultimately arrive at a lens $\mathfrak{L}'$ which differs from the original lens only in the fact that the edges a and b have interchanged their roles. If $2\pi q'/p$ is the screw angle of $\mathfrak{L}'$, then in the cyclic sequence $\mathfrak{T}_1, \mathfrak{T}_{1+q}, \ldots, \mathfrak{T}_2, \ldots$ in which the tetrahedra are arranged about the axis a of $\mathfrak{L}'$, the tetrahedra $\mathfrak{T}_1$ and $\mathfrak{T}_2$, of which the latter coincides with $\mathfrak{T}_{2+xp}$, will differ by q' places on one side and will differ by $p - q'$ places on the other side. Consequently, the difference of indexes of $\mathfrak{T}_{2+xp}$ and $\mathfrak{T}_1$ in the cyclic sequence will be equal to qq' or $q(p - q')$, respectively. But this difference is also equal to $(2 + xp) - 1$. Thus,

$$(2 + xp) - 1 = qq' \quad \text{or} \quad (2 + xp) - 1 = q(p - q'),$$

which gives

$$qq' \equiv \pm 1 \pmod{p}.$$

Thus the lens spaces $(7, 2)$ and $(7, 3)$ are homeomorphic for example because $2 \cdot 3 \equiv -1$ (mod 7). But one cannot decide from Theorem II whether the lens spaces $(p, 1)$ and $(p, 2)$ are homeomorphic. We shall later (§77) introduce an invariant which is not associated to the fundamental group and which permits certain lens spaces to be distinguished. It can show, for example, that $(5, 1)$ and $(5, 2)$ are distinct spaces. On the other hand, Theorem II does not distinguish between $(7, 1)$ and $(7, 2)$.

It should be noted that every lens space can be decomposed into two full rings having a torus as their common boundary. Bore a full cylinder $\mathfrak{B}$ out of the lens by boring along (and concentric with) the lens axis b. After identification of the lens caps, $\mathfrak{B}$ will close to form a full ring. The same is true for the complementary space $\mathfrak{A}$ which remains after boring out $\mathfrak{B}$. To see this we need only to decompose the lens $\mathfrak{L}$ into p tetrahedra as previously and to assemble the lens $\mathfrak{L}'$ from them. The complementary space $\mathfrak{A}$ then becomes a cylinder in $\mathfrak{L}'$ which surrounds the axis a of $\mathfrak{L}'$. When equivalent points are identified, $\mathfrak{A}$ will close to form a full ring (cf. §63).

As another example we shall investigate the *spherical dodecahedron space* (Kneser [8, p. 256]). This space arises from a dodecahedron when one twists opposite lying pentagons by $\pi/5$ radians relative to one another and then identifies them. The edge network of the dodecahedron, which completely determines the space, is drawn in Fig. 112. There exist $\alpha^0 = 5$ nonequivalent vertices O, P, Q, R, S. There are ten nonequivalent edges, each formed by identifying three equivalent edges. The Euler characteristic is $N = -5 + 10 - 6 + 1 = 0$; thus we are dealing with a manifold. We select O as the initial point of the closed paths and we select the paths $a, h, f^{-1}, f^{-1}d$ as auxiliary paths leading to the vertices P, Q, R, S, respectively. The generating path classes of the fundamental group will then be represented by the closed paths

$$
\begin{aligned}
&A = aa^{-1}, &\quad &B = abh^{-1}, \\
&C = hcf, &\quad &D = f^{-1}d(d^{-1}f), \\
&E = (f^{-1}d)e, &\quad &F = f^{-1}f, \\
&G = (f^{-1}d)ga^{-1}, &\quad &H = hh^{-1}, \\
&J = aif, &\quad &K = hk(d^{-1}f).
\end{aligned}
$$

The relations of type (I) follow after one writes the right-hand sides of the above equations in capital letters instead of lower case letters. We then get $A = D = F = H = 1$ and the remaining relations become trivial.

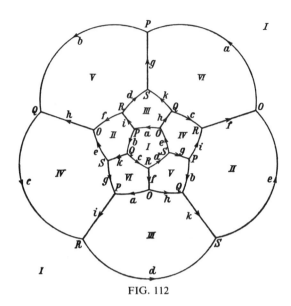

FIG. 112

By running around the pentagons we get the following six relations of type (II):

$$\left.\begin{array}{r} ABCDE = 1 \\ BKEF^{-1}J^{-1} = 1 \\ AJDK^{-1}H^{-1} = 1 \\ CJ^{-1}G^{-1}EH = 1 \\ BH^{-1}F^{-1}DG = 1 \\ AG^{-1}K^{-1}CF = 1 \end{array}\right\}
\quad \text{or} \quad
\left\{\begin{array}{r} BCE = 1 \\ BKEJ^{-1} = 1 \\ J = K \\ CJ^{-1}G^{-1}E = 1 \\ B = G^{-1} \\ G^{-1}K^{-1}C = 1 \end{array}\right.$$

Elimination of G and K gives

$$BCE = 1,$$
$$BJEJ^{-1} = 1,$$
$$CJ^{-1}BE = 1,$$
$$BJ^{-1}C = 1.$$

From the first and fourth of these relations we get

$$E = C^{-1}B^{-1}, \qquad J = CB.$$

Using these to eliminate E and J from the second and third relations, we get

$$BCBC^{-1} \cdot B^{-2}C^{-1} = 1 \tag{I}$$

and

$$CB^{-1}C^{-1}BC^{-1}B^{-1} = 1. \tag{II}$$

We determine the first homology group from these two relations by making relations (I) and (II) Abelian! As always, we use additive notation for Abelian groups, and we denote the elements of the homology group by means of symbols with bars. We get

$$\bar{C} = 0, \tag{Ī}$$

$$-\bar{C} - \bar{B} = 0; \tag{ĪĪ}$$

thus $\bar{B} = \bar{C} = 0$. That is, the first homology group consists of the null element alone. Since the dodecahedron space is orientable, we have the following values for the Betti numbers:

$$p^0 = 1, \qquad p^1 = p^2 = 0, \qquad p^3 = 1.$$

There are no torsion coefficients.

The numbers above are just the numerical invariants of the 3-sphere. Thus the homology groups are not in themselves sufficient to distinguish whether the 3-sphere does or does not coincide with the dodecahedron space. To decide this we examine whether the fundamental groups of these spaces differ. To do so we transform the somewhat untransparent relations (I) and (II) further. We set (II) into (I) at the position indicated by the dot in (I). In place of (I) we get the relation

$$BCBC^{-1} \cdot CB^{-1}C^{-1}BC^{-1}B^{-1} \cdot B^{-2}C^{-1} = 1,$$

or, after shortening this,

$$B^2C^{-1}B^{-3}C^{-1} = 1. \tag{I'}$$

By introducing a new generator U into (I) and (II) where U is defined by $C = U^{-1}B$, we get

$$B^2 \cdot B^{-1}U \cdot B^{-3} \cdot B^{-1}U = 1,$$

$$U^{-1}B \cdot B^{-1} \cdot B^{-1}U \cdot B \cdot B^{-1}U \cdot B^{-1} = 1$$

or

$$B^4 = UBU, \qquad U^2 = BUB$$

or also

$$B^5 = (BU)^2 = U^3. \tag{III}$$

We recognize from the relations (III) that the dodecahedron space is not homeomorphic to the 3-sphere. That is because the fundamental group does not consist of the unit element alone. Instead, the relations (III) are satisfied by the icosahedral group, if one interprets B as a rotation of $2\pi/5$ radians about a vertex of the icosahedron and interprets U as a rotation of $2\pi/3$ radians, having the same sense of rotation, about the midpoint of a triangle adjoining that vertex. The icosahedral group is therefore either the group (III) itself or a factor group of (III). In either case the fundamental group does not consist of just the unit element alone. It is possible to show, by the way, that (III) is of order 120 and is the "binary icosahedral group."[*]

The spherical dodecahedron space is a manifold which has the same homology groups as a 3-sphere without, however, being homeomorphic to it. Such a manifold is called a *Poincaré space.* Infinitely many Poincaré spaces are known. But the spherical dodecahedron space is the only one known which has a finite fundamental group.[33]

The homology groups are not sufficient to characterize the 3-sphere. Whether the 3-sphere is characterized by its fundamental group is the content of the "Poincaré conjecture," which remains unproven to this day. Since the fundamental group of the 3-sphere consists of the unit element alone, we can also state the problem as follows: Aside from the 3-sphere do there exist other 3-dimensional closed manifolds such that each closed path can be contracted to a point (is null homotopic)?[**]

[*] *Jahresber. Deutsch. Math.-Verein.* **42** (1932), problem 84, p. 3.

[**] *Editor's Note*: As of January 1979, this famous problem is still open! However, new (unpublished) results of W. Thurston have established the following weak version: If a simply connected 3-manifold M is a cyclic branched covering space of S^3, then M is in fact homeomorphic to S^3.

Problems

1. The *hyperbolic dodecahedron space* is formed by closing the full dodecahedron as follows. One twists opposite lying faces by $3\pi/5$ radians (not $\pi/5$ radians as in the case of spherical dodecahedron space) relative to one another and identifies them. Show that the hyperbolic dodecahedron space is a manifold and prove, by presenting its fundamental group, that it has three torsion coefficients of value 5 and has the Betti number $p^1 = 0$.

2. Prove that the relations of the fundamental group of the octahedron space (§61) are

$$abc = adb = acd = bdc = 1.$$

Prove that the tetrahedron group is a factor group of this group. (The relations are valid in the tetrahedral group when one interprets a, b, c, d as the four rotations by $2\pi/3$ radians about the four vertices of the tetrahedron.)

3. One can derive a new 3-dimensional manifold from two 3-dimensional manifolds $\mathfrak{M}_1$ and $\mathfrak{M}_2$ by the process of "connected sums." One bores a small ball out of each manifold and glues together the two resulting "manifolds with boundary" along their bounding spherical surfaces. This can be done in two essentially different ways, either by preserving or by reversing orientation. Determine the fundamental group of the resulting closed manifold and the first homology group in each case, assuming that the fundamental groups of $\mathfrak{M}_1$ and $\mathfrak{M}_2$ are known.

4. Using the results of Problem 3 construct a 3-dimensional manifold having an arbitrarily given first homology group.

63. The Heegaard Diagram

We have previously presented general theorems dealing with 3-dimensional manifolds but we have only become acquainted with individual examples of manifolds. In contrast to the situation in 2 dimensions, the problem of cataloging all 3-dimensional manifolds is unsolved. We can look upon the *Heegaard diagram*, which we now explain, as a method of proceeding to its solution.

A *handle body of genus h* arises from a 3-dimensional ball when one carries out identifications of type (I) (§61) pairwise on $2h$ distinct disks on the surface of the ball. We can perform the identification in ordinary 3-space by deforming the ball in a manner so that associated disks come into coincidence with one another. The body arising in this way can also be regarded as a full sphere having h full handles attached. Its surface is an oriented surface of genus h. The h circles which arise from the boundaries of the pairwise identified surface elements are called the meridian circles of the handle body. A handle body of genus 1 is an ordinary solid torus. We have the following theorem:

THEOREM. *One can obtain each orientable 3-dimensional manifold $\mathfrak{M}^3$ by gluing together two handle bodies of the same genus along their surfaces, that is, by mapping their surfaces topologically one onto the other and identifying corresponding points.*

Proof. Consider the 3-dimensional manifold to be simplicially decomposed and bore out all of the α^0 vertices of the decomposition by the procedure of boring out a small ball surrounding each vertex. Likewise, bore out the α^1

edges by boring out small full cylinders about them, where the cylinders connect the balls surrounding the endpoints of the edges. The subspace $\overline{\mathfrak{M}}_1^3$ which remains after one has bored out the vertices and edges is composed of α^3 simplexes which are truncated along their edges. This space is a handle body. For one can construct it by starting with a truncated simplex $\mathfrak{E}_1^3$, joining a neighboring simplex $\mathfrak{E}_2^3$ along one face, then joining an additional simplex $\mathfrak{E}_3^3$ to either $\mathfrak{E}_1^3$ or $\mathfrak{E}_2^3$ along a face and so forth. If this exhausts all of the simplexes, then a topological full sphere will result, in which certain disjoint surface elements of its surface still remain to be identified pairwise. It is clear that the association of these surface elements must be of type (I) since in the other cases $\overline{\mathfrak{M}}_1^3$ would be nonorientable and consequently $\mathfrak{M}^3$ would also be nonorientable. One can show in like manner that the bored-out part $\overline{\mathfrak{M}}_2^3$, consisting of α^0 full spheres and α^1 full cylinders is a handle body.

From now on we can, then, represent each orientable 3-dimensional manifold by means of an orientable surface $\mathfrak{M}^2$ of genus h; this is the surface common to $\overline{\mathfrak{M}}_1^3$ and $\overline{\mathfrak{M}}_2^3$ and we must mark on $\mathfrak{M}^2$ how the handle bodies are placed upon it. We do this by constructing the meridian circles of $\overline{\mathfrak{M}}_1^3$ and of $\overline{\mathfrak{M}}_2^3$ on $\mathfrak{M}^2$. These form two systems Σ_1 and Σ_2, each of h circles which are free of double points and which are disjoint within each system (but not necessarily between the two systems). The systems of meridian circles have the property that $\mathfrak{M}^2$ becomes a 2-sphere having $2h$ holes when it is cut apart along the h circles of Σ_1 (or of Σ_2).* If one is given an arbitrary orientable surface of genus h, containing two such systems of h circles Σ_1 and Σ_2, then a 3-dimensional manifold will always be determined. For one can cut apart $\mathfrak{M}^2$ along the circles of Σ_1 to form a 2-sphere with h holes, close the holes with surface elements (caps), and fill in the resulting 2-sphere to form a closed 3-ball. Pairwise identification of the caps results in a handle body whose surface is $\mathfrak{M}^2$ and whose meridian circles are the circles of Σ_1. One constructs the handle body for the system Σ_2 in like manner. A closed surface $\mathfrak{M}^2$ together with the two systems Σ_1 and Σ_2 is called a Heegaard diagram of the manifold $\mathfrak{M}^3$.

One can easily find the Heegaard diagrams on the torus. They consist of any two essential cuts (double-point-free closed edge sequences of a simplicial decomposition) which are not null homotopic. The lens spaces are among the manifolds which have their Heegaard diagrams on the torus. As we saw in §62, each lens space can be decomposed into two full rings when the lens axis is bored out. The topological product of a circle with the 2-sphere also has its Heegaard diagram lying on the torus, because this topological product arises by identifying corresponding surface points of two full rings which have been mapped topologically one onto the other; one can thereby say that it arises by means of a doubling of the full ring (§36). In this example the two meridian

---

*The cutting apart makes sense when the circles of Σ_1 and Σ_2 are composed of edges of a simplicial decomposition. We assume this to be the case.

circles of the Heegaard diagram coincide. One can prove that this exhausts all of the manifolds whose Heegaard diagrams lie on the torus.[34]

Little is known about Heegaard diagrams of higher genus.* The spherical dodecahedron space has a Heegaard diagram of genus 2. H. Poincaré first studied this space by means of its Heegaard diagram.[33]

One can also define the Heegaard diagram for nonorientable manifolds. One needs only to introduce nonorientable handle bodies. These arise from a full sphere when one pairwise associates disjoint polygons on its surface, where at least one association is of type two. The surface which results is a nonorientable surface having an even characteristic N and is thus also of even genus $k = 2k'$ [from §38 the characteristic remains the same regardless of whether a type (I) or a type (II) association is made]. In the same way as for orientable manifolds, it can be proved that each closed nonorientable 3-dimensional manifold can be decomposed into two nonorientable handle bodies.

The construction of 3-dimensional manifolds has been reduced to a 2-dimensional problem by means of the Heegaard diagram. This problem is the enumeration of all Heegaard diagrams. Even if the diagrams could all be enumerated, the homeomorphism problem in 3 dimensions would not be solved because a criterion is still lacking for deciding when two different Heegaard diagrams generate the same manifold. The enumeration has been carried out successfully in the simplest case, that of Heegaard diagrams of genus 1, but the problem of coincidence of manifolds, that is, the homemorphism problem for lens spaces, has not been solved even here.**

Another way to attempt the enumeration of all 3-dimensional manifolds would be to construct all polyhedra having pairwise association of faces. This also is a 2-dimensional problem and it has met with as little success at solution as the problem of enumerating the Heegaard diagrams.

It is known from the theory of functions of complex variables that one can obtain any closed orientable surface as a branched covering surface of the 2-sphere, where the branching occurs at finitely many points. Corresponding to this result, it is possible to describe each closed orientable 3-dimensional manifold as a branched covering of the 3-sphere.[35] In this case the branching occurs along closed curves (knots) which lie in the 3-sphere. Here also the enumeration and distinguishing of individual covering spaces leads to unanswered questions. On occasion the same manifold can be derived as branched coverings of the 3-sphere with quite distinct knots as branch sets; as an example, three different branch sets are known for the spherical dodecahedron space.[33]

---

* *Editor's Note*: As of January 1979, the classification problem for 3-dimensional manifolds which admit Heegaard diagrams of genus $g > 1$ remains open. Surprisingly, related techniques of "handle decompositions" of manifolds in dimension > 4 have proved to be of fundamental importance (for example, see Smale [1]).

** See the Editor's notes in §61 and §63.

Problem

Show that if a 3-dimensional manifold can be represented by a polyhedron having h pairs of associated faces, then the manifold has a Heegaard diagram of genus h. (To what extent the converse is valid is still unsettled.)*

64. 3-Dimensional Manifolds with Boundary

A 3-dimensional manifold with boundary is a finite, pure, connected complex $\Re^2$ having a boundary and admitting a simplicial decomposition with the following properties:

1. The neighborhood complex of each interior point of $\Re^3$ is a 2-sphere.
2. The neighborhood complex of each boundary point relative to $\Re^3$ is a disk.

By examining the homology groups at the points of $\Re^3$ it is easy to prove that each arbitrary simplicial decomposition of $\Re^3$ will also possess these properties.

The simplest examples of 3-dimensional manifolds with boundary are the closed 3-ball, the hollow 3-ball, and the solid torus (full ring).

It is also possible to define a 3-dimensional manifold with boundary as a pure complex having a boundary whose double is a closed 3-dimensional manifold.

The boundary $\Re^2$ of $\Re^3$ consists of one or more closed surfaces.

This is true because the neighborhood complex of a boundary point Q relative to $\Re^3$ is a disk and hence the neighborhood complex of Q relative to $\Re^2$ is a circle. Thus if $\Re^2$ is decomposed into isolated subcomplexes $\Re_1^2, \Re_2^2, \ldots, \Re_\nu^2$, then each of them will be a finite, connected, homogeneous 2-dimensional complex, that is, a closed surface (§39).

We now ask the questions: Can an arbitrary closed surface occur as the boundary of a 3-dimensional manifold with boundary, and to what extent are the properties of $\Re^3$ determined by its boundary surfaces?

If $\Re^3$ is orientable, then by coherently orienting all 3-simplexes we obtain a 3-chain whose boundary is a closed 2-chain on $\Re^2$. Every 2-simplex of $\Re^2$ appears in this 2-chain. Thus there must exist nonvanishing closed 2-chains on the boundary surfaces $\Re_1^2, \Re_2^2, \ldots, \Re_r^2$. This is possible only for oriented surfaces. We then have:

THEOREM I. *The boundary of an orientable 3-dimensional manifold with boundary consists of orientable surfaces.*

The converse of this theorem is not valid; for example, a nonorientable manifold bounded by a 2-sphere is obtained by removing a 3-simplex from any orientable closed manifold.

---

* See the Editor's note in §62.

COROLLARY. *It is not possible to embed the projective plane $\mathfrak{P}^2$ in Euclidean 3-space $\mathfrak{R}^3$ or in the 3-sphere $\mathfrak{S}^3$ (which is the completion of $\mathfrak{R}^3$ by addition of an improper point).*

More precisely: $\mathfrak{P}^2$ cannot be a subcomplex of a simplicial decomposition of the 3-sphere.[36] This follows from the fact that $\mathfrak{P}^2$ is a pure complex having no boundary (§12). Considered as a subcomplex of $\mathfrak{S}^3$, $\mathfrak{P}^2$ would be a closed 2-chain mod 2 (§23). But the second connectivity number of $\mathfrak{S}^3$ is $q^2 = 0$; therefore each closed 2-chain mod 2 has a boundary (§24) and there would thus exist a 3-dimensional subcomplex $\mathfrak{R}^3$ having boundary $\mathfrak{P}^2$. The subcomplex $\mathfrak{R}^3$ would be a 3-dimensional manifold with boundary since any point, Q of $\mathfrak{R}^3$ which did not lie on $\mathfrak{P}^2$ would have a 2-sphere as its neighborhood complex relative to $\mathfrak{R}^3$. But if Q did lie on $\mathfrak{P}^2$, then its neighborhood complex relative to $\mathfrak{S}^3$ would be decomposed by $\mathfrak{P}^2$ into two surfaces with boundary having a common boundary circle. These surfaces with boundary would both be disks since together they form a 2-sphere. Since $\mathfrak{R}^3$ would be orientable, because $\mathfrak{S}^3$ is orientable, this would result in a contradiction to Theorem I.

When one forms the double $\mathfrak{R}^3_2$ of $\mathfrak{R}^3$, the neighborhood complex of each boundary point also experiences a doubling. The double of a disk is a 2-sphere; consequently, $\mathfrak{R}^3_2$ is homogeneous and is therefore a closed 3-dimensional manifold. Its Euler characteristic must vanish (§59). Thus if N denotes the Euler characteristic of $\mathfrak{R}^3$ and $\overline{N}$ denotes the Euler characteristic of its boundary, $\mathfrak{R}^2$, we have

$$2N - \overline{N} = 0 \tag{1}$$

and thus have

THEOREM II. *The Euler characteristic of a 3-dimensional manifold uniquely determines the Euler characteristic of its boundary, and conversely.*

There also follows from (1)

THEOREM III. *The Euler characteristic of the boundary of a 3-dimensional manifold with boundary is always an even integer. Thus the projective plane, or more generally a nonorientable surface of odd genus, cannot be the boundary of a 3-dimensional manifold.**

On the other hand, it is quite possible that two projective planes taken together form the boundary. One can see this from the example of the topological product of the line interval and the projective plane.

If $\mathfrak{R}^3$ is orientable and if its bounding surfaces $\mathfrak{R}^2_1, \mathfrak{R}^2_2, \ldots, \mathfrak{R}^2_r$ are

---

Editor's Note: We see here the beginnings of "cobordism theory," which concerns itself, broadly speaking, with whether an n-manifold bounds a particular type of $(n + 1)$-manifold. Theorem III establishes that the cobordism group of nonoriented, 2-dimensional (possibly disconnected) manifolds is nontrivial.

respectively of genus $h_1, h_2, \ldots, h_r$, then from §38

$$\overline{N} = \sum_{i=1}^{r} (2h_i - 2) = 2 \sum_{i=1}^{r} h_i - 2r.$$

Furthermore, in

$$N = -p^0 + p^1 - p^2 + p^3$$

the Betti number $p^0 = 1$ and $p^3 = 0$ since there exist no closed nonzero simplicial 3-chains in $\mathfrak{R}^3$. Substituting these values for $\overline{N}$ and N into (1) we get

$$p^1 = 1 + p^2 + \frac{1}{2} \left(2 \sum_{i=1}^{r} h_i - 2r \right) = p^2 - (r - 1) + \sum_{i=1}^{r} h_i.$$

But $p^2 \geqq r - 1$, because the boundary surfaces $\mathfrak{R}_1^2, \mathfrak{R}_2^2, \ldots, \mathfrak{R}_{r-1}^2$ represent $r - 1$ 2-chains which are homologously independent after one has oriented them coherently, since when taken together they do not bound. Consequently,

$$p^1 \geqq \sum_{i=1}^{r} h_i$$

and we have

THEOREM IV. *The first Betti number of an orientable 3-dimensional manifold with boundary is equal to or greater than the total number of handles of the boundary surfaces.*

Equality will hold for example for the handle body of genus h.

As a consequence of this theorem, an oriented 3-dimensional manifold with boundary for which each closed path is null homologous can only have 2-spheres as boundary surfaces. In particular, this is true for each simply connected manifold of this type.[37]

65. Construction of 3-Dimensional Manifolds out of Knots

As an example of the many possible methods[38] for constructing 3-dimensional manifolds we shall discuss the procedure given by Dehn [1], of boring out and subsequently filling in a knot. It can be described, briefly, as follows: A knot is bored out of the 3-sphere, which is Euclidean 3-space $\mathfrak{R}^3$ completed by adding a point at infinity. The resulting manifold with boundary, the exterior space of the knot, is formed into a new closed manifold. This is done by attaching a full ring to it, the closure ring, by identifying the boundary surfaces of the exterior space and the ring in a topological mapping.

We first explain the concepts "knot" and "boring out." We assume that the 3-sphere $\mathfrak{S}^3$ has been simplicially decomposed in some manner and we also denote the decomposition by $\mathfrak{S}^3$. A *knot* $\mathfrak{k}$ is a 1-dimensional subcomplex of this simplicial decomposition, such that each vertex of the subcomplex is

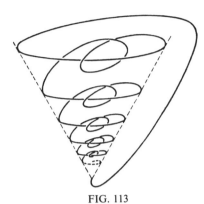

FIG. 113

incident with exactly two edges. A knot is then a topological circle, that is, the topological image of a circle. But not every topological circle in $\mathfrak{S}^3$ is a knot. As an example, a topological circle having infinitely many loops which successively diminish in size and accumulate at a point, that is, an "infinitely knotted circle" as appears in Fig. 113, is not a knot, because it can not be constructed from edges of a simplicial decomposition of $\mathfrak{S}^3$.

A knot is also defined on occasion as a closed double-point-free sequence of edges in $\mathfrak{R}^3$, consisting of finitely many straight line segments. Such a sequence of edges is a knot in the sense of our definition, since we can easily make the straight line segments edges of a simplicial decomposition.

Two knots are said to be *equivalent* if there exists an orientation preserving homeomorphism of $\mathfrak{S}^3$ onto itself which transforms one knot to the other.[39]

In order to explain the term "boring out" we shall make an additional assumption about the simplicial decomposition of $\mathfrak{S}^3$. We shall assume that whenever all vertices of an i-simplex lie on $\mathfrak{k}$, then the i-simplex itself belongs to $\mathfrak{k}$; that is, we assume that no triangle will have all of its vertices lying on $\mathfrak{k}$, so that some triangle edge will not be a chord of $\mathfrak{k}$. This property can always be achieved by means of a normal subdivision.

If, then $P_1, P_2, \ldots, P_r$ are the vertices of $\mathfrak{k}$ listed in a cyclic ordering and $\mathfrak{St}_1, \mathfrak{St}_2, \ldots, \mathfrak{St}_r$ are the simplicial stars of the normal subdivision, $\dot{\mathfrak{S}}^3$ of $\mathfrak{S}^3$, then the following holds from our recently stated assumption concerning $\mathfrak{S}^3$: only successive stars $\mathfrak{St}_\rho$ and $\mathfrak{St}_{\rho+1}$ (where $\rho + 1$ is to be reduced mod r if necessary) have points in common and these common points will always form exactly one surface element, $\mathfrak{E}_{\rho, \rho+1}$.* In addition, the surface elements $\mathfrak{E}_{\rho, \rho+1}$ have no points in common with one another. That is, as point sets they are disjoint. This implies that the subcomplex composed of the simplicial stars $\mathfrak{St}_1, \mathfrak{St}_2, \ldots, \mathfrak{St}_r$ is a full ring $\mathfrak{B}_0$ having $\mathfrak{k}$ as its core. For we can consider that $\mathfrak{B}_0$ is constructed by starting with the star $\mathfrak{St}_1$, attaching the following

---

*When one considers the simplicial decomposition of $\mathfrak{S}^3$ to be a cellular division (§68), the 3-stars $\mathfrak{St}_\rho$ and the surface elements $\mathfrak{E}_{\rho, \rho+1}$ are the cells dual to the vertices and edges respectively of $\mathfrak{k}$.

star along a surface element and successively adding stars in this manner so that a 3-ball results at each step until finally, as the last step, disjoint surface elements on the boundary of a 3-ball are identified. The process of boring out the knot $\mathfrak{k}$ consists of removing the interior points of the full ring $\mathfrak{B}_0$. The residual space $\mathfrak{A}$ is a 3-dimensional manifold with boundary whose boundary is a torus $\mathfrak{T}$. We will determine the homology group $\mathfrak{H}_{\mathfrak{A}}$ for dimension 1 of the "*exterior space*" $\mathfrak{A}$. We select a meridian circle m_0 on the torus $\mathfrak{T}$, for example a boundary path of the surface element $\mathfrak{E}_{r_0}$, and we select a longitude circle b_0 which is a path in $\mathfrak{B}_0$ homologous to the oriented core $\mathfrak{k}$. The circle b_0 is then determined only up to a multiple of m_0. We now apply Theorem II of §52 to the two subcomplexes $\mathfrak{B}_0$ and $\mathfrak{A}$ of $\dot{\mathfrak{S}}^3$. The first homology group of $\mathfrak{B}_0$ has generators m_0 and b_0 and the one relation $m_0 \sim 0$.*
If we also consider the homology group $\mathfrak{H}_{\mathfrak{A}}$ to be presented in terms of its generators and relations, then we may assume that m_0 and b_0 appear among its generators. The relations of the homology group $\mathfrak{H}_{\mathfrak{S}}$ of $\mathfrak{S}^3$ are obtained by adding the relations of $\mathfrak{B}_0$, that is, $m_0 \sim 0$, to the set of relations of $\mathfrak{H}_{\mathfrak{A}}$. The latter relations are not yet known to us (here we disregard the commutation relations since they appear implicitly when additive notation is used for the homologies). The group $\mathfrak{H}_{\mathfrak{S}}$ consists of the null element alone. If we then set $m_0 \sim 0$ in the homology group $\mathfrak{H}_{\mathfrak{A}}$, all elements become 0. This is possible only if $\mathfrak{H}_{\mathfrak{A}}$ is a cyclic group generated by m_0. *The cyclic group $\mathfrak{H}_{\mathfrak{A}}$ is a free cyclic group.*** For from Theorem IV of §64 the first Betti number of $\mathfrak{A}$ is equal to or greater than the total number of handles of the boundary surfaces and is therefore at least 1; thus the group $\mathfrak{H}_{\mathfrak{A}}$ is an infinite group. In particular, $b_0 \sim xm_0$ in $\mathfrak{A}$, where x is a uniquely determined integer. If $x \neq 0$, we replace b_0 by the longitude circle $b_0^* \sim b_0 - xm_0$, which is then null homologous in $\mathfrak{A}$.

We obtain a closed manifold $\mathfrak{M}^3$ from $\mathfrak{A}$ by mapping the ring boundary surface $\mathfrak{T}'$ of a full ring $\mathfrak{B}'$ topologically onto the torus boundary surface $\mathfrak{T}$ of $\mathfrak{A}$ and identifying corresponding points.† The image m on $\mathfrak{T}$ of a meridian circle m' of $\mathfrak{B}'$ is a path on $\mathfrak{T}$ and is thus homologous to a linear combination of m_0 and b_0^*:

$$m \sim \alpha m_0 + \beta b_0^* \qquad \text{(on } \mathfrak{T}). \qquad (1)$$

The integers α and β are relatively prime to one another, by the way, since m is free of double points. The first homology group $\mathfrak{H}_{\mathfrak{M}}$ of $\mathfrak{M}^3$ is obtained from $\mathfrak{H}_{\mathfrak{A}}$ using Theorem IV of §64 by setting the meridian circle m of the closure

---

*The generators of the homology group are the homology classes determined by m_0 and b_0. Thus, rigorously, we must substitute the homology class of m_0 for m_0 and replace the symbol $\sim$ by the equality sign relating group elements.

** A simpler proof of this fact is found in §77.

†The claim that a complex always results is equivalent to claiming that the simplicial decomposition of $\mathfrak{T}$ after being mapped topologically onto $\mathfrak{T}'$ can be extended to give a simplicial decomposition of $\mathfrak{B}'$. We shall not give the proof of this fact here. We shall assume instead that the topological mapping of $\mathfrak{T}$ onto $\mathfrak{T}'$ has this property.

ring homologous to 0:

$$\alpha m_0 + \beta b_0^* \sim 0 \qquad (\text{in } \mathfrak{M}^3).$$

Thus

$$\alpha m_0 \sim 0$$

is obtained as the defining relation of $\mathfrak{H}_{\mathfrak{M}}$ since $b_0^* \sim 0$ in $\mathfrak{A}$. Thus $\mathfrak{H}_{\mathfrak{M}}$ is a cyclic group of order α.

In the special case that $\alpha = 0$ in (1) we obtain a Poincaré space or the 3-sphere; there are infinitely many ways to effect the closure, depending upon the choice of β.

We shall illustrate this procedure for the case of the trefoil knot! From §52, the relations of the fundamental group of $\mathfrak{A}$ are

$$A^2 = B^3. \tag{2}$$

Choose a point of $\mathfrak{X}$ as the initial point of the closed paths. We can choose a path

$$m_0 = BA^{-1}$$

(Fig. 109) as a meridian circle and can choose a path

$$b_0 = A^2$$

as a longitude circle for, in the notation of §52, A^2 is the middle circle of the annulus $\mathfrak{D}$. To find the null homologous longitude circle b_0^* in the exterior space we form the expression

$$b_0^* = A^2(BA^{-1})^{-x} \sim A^{2+x}B^{-x},$$

where the integer x is not yet determined. We must then set $x = 0$ because of (2):

$$b_0^* \sim A^{-4}B^6 \sim (A^{-2}B^3)^2 \sim 0 \qquad (\text{in } \mathfrak{A}). \tag{3}$$

To obtain a Poincaré space we choose a path

$$m = m_0(b_0^*)^\beta = BA^{-1}\left[A^2(BA^{-1})^6\right]^\beta \tag{4}$$

as the meridian circle of the closure ring. The path m is null homotopic in the closure ring and therefore also in $\mathfrak{M}^3$ so that, besides (2), the relation

$$(BA^{-1})\left[A^2(BA^{-1})^6\right]^\beta = 1 \tag{5}$$

also holds in $\mathfrak{M}^3$. It follows from Theorem I of §52 that these two relations are the only relations of the fundamental group of $\mathfrak{M}^3$. One obtains the simplest closing of $\mathfrak{A}$ when $\beta = 0$. One then gets $m = m_0$ and the 3-sphere with which we started results. For the case $\beta = -1$ we get for the relations (2) and (5):

$$A^2 = B^3, \qquad BA^{-1}(BA^{-1})^{-6}A^{-2} = 1 \qquad \text{or} \qquad A^2 = (BA^{-1})^{-5}$$

Introducing a new generator C in place of A by means of the equation $A = CB$ we get

$$(CB)^2 = B^3 = C^5.$$

Apart from notation, these are the relations of the spherical dodecahedron space. To prove that not only the fundamental groups coincide but also the spaces themselves, requires methods beyond the scope of this book.[33]

Poincaré spaces arise in a like manner for other values of β and, in fact, these spaces are pairwise nonhomeomorphic since their fundamental groups can be proven to be distinct.

Because of their clear geometric significance, homogeneous complexes play a distinctive role among the complexes. We have given the name "manifolds" to the homogeneous complexes in 2 and 3 dimensions and we have attempted to gain a complete view of their properties. Our attempt was successful in 2 dimensions. In 3 dimensions we did not get further than a presentation of more or less systematically arranged examples. The complete classification of n-dimensional manifolds is a hopeless task at the present time.* Consequently, we let the homeomorphism problem recede into the background and we shift our emphasis to presenting general theorems which will all be based upon the possibility of providing manifolds with dual cellular divisions. Among the theorems which we treat here are the Poincaré duality theorem and the theory of intersection and linking numbers.

It is of lesser significance that we shall find it convenient to define manifolds of more than 3 dimensions in a more general way than by their homogeneity. We begin by introducing an auxiliary concept, the star complex, which prepares the way for such a definition and for the introduction of dual cellular divisions.

66. Star Complexes

Stated briefly, a star complex is a finite simplicial complex whose simplexes of dimensions 0 through n are partitioned into stars in such a way that the outer boundary of an i-star consists entirely of $(i - 1)$-stars. The precise definition is given by using mathematical induction over the dimension of the star complex as follows:

A 0-star complex $\Re_a^0$ consists of finitely many points

$$\mathfrak{a}_1^0, \mathfrak{a}_2^0, \ldots, \mathfrak{a}_{\alpha^0}^0$$

which are called the null stars of $\Re_a^0$. One obtains a 1-star complex $\Re_a^1$, by adding finitely many 1-stars

$$\mathfrak{a}_1^1, \mathfrak{a}_2^1, \ldots, \mathfrak{a}_{\alpha^1}^1$$

---

* *Editor's note*: It is now known from a result of Markov [1] that no algorithm exists for solving the homeomorphism problem in dimensions $\geqslant 4$.

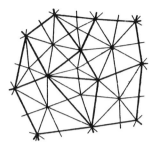

FIG. 114

to the collection of stars, where the outer boundary of each α_{ν}^{1} consists of certain 0-stars of $\Re_{a}^{0}$. If $\Re_{a}^{n-1}$ is an $(n-1)$-star complex having the stars

$$\alpha_{1}^{k}, \alpha_{2}^{k}, \ldots, \alpha_{\alpha^{k}}^{k}$$

of dimension k $(k = 0, \ldots, n-1)$, then an n-star complex, $\Re_{a}^{n}$ is obtained by adding finitely many n-stars $\alpha_{1}^{n}, \alpha_{2}^{n}, \ldots, \alpha_{\alpha^{n}}^{n}$ to the collection of stars, where the outer boundary of each α_{ν}^{n} consists of certain $(n-1)$-stars of $\Re_{a}^{n-1}$.

The simplicial complex which is formed by the collection of all of the stars of $\Re_{a}^{n}$ is called the *normal subdivision* of $\Re_{a}^{n}$ and we will denote it by $\dot{\Re}_{a}^{n}$. A portion of a 2-star complex is illustrated in Fig. 114; the 2-stars consist of the polygons whose boundaries appear in darker outline.

EXAMPLE. We can obtain a star complex from any finite complex $\Re^{n}$ by first decomposing it simplicially in an arbitrary way and then normally subdividing the resulting simplicial complex. The k-stars are the normally subdivided k-simplexes of $\Re^{n}$.

In $\Re_{a}^{n}$, if an $(n-1)$-star α_{ι}^{k-1} lies on the outer boundary of a k-star α_{κ}^{k}, then α_{ι}^{k-1} and α_{κ}^{k} are said to be *directly incident*. If there exists a sequence of stars of increasing dimension $\alpha^{i}, \alpha^{i+1}, \ldots, \alpha^{k-1}, \alpha^{k}$ between two stars α^{i} and α^{k} $(i < k)$ such that each star is directly incident with the following star, then α^{i} and α^{k} are said to be incident via intermediate stars or just *incident*.

It follows from the definition of a star complex that *one can arbitrarily prescribe the number of stars of dimensions 0 through n $\alpha^{0}, \alpha^{1}, \ldots, \alpha^{n}$ as well as the direct incidences, subject only to the requirement that each k-star $(k > 0)$ be incident with at least one $(k-1)$-star.* If P_{κ}^{k} is the center point* of the star α_{κ}^{k}, then one can construct the star complex rectilinearly in a Euclidean space of dimension $\alpha^{0} + \alpha^{1} + \cdots + \alpha^{n}$ as follows. One selects $\alpha^{0} + \alpha^{1} + \cdots + \alpha^{n}$ linearly independent points

$$P_{1}^{0}, P_{2}^{0}, \ldots, P_{\alpha^{0}}^{0}, \ldots, P_{1}^{n}, P_{2}^{n}, \ldots, P_{\alpha^{n}}^{n}$$

in this space and projects from the points $P_{1}^{1}, P_{2}^{1}, \ldots, P_{\alpha^{1}}^{1}$ to those points

---

*Here and in the treatment to follow, the upper index on a vertex refers to the dimension of the star having the vertex as center point.

$P_1^0, P_2^0, \ldots, P_{\alpha^1}^0$ which are to comprise the outer boundaries of the stars $\mathfrak{a}_1^1, \mathfrak{a}_2^1, \ldots, \mathfrak{a}_{\alpha^1}^1$ and so forth.

We do not require that each k-star be incident with a $(k + 1)$-star. If this turns out to be the case, however, then the star complex is said to be *pure*. The star complex $\mathfrak{R}_a^n$ is obviously pure if and only if its normal subdivision $\dot{\mathfrak{R}}_a^n$ is a pure simplicial complex. On the other hand, a k-star $\mathfrak{a}^k$ of $\mathfrak{R}_a^n$ will be a pure k-dimensional subcomplex of $\dot{\mathfrak{R}}_a^n$ in every case. That is, each i-simplex $(i < k)$ of $\mathfrak{a}^k$ is always a face of at least one k-simplex.

We now express the properties of an n-star complex and of its normal subdivision in terms of the incidences between its stars. For this purpose we use

THEOREM I. *If $\mathfrak{a}^0, \mathfrak{a}^1, \ldots, \mathfrak{a}^k$ is a sequence of stars such that each star is directly incident with the star following it, then their center points $P^0, P^1, \ldots, P^k$ form the vertices of a k-simplex belonging to $\mathfrak{a}^k$ and each k-simplex of the star $\mathfrak{a}^k$ can be derived from such a sequence.*

Proof. The theorem is trivially valid for $k = 0$. Assume that the theorem has already been proved for the dimension $k - 1$. Then it is also true for the dimension k. For if $\mathfrak{a}^0, \mathfrak{a}^1, \ldots, \mathfrak{a}^{k-1}$ is a sequence of directly incident stars, then their center points $P^0, P^1, \ldots, P^{k-1}$ are the vertices of a simplex belonging to $\mathfrak{a}^{k-1}$ by assumption. Because $\mathfrak{a}^{k-1}$ and $\mathfrak{a}^k$ are directly incident, the simplex $(P^0, P^1, \ldots, P^{k-1})$ will lie on the outer boundary of $\mathfrak{a}^k$, so that the vertices $P^0, P^1, \ldots, P^{k-1}, P^k$ span a simplex belonging to $\mathfrak{a}^k$.

If, conversely, $\mathfrak{E}^k$ is an arbitrary k-simplex of $\mathfrak{a}^k$, then P^k is one of its vertices and the face opposite to P^k, $\mathfrak{E}^{k-1}$, lies on the outer boundary of $\mathfrak{a}^k$ and belongs to a star there which is incident with $\mathfrak{a}^k$. By our initial induction assumption, the vertices of $\mathfrak{E}^{k-1}$ are the center points of a sequence of successively incident stars $\mathfrak{a}^0, \mathfrak{a}^1, \ldots, \mathfrak{a}^{k-1}$. The sequence $\mathfrak{a}^0, \mathfrak{a}^1, \ldots, \mathfrak{a}^{k-1}, \mathfrak{a}^k$, where the center points of the stars are the vertices of $\mathfrak{E}^k$, then has the same property.

As a generalization of Theorem I we have

THEOREM II. *If $\mathfrak{a}^i, \mathfrak{a}^k, \ldots, \mathfrak{a}^l$ $(i < k < \cdots < l)$ is a sequence of stars of increasing dimension such that any two successive stars are incident (but not necessarily directly incident), then their center points $P^i, P^k, \ldots, P^l$ are the vertices of a simplex of $\dot{\mathfrak{R}}_a^n$ and one obtains all simplexes of $\dot{\mathfrak{R}}_a^n$ in this way.*

Proof. By inserting additional stars one can complete the sequence $\mathfrak{a}^i, \mathfrak{a}^k, \ldots, \mathfrak{a}^l$ to form a sequence of stars $\mathfrak{a}^0, \mathfrak{a}^1, \ldots, \mathfrak{a}^{l-1}, \mathfrak{a}^l$ in which any two successive stars are directly incident. From Theorem I the center points of these stars make up the vertices of a simplex belonging to $\mathfrak{a}^l$ and the center points of the stars given originally therefore span a face of this simplex.

Conversely, let $\mathfrak{E}^i$ be an arbitrary simplex of $\dot{\mathfrak{R}}_a^n$. Let us consider the subcomplexes $\mathfrak{R}_a^0, \mathfrak{R}_a^1, \ldots, \mathfrak{R}_a^n$ which are formed, respectively, by all the

0-stars, all of the 0-stars and 1-stars, and so forth until finally, all of the stars of $\dot{\Re}_a^n$. Let $\dot{\Re}_a^h$ be the first of these subcomplexes in which $\mathfrak{E}^j$ appears. A vertex of $\mathfrak{E}^j$ will then be the center point of an h-star $\mathfrak{a}^h$, while the opposite face of $\mathfrak{E}^j$ will belong to the outer boundary of $\mathfrak{a}^h$. The simplex $\mathfrak{E}^j$ is a face of a simplex $\mathfrak{E}^h$ of $\mathfrak{a}^h$ (or else $\mathfrak{E}^h = \mathfrak{E}^j$) whose vertices are the center points of successive incident stars. This follows from Theorem I. The vertices of $\mathfrak{E}^j$ are therefore the center points of incident stars (either directly or via intermediate stars).

If one knows, then, how many stars of each dimension appear in $\dot{\Re}_a^n$ and which stars are directly incident with one another, the intermediate stars can be found, and by using Theorem II the simplexes of $\dot{\Re}_a^n$ can be constructed. From Theorem I one can determine which simplexes of $\dot{\Re}_a^n$ form a star of $\Re_a^n$. If it is possible to establish a one-to-one correspondence between the stars of two star complexes $\Re_a^n$ and $'\Re_a^n$ in such a way that stars which are directly incident in $\Re_a^n$ correspond to stars which are directly incident in $'\Re_a^n$ and vice versa, then a one-to-one correspondence will have been established at the same time between the normal subdivisions $\dot{\Re}_a^n$ and $'\dot{\Re}_a^n$. We say that two such star complexes $\Re_a^n$ and $'\Re_a^n$ are *isomorphic*.

From Theorem II there also follows

THEOREM III. *If*

$$P^i, P^k, \ldots, P^l \qquad (i < k < \cdots < l)$$

are the vertices of a simplex of $\dot{\Re}_a^n$ *and*

$$P^l, P^m, \ldots, P^z \qquad (l < m < \cdots < z)$$

are the vertices of a second simplex of $\dot{\Re}_a^n$, *then*

$$P^i, P^k, \ldots, P^l, P^m, \ldots, P^z$$

are also the vertices of a simplex of $\dot{\Re}_a^n$.

For the stars $\mathfrak{a}^i, \mathfrak{a}^k, \ldots, \mathfrak{a}^l, \mathfrak{a}^m, \ldots, \mathfrak{a}^z$ are successively incident with one another, even if not directly incident.

We now turn our attention to the pure star complexes $\Re_a^n$. For such a complex there holds an important theorem which states the existence of a dual star complex. It is this property which motivates our study of the star complexes.

THEOREM IV. *For each pure star complex* $\Re_a^n$ *there exists a dual star complex* $\Re_b^n$ *which is determined up to an isomorphism.*

The dual complex is characterized by the following properties: In one-to-one correspondence with each k-star $\mathfrak{a}^k$ *of* $\Re_a^n$ *there exists a "dual"* $(n - k)$-*star* $\mathfrak{b}^{n-k}$ *of* $\Re_b^n$ *and incident stars correspond to incident stars.*

Proof. In the star complex $\Re_b^n$ which is to be constructed the number of stars of each dimension is already given, as well as the direct incidences of the

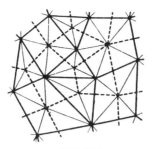

FIG. 115

stars. The single requirement to be fulfilled, that each k-star be incident with at least one $(k - 1)$-star, is satisfied since each $(n - k)$-star of $\Re_a^n$ is incident with at least one $(n - k + 1)$-star, because $\Re_a^n$ is a pure complex.

In Fig. 115 the 2-stars of the original star complex (the same as in Fig. 114) are shown with their boundaries in darker outline. The 2-stars of the dual complex are the polygons whose boundaries are shown in broken lines.

When one forms the star complex dual to $\Re_b^n$, one again obtains $\Re_a^n$. The relation between $\Re_a^n$ and $\Re_b^n$ is therefore reciprocal.

Since a one-to-one correspondence exists between the stars $\mathfrak{a}_\kappa^k$ of $\Re_a^n$ and the dual stars $\mathfrak{b}_\kappa^{n-k}$ of $\Re_b^n$, there also exists a one-to-one correspondence between the center points P_κ^k and Q_κ^{n-k}, that is, between the vertices of the normal subdivisions $\dot{\Re}_a^n$ and $\dot{\Re}_b^n$. If $P^i P^k \cdots P^l$ $(i < k \cdots < l)$ is now a simplex of $\dot{\Re}_a^n$, then by Theorem II the corresponding stars $\mathfrak{a}^i, \mathfrak{a}^k, \ldots, \mathfrak{a}^l$ are successively incident. The same is also true for the dual stars $\mathfrak{b}^{n-i}$, $\mathfrak{b}^{n-k}, \ldots, \mathfrak{b}^{n-l}$. Consequently, from Theorem II their center points likewise form the vertices of a simplex of $\dot{\Re}_b^n$. That is, in the mapping $P_\kappa^k \leftrightarrow Q_\kappa^{n-k}$ the simplexes of $\dot{\Re}_a^n$ and $\dot{\Re}_b^n$ transform to one another. *We can therefore regard the dual star complexes $\dot{\Re}_a^n$ and $\dot{\Re}_b^n$ as being different stellar divisions of one and the same simplicial complex $\dot{\Re}^n = \dot{\Re}_a^n = \dot{\Re}_b^n$.* We shall take this viewpoint from now on.

We now adopt the convention that the vertices of any simplex of $\dot{\Re}^n$ are always to be listed in order of increasing upper index. From Theorem II all of these upper indices are different. That is, the vertices are center points of stars of different dimensions. Each simplex of $\dot{\Re}^n$ accordingly has a definite first vertex and a definite last vertex. Since from Theorem I a k-simplex of $\dot{\Re}^n$ which belongs to the k-star $\mathfrak{a}^k$ of $\Re_a^n$ will be given by a sequence of $k + 1$ successively incident stars $\mathfrak{a}^0, \mathfrak{a}^1, \ldots, \mathfrak{a}^k$ and a simplex of the dual star $\mathfrak{b}^{n-k}$ will be correspondingly given by a sequence $\mathfrak{a}^k, \mathfrak{a}^{k+1}, \ldots, \mathfrak{a}^n$, the location of a star with respect to its dual star can be specified by the following theorem:

THEOREM V. *The star $\mathfrak{a}^k$ consists of all of the k-simplexes which have its center point P^k as their last vertex; its dual star $\mathfrak{b}^{n-k}$ consists of all of the $(n - k)$-simplexes which have P^k as their first vertex.*

Two dual stars will therefore have only the center point in common, while two nondual stars $\mathfrak{a}^k$ and $\mathfrak{b}^{n-k}$ will be disjoint, that is, will have no point in common.

More generally we have

THEOREM VI. *The intersection of two stars $\mathfrak{a}^k$ having center point P^k and $\mathfrak{b}^{n-i}$ having center point P^i consists of all of the $(k-i)$-simplexes of $\mathfrak{R}^n$ which have P^i as first vertex and P^k as last vertex.*

The stars will have a nonempty intersection, then, only if $i \leqq k$ and the center point of $\mathfrak{b}^{n-i}$ belongs to $\mathfrak{a}^k$.

Proof. (α) If $(P^i P^{i+1} \cdots P^{k-1} P^k)$ is a $(k-i)$-simplex, then it will belong to an n-simplex $(P^0 \cdots P^i \cdots P^k \cdots P^n)$. According to Theorem V, $(P^0 \cdots P^k)$ belongs to $\mathfrak{a}^k$ and $(P^i \cdots P^n)$ belongs to $\mathfrak{b}^{n-i}$ and thus the common face $(P^i P^{i+1} \cdots P^{k-1} P^k)$ belongs to the intersection of $\mathfrak{a}^k$ and $\mathfrak{b}^{n-i}$.

(β) We must also show that each simplex of the intersection of $\mathfrak{a}^k$ and $\mathfrak{b}^{n-i}$ is a face of a $(k-i)$-simplex $(P^i \cdots P^k)$. If $\mathfrak{E} = (P^r \cdots P^s)$ is such a common simplex [not necessarily $(s-r)$-dimensional] of $\mathfrak{a}^k$ and $\mathfrak{b}^{n-i}$, then there exists an $(s-r)$-dimensional simplex $\mathfrak{E}^{s-r}$ which has P^r as its first vertex and P^s as its last vertex, for $\mathfrak{E}$ is even a face of an n-dimensional simplex. Furthermore, there exists a simplex $\mathfrak{E}^{k-s}$ which has P^s as first vertex and P^k as last vertex (for P^s belongs to $\mathfrak{a}^k$) and a simplex $\mathfrak{E}^{r-i}$ having P^i as first vertex and P^r as last vertex (for P^r belongs to $\mathfrak{b}^{n-i}$). From Theorem

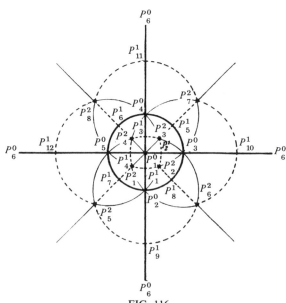

FIG. 116

III, * the three simplexes $\mathfrak{E}^{r-i}$, $\mathfrak{E}^{s-r}$ and $\mathfrak{E}^{k-s}$ are faces of a $(k-i)$-simplex having P^i as first vertex and P^k as last vertex.

We shall illustrate these concepts for a simplicial complex $\mathfrak{R}^2_a$, the octahedron. When its eight triangles are normally subdivided a star complex $\mathfrak{R}^2$ arises, composed of eight 2-stars having center points $P^2_1, P^2_2, \ldots, P^2_8$. The midpoints of the edges of the octahedron are the center points $P^1_1, P^1_2, \ldots, P^1_{12}$ of the twelve 1-stars of the complex, while the vertices of the octahedron are the six 0-stars. The dual star of the 0-star P^0_1 is made up of the eight triangles which have P^0_1 as first vertex. Its outer boundary is made up of four 1-stars, that is, a quadrilateral belonging to the dual star complex $\mathfrak{R}^2_b$, the cube shown dotted in Fig. 116. The figure shows the octahedral division and the dual cubic division in stereographic projection.

67. Cell Complexes

A k-star is called a k-cell if its outer boundary has the same homology groups as the $(k-1)$-sphere and, for $k > 1$, satisfies the additional requirement that it be a closed pseudomanifold. A 0-cell is an individual point. A 1-cell consists of two intervals incident at a point. In the case of dimension 2 the outer boundary is a circle and the 2-cell is a disk. In the case of dimension 3 the outer boundary is a 2-sphere and the 3-cell is a 3-ball (see Problem 1 of §39). For dimension 4 and higher dimensions not all cells can be catalogued. There exist cells which cannot be mapped topologically onto a closed n-ball. An example of this is a 4-dimensional simplicial star whose outer boundary is a 3-dimensional Poincaré space.

A star complex is called a *cell complex* if all of its stars are cells.

Since each normally subdivided k-simplex is a k-cell, one can make each finite simplicial complex into a cell complex by normally subdividing it. Consequently one can regard the cells as being a generalization of the simplexes. They play the same role in a cell complex as the simplexes in a simplicial complex. The reason why we cannot restrict ourselves to the use of simplexes in our investigations will first be shown in the next section, in our investigations of manifolds.

For $k > 1$ the outer boundary of a k-cell $\mathfrak{a}^k$ is a $(k-1)$-dimensional orientable pseudomanifold. For as in the case of a $(k-1)$-sphere, its $(k-1)$th homology group is the free cyclic group and according to §24 this is equivalent to orientability. From §24, then, $\mathfrak{a}^k$ is itself an oriented pseudomanifold with boundary. Consequently, one can orient its k-simplexes coherently (in two opposite ways). By doing this we make $\mathfrak{a}^k$ into an *oriented k-cell*, which we denote by a^k. In the case $k = 1$ it is also possible to coherently orient the two 1-simplexes composing the 1-cell. We orient 0-cells as we do 0-simplexes (§9). From now on we shall consider an oriented k-cell $(k \cong 0)$ to be a set of oriented simplexes, which we can regard as a particular

---

* *Editor's note*: The German original refers here to Theorem IV; however, it seems clear to us that Theorem III was intended.

k-chain such that each simplex appears with multiplicity 1. The boundary chain of an oriented cell a^k consists of the coherently oriented outer boundary of $\mathfrak{a}^k$ (§24). This consequently provides an orientation of the $(k - 1)$-cells which compose the outer boundary of a k-cell in a cell complex; we call this orientation the *orientation induced by the oriented k-cell*.

From now on we shall consider the cells of the cell complex $\mathfrak{R}^n$ to have a fixed orientation; for each k-cell we arbitrarily choose one of the two possible orientations. Denote the cells oriented in this manner by a_κ^k ($k = 0, 1, \ldots, n$, $\kappa = 1, 2, \ldots, \alpha^k$). The oriented cells of dimension k are linearly independent of one another, since no two k-cells have a common k-simplex.

We define a *k-cell chain* of a cell complex to be a formal sum

$$q_1 a_1^k + q_2 a_2^k + \cdots + q_{\alpha^k} a_{\alpha^k}^k, \tag{1}$$

where the cells a_κ^k are regarded as simplicial chains of the normal subdivision $\dot{\mathfrak{R}}^n$ of $\mathfrak{R}^n$. A cell chain is then a particular simplicial chain, one such that the simplexes can be formed into cells according to (1). If such a grouping of simplexes into cells is possible, then it can be done in only one way; because of the linear independence of the k-cells, the coefficients $q_1, q_2, \ldots, q_{\alpha^k}$ are determined uniquely by the chain.

A cell chain is said to be closed or null homologous when the simplicial chain is, respectively, closed or null homologous. The boundary of a cell chain is again a cell chain. For the boundary of an individual cell a^{k+1} is a cell chain:

$$\mathfrak{R} \partial a_\lambda^{k+1} = \sum_{\kappa=1}^{\alpha^k} \varepsilon_{\kappa\lambda}^k a_\kappa^k \qquad (\lambda = 1, 2, \ldots, \alpha^{k+1}). \tag{2}$$

Here $\varepsilon_{\kappa\lambda}^k = 1$ if the orientation of a_κ^k is that induced by a_κ^{k+1} and $\varepsilon_{\kappa\lambda}^k = -1$ if its orientation is opposite to the induced orientation; $\varepsilon_{\kappa\lambda}^k = 0$ when a_κ^k is not incident with a_κ^{k+1}.

For each closed simplicial k-chain on a cell complex $\mathfrak{R}^n$ there exists a homologous cell chain. More generally, we shall prove

THEOREM I. *If U^k is a simplicial chain on a cell complex $\mathfrak{R}^n$, and if $\mathfrak{R} \partial U^k$ is a $(k - 1)$-dimensional cell chain,* which can also be the chain 0, then there exists a cell chain homologous to U^k.*

Proof. First let $k = 0$. Let E^0 be a 0-simplex on U^0. Then E^0 is either itself a 0-cell or is the center point of an l-cell upon whose outer boundary a 0-cell lies which is obviously homologous to E^0. Each 0-simplex is thus homologous to a 0-cell and therefore each 0-chain is homologous to a 0-cell chain.

If $k > 0$, then we examine the subchain V^k of all k-simplexes of U^k which are incident with the center point P^n of an n-cell $\mathfrak{a}^n$. Since $\mathfrak{R} \partial U^k$ consists

---

* For $k = 0$ of course we omit the requirement that the boundary be a cell chain.

entirely of $(k - 1)$-cells, no boundary simplex of U^k, and therefore also of V^k, can be incident with P^n. Thus $\Re \partial V^k$ lies on the outer boundary of $\mathfrak{a}^n$.

(a) $k = n$. In this case V^n is an n-chain on the cell $\mathfrak{a}^n$ and its boundary lies on the outer boundary of $\mathfrak{a}^n$. Thus from §24 V^n is a multiple of a^n. The chain U^n is composed entirely of n-cells.

(b) $0 < k < n$. In this case $\Re \partial V^k$ is a closed $(k - 1)$-chain of the outer boundary of $\mathfrak{a}^n$ and it is therefore null homologous on the outer boundary of $\mathfrak{a}^n$, because the outer boundary has the same homology groups as an $(n - 1)$-sphere. (This is also true for $k = 1$, because the sum of the coefficients of the 0-chain $\Re \partial V^k$ vanishes, as always, for the boundary of a 1-chain.) Thus on the outer boundary of $\mathfrak{a}^n$ there exists a chain $'V^k$ having boundary $\Re \partial V^k$. $V^k - 'V^k$ is then a closed chain on $\mathfrak{a}^n$ and is therefore null homologous when regarded as a chain on that n-star (§19). If one then replaces the subchain V^k in U^k by $'V^k$, one obtains a chain homologous to U^k which no longer contains any interior simplex of $\mathfrak{a}^n$. One proceeds in this manner with all n-cells and thereby obtains a chain $U_1^k \sim U^k$ which lies on the subcomplex $\Re^{n-1}$ of $\Re^n$ formed by the cells of dimensions 0 through $n - 1$. If $k = n - 1$, then U_1^k is already a cell chain, from (a). Otherwise, one again applies the procedure (b) to $\Re^{n-1}$ and replaces U_1^k by a homologous chain U_2^k which lies on the complex $\Re^{n-2}$ of all 0-, 1-, . . . , $(n - 2)$-cells of $\Re^n$. This push-out procedure is repeated until one finally arrives at a chain $U_{n-k}^k \sim U^k$ which lies on the subcomplex $\Re^k$ of all cells up to dimension k, and to which one can, finally, apply (a).

We have thereby achieved the following important result:

THEOREM II. *The oriented cells of a cell complex form a system of blocks.*

For the conditions (Bl1) through (Bl4) of §22 are satisfied by the cells. One can then use the cell chains to calculate the homology groups. The matrices $[\varepsilon_{\kappa\lambda}^k]$ resulting from the boundary relations (2) are to be used as incidence matrices here. The cell complex is completely determined (up to isomorphism) from these incidence matrices, since a star complex is determined by giving its stars and their direct incidences.

Just as we can form simplicial chains from oriented simplexes and form simplicial chains mod 2 from nonoriented simplexes, we can form *cells mod 2* from nonoriented cells and *cell chains* mod 2 from cell chains. A k-cell mod 2 (which we also call a nonoriented k-cell) is the simplicial k-chain mod 2 which is formed from all of the nonoriented k-simplexes of the cell. A k-cell chain mod 2 is a linear combination of k-cells mod 2 whose coefficients are residue classes mod 2, that is, $\breve{0}$ or $\breve{1}$. The theorems concerning cell chains can be carried over to cell chains mod 2 as in the case of simplicial chains. One has only to replace the coefficients in the domain of integers by their residue classes mod 2 in all relations. The cells mod 2 form a block system mod 2.

68. Manifolds

In dimensions 2 and 3 we have defined a manifold to be a homogeneous complex, that is, a complex such that each point possesses a neighborhood homeomorphic to the interior of a 2- or 3-dimensional ball, respectively. In spaces of higher dimension this definition is unsuitable in the present state of development of topology. This is due to the fact that the definition cannot be interpreted combinatorially. We do not have any procedure for deciding whether a simplicial complex of more than 3 dimensions is or is not homogeneous when it is given by its schema. Specifically, we do not know whether we may conclude from the homogeneity of an n-dimensional complex $\Re^n$ that the $(n-1)$-dimensional neighborhood complex of a vertex of a simplicial decomposition is homeomorphic to the $(n-1)$-sphere. But even if this were the case, we would still have to decide whether a given $(n-1)$-dimensional simplicial complex is an $(n-1)$-sphere. This "sphere problem" has not yet been solved for more than 2 dimensions. It is, however, possible to prove a large number of theorems which relate to the homology properties (but not homotopy properties) of homogeneous complexes without fully exploiting the homogeneity of the complex. These theorems are still valid for an arbitrary complex if it only behaves like a homogeneous complex with respect to the homology properties in neighborhoods of each of its points. It is sufficient to require that the homology groups at each point be the same as those of the $(n-1)$-sphere.

Correspondingly, we give this definition: *a (closed) n-dimensional manifold $\mathfrak{M}^n$ $(n > 0)$ is a connected, finite n-dimensional complex which at every point has the same homology groups as the $(n-1)$-sphere.*[40]

The 0th and $(n-1)$th homology groups at a point are then free cyclic groups. All others consist of the null element alone, except for the case $n = 1$; in this exceptional case the 0th homology group is the free group having two generators.

As in the case of surfaces we can also define infinite manifolds and manifolds with boundary,[41] besides the closed manifolds, which are the only ones to be considered here. The closed manifolds are finite manifolds without boundary.

THEOREM I. *All connected finite homogeneous complexes are manifolds.*

This follows from Theorem II of §33.

In particular, the 2- and 3-dimensional manifolds of Chapters VI and IX are also manifolds in the sense of the definition given above. The same is true for the n-sphere $\mathfrak{S}^n$ and the projective n-space $\mathfrak{P}^n$ for $n > 0$. There exists only one 1-dimensional manifold. It is homeomorphic to the circle. For the number of free generators of the 0th homology group at a vertex of a 1-dimensional simplicial complex is equal to the number of 1-simplexes leaving the vertex; since the 0th homology group is to be the free group having two generators, exactly two 1-simplexes must be incident with each vertex. From now on we shall not consider the trivial case $n = 1$ in our investigations.

If a complex $\mathfrak{K}^n$ is given together with a simplicial decomposition, then one can always decide whether it is a manifold. The ith homology group is the same at all inner points of a k-simplex. Thus one needs only to examine the homology groups at the midpoints of all of the k-simplexes, that is, at the vertices of the normal subdivision of $\mathfrak{K}^n$.

We shall now show that:

I.   Every manifold $\mathfrak{M}^n$ is a pure complex. That is, in an arbitrary simplicial decomposition each k-simplex ($k < n$) is incident with at least one n-simplex.

Otherwise, at an inner point of the k-simplex the $(n - 1)$th homology group would not be the free cyclic group but would consist only of the null element.

II.   Each $(n - 1)$-simplex is incident with exactly two n-simplexes.

For if an $(n - 1)$-simplex were incident with $\nu \neq 2$ n-simplexes, then the $(n - 1)$th homology group at an inner point of the $(n - 1)$-simplex would be the free Abelian group of $\nu - 1 \neq 1$ generators (§32, Example 2).

This implies that the neighborhood complex $\mathfrak{A}^{n-1}$ of a vertex P, that is, the outer boundary of the simplicial star having P as center point, is a pure $(n - 1)$-dimensional complex such that each $(n - 2)$-simplex is incident with exactly two $(n - 1)$-simplexes. Furthermore, since $\mathfrak{A}^{n-1}$ has the same homology groups as the $(n - 1)$-sphere, the $(n - 1)$th connectivity number of $\mathfrak{A}^{n-1}$ is $q^{n-1} = 1$. The neighborhood complex $\mathfrak{A}^{n-1}$ is therefore an $(n - 1)$-dimensional pseudomanifold (cf. §24).

There also follows from this:

III.   In a manifold $\mathfrak{M}^n$ one can connect any two n-simplexes with one another by a sequence of successively incident sequences of dimension n and dimension $n - 1$.

To show this we first assume the statement to be false. We can then decompose $\mathfrak{M}^n$ into two subcomplexes which have as their intersection a complex $\mathfrak{K}^k$ of dimension at most $n - 2$. In one subcomplex we place all n-simplexes which can be connected with a particular n-simplex in the abovementioned way. In the other subcomplex we have the remaining n-simplexes. On the neighborhood complex $\mathfrak{A}^{n-1}$ of a vertex of $\mathfrak{K}^k$, the two subcomplexes then appear as $(n - 1)$-dimensional subcomplexes whose intersection is either empty or is $(n - 3)$-dimensional at most. But we cannot, then, connect each two $(n - 1)$-simplexes on $\mathfrak{A}^{n-1}$ sequentially, which contradicts the fact that $\mathfrak{A}^{n-1}$ is a pseudomanifold!

Since the properties I through III shown here coincide with the properties (PM1) through (PM3) defining a pseudomanifold, we have

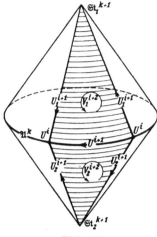

FIG. 117

THEOREM II. *Every manifold is a pseudomanifold.*

The theorems proved earlier with regard to pseudomanifolds are, therefore, also valid for manifolds. In particular, we can classify manifolds as being either orientable or nonorientable.

To obtain additional properties of manifolds we prove the following lemma:

LEMMA. *Given a complex* $\mathfrak{B}^{k+1}$ *($k \geqq 0$) which consists of two simplicial stars* $\mathfrak{St}_1^{k+1}$ *and* $\mathfrak{St}_2^{k+1}$ *having a common outer boundary* $\mathfrak{A}^k$, *and such that the homology groups of* $\mathfrak{B}^{k+1}$ *are those of the $(k + 1)$-sphere, then the homology groups of* $\mathfrak{A}^k$ *are those of the k-sphere.*

Figure 117 illustrates the case $k = 1$.

Proof. Let U^i be a closed i-chain on $\mathfrak{A}^k$, $0 \leqq i \leqq k$.* For the case $i = 0$ we also assume that U^i has a sum of coefficients of value 0. In the figure, U^i consists of two points having multiplicities $+1$ and -1. Then $U^i \sim 0$ in $\mathfrak{St}_1^{k+1}$ and also in $\mathfrak{St}_2^{k+1}$, from §19. There thus exist chains U_1^{i+1} and U_2^{i+1} in $\mathfrak{St}_1^{k+1}$ and $\mathfrak{St}_2^{k+1}$ such that

$$\mathfrak{R}\, \partial U_1^{i+1} = U^i, \tag{1}$$

$$\mathfrak{R}\, \partial U_2^{i+1} = U^i. \tag{2}$$

The chain

$$U_1^{i+1} - U_2^{i+1} \tag{3}$$

is then closed.

---

* All chains in this proof are assumed to be nonsingular, that is, simplicial. We can omit the trivial case $k = 0$ in the proof.

Now assume

I. $i < k$. The chain (3) is null homologous on $\mathfrak{B}^{k+1}$ since its dimension $i + 1$ is larger than 0 and smaller than $k + 1$ and $\mathfrak{B}^{k+1}$ has the same homology groups as the $(k + 1)$-sphere. Accordingly, let

$$\mathfrak{R} \partial V^{i+2} = U_1^{i+1} - U_2^{i+1}. \tag{4}$$

We decompose V^{i+2} into the sum

$$V^{i+2} = V_1^{i+2} + V_2^{i+2}, \tag{5}$$

where V_1^{i+2} consists of all of the simplexes of V^{i+2} which are incident with the center point of $\mathfrak{St}_1^{k+1}$. Then from (5) and (4)

$$\mathfrak{R} \partial V_1^{i+2} = \mathfrak{R} \partial V^{i+2} - \mathfrak{R} \partial V_2^{i+2} = U_1^{i+1} - U_2^{i+1} - \mathfrak{R} \partial V_2^{i+2}. \tag{6}$$

The left side of (6) belongs to $\mathfrak{St}_1^{k+1}$ and so does the chain U_1^{i+1} on the right side. Thus the chain $U_2^{i+1} + \mathfrak{R} \partial V_2^{i+2} = U^{i+1}$ must also belong to $\mathfrak{St}_1^{k+1}$. On the other hand, it belongs to $\mathfrak{St}_2^{k+1}$, as indicated by the index 2, and must lie on $\mathfrak{A}^k$. It follows from (6) that $U_1^{i+1} - U^{i+1} \sim 0$. Chains U_1^{i+1} and U^{i+1} therefore have the same boundary U^i. We have thereby constructed a chain U^{i+1} lying on $\mathfrak{A}^k$ which has the given chain U^i as its boundary and we have shown that the homology groups of the dimensions 1 through $k - 1$ of $\mathfrak{A}^k$ consist only of the null element, while the 0th homology group is the free cyclic group.

II. $i = k$. Since the $(k + 1)$th homology group of $\mathfrak{B}^{k+1}$ is the free cyclic group, there exists a chain B^{k+1} such that each closed $(k + 1)$-chain is a multiple of B^{k+1}. In particular, this is true for the chain (3):

$$U_1^{k+1} - U_2^{k+1} = m B^{k+1}.$$

The chain B^{k+1} can be decomposed uniquely into two chains B_1^{k+1} and B_2^{k+1} which belong to $\mathfrak{St}_1^{k+1}$ and $\mathfrak{St}_2^{k+1}$, respectively. It then follows that $U_1^{k+1} = m B_1^{k+1}$, so that $U^k = \mathfrak{R} \partial U_1^{k+1} = m \mathfrak{R} \partial B_1^{k+1}$. The chain U^k is thus the m-fold multiple of the closed chain $\mathfrak{R} \partial B_1^{k+1}$, that is, the kth homology group of $\mathfrak{A}^k$ is the free cyclic group.

An important theorem now follows:

THEOREM III. *If P is a vertex of a simplicially decomposed manifold $\mathfrak{M}^n$, then the outer boundary $\mathfrak{A}^{n-1}$ of the simplicial star $\mathfrak{St}^n$ about P is an $(n - 1)$-dimensional manifold $(n \geqq 2)$.*

Proof. I. $\mathfrak{A}^{n-1}$ is connected, finite and $(n - 1)$-dimensional. This is because $\mathfrak{A}^{n-1}$ is an $(n - 1)$-dimensional pseudomanifold.

II. To show that $\mathfrak{A}^{n-1}$ has the same homology groups as the $(n - 2)$-sphere at each point A let us ignore all simplexes of $\mathfrak{M}^n$ which do not belong to $\mathfrak{St}^n$ and let us also assume that A is a vertex of the simplicial decomposition of $\mathfrak{St}^n$. This is in reality no restriction, since we can subdivide

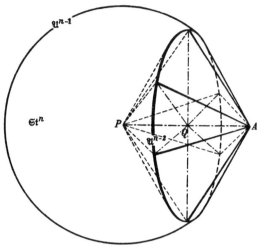

FIG. 118

$\mathfrak{A}^{n-1}$ so that A becomes a vertex and we can complete this subdivision of $\mathfrak{A}^{n-1}$ to a subdivision of $\mathfrak{St}^n$ by projecting out from P. Let $\mathfrak{St}^{n-1}$ denote the simplicial star on $\mathfrak{A}^{n-1}$ having A as center point and let $\mathfrak{A}^{n-2}$ be its outer boundary (Fig. 118). An n-simplex which contains the interval (PA) will then have vertices: $P, A, A_0, A_1, \ldots, A_{n-2}$, where

$$(A_0 A_1 \cdots A_{n-2})$$

is a $(n-2)$-simplex of $\mathfrak{A}^{n-2}$. Using the midpoint Q of (PA), we decompose the n-simplex into two n-simplexes

$$(PQA_0 A_1 \cdots A_{n-2}) \qquad \text{and} \qquad (AQA_0 A_1 \cdots A_{n-2}),$$

and we do the same with all of the other n-simplexes of $\mathfrak{St}^n$ incident with (PA). By means of this subdivision of $\mathfrak{St}^n$, Q becomes a vertex, and the outer boundary $\mathfrak{B}$ of the simplicial star having Q as center point is composed of two parts: the totality of $(n-1)$-simplexes $(PA_0 A_1 \cdots A_{n-2})$ and the totality of $(n-1)$-simplexes $(AA_0 A_1 \cdots A_{n-2})$, where all of the vertices of $(n-2)$-simplexes of $\mathfrak{A}^{n-2}$ are substituted into these expressions. That is, the neighborhood complex of Q consists of two simplicial stars having a common outer boundary $\mathfrak{A}^{n-2}$. Now from the definition of a manifold, the homology groups of the neighborhood complex $\mathfrak{B}^{n-1}$ of Q are the same as those of the $(n-1)$-sphere. From the lemma, then, the homology groups of $\mathfrak{A}^{n-2}$, that is, the homology groups of the outer boundary of $\mathfrak{A}^{n-1}$ at the point A, are the same as those of the $(n-2)$-sphere.

We have already seen that each connected, finite homogeneous complex is a manifold. We now show as a converse

THEOREM IV. *Each closed manifold is homogeneous for the dimensions $n = 1, 2, 3$ and only for these dimensions.*

The only 1-dimensional manifold, the circle, is obviously homogeneous.

If P is a point of a 2-dimensional manifold, we simplicially decompose the manifold with P as a vertex. From Theorem III, the neighborhood complex of P is a 1-dimensional manifold, and so is homeomorphic to a circle. Thus the simplicial star having center point P is homeomorphic to a disk. For 2 dimensions, then, a closed manifold is a finite, connected homogeneous complex. That is, it is the same as a closed surface (§39). We have already given a complete classification of the closed surfaces in Chapter VI.

In a 3-dimensional manifold $\mathfrak{M}^3$ the neighborhood complex of a point is a 2-dimensional manifold which has the same homology groups as the 2-sphere. From the fundamental theorem of surface topology, the 2-sphere is the only closed surface possessing this property. The simplicial star about a point of $\mathfrak{M}^3$ is thus homeomorphic to the 3-ball. That is, $\mathfrak{M}^3$ is homogeneous and is therefore a complex, as we considered it in Chapter IX.

This method of proof fails in more than 3 dimensions, because there exist 3-dimensional manifolds, the Poincaré spaces, which have the same homology groups as the 3-sphere but do not coincide with it. In fact, 4-dimensional nonhomogeneous manifolds do exist. An example is a complex $\mathfrak{K}^4$ consisting of two simplicial stars $\mathfrak{S}t_1^4$ and $\mathfrak{S}t_2^4$ which have the spherical dodecahedral space as a common outer boundary. All points, except for the center points P_1 of $\mathfrak{S}t_1^4$ and P_2 of $\mathfrak{S}t_2^4$, have the 3-sphere as a neighborhood complex. But the homology groups are the same as those of the 3-sphere, even at P_1 and P_2. For the neighborhood complex of P_1 and of P_2 is the spherical dodecahedral space, whose homology groups are the same as those of the 3-sphere (§62). Consequently, $\mathfrak{K}^4$ is a manifold. It is not, however, homogeneous. The points P_1 and P_2 have no neighborhoods homeomorphic to the interior of the 3-ball, since the fundamental group at P_1 and at P_2 is the binary icosahedral group (from §62) and not the unit element alone.

After having discussed the scope of our concept of a manifold, we now go to the most important property of manifolds, which is *the existence of dual cellular divisions*. By means of simplicial decomposition and an appropriate joining of its simplexes into stars we can make a given manifold $\mathfrak{M}^n$ into a star complex. For example, we get such a stellar division by normally subdividing an arbitrary simplicial decomposition of $\mathfrak{M}^n$ (§66). It is then true that *each division of a manifold into stars is also a division into cells.*

Proof. First of all, all n-stars are cells. For the outer boundary of an n-star is a pseudomanifold (from earlier in this section), and has the same homology groups as the $(n - 1)$-sphere since the homology groups at its center point are those of the $(n - 1)$-sphere. By the same reasoning, the $(n - 1)$-stars of an $(n - 1)$-dimensional manifold are cells. In particular, the $(n - 1)$-stars which form the outer boundary of an n-cell of $\mathfrak{M}^n$ are cells, for by Theorem III this outer boundary is an $(n - 1)$-dimensional manifold. Since each $(n - 1)$-star is incident with at least one n-star, for a manifold is a pure complex, all $(n - 1)$-stars are cells. One proceeds further by induction until one has reached the 0-stars, whereby the theorem has been proved.

Along with each stellar division, its dual stellar division is also a cellular division. Thus *each manifold admits dual cellular divisions.*

In the sections to follow we will denote

a given cellular division of $\mathfrak{M}^n$ by $\mathfrak{M}^n_a$,
the dual cellular division by $\mathfrak{M}^n_b$, and
the common normal subdivision by $\dot{\mathfrak{M}}^n$;

we denote the oriented cells of $\mathfrak{M}^n_a$ by a^k_α, the dual cells of $\mathfrak{M}^n_b$ by b^{n-k}_κ, the common center point of a^k_α and b^{n-k}_κ by P^k_κ $(k = 0, 1, \ldots, n, \kappa = 1, 2, \ldots, \alpha^k)$.

69. The Poincaré Duality Theorem

We are given a manifold $\mathfrak{M}^n$ together with dual cellular divisions $\mathfrak{M}^n_a$ and $\mathfrak{M}^n_b$. Assume that they are orientable and have been given fixed orientations. That is, the n-simplexes of the normal subdivision $\dot{\mathfrak{M}}^n$ have been coherently oriented with one of the two possible opposite orientations. If a^k_κ and b^{n-k}_κ are oriented dual cells we define the *intersection number* $\mathbb{S}(a^k_\kappa, b^{n-k}_\kappa)$ by means of the following rule.

RULE. Select a k-simplex from the normal subdivision of a^k_κ,

$$E^k = \xi(P^0 P^1 \cdots P^k)$$

whose orientation is given by that of a^k_κ. Likewise, select an $(n - k)$-simplex

$$E^{n-k} = \eta(P^k P^{k+1} \cdots P^n)$$

from the normal subdivision of b^{n-k}_κ. Then from Theorem III, Section 66,

$$E^n = \zeta(P^0 P^1 \cdots P^k \cdots P^n)$$

is an n-simplex whose sign ζ is determined by the orientation of $\dot{\mathfrak{M}}^n$. The intersection number $\mathbb{S}(a^k_\kappa, b^{n-k}_\kappa)$ is defined[42] as the product

$$\mathbb{S}(a^k_\kappa, b^{n-k}_\kappa) = \xi\eta\zeta. \tag{1}$$

It will always have value ± 1. In Fig. 119, $n = 2$, $k = 1$, $\xi = 1$, $\eta = \zeta = -1$. In this definition it is unimportant whether one arranges the vertices of E^k and E^{n-k} according to increasing upper index, so long as P^k stands in last place in E^k and in first place in E^{n-k} and the ordering of the vertices in E^n is the same as in E^k and E^{n-k}. For if one were to interchange two vertices $(\neq P^k)$ in E^k, both ξ and ζ would change sign while η would remain unchanged.

The definition is independent of the particular choice of simplexes E^k and E^{n-k} from the normal subdivisions of the cells a^k_κ and b^{n-k}_κ, respectively. For if one replaces E^k by a simplex $'E^k$ which has all vertices in common with E^k except for one vertex, let us say P^0 which is replaced by $'P^0$, then $'E^k = -\xi('P^0 P^1 \cdots P^k)$. For E^k and $'E^k$ induce opposite orientations in the common $(k - 1)$-dimensional face, because of the assumed coherent

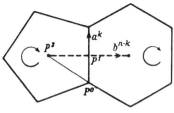

FIG. 119

orientation within a_κ^k. In place of E^n there appears

$$'E^n = -\zeta('P^0 P^1 \cdots P^k \cdots P^n),$$

whose sign is also given correctly, because E^n and $'E^n$ induce opposite orientations in the common $(n-1)$-simplex. Consequently, we get the intersection number

$$(-\xi)\eta(-\zeta) = \xi\eta\zeta,$$

as before. By repeating this procedure of replacing a simplex by a neighboring simplex, one can eventually get to any desired simplex of a_κ^k, for the outer boundary of a_κ^k is a pseudomanifold. The intersection number is preserved at each step of the procedure. One can apply the same considerations to the cell b_κ^{n-k}.

On the other hand, it is obvious that the intersection number reverses sign when one reverses the orientation of either $a_\kappa^k, b_\kappa^{n-k}$, or $\mathfrak{M}^n$.

In the case $k = 0$, a_κ^k consists of a single vertex P^0 which is also the center point of the dual cell b_κ^n. In this case,

$$E^0 = \xi(P^0) \qquad \text{and} \qquad E^n = \eta(P^0 P^1 \cdots P^n).$$

If we assume, in addition, that a_κ^0 is oriented with the sign $+1$, then $\xi = 1$ and

$$\mathbb{S}(a_\kappa^0, b_\kappa^n) = \eta\zeta = +1 \quad \text{or} \quad -1, \quad \text{respectively},$$

according to whether the orientation of b_κ^n does or does not agree with that of $\mathfrak{M}^n$. Here we can also regard the intersection number as a "covering number" of the point P^0 by the cell b_κ^n.

When one interchanges the two dual cells, the intersection number may change sign under certain circumstances. We have

$$\mathbb{S}(a_\kappa^k, b_\kappa^{n-k}) = (-1)^{k(n-k)}\mathbb{S}(b_\kappa^{n-k}, a_\kappa^k). \tag{2}$$

Proof.

$$E^k = \xi(P^0 P^1 \cdots P^k) = \xi(-1)^k(P^k P^0 \cdots P^{k-1}),$$

$$E^{n-k} = \eta(P^k P^{k+1} \cdots P^n) = \eta(-1)^{n-k}(P^{k+1} \cdots P^n P^k),$$

$$E^n = \zeta(P^0 P^1 \cdots P^{k-1} P^k P^{k+1} \cdots P^n)$$

$$= \zeta(-1)^{k+(n-k)(k+1)}(P^{k+1} \cdots P^n P^k P^0 \cdots P^{k-1}).$$

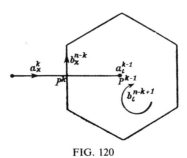

FIG. 120

From our previous rule, $\mathfrak{S}(b_\kappa^{n-k}, a_\kappa^k)$ is the product of the signs appearing on the right-hand sides of the last three equations:

$$\mathfrak{S}\left(b_\kappa^{n-k}, a_\kappa^k\right) = \xi\eta\zeta(-1)^{k(n-k)} = (-1)^{k(n-k)}\mathfrak{S}\left(a_\kappa^k, b_\kappa^{n-k}\right).$$

Now along with a_κ^k and b_κ^{n-k} we shall look at two dual cells a_ι^{k-1} and b_ι^{n-k+1} incident with them. We orient the latter cells so that we have (Fig. 120)

$$\mathfrak{R}\partial a_\kappa^k = a_\iota^{k-1} + \cdots, \tag{3}$$

$$\mathfrak{R}\partial b_\iota^{n-k+1} = b_\kappa^{n-k} + \cdots. \tag{4}$$

Let P^k and P^{k-1} be the center points of a_κ^k and a_ι^{k-1}, let

$$E^k = \xi(P^0 P^1 \cdots P^k)$$

be an oriented simplex of a_κ^k, and let

$$E^{n-k+1} = \eta(P^{k-1} P^k \cdots P^n)$$

be an oriented simplex of b_ι^{n-k+1}. Then

$$E^{k-1} = \xi(-1)^k (P^0 P^1 \cdots P^{k-1})$$

is a simplex of a_ι^{k-1} and

$$E^{n-k} = \eta(P^k P^{k+1} \cdots P^n)$$

is a simplex of b_κ^{n-k}. If the orientation of $\mathfrak{M}^n$ is again given by

$$E^n = \zeta(P^0 P^1 \cdots P^n),$$

then by our rule

$$\mathfrak{S}\left(a_\kappa^k, b_\kappa^{n-k}\right) = \xi\eta\zeta$$

and

$$\mathfrak{S}\left(a_\iota^{k-1}, b_\iota^{n-k+1}\right) = (-1)^k \xi\eta\zeta,$$

so that

$$\mathfrak{S}\left(a_\kappa^k, b_\kappa^{n-k}\right) = (-1)^k \mathfrak{S}\left(a_\iota^{k-1}, b_\iota^{n-k+1}\right). \tag{5}$$

More generally, if we assume that the orientations of the cells a_ι^{k-1} and b_ι^{n-k+1} are not fixed by Eqs. (3) and (4), but are prescribed arbitrarily, then instead of Eqs. (3) and (4) there appear the equations

$$\mathcal{R}\partial a_\kappa^k = {}^{(a)}\varepsilon_{\iota\kappa}^{k-1}a_\iota^{k-1} + \cdots = \rho a_\kappa^{k-1} + \cdots, \tag{6}$$

$$\mathcal{R}\partial b_\iota^{n-k+1} = {}^{(b)}\varepsilon_{\kappa\iota}^{n-k}b_\kappa^{n-k} + \cdots = \sigma b_\kappa^{n-1} + \cdots. \tag{7}$$

We then have

$$\mathbb{S}\big(a_\kappa^k, b_\kappa^{n-k}\big) = \rho\sigma(-1)^k \mathbb{S}\big(a_\iota^{k-1}, b_\iota^{n-k+1}\big). \tag{8}$$

For upon reversing the orientation of a_ι^{k-1} or b_ι^{n-k+1} the sign on the right-hand side of (5) reverses.

Let us now assume that the cells a_κ^k ($k = 0, 1, \ldots, n$, $\kappa = 1, 2, \ldots, \alpha^k$) are oriented arbitrarily, but that the cells b_κ^{n-k} of the dual cellular division are oriented so that the intersection numbers are

$$\mathbb{S}\big(a_\kappa^k, b_\kappa^{n-k}\big) = 1.$$

It then follows from (8) that $\rho\sigma(-1)^k = 1$ or

$$\rho = (-1)^k\sigma. \tag{9}$$

Now ρ is nothing other than the coefficient ${}^{(a)}\varepsilon_{\iota\kappa}^{k-1}$ in the incidence matrix $\mathbf{E}_a^{k-1}$ of the cellular division $\mathfrak{M}_a^n$ and σ is the coefficient ${}^{(b)}\varepsilon_{\kappa\iota}^{n-k}$ in the incidence matrix $\mathbf{E}_b^{n-k}$ of the cellular division $\mathfrak{M}_b^n$. Equation (9) then states that the matrix $\mathbf{E}_a^{k-1}$ is equal to the transpose* of the incidence matrix $\mathbf{E}_b^{n-k}$ multiplied by $(-1)^k$, that is,

$$\mathbf{E}_a^{k-1} = (-1)^k\bar{\mathbf{E}}_b^{n-k}, \tag{10}$$

or that the following boundary relations hold (we can now again omit the superscripted indexes ${}^{(a)}$ and ${}^{(b)}$ of $\varepsilon_{\iota\kappa}^{k-1}$), that is,

$$\mathcal{R}\partial a_\kappa^k = \sum_{\iota=1}^{\alpha^{k-1}} \varepsilon_{\iota\kappa}^{k-1}a_{\iota\kappa}^{k-1},$$

$$\mathcal{R}\partial b_\iota^{n-k+1} = \sum_{\kappa=1}^{\alpha^k} (-1)^k\varepsilon_{\iota\kappa}^{k-1}b_\kappa^{n-k}. \tag{11}$$

The result is

THEOREM I. *Given dual cellular divisions $\mathfrak{M}_a^n$ and $\mathfrak{M}_b^n$, if one orients the cells of $\mathfrak{M}_b^n$ in a way such that the intersection number of each cell of $\mathfrak{M}_a^n$ with the corresponding dual cell of $\mathfrak{M}_b^n$ in this order is equal to 1, then the incidence matrix $\mathbf{E}_a^{k-1}$ of $\mathfrak{M}_a^n$ is equal to the transpose of the incidence matrix $\mathbf{E}_b^{n-k}$ of $\mathfrak{M}_b^n$ multiplied by $(-1)^k$.*

---

*A matrix is transposed by interchanging rows and columns or, what is the same thing, mirroring them about the main diagonal. We denote the transpose of a matrix by a bar over the matrix, slanting in the direction of the main diagonal.

We arrive at the Poincaré duality theorem by calculating the homology groups of $\mathfrak{M}^n$ in two different ways, once from the cellular division $\mathfrak{M}_a^n$ and once from the cellular division $\mathfrak{M}_b^n$, and then using the matrix equation (10). For the kth Betti number we then get

$$p^k = \alpha_a^k - \gamma_a^k - \gamma_a^{k-1} = \alpha_b^k - \gamma_b^k - \gamma_b^{k-1} \qquad (k = 0, 1, \ldots, n).$$

Here α_a^k denotes the number of k-cells of $\mathfrak{M}_a^n$, γ_a^k the rank of $\mathbf{E}_a^k$, and $\gamma_a^0 = \gamma_a^n = 0$ (§22). The numbers indicated with the index b have a corresponding significance. Since there is exactly one dual $(n - k)$-cell corresponding to each k-cell, $\alpha_a^k = \alpha_b^{n-k}$. From (10), $\gamma_a^{k-1} = \gamma_b^{n-k}$. Thus

$$p^k = \alpha_b^{n-k} - \gamma_b^{n-k-1} - \gamma_b^{n-k} = p^{n-k}.$$

The torsion coefficients for the dimension k are, on the one hand, the invariant factors of $\mathbf{E}_a^k$ which are not equal to 1; on the other hand, they are the invariant factors of $\mathbf{E}_b^k = (-1)^{n-k} \mathbf{E}_a^{n-k-1}$ and therefore coincide with the torsion coefficients for the dimension $n - k - 1$. We thus have

THEOREM II (POINCARÉ DUALITY THEOREM). *The kth Betti number of an orientable closed manifold is the same as the $(n - k)$th Betti number $(k = 0, 1, \ldots, n)$; the k-dimensional torsion coefficients are the same as the $(n - k - 1)$-dimensional torsion coefficients $(k = 0, 1, \ldots, n - 1)$.*

It follows, in particular, that there are no $(n - 1)$-dimensional torsion coefficients for an orientable closed manifold, a fact already known to us. For there are no 0-dimensional torsion coefficients (§18).

This duality theorem depends upon the orientability of the manifold in an essential way. The matrix equation (10) cannot be obtained for any orientation of the n-cells, in the case of a nonorientable manifold, because $\mathbf{E}_a^0$ has rank $\alpha_a^0 - 1$ and $\mathbf{E}_b^{n-1}$ has rank $\alpha_b^n = \alpha_a^0$ (§24).

It is also possible to find a duality theorem for nonorientable manifolds. This theorem can be derived in a simpler manner, because we need not consider the cell orientations. In this case Eq. (10) does not hold between the incidence matrices $\mathbf{E}_a^{k-1}$ and $\mathbf{E}_b^{n-k}$. Instead we have the congruence

$$\mathbf{E}_a^{k-1} \equiv \grave{\mathbf{E}}_b^{n-k} \pmod 2.$$

The elements of $\mathbf{E}_a^{k-1}$ and $\grave{\mathbf{E}}_b^{n-k}$ differ at most in sign. In other words, the incidence matrices mod 2, $\check{\mathbf{E}}_a^{k-1}$ and $\check{\mathbf{E}}_b^{n-k}$, transform to one another upon changing rows and columns and therefore have the same rank $\delta_a^{k-1} = \delta_b^{n-k}$. The kth connectivity number is given by

$$q^k = \alpha_a^k - \delta_a^k - \delta_a^{k-1} = \alpha_b^k - \delta_b^k - \delta_b^{k-1}$$
$$= \alpha_a^{n-k} - \delta_a^{n-k-1} - \delta_a^{n-k} = q^{n-k},$$

as a result of formula (8) of §23. We therefore have

THEOREM III (DUALITY THEOREM mod 2). *The kth connectivity number of an orientable or nonorientable closed manifold is equal to the $(n - k)th$ connectivity number.*

From Eq. (12) of §23 there follows

THEOREM IV. *The Euler characteristic of a manifold of odd dimension is* $N = 0$.

70. Intersection Numbers of Cell Chains

The topological content of the duality theorems just derived formally from the incidence matrices becomes apparent only when one goes from Betti numbers to Betti bases and proves that the k-dimensional Betti basis stands in a duality relation with the $(n - k)$-dimensional Betti basis. To demonstrate this we must define the intersection number of two dual cell chains and use it to construct dual Betti bases (§71). Let

$$A^k = \xi_1 a_1^k + \xi_2 a_2^k + \cdots + \xi_{\alpha^k} a_{\alpha^k}^k$$

and

$$B^{n-k} = \eta_1 b_1^{n-k} + \eta_2 b_2^{n-k} + \cdots + \eta_{\alpha^k} b_{\alpha^k}^{n-k}$$

be two chains on dual cellular divisions. Since $\alpha_b^{n-k} = \alpha_a^k$ and is equal, for example, to α^k, the indices run from 1 to the common value α^k in both chains. The two chains have at most finitely many points in common, for a cell a_κ^k is disjoint with every cell b_λ^{n-k} except for the dual cell b_κ^{n-k}, which has a common center point with a_κ^k. The intersection number of two dual cells was defined at the beginning of the last section. For two nondual cells of the dual divisions we define their intersection number to be 0. The intersection number of the chains A^k and B^{n-k} is then the sum of the intersection numbers of the individual cells. More precisely,

$$\mathbb{S}(A^k, B^{n-k}) = \sum_{\kappa=1}^{\alpha^k} \sum_{\lambda=1}^{\alpha^k} \xi_\kappa \eta_\lambda \mathbb{S}(a_\kappa^k, b_\lambda^{n-k}) = \sum_{\kappa=1}^{\alpha^k} \xi_\kappa \eta_\kappa \mathbb{S}(a_\kappa^k, b_\kappa^{n-k}). \quad (1)$$

The above equation describes the intersection number of two chains as a bilinear form on the two vectors representing the chains.

The given fixed orientation of $\mathfrak{M}^n$ determines a particular orientation for all of the n-cells of $\mathfrak{M}_b^n$. Call the sum of all of the n-cells oriented in this way M_b^n. Setting $k = 0$ and $B^{n-k} = M_b^n$ into formula (1), it follows from §69 that *the intersection number of a chain A^0 with the oriented chain M_b^n of the manifold is equal to the sum of the coefficients of A^0.*

We obviously have the rules of computation

$$\mathbb{S}(A_1^k + A_2^k, B^{n-k}) = \mathbb{S}(A_1^k, B^{n-k}) + \mathbb{S}(A_2^k, B^{n-k}) \quad (2)$$

and, using formula (2) of §69,

$$\mathbb{S}(A^k, B^{n-k}) = (-1)^{(n-k)k} \mathbb{S}(B^{n-k}, A^k). \quad (3)$$

It is also possible to extend formula (5) of §69,

$$\mathfrak{S}\left(a_{\kappa}^{k}, b_{\kappa}^{n-k}\right) = (-1)^{k}\mathfrak{S}\left(a_{\iota}^{k-1}, b_{\iota}^{n-k+1}\right)$$

from the case of two dual cells to the case of arbitrary chains of dimensions k and $n - k$. Remembering that all boundary cells of b_{ι}^{n-k+1} are disjoint with a_{κ}^{k}, with the exception of b_{κ}^{n-k}, we can replace b_{κ}^{n-k} on the left-hand side of this formula by $\mathfrak{R}\partial b_{\iota}^{n-k+1}$ and, correspondingly, substitute $\mathfrak{R}\partial a_{\kappa}^{k}$ for a_{ι}^{k-1} on the right-hand side. We get

$$\mathfrak{S}\left(a_{\kappa}^{k}, \mathfrak{R}\partial b_{\iota}^{n-k+1}\right) = (-1)^{k}\mathfrak{S}\left(\mathfrak{R}\partial a_{\kappa}^{k}, b_{\iota}^{n-k+1}\right). \tag{4}$$

This formula has been proved under the assumption that the cell a_{ι}^{k-1}, the dual to b_{ι}^{n-k+1}, is incident to a_{κ}^{k}. The formula is also valid when this is not the case, as both sides then vanish. For the left-hand side is nonzero only if the cell b_{κ}^{n-k}, the dual to a_{κ}^{k}, appears in the boundary of b_{ι}^{n-k+1}; that is, if b_{ι}^{n-k+1} and b_{κ}^{n-k} are incident and consequently the dual cells a_{ι}^{k-1} and a_{κ}^{k} are incident. In like manner, the right-hand side is nonzero only if the cell a_{ι}^{k-1}, the dual to b_{ι}^{n-k+1}, appears in the boundary of a_{κ}^{k}. Formula (4) is therefore valid for two arbitrary cells a_{κ}^{k} and b_{ι}^{n-k+1}. Consequently, from (2), the formula

$$\mathfrak{S}(A^{k}, \mathfrak{R}\partial B^{n-k+1}) = (-1)^{k}\mathfrak{S}(\mathfrak{R}\partial A^{k}, B^{n-k+1}) \tag{5}$$

is valid for two arbitrary cell chains A^{k} and B^{n-k+1} on the dual cellular divisions $\mathfrak{M}_{a}^{n}$ and $\mathfrak{M}_{b}^{n}$.

We shall make immediate application of this important formula. If A^{k} is closed, that is, $\mathfrak{R}\partial A^{k} = 0$, then the right-hand side of (5) is equal to zero. But $\mathfrak{R}\partial B^{n-k+1}$ is an arbitrary null homologous $(n - k)$-chain on the left-hand side. We therefore have

THEOREM I. *A null homologous cell chain B^{n-k} of dimension $n - k$ has the intersection number zero with respect to each closed k-chain A^{k} of the dual cellular division.*

The theorem is also valid when $B^{n-k} \approx 0$ (that is, when B^{n-k} is only division–null homologous). In that case

$$cB^{n-k} \sim 0 \qquad (c \neq 0)$$

and also

$$\mathfrak{S}(A^{k}, cB^{n-k}) = c\mathfrak{S}(A^{k}, B^{n-k}).$$

From Theorem I the left-hand side of the above equation is equal to 0, and thus so is the right-hand side. That implies $\mathfrak{S}(A^{k}, B^{n-k}) = 0$, since $c \neq 0$, and leads to the following theorem:

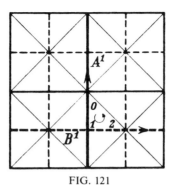

FIG. 121

THEOREM II. *If* $A^k \approx {'A}^k$ *and* $B^{n-k} \approx {'B}^{n-k}$ *are closed chains on dual cellular divisions, then*

$$\mathfrak{S}(A^k, B^{n-k}) = \mathfrak{S}({'A}^k, {'B}^{n-k}).$$

That is, the intersection number of two closed chains remains unchanged when one replaces each chain by a division–null homologous chain of the same cellular division.

As an example we shall consider a cellular division $\mathfrak{M}_a^2$ of the torus into four squares. These are the squares in darker outline in Fig. 121. The opposite sides of the large square are to be identified. The dual cellular division $\mathfrak{M}_b^2$ also consists of 4 squares. These are shown dotted in the figures. The common normal subdivision $\mathfrak{M}^2$ consists of 4×8 triangles, which we assume to be coherently oriented as indicated by the circular arrow in one of the triangles of the figure. We wish to find the intersection number of the cell chain A^1 and B^1 which are closed 1-chains belonging to the dual cellular divisions. These chains (indicated by double lines in the figure) each consists of two 1-cells. Since we have chosen $n = 2$ and $k = 1$, $n - k = 1$. The intersection number will be determined from our rule (§69). On A^1 we select the 1-simplex* $-(P^0P^1)$ and on B^1 we select the 1-simplex $+(P^1P^2)$. We form from them the 2-simplex $(P^0P^1P^2)$; in the coherent orientation we are using, this has a positive sign $+1$. The intersection number is therefore

$$\mathfrak{S}(A^1, B^1) = (-1)(+1)(+1) = -1.$$

In defining intersection numbers, it is essential to assume that the manifold is orientable. If we wish to extend our theory to nonorientable manifolds, then we must restrict our considerations to chains mod 2. For two cell chains mod 2, $\mathfrak{A}^k$ and $\mathfrak{B}^{n-k}$ lying on dual cellular divisions, their *intersection number* mod 2 is a residue class of the integers mod 2. It is defined as

$$\mathfrak{S}(\mathfrak{A}^k, \mathfrak{B}^{n-k}) = \breve{0} \quad \text{or} \quad \breve{1}, \quad \text{respectively,}$$

according to whether the number of points common to $\mathfrak{A}^k$ and $\mathfrak{B}^{n-k}$ is even or odd respectively. Formulas (1) through (5) are also valid for chains mod 2 · when one replaces the integers appearing in them by their residue classes mod 2. One can then, of course, omit the sign factor appearing in the formulas.

*In the figure, the points are indicated only by their indices.

71. Dual Bases

We are given two arrays of "variables" or "indeterminate quantities"

$$x_1, x_2, \ldots, x_m \qquad \text{and} \qquad y_1, y_2, \ldots, y_m$$

which are transformed by two integer-valued unimodular linear transformations

$$\left.\begin{aligned} x_1 &= \alpha_{11}\bar{x}_1 + \alpha_{12}\bar{x}_2 + \cdots + \alpha_{1m}\bar{x}_m \\ &\vdots \\ x_m &= \alpha_{m1}\bar{x}_1 + \alpha_{m2}\bar{x}_2 + \cdots + \alpha_{mm}\bar{x}_m \end{aligned}\right\}, \qquad (\alpha)$$

$$\left.\begin{aligned} y_1 &= A_{11}\bar{y}_1 + A_{12}\bar{y}_2 + \cdots + A_{1m}\bar{y}_m \\ &\vdots \\ y_m &= A_{m1}\bar{y}_1 + A_{m2}\bar{y}_2 + \cdots + A_{mm}\bar{y}_m \end{aligned}\right\} \qquad (\mathbf{A})$$

to new arrays of variables (indicated by bars). The two transformations and also the arrays of variables are said to be *contragredient* if the transformation matrices α and $\mathbf{A}$ are related so that the one is the transpose of the inverse of the other:

$$\mathbf{A} = \grave{\alpha}^{-1}.$$

The element A_{ik} is the cofactor of α_{ik} divided by the determinant $|\alpha| = \pm 1$. The relation of contragredience is obviously a reciprocal one.

Since $\mathbf{A}^{-1} = \grave{\alpha}$, the solution of the equations ($\mathbf{A}$) is

$$\left.\begin{aligned} \bar{y}_1 &= \alpha_{11}y_1 + \alpha_{21}y_2 + \cdots + \alpha_{m1}y_m \\ &\vdots \\ \bar{y}_m &= \alpha_{1m}y_1 + \alpha_{2m}y_2 + \cdots + \alpha_{mm}y_m \end{aligned}\right\}. \qquad (\mathbf{A}^{-1})$$

Thus

$$\sum_{i=1}^{m} x_i y_i = \sum_i \sum_k \alpha_{ik}\bar{x}_k y_i = \sum_k \bar{x}_k \left(\sum_i \alpha_{ik} y_i\right) = \sum_{k=1}^{m} \bar{x}_k \bar{y}_k.$$

Two contragredient transformations will then leave the "unit bilinear form" invariant. We could have also defined contragredience by means of this property.

We shall speak of *cogredient* transformations of the two arrays of variables, on the other hand, when they both transform in the same way.

We have already encountered examples of cogredient and contragredient transformations in §21.

Once again let $\mathfrak{M}^n$ be an orientable manifold having dual cellular divisions $\mathfrak{M}_a^n$ and $\mathfrak{M}_b^n$. Let the dual cells a_κ^k and b_κ^{n-k} be oriented so that

$\mathcal{S}(a_\kappa^k, b_\kappa^{n-k}) = +1$. By carrying out the unimodular transformation $(\boldsymbol{\alpha})$ on the cells

$$a_1^k, a_2^k, \ldots, a_{\alpha^k}^k$$

one thereby goes to a new basis

$$\bar{a}_1^k, \bar{a}_2^k, \ldots, \bar{a}_{\alpha^k}^k$$

of the lattice of all k-cell chains of $\mathfrak{M}_a^n$. The coefficients $\xi_1, \xi_2, \ldots, \xi_{\alpha^k}$ appearing in a chain

$$A^k = \xi_1 a_1^k + \xi_2 a_2^k + \cdots + \xi_{\alpha^k} a_{\alpha^k}^k$$

will transform according to the contragredient transformation $(\mathbf{A})$. For we have

$$A^k = \sum_{\kappa=1}^{\alpha^k} \xi_\kappa a_\kappa^k = \sum_{\kappa=1}^{\alpha^k} \bar{\xi}_\kappa \bar{a}_\kappa^k.$$

The dual cells

$$b_1^{n-k}, b_2^{n-k}, \ldots, b_{\alpha^k}^{n-k}$$

are now to be transformed contragrediently to the a_κ^k, that is, according to the equations $(\mathbf{A})$. The coefficients η_κ in the chain

$$B^{n-k} = \sum_{\kappa=1}^{\alpha^k} \eta_\kappa b_\kappa^{n-k} = \sum_{\kappa=1}^{\alpha^k} \bar{\eta}_\kappa \bar{b}_\kappa^{n-k}$$

will then also be transformed contragrediently to the b_κ^{n-k} and, thereby, contragrediently to the ξ_κ. The bilinear form $\sum \xi_\kappa \eta_\kappa$ will thus remain invariant:

$$\sum_{\kappa=1}^{\alpha^k} \xi_\kappa \eta_\kappa = \sum_{\kappa=1}^{\alpha^k} \bar{\xi}_\kappa \bar{\eta}_\kappa.$$

The left-hand side of the above equation is equal to the intersection number $\mathcal{S}(A^k, B^{n-k})$. This shows that when one uses new bases $\bar{a}_\kappa^k$ and $\bar{b}_\kappa^{n-k}$, the intersection number of two arbitrary chains $\sum \bar{\xi}_\kappa \bar{a}_\kappa^k$ and $\sum \bar{\eta}_\kappa \bar{b}_\kappa^{n-k}$ is still given by the unit bilinear form.

Two cell bases $\bar{a}_\kappa^k$ and $\bar{b}_\kappa^{n-k}$ which have this property are said to be dual to each other. It follows that for each arbitrary cellular basis $\bar{a}_1^k, \bar{a}_2^k, \ldots, \bar{a}_{\alpha^k}^k$ there exists one and only one dual basis $\bar{b}_1^{n-k}, \bar{b}_2^{n-k}, \ldots, \bar{b}_{\alpha^k}^{n-k}$ and each pair of dual cellular bases can be obtained from one particular pair by means of contragredient unimodular transformations.

Another way of characterizing the dual bases is to require that *the matrix of intersection numbers for the dual bases is the unit matrix having α^k rows*:

$$\mathcal{S}\left(\bar{a}_\mu^k, \bar{b}_\nu^{n-k}\right) = \delta_{\mu\nu}.$$

We shall now examine how the boundary relations (11) of §69:

$$\mathcal{R}\partial a_\kappa^k = \sum_{\iota=1}^{\alpha^{k-1}} \varepsilon_{\iota\kappa}^{k-1} a_\iota^{k-1}, \tag{1}$$

$$\mathcal{R}\partial b_\iota^{n-k+1} = (-1)^k \sum_{\kappa=1}^{\alpha^k} \varepsilon_{\iota\kappa}^{k-1} b_\kappa^{n-k} \tag{2}$$

are changed by a contragredient transformation of dual bases. When we transform the array of cells a_κ^k by an integer-valued unimodular transformation, the $\varepsilon_{\iota\kappa}^{k-1}$ in Eq. (1) will transform cogrediently to the a_κ^k for a fixed index ι; on the other hand, the $\varepsilon_{\iota\kappa}^{k-1}$ in Eq. (2) will transform contragrediently to the b_κ^{n-k}. Since the a_κ^k and the b_κ^{n-k} are to be transformed by contragredient transformations, however, the coefficients $\varepsilon_{\iota\kappa}^{k-1}$ in (2) will transform cogrediently to the a_κ^k, that is, just as in Eq. (1). When we replace the dual bases a_ι^{k-1} and b_ι^{n-k+1} by new dual bases, using two contragredient transformations, the $\varepsilon_{\iota\kappa}^{k-1}$ will likewise transform (for fixed index κ) cogrediently to the b_ι^{n-k+1} in both equations (and hence contragrediently to the a_ι^{k-1}). After the two transformations, the coefficients $\bar{\varepsilon}_{\iota\kappa}^{k-1}$ will still be identical in the two transformed boundary relations

$$\mathcal{R}\partial\bar{a}_\kappa^k = \sum_{\iota=1}^{\alpha^{k-1}} \bar{\varepsilon}_{\iota\kappa}^{k-1} \bar{a}_\iota^{k-1}, \tag{$\bar{1}$}$$

$$\mathcal{R}\partial\bar{b}_\iota^{n-k+1} = (-1)^k \sum_{\kappa=1}^{\alpha^k} \bar{\varepsilon}_{\iota\kappa}^{k-1} \bar{b}_\kappa^{n-k}. \tag{$\bar{2}$}$$

From §22, one can simultaneously bring the cell incidence matrices

$$\mathbf{E}^0, \mathbf{E}^1, \dots, \mathbf{E}^{n-1}$$

of the cellular division $\mathfrak{M}_a^n$ to normal form, by means of integer-valued unimodular transformations of the 0- through n-dimensional a-bases. We shall choose our notation so that the matrices $[\bar{\varepsilon}_{\iota\kappa}^{k-1}]$ already appear in normal form.

The basis chains $\bar{a}_\kappa^k$ then decompose into three classes:

$$\begin{aligned} {}^1A_\lambda^k & \quad (\lambda = 1, 2, \dots, \gamma^k = \gamma^{n-k-1}), \\ {}^2A_\mu^k & \quad (\mu = 1, 2, \dots, p^k = p^{n-k}), \\ {}^3A_\nu^k & \quad (\nu = 1, 2, \dots, \gamma^{k-1} = \gamma^{n-k}). \end{aligned} \tag{3}$$

In §22 these three classes were denoted by $\bar{A}_\lambda^k, \bar{B}_\mu^k, \bar{C}_\nu^k$. The cell chains ${}^1A_\lambda^k$ are division–null homologous. The ${}^2A_\mu^k$ form a Betti basis. The ${}^3A_\nu^k$ are nonclosed chains. The boundary relations ($\bar{1}$) now become

$$\mathcal{R}\partial {}^1A_\lambda^k = 0, \qquad \mathcal{R}\partial {}^2A_\mu^k = 0, \qquad \mathcal{R}\partial {}^3A_\nu^k = c_\nu^{k-1}\, {}^1A_\nu^{k-1}. \tag{4}$$

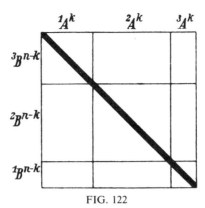

FIG. 122

The chains of the basis dual to (3) will be denoted, in order, by

$$
\begin{aligned}
{}^{3}B_{\lambda}^{n-k} \quad & (\lambda = 1, 2, \ldots, \gamma^{n-k-1} = \gamma^{k}), \\
{}^{2}B_{\mu}^{n-k} \quad & (\mu = 1, 2, \ldots, p^{n-k} = p^{k}), \\
{}^{1}B_{\nu}^{n-k} \quad & (\nu = 1, 2, \ldots, \gamma^{n-k} = \gamma^{k-1}).
\end{aligned}
\tag{5}
$$

Because the coefficients $\bar{\varepsilon}$ appearing in $(\bar{1})$ and $(\bar{2})$ have been proved to be the same, the boundary relations $(\bar{2})$ can now be written

$$
\begin{aligned}
\Re \partial\,{}^{3}B_{\nu}^{n-k+1} &= (-1)^{k} c_{\nu}^{k-1}\,{}^{1}B_{\nu}^{n-k}, \\
\Re \partial\,{}^{2}B_{\zeta}^{n-k+1} &= 0, \\
\Re \partial\,{}^{1}B_{\vartheta}^{n-k+1} &= 0.
\end{aligned}
\tag{6}
$$

It follows from this, when one uses the $\bar{b}$-basis, that the incidence matrices likewise appear simultaneously in normal form. The cell chains ${}^{3}B_{\lambda}^{n-k}$ are nonclosed chains, the ${}^{2}B_{\mu}^{n-k}$ form a Betti basis, and the ${}^{1}B_{\nu}^{n-k}$ are division–null homologous.

Since the matrix of intersection numbers of the dual bases (Fig. 122) is a unit matrix, the middle submatrix of intersection numbers

$$
\mathcal{S}\left({}^{2}A_{\rho}^{k}, {}^{2}B_{\sigma}^{n-k}\right) = \delta_{\rho\sigma} \qquad (\rho, \sigma = 1, 2, \ldots, p^{k})
\tag{7}
$$

is also a unit matrix.

If we define two *dual Betti bases* ${}^{2}A_{\mu}^{k}$ and ${}^{2}B_{\mu}^{n-k}$ as Betti bases on dual cellular divisions such that their intersection matrix (7) is the unit matrix, then we have demonstrated the existence of dual Betti bases.

From here it is only one step to our result:

THEOREM I. *For each given k-dimensional Betti basis there exists a dual $(n-k)$-dimensional Betti basis, uniquely determined up to division–null homologous chains.*

For one can obtain the most general Betti basis from the particular Betti basis $^2A_\mu^k$ by means of an integer-valued unimodular transformation and by addition of division–null homologous chains. But the transformation of the particular Betti basis $^2B_\mu^{n-k}$ is determined uniquely by the transformation of its dual basis $^2A_\mu^k$, namely, as the contragredient transformation, when one requires that the intersection matrix be again the unit matrix (7) after the transformation. Since the intersection numbers do not change upon addition of division–null homologous chains, one can add division–null homologous chains to the new k-dimensional Betti basis as well as to the new $(n - k)$-dimensional Betti basis.

The existence of dual Betti bases discloses the underlying reason for the Poincaré duality relation $p^k = p^{n-k}$ between the Betti numbers. In a similar way, we can provide a topological foundation for the relation between the k-dimensional and $(n - k)$-dimensional torsion coefficients, as stated in the duality theorem of §69.

The first ρ^k chains among the chains $^1A_\lambda^k$ of (3), that is, the chains

$$^1A_1^k, {}^1A_2^k, \ldots, {}^1A_{\rho^k}^k,$$

form a torsion basis for the dimension k (§20) and the chains

$$^1B_1^{n-k-1}, {}^1B_2^{n-k-1}, \ldots, {}^1B_{\rho^k}^{n-k-1}$$

form a torsion basis for the dimension $n - k - 1$. The matrix of intersection numbers

$$\left({}^1A_\sigma^k, {}^3B_\tau^{n-k}\right) = \delta_{\sigma\tau} \qquad (\sigma, \tau = 1, 2, \ldots, \rho^k)$$

is a unit matrix, when regarded as a submatrix of the matrix of Fig. 122. Using the abbreviated notation

$$^1A_\sigma^k = A_\sigma^k, \qquad (-1)^{k+1}\,{}^1B_\tau^{n-k-1} = B_\tau^{n-k-1}, \qquad {}^3B_\tau^{n-k} = B_\tau^{n-k},$$

we have

THEOREM II. *There exists a torsion basis for the dimension k*

$$A_1^k, A_2^k, \ldots, A_{\rho^k}^k$$

and a torsion basis for the dimension $n - k$

$$B_1^{n-k-1}, B_2^{n-k-1}, \ldots, B_{\rho^k}^{n-k-1}$$

having the following property. If

$$B_1^{n-k}, B_2^{n-k}, \ldots, B_{\rho^k}^{n-k}$$

are arbitrary chains for which the relations

$$\Re\,\partial B_1^{n-k} = c_1^k B_1^{n-k-1}, \qquad \Re\,\partial B_2^{n-k} = c_2^k B_2^{n-k-1}, \qquad \Re\,\partial B_{\rho^k}^{n-k} = c_{\rho^k}^k B_{\rho^k}^{n-k-1}$$

hold (the c_τ^k are the k-dimensional torsion coefficients), then the matrix of *intersection numbers*

$$\mathfrak{S}\left(A_\sigma^k, B_\tau^{n-k}\right) = \delta_{\sigma\tau}$$

is the unit matrix having ρ^k *rows.*[43]

We have proved the theorem only for the case that the $(n-k)$-chain B_τ^{n-k} spanning $c_\tau^k B_\tau^{n-k-1}$ is the particular chain $^3B_\tau^{n-k}$. It is valid, however, for any chain B_τ^{n-k} spanning $c_\tau^k B_\tau^{n-k-1}$. For by Theorem I of §70 we have

$$\mathfrak{S}\left(A_\sigma^k, B_\tau^{n-k} - {}^3B_\tau^{n-k}\right) = 0,$$

since A_σ^k is division–null homologous and $B_\tau^{n-k} - {}^3B_\tau^{n-k}$ is a closed chain. Thus $\mathfrak{S}(A_\sigma^k, B_\tau^{n-k}) = \mathfrak{S}(A_\sigma^k, {}^3B_\tau^{n-k})$.

We can also carry out the developments of this section for chains mod 2 and we obtain, corresponding to Theorem I:

THEOREM III. *For each connectivity basis*

$$\mathfrak{A}_1^k, \mathfrak{A}_2^k, \ldots, \mathfrak{A}_{q^k}^k$$

of the dimension k *there exists a dual connectivity basis of the dimension* $n-k$

$$\mathfrak{B}_1^{n-k}, \mathfrak{B}_2^{n-k}, \ldots, \mathfrak{B}_{q^k}^{n-k}$$

such that the matrix of intersection numbers mod 2 *is the unit matrix having* q^k *rows.*

Examples to illustrate this section are to be found in §75.

72. Cellular Approximations

Until now, we have defined intersection numbers only for cell chains in dual cellular divisions and have investigated the intersection numbers only for this case. We intend next to make our results independent of any particular cellular division and to define intersection numbers for arbitrary singular chains. This is done by approximating the singular chains. The approximation is not a simplicial approximation, as previously, but proceeds using cellular divisions. We shall now explain this process of cellular approximation. The investigations to follow will be valid in an arbitrary cell complex $\Re^n$. We shall not assume that $\Re^n$ is a manifold.

The underlying fundamental theorem is the following approximation theorem, which describes the extension of the simplicial approximation theorem of Chapter IV to cell complexes.

THEOREM I. *If* A^k *is a singular chain on a cell complex* $\Re^n$ *and if the boundary* * A^{k-1} *of* A^k *is a cell chain on* $\Re^n$ *(in particular it can be the* $(k-1)$*-chain* 0*), then there exists a cell chain* $\overline{A}^k$ *homologous to* A^k.

---

* For the case $k = 0$ this requirement is of course eliminated.

Proof. Since the boundary of A^k is a cell chain for $k > 0$, and is therefore also a simplicial chain, by the approximation theorem of §28 there exists a homologous simplicial chain $'A^k$ (which is thus a chain on the normal subdivision $\dot{\Re}^n$ of the cell complex $\Re^n$). By §67 there exists a cell chain homologous to this simplicial chain. The theorem remains valid when $k = 0$; the proof is simpler in this case, since one does not have to pay attention to the boundary.

COROLLARY. *If $\langle A^k \rangle$ is the smallest cell subcomplex containing A^k, then one may assume that $\langle A^k \rangle$ also contains $\overline{A}^k$ and that $A^k \sim \overline{A}^k$ on $\langle A^k \rangle$.*

The smallest cell subcomplex $\langle A^k \rangle$ is defined as the intersection of all cell subcomplexes to which A^k belongs. It is empty if and only if A^k is the k-chain 0.* The corollary follows immediately from Theorem I. One removes all cells from $\Re^n$ which do not belong to $\langle A^k \rangle$ and then regards A^k as being a chain of the cellular complex $\langle A^k \rangle$ and, finally, applies Theorem I to $\langle A^k \rangle$ instead of to $\Re^n$.

We will now make a cellular approximation of an arbitrary "singular complex" $\mathfrak{C}^m$ in $\Re^n$. By a singular complex we mean a set of finitely many singular nondegenerate simplexes for which, along with each singular simplex, each of its nondegenerate faces is also included in this set. We make the cellular approximation of $\mathfrak{C}^m$ as follows: First, we arbitrarily orient all singular simplexes of $\mathfrak{C}^m$. Let these singular simplexes be called

$$X_1^0, X_2^0, \ldots, X_{\beta^0}^0, \ldots, X_1^m, X_2^m, \ldots, X_{\beta^m}^m.$$

To each X_κ^k we assign a k-cell chain, Ap X_κ^k as its *approximation* and we also assign a singular $(k + 1)$-chain Verb X_κ^k, which is called the *connection chain* (German: Verbindungskette) of X_κ^k with its approximation. To each singular chain

$$U^k = \sum u_\kappa X_\kappa^k$$

there will then correspond a particular approximation

$$\text{Ap } U^k = \sum u_\kappa \text{ Ap } X_\kappa^k$$

and a singular connection chain

$$\text{Verb } U^k = \sum u_\kappa \text{ Verb } X_\kappa^k.$$

We select the chains Ap X_κ^k and Verb X_κ^k so that the following conditions will be satisfied for each chain U^k ($k = 0, 1, \ldots, m$):

(a) Ap U^k and Verb U^k lie on the smallest cellular subcomplex containing U^k.

---

*In this case, $\langle A^k \rangle$ is not a subcomplex in the sense of §12 because there we excluded the empty subcomplex.

(b) $\mathfrak{R}\partial \operatorname{Ap} U^k = \operatorname{Ap} \mathfrak{R}\partial U^k.$

(c) $\mathfrak{R}\partial \operatorname{Verb} U^k = U^k - \operatorname{Ap} U^k - \operatorname{Verb} \mathfrak{R}\partial U^k.*$

If a cell chain has been assigned to each singular simplex in this way, then we say that the singular complex $\mathfrak{C}^m$ in $\mathfrak{R}^n$ has been given a cellular approximation.

It is clear that if one requires conditions (a), (b), and (c) to hold on all singular simplexes $U^k = X^k$, then this is sufficient to provide a cellular approximation. For these conditions will then be satisfied automatically on all arbitrary chains U^k.

We now claim

THEOREM II. *Any singular complex $\mathfrak{C}^m$ can be given a cellular approximation.*

We will also show that: *Given a cellular approximation of a singular subcomplex,** $\mathfrak{c}$, of $\mathfrak{C}^m$ the approximation can be extended to give an approximation of $\mathfrak{C}^m$.*

Proof. We select a simplex X^k of smallest possible dimension from the set of all simplexes of $\mathfrak{C}^m$ which do not belong to $\mathfrak{c}$. This will either be a 0-simplex or else, in the case $k > 0$, all nondegenerate faces of X^k will belong to $\mathfrak{c}$. We will construct the approximation and the connection chains of X^k.

First let $k > 0$. Then the approximations and the connection chains for the nondegenerate faces of X^k, which in fact belong to $\mathfrak{c}$, are already known. In particular,

$$X^k - \operatorname{Verb} \mathfrak{R}\partial X^k \qquad (1)$$

is a well-determined singular chain. Its boundary is

$$\mathfrak{R}\partial X^k - \mathfrak{R}\partial \operatorname{Verb} \mathfrak{R}\partial X^k.$$

If we temporarily set $\mathfrak{R}\partial X^k$ equal to U^{k-1} we can apply formula (c) for the dimension $k - 1$ to the second term of the above expression. The boundary of the singular chain is then

$$\mathfrak{R}\partial X^k - \left(\mathfrak{R}\partial X^k - \operatorname{Ap} \mathfrak{R}\partial X^k - \operatorname{Verb} \mathfrak{R}\partial \mathfrak{R}\partial X^k \right) = \operatorname{Ap} \mathfrak{R}\partial X^k. \qquad (2)$$

But as an approximation, this chain is a cell chain (Fig. 123). If we then temporarily denote the chain (1) by A^k, we can apply Theorem I and obtain a cell chain $\overline{A}^k \sim A^k$ which, like A^k, has (2) as its boundary. We now define $\operatorname{Ap} X^k$ to be the cell chain $\overline{A}^k$. Using $\mathfrak{R}\partial \overline{A}^k = \mathfrak{R}\partial A^k$ we get

$$\mathfrak{R}\partial \operatorname{Ap} X^k = \operatorname{Ap} \mathfrak{R}\partial X^k. \qquad (3)\cdot$$

Furthermore, if $\langle X^k \rangle$ is the smallest cellular complex of $\mathfrak{R}^n$ which contains

---

*In the case $k = 0$, condition (b) is eliminated and one is to set in (c) $\operatorname{Verb} \mathfrak{R}\partial U^k = 0$. The structure of formula (c) is exactly the same as the connection formulas of §29.

** This is a singular complex whose simplexes are, at the same time, simplexes of $\mathfrak{C}^m$.

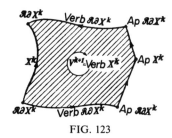

FIG. 123

X^k, then $\mathcal{R}\partial X^k$ will also belong to $\langle X^k \rangle$ and, because of (a), so will Verb $\mathcal{R}\partial X^k$, and hence also the chain (1). From the corollary to Theorem I its approximation $\overline{A}^k = \text{Ap } X^k$ will also belong to $\langle X^k \rangle$. According to Theorem I we also have $A^k \sim \overline{A}^k$ on X^k. That is, there exists a singular connection chain V^{k+1} on $\langle X^k \rangle$ which has the boundary $A^k - \overline{A}^k$. Setting

$$V^{k+1} = \text{Verb } X^k$$

we get

$$\mathcal{R}\partial \text{ Verb } X^k = \mathcal{R}\partial V^{k+1} = A^k - \overline{A}^k = (X^k - \text{Verb } \mathcal{R}\partial X^k) - \text{Ap } X^k.$$

This completes the construction of the chains $\text{Ap } X^k$ and $\text{Verb } X^k$ and proves that the requirements (a), (b), and (c) are satisfied for $U^k = X^k$. Since we have assumed that (a), (b), and (c) hold for all simplexes of c, these conditions hold for each singular chain which can be formed from the simplexes of c together with the additional simplex X^k. In this way one simplex after the other can be added on until one has finally approximated the whole of $\mathfrak{C}^m$.

We must still treat the simple case that our selected simplex is 0-dimensional. The smallest cellular subcomplex $\langle X^0 \rangle$ is then the cell of smallest dimension on which X^0 lies. From the corollary to Theorem I there exists a cellular approximation $\text{Ap } X^0 \sim X^0$ on $\langle X^0 \rangle$ and hence a singular connection chain $\text{Verb } X^0$ having the boundary $X^0 - \text{Ap } X^0$. This completes our proof.

Let us once again assume that A^k is an arbitrary singular k-chain. Its k-simplexes, taken together with all of their nondegenerate faces of dimensions $k - 1$ through 0, form a singular complex $\mathfrak{A}^k$. We shall make a cellular approximation of A^k by approximating $\mathfrak{A}^k$.

If

$$A_1^{k_1}, A_2^{k_2}, \ldots, A_r^{k_r} \tag{4}$$

are singular chains having arbitrary dimensions, then their simplexes and the nondegenerate faces of their simplexes likewise form a singular complex $\mathfrak{C}$. We say that the chains (4) are *approximated simultaneously* when $\mathfrak{C}$ is approximated. Each linear relation which holds between the chains (4) and the boundaries of these chains will also hold for their approximations. That is, if

$$\phi\left(A_1^{k_1}, A_2^{k_2}, \ldots, A_r^{k_r}, \mathcal{R}\partial A_1^{k_1}, \mathcal{R}\partial A_2^{k_2}, \ldots, \mathcal{R}\partial A_r^{k_r}\right) = 0$$

is a linear relation with integer coefficients, then

$$\phi\left(\operatorname{Ap} A_1^{k_1}, \operatorname{Ap} A_2^{k_2}, \ldots, \operatorname{Ap} A_r^{k_r}, \mathfrak{R}\partial \operatorname{Ap} A_1^{k_1}, \mathfrak{R}\partial \operatorname{Ap} A_2^{k_2}, \ldots, \mathfrak{R}\partial \operatorname{Ap} A_r^{k_r}\right) = 0$$

also holds. For $\operatorname{Ap}(A + B) = \operatorname{Ap} A + \operatorname{Ap} B$ and $\operatorname{AP} \mathfrak{R}\partial = \mathfrak{R}\partial \operatorname{Ap}$. If one does not approximate the chains (4) simultaneously, but instead approximates each one individually, then the relations of course will not be preserved. For example, if $A_1^{k_1}$ and $A_2^{k_2}$ have the same boundary, this will not necessarily be true for the chains $\operatorname{Ap} A_1^{k_1}$ and $\operatorname{Ap} A_2^{k_2}$ if they are approximated individually.

If

$$A_1^{k_1}, A_2^{k_2}, \ldots, A_s^{k_s} \tag{5}$$

is a subsystem of (4) and if (5) has already been approximated simultaneously, then one can extend this approximation to a simultaneous approximation of the whole system (4) because the singular complex belonging to the chains (5) is a subcomplex of $\mathfrak{C}$.

Our considerations can be carried over in the same sense to chains mod 2.

73. Intersection Numbers of Singular Chains

We now study an orientable manifold $\mathfrak{M}^n$ which has been given an orientation determined by an orienting chain O^n (§36). The n-simplexes of a simplicial decomposition of $\mathfrak{M}^n$ can then be oriented in one and only one way such that the resulting n-chain is $\sim O^n$. It is our present goal to define the intersection number $\mathfrak{S}(A^k, B^{n-k})$ of two arbitrary singular chains A^k and B^{n-k}. We must of course impose the following restriction:

(R)   The boundary A^{k-1} of A^k is disjoint with B^{n-k} and the boundary B^{n-k-1} of B^{n-k} is disjoint with A^k (considered as point sets).

This restriction was satisfied automatically in our previous investigation of cell chains A^k and B^{n-k} on dual cellular divisions because two cells of dimension k and $n - k$, respectively, on dual divisions are either disjoint or have just a center point in common. We need the restriction (R) for singular chains. This is because the definition makes use of cellular approximation and one can arbitrarily approximate two singular 1-chains A^1 and B^1 which violate the restriction (R), for example as shown in Fig. 124, by either two intersecting chains or two disjoint chains, regardless of how fine the cellular division of the embedding manifold may be.

Along with the previously given singular chains A^k and B^{n-k} which satisfy the condition (R), we now consider two additional singular chains $'A^k$ and $'B^{n-k}$, which also satisfy (R) (Fig. 125). We denote the boundaries of these chains by

$$\mathfrak{R}\partial A^k = A^{k-1}, \qquad \mathfrak{R}\partial \, 'A^k = \, 'A^{k-1}, \tag{1a}$$

$$\mathfrak{R}\partial B^{n-k} = B^{n-k-1}, \qquad \mathfrak{R}\partial \, 'B^{n-k} = \, 'B^{n-k-1}. \tag{1b}$$

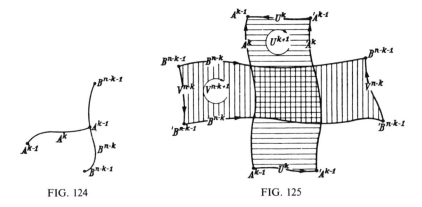

FIG. 124   FIG. 125

Furthermore, we require the existence of singular "connection chains" $U^k, U^{k+1}, V^{n-k}, V^{n-k+1}$ which are to satisfy the following connection formulas:*

$$\mathfrak{R}\,\partial U^k = A^{k-1} - {}'A^{k-1} \atop \mathfrak{R}\,\partial U^{k+1} = A^k - {}'A^k - U^k \Big\},\qquad (2a)$$

$$\mathfrak{R}\,\partial V^{n-k} = B^{n-k-1} - {}'B^{n-k-1} \atop \mathfrak{R}\,\partial V^{n-k-1} = B^{n-k} - {}'B^{n-k} - V^{n-k} \Big\}.\qquad (2b)$$

Finally, we require not only that the chains A^k and B^{n-k-1} be disjoint but also that each of the chains

$$A^k, {}'A^k, U^{k+1} \qquad (3a)$$

be disjoint with each of the chains

$$B^{n-k-1}, {}'B^{n-k-1}, V^{n-k} \qquad (4b)$$

and likewise that the chains

$$B^{n-k}, {}'B^{n-k}, V^{n-k+1} \qquad (3b)$$

be disjoint with each of the chains

$$A^{k-1}, {}'A^{k-1}, U^k. \qquad (4a)$$

If such connection chains exist, then we may say that the two pairs of chains A^k, B^{n-k} and ${}'A^k, {}'B^{n-k}$ are *connected* to one another. For example, if we carry out a sufficiently small deformation on the manifold $\mathfrak{M}^n$ which transforms the chains A^k, B^{n-k} to the chains ${}'A^k, {}'B^{n-k}$, then these pairs are connected. For during the deformation A^k, A^{k-1}, B^{n-k}, and B^{n-k-1} sweep out certain connecting chains $U^{k+1}, U^k, V^{n-k+1}, V^{n-k}$ for which the formulas (2a) and (2b) hold (cf. §31). If the deformation is small enough, the chains (3a) will be disjoint with the chains (4b) and the chains (3b) will be disjoint with the chains (4a).

---

*In Fig. 125, A^{k-1} consists of two points having opposite orientations (signs). The same is true for ${}'A^{k-1}, B^{n-k-1}$, and ${}'B^{n-k-1}$.

Also, when we approximate the chains A^k and B^{n-k} in sufficiently fine dual cellular divisions $\mathfrak{M}_a^n$ and $\mathfrak{M}_b^n$, we shall obtain a pair of chains connected with these chains. In particular, let $\langle A^k \rangle$ and $\langle A^{k-1} \rangle$ be the smallest cellular complexes of $\mathfrak{M}_a^n$ to which A^k and A^{k-1} belong and, in the same way, let $\langle B^{n-k} \rangle$ and $\langle B^{n-k-1} \rangle$ be the smallest cellular subcomplexes of $\mathfrak{M}_b^n$ upon which B^{n-k} and B^{n-k-1} lie. If the cells of $\mathfrak{M}_a^n$ and $\mathfrak{M}_b^n$ are then so small that $\langle A^k \rangle$ is disjoint with $\langle B^{n-k-1} \rangle$ and $\langle B^{n-k} \rangle$ is disjoint with $\langle A^{k-1} \rangle$, considered as point sets, the cellular approximations of A^k, A^{k-1}, B^{n-k}, B^{n-k-1}, and the singular connection chains belonging to them will likewise lie on these smallest subcomplexes and will satisfy the disjointness conditions which we require for connected pairs of chains. We see in the following manner that the cellular divisions $\mathfrak{M}_a^n$ and $\mathfrak{M}_a^n$ can always chosen to be sufficiently fine. Consider $\mathfrak{M}^n$ to lie in a Euclidean space, in which the distance between two points of $\mathfrak{M}^n$ is the usual Euclidean distance. The points of A^k and the points of B^{n-k-1} form two closed sets of the Euclidean space and these sets have a positive distance δ' between them. In like manner, let $\delta'' > 0$ be the distance between A^{k-1} and B^{n-k}. If δ is the smaller of the two distances δ' and δ'', then we need only to make the cells of $\mathfrak{M}_a^n$ and $\mathfrak{M}_a^n$ smaller in diameter than $\delta/2$ to insure that $\langle A^k \rangle$ is disjoint with $\langle B^{n-k-1} \rangle$ and $\langle A^{k-1} \rangle$ is disjoint with $\langle B^{n-k} \rangle$. The existence of cellular divisions having cells this small follows because $\mathfrak{M}^n$ can be given an arbitrarily fine simplicial decomposition and each (normally subdivided) simplicial decomposition is a cellular division.

We can now define the intersection number of two singular chains A^k and B^{n-k} which satisfy the restriction (R) as follows:

Replace A^k and B^{n-k} by two cell chains $'A^k$ and $'B^{n-k}$ lying on dual cellular divisions $'\mathfrak{M}_a$ and $'\mathfrak{M}_b$ so that $'A^k, 'B^{n-k}$ form a pair of chains connected with A^k, B^{n-k}; we define the intersection number of A^k and B^{n-k} to be the intersection number

$$\mathfrak{S}('A^k, 'B^{n-k})$$

which was defined in §70.

74. Invariance of Intersection Numbers

We must now demonstrate that our definition of the intersection number is independent of the particular choice of cellular divisions $'\mathfrak{M}_a$ and $'\mathfrak{M}_b$ and is independent of the pairs of cell chains connected with A^k, B^{n-k} on these cellular divisions. Until we have proven this, the symbol $\mathfrak{S}(A^k, B^{n-k})$ will only make sense for two cell chains on dual division.*

---

*For simplicity we shall temporarily omit the dimensionality index n of $\mathfrak{M}$. In Theorems I through III we assume that $\mathfrak{M}$ is embedded in a Euclidean space. When we say that a cell is $< \delta$, we shall mean that its diameter in this Euclidean space is $< \delta$.

THEOREM I. *If A^k, B^{n-k} and $'A^k, 'B^{n-k}$ are connected pairs of singular chains and $\bar{A}^k, '\bar{A}^k$ are simultaneous approximations of A^k and $'A^k$ in a cellular division $\mathfrak{M}_a$, while $\bar{B}^{n-k}, '\bar{B}^{n-k}$ are simultaneous approximations of B^{n-k} and $'B^{n-k}$ on the dual cellular division $\mathfrak{M}_b$, then*

$$\mathbb{S}(\bar{A}^k, \bar{B}^{n-k}) = \mathbb{S}('\bar{A}^k, '\bar{B}^{n-k})$$

whenever the cells of $\mathfrak{M}_a$ and $\mathfrak{M}_b$ are smaller than an appropriately chosen δ.

Proof. We choose δ to be so small that not only the chains (3a), (4b) and (3b), (4a) of the preceding section are disjoint, but also the smallest cellular subcomplexes on which these chains lie are disjoint. Here, subcomplexes on $\mathfrak{M}_a$ are to be taken for the chains (3a) and (4a) and subcomplexes on $\mathfrak{M}_b$ are to be taken for the chains (3b) and (4b). We can extend the given simultaneous approximation of the chains $A^k, 'A^k$ in the cellular division $\mathfrak{M}_a$ to a simultaneous approximation of all of the chains (3a) and (4a). Likewise, we can extend the simultaneous approximations of the chains B^{n-k} and $'B^{n-k}$ in the cellular division $\mathfrak{M}_b$ to a simultaneous approximation of all of the chains (3b) and (4b). If we denote the approximations by means of bars written above the corresponding symbols, then formulas $(\bar{1}a), (\bar{1}b), (\bar{2}a), (\bar{2}b)$ which are obtained from (1a), (1b), (2a), (2b) by putting bars above the chains are valid according to §72. The chains $(\bar{3}a)$ corresponding to the chains (3a) are disjoint with the chains $(\bar{4}b)$ and the chains $(\bar{3}b)$ are disjoint with $(\bar{4}a)$.

From now on we will need to deal only with chains having bars written above them. Of these, $(\bar{3}a)$ and $(\bar{4}a)$ are cell chains on $\mathfrak{M}_a$ and $(\bar{3}b), (\bar{4}b)$ are cell chains on the dual division $\mathfrak{M}_b$. The chains with bars written above them are located arbitrarily closely to the unbarred chains and we can illustrate them, as before, by Fig. 125.

The considerations which follow are purely combinatorial in nature since the chains in question lie on dual cellular divisions. The proof that the two intersection numbers $\mathbb{S}(\bar{A}^k, \bar{B}^{n-k})$ and $\mathbb{S}('\bar{A}^k, '\bar{B}^{n-k})$ are the same will be carried out in two steps. By using formula (5) of §70 we get

$$\mathbb{S}(\bar{A}^k, \mathfrak{R}\partial\bar{V}^{n-k+1}) = (-1)^k \mathbb{S}(\mathfrak{R}\partial\bar{A}^k, \bar{V}^{n-k+1}) = 0,$$

since $\bar{A}^{k-1} (= \mathfrak{R}\partial\bar{A}^k)$ and $\bar{V}^{n-k+1}$ are disjoint. On the other hand, upon substituting the value of $\mathfrak{R}\partial\bar{V}^{n-k+1}$ given in formula $(\bar{2}b)$ we get

$$\mathbb{S}(\bar{A}^k, \mathfrak{R}\partial\bar{V}^{n-k+1}) = \mathbb{S}(\bar{A}^k, \bar{B}^{n-k} - '\bar{B}^{n-k} - \bar{V}^{n-k}) = \mathbb{S}(\bar{A}^k, \bar{B}^{n-k} - '\bar{B}^{n-k})$$

since $\bar{A}^k$ and $\bar{V}^{n-k}$ are disjoint. Thus

$$\mathbb{S}(\bar{A}^k, \bar{B}^{n-k}) = \mathbb{S}(\bar{A}^k, '\bar{B}^{n-k}). \tag{1}$$

In like manner,

$$\mathbb{S}(\mathfrak{R}\partial\bar{U}^{k+1}, '\bar{B}^{n-k}) = (-1)^{k+1}\mathbb{S}(\bar{U}^{k+1}, \mathfrak{R}\partial'\bar{B}^{n-k}) = 0$$

since $\bar{U}^{k+1}$ and $'\bar{B}^{n-k-1}$ are disjoint. On the other hand, upon substituting the value of $\mathcal{R} \partial \overline{U}^{k+1}$ given in formula $(\overline{2}a)$ one gets

$$\mathcal{S}(\mathcal{R} \partial \overline{U}^{k+1}, {}'\overline{B}^{n-k}) = \mathcal{S}(\overline{A}^k - {}'\overline{A}^k - \overline{U}^k, {}'\overline{B}^{n-k}) = \mathcal{S}(\overline{A}^k - {}'\overline{A}^k, {}'\overline{B}^{n-k})$$

since $\overline{U}^k$ and $'\overline{B}^{n-k}$ are disjoint. Thus

$$\mathcal{S}(\overline{A}^k, {}'\overline{B}^{n-k}) = \mathcal{S}({}'\overline{A}^k, {}'\overline{B}^{n-k}). \tag{2}$$

Our desired result,

$$\mathcal{S}(\overline{A}^k, \overline{B}^{n-k}) = \mathcal{S}({}'\overline{A}^k, {}'\overline{B}^{n-k}),$$

follows from (1) and (2).

The transition from one cellular division to another is accomplished by means of the following theorem:

THEOREM II^k. *Let a^k and b^{n-k} be oriented dual cells on dual cellular divisions $\mathfrak{M}_a$ and $\mathfrak{M}_b$ of $\mathfrak{M}$ and let these cells have intersection number*

$$\eta = \mathcal{S}(a^k, b^{n-k}) = \pm 1.$$

When one approximates a^k and b^{n-k} in two new dual cellular divisions $\overline{\mathfrak{M}}_a$ and $\overline{\mathfrak{M}}_b$ whose cells are smaller than an appropriately chosen δ, the approximations $\bar{a}^k$ and $\bar{b}^{n-k}$ have the same intersection number*

$$\eta = \mathcal{S}(\bar{a}^k, \bar{b}^{n-k}).$$

Here a^k is a coherently oriented simplicial star of the normal subdivision $\dot{\mathfrak{M}}$ of $\mathfrak{M}_a$ and thus represents a singular k-chain with respect to $\mathfrak{M}_a$ and can be given a cellular approximation in the sense of §72. Thus $\bar{a}^k$ is a cell chain on $\overline{\mathfrak{M}}_a$. Likewise, $\bar{b}^{n-k}$ is a cell chain on the dual division $\overline{\mathfrak{M}}_b$.

The proof of Theorem II^k is achieved by induction over all values of k. We first prove Theorem II^0. Here we are dealing with dual cells a^0 and b^n. We coherently orient both of the cellular divisions $\mathfrak{M}_b$ and $\overline{\mathfrak{M}}_b$ so that the resulting chains M_b^n and $\overline{M}_b^n$ become homologous to an orienting chain O^n:

$$M_b^n \sim \overline{M}_b^n \sim O^n. \tag{3}$$

Now

$$M_b^n = \varepsilon_b b^n + {}'b^n. \tag{4}$$

Here $\varepsilon_b = \pm 1$ is the sign with which b^n appears in M_b^n and $'b^n$ is the sum of those n-cells of M_b^n which differ from b^n. We can extend the given approximation of b^n in $\overline{\mathfrak{M}}_b$ to a simultaneous approximation of b^n and $'b^n$. In the approximation let $'b^n$ transform to $'\bar{b}^n$. The approximation of $M_b^n = \varepsilon_b b^n + {}'b^n$ is then the chain $\varepsilon_b \bar{b}^n + {}'\bar{b}^n$ and since, on being

---

*Note that the symbol $\mathcal{S}$ appears here in two different contexts: once as the intersection number with respect to the dual divisions $\mathfrak{M}_a$ and $\mathfrak{M}_b$ and the other time as the intersection number with respect to $\overline{\mathfrak{M}}_a$ and $\overline{\mathfrak{M}}_b$.

approximated, a closed chain transforms to a homologous chain and furthermore $\overline{M}_b^n$ is the only cell chain on $\overline{\mathfrak{M}}_b$ homologous to M_b^n,

$$\overline{M}_b^n = \varepsilon_b \overline{b}^n + {'\overline{b}}^n. \tag{$\overline{4}$}$$

We now select the cells of $\overline{\mathfrak{M}}_a$ and $\overline{\mathfrak{M}}_b$ to be so small that not only a^0 and ${'b}^n$ but also even the smallest subcomplexes of $\overline{\mathfrak{M}}_a$ and $\overline{\mathfrak{M}}_b$ to which a^0 and ${'b}^n$, respectively, belong are disjoint. The same will then be true for the chains $\overline{a}^0$ and ${'\overline{b}}^n$ since these chains belong, as approximations, to these smallest subcomplexes. If $\varepsilon_a = \pm 1$ denotes the sum of coefficients of a^0, then it is also the sum of coefficients of $\overline{a}^0$ for, as an approximation, $\overline{a}^0$ is homologous to a^0. We then have

$$\mathfrak{S}(a^0, b^n) = \varepsilon_b \mathfrak{S}(a^0, \varepsilon_b b^n) = \varepsilon_b \mathfrak{S}(a^0, \varepsilon_b b^n + {'b}^n)$$

$$= \varepsilon_b \mathfrak{S}(a^0, M_b^n) = \varepsilon_b \varepsilon_a \qquad (\S 70). \tag{5}$$

Similarly,

$$\mathfrak{S}(\overline{a}^0, \overline{b}^n) = \varepsilon_b \mathfrak{S}(\overline{a}^0, \varepsilon_b \overline{b}^n),$$

and furthermore, since $\overline{a}^0$ and ${'\overline{b}}^n$ are disjoint,

$$\mathfrak{S}(\overline{a}^0, \overline{b}^n) = \varepsilon_b \mathfrak{S}(\overline{a}^0, \varepsilon_b \overline{b}^n + {'\overline{b}}^n) = \varepsilon_b \mathfrak{S}(\overline{a}^0, \overline{M}_b^n) = \varepsilon_b \varepsilon_a. \tag{$\overline{5}$}$$

From (5) and ($\overline{5}$) we get

$$\mathfrak{S}(a^0, b^n) = \mathfrak{S}(\overline{a}^0, \overline{b}^n),$$

which proves Theorem II^0.

We now assume that Theorem II^{k-1} has already been proved. Let a^{k-1} be a cell on $\mathfrak{M}_a$ which is incident with a^k. The cell b^{n-k+1} dual to a^{k-1} is then incident with b^{n-k}. Choose the orientations of a^{k-1} and b^{n-k+1} so that a^{k-1} appears with multiplicity $+1$ in the boundary of a^k and b^{n-k} appears with multiplicity $+1$ in the boundary of b^{n-k+1} (Fig. 126):

$$\mathfrak{R}\partial a^k = a^{k-1} + {'a}^{k-1},$$
$$\mathfrak{R}\partial b^{n-k+1} = b^{n-k} + {'b}^{n-k}.$$

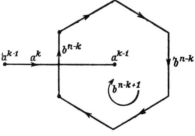

FIG. 126

The approximation of b^{n-k} as $\bar{b}^{n-k}$ can be extended to a simultaneous approximation of b^{n-k}, $'b^{n-k}$, and b^{n-k+1}. Let the approximations of these chains be called $\bar{b}^{n-k}$, $'\bar{b}^{n-k}$, and $\bar{b}^{n-k+1}$. We now choose the cellular divisions $\overline{\mathfrak{M}}_a$ and $\overline{\mathfrak{M}}_b$ to be sufficiently fine so that the chains $'a^{k-1}$ and b^{n-k+1}, on the one hand, and $'b^{n-k}$ and a^k, on the other hand, are disjoint and are also disjoint even after the approximation. Thus we have

(α) $'\bar{a}^{k-1}$ is disjoint with $\bar{b}^{n-k+1}$, and
(β) $'\bar{b}^{n-k}$ is disjoint with $\bar{a}^k$.

In addition, from Theorem II^{k-1} we can also require

(γ) $\mathfrak{S}(a^{k-1}, b^{n-k+1}) = \mathfrak{S}(\bar{a}^{k-1}, \bar{b}^{n-k+1})$.

Now by (5) of §69,

$$\mathfrak{S}(a^k, b^{n-k}) = (-1)^k \mathfrak{S}(a^{k-1}, b^{n-k+1}). \tag{6}$$

Correspondingly,

$$\mathfrak{S}(\bar{a}^k, \bar{b}^{n-k}) = \mathfrak{S}(\bar{a}^k, \bar{b}^{n-k} + '\bar{b}^{n-k}) \qquad (\beta)$$

$$= \mathfrak{S}(\bar{a}^k, \mathfrak{R}\partial\bar{b}^{n-k+1})$$

$$= (-1)^k \mathfrak{S}(\mathfrak{R}\partial\bar{a}^k, \bar{b}^{n-k+1}) \qquad [(5), §70]$$

$$= (-1)^k \mathfrak{S}(\bar{a}^{k-1} + '\bar{a}^{k-1}, \bar{b}^{n-k+1})$$

$$= (-1)^k \mathfrak{S}(\bar{a}^{k-1}, \bar{b}^{n-k+1}) \qquad (\alpha). \tag{$\bar{6}$}$$

By (γ), the right-hand side of (6) is equal to the right-hand side of ($\bar{6}$) and thus the left-hand side of (6) is equal to the left-hand side of ($\bar{6}$), which was to be proven.

Theorem II^k can easily be extended to cell chains:

THEOREM III. *If*

$$A^k = \sum \xi_\mu a_\mu^k \qquad and \qquad B^{n-k} = \sum \eta_\nu b_\nu^{n-k}$$

are cell chains on dual cellular divisions $\mathfrak{M}_a$ and $\mathfrak{M}_b$, respectively, of the orientable manifold $\mathfrak{M}$, and $\bar{A}^k$ and $\bar{B}^{n-k}$ are their approximations in two other dual cellular divisions $\overline{\mathfrak{M}}_a$ and $\overline{\mathfrak{M}}_b$, then

$$\mathfrak{S}(A^k, B^{n-k}) = \mathfrak{S}(\bar{A}^k, \bar{B}^{n-k})$$

whenever the cells of $\overline{\mathfrak{M}}_a$ and $\overline{\mathfrak{M}}_b$ are smaller than an appropriately chosen δ.

Proof. Choose $\overline{\mathfrak{M}}_a$ and $\overline{\mathfrak{M}}_b$ sufficiently fine so that when one approximates the cells a_μ^k of A^k by $\bar{a}_\mu^k$ and b_ν^{n-k} of B^{n-k} by $\bar{b}_\nu^{n-k}$ one will have

$$\mathfrak{S}(\bar{a}_\mu^k, \bar{b}_\nu^{n-k}) = \mathfrak{S}(a_\mu^k, b_\nu^{n-k}).$$

This is possible when $\mu = \nu$ by virtue of Theorem II^k. When $\mu \neq \nu$ the cells a_μ^k and b_ν^{n-k} are disjoint and, for a sufficiently fine division, their approximations are disjoint. One then has

$$\mathcal{S}(A^k, B^{n-k}) = \mathcal{S}\left(\sum \xi_\mu a_\mu^k, \sum \eta_\nu b_\nu^{n-k}\right) = \sum \xi_\mu \eta_\nu \mathcal{S}\left(a_\mu^k, b_\nu^{n-k}\right)$$

and

$$\mathcal{S}(\overline{A}^k, \overline{B}^{n-k}) = \mathcal{S}\left(\sum \xi_\mu \overline{a}_\mu^k, \sum \eta_\nu \overline{b}_\nu^{n-k}\right) = \sum \xi_\mu \eta_\nu \mathcal{S}\left(\overline{a}_\mu^k, \overline{b}_\nu^{n-k}\right).$$

Since the right-hand sides are the same, we get

$$\mathcal{S}(A^k, B^{n-k}) = \mathcal{S}(\overline{A}^k, \overline{B}^{n-k}).$$

The invariance of the intersection numbers now comes about as follows: Corresponding to the given singular chains A^k and B^{n-k} choose the pair of chains $'A^k, 'B^{n-k}$ on the dual cellular divisions $'\mathfrak{M}_a$ and $'\mathfrak{M}_b$, where the pair is connected with A^k, B^{n-k}. Likewise, on the dual cellular divisions $''\mathfrak{M}_a$ and $''\mathfrak{M}_b$ choose the pair of chains $''A^k, ''B^{n-k}$ connected with A^k, B^{n-k}. The chains $A^k, 'A^k, ''A^k$ will then transform to chains $\overline{A}^k, '\overline{A}^k, ''\overline{A}^k$ by means of a simultaneous approximation in a third cellular division $\overline{\mathfrak{M}}_a$. Likewise, the chains B^{n-k}, $'B^{n-k}$, and $''B^{n-k}$ will transform by a simultaneous approximation in the dual cellular division $\overline{\mathfrak{M}}_b$ to the chains $\overline{B}^{n-k}$, $'\overline{B}^{n-k}$, and $''\overline{B}^{n-k}$. From Theorem I, when $\overline{\mathfrak{M}}_a$ and $\overline{\mathfrak{M}}_b$ are sufficiently fine

$$\mathcal{S}(\overline{A}^k, \overline{B}^{n-k}) = \mathcal{S}('\overline{A}^k, '\overline{B}^{n-k}),$$

and

$$\mathcal{S}(\overline{A}^k, \overline{B}^{n-k}) = \mathcal{S}(''\overline{A}^k, ''\overline{B}^{n-k});$$

thus

$$\mathcal{S}('\overline{A}^k, '\overline{B}^{n-k}) = \mathcal{S}(''\overline{A}^k, ''\overline{B}^{n-k}). \tag{7}$$

Furthermore, from Theorem III

$$\mathcal{S}('\overline{A}^k, '\overline{B}^{n-k}) = \mathcal{S}('A^k, 'B^{n-k})$$

since $'A^k$ and $'B^{n-k}$ are cell chains on dual cellular divisions, and likewise

$$\mathcal{S}(''\overline{A}^k, ''\overline{B}^{n-k}) = \mathcal{S}(''A^k, ''B^{n-k}).$$

Setting this into (7) we get

$$\mathcal{S}('A^k, 'B^{n-k}) = \mathcal{S}(''A^k, ''B^{n-k}).$$

$$
\begin{array}{ccc}
\boxed{''A^k, 'B^{n-k}} & \boxed{A^k, B^{n-k}} & \boxed{'A^k, 'B^{n-k}} \\
\updownarrow & \updownarrow & \updownarrow \\
\boxed{''\overline{A}^k, ''\overline{B}^{n-k}} \longleftrightarrow & \boxed{\overline{A}^k, \overline{B}^{n-k}} \longleftrightarrow & \boxed{'\overline{A}^k, '\overline{B}^{n-k}}
\end{array}
$$

Let us review the steps leading to the proof, with the aid of the accompanying schematic. The singular chains A^k, B^{n-k} were replaced in two different dual cellular divisions by the pairs of chains connected with them: $'A^k, 'B^{n-k}$ and $''A^k, ''B^{n-k}$, respectively. In these cellular divisions the intersection numbers were determined combinatorially by the method of §70. The equality of the intersection numbers found in this way was demonstrated by making a cellular approximation of all six chains in a third cellular division (indicated by bars in the lower row). The equality was shown, for the directions of the vertical arrows, with the aid of Theorem III; the equality was shown, for the directions of the horizontal arrows, with the aid of Theorem I of this section.[44]

It is now easy to extend the theorems proven for cell chains in §70 to the case of singular chains. In this regard, we shall make the implicit assumption that whenever two singular chains A^k and B^{n-k} appear in a symbol $\mathcal{S}(A^k, B^{n-k})$, they satisfy the condition (R) of §73.

If two singular chains A^k and B^{n-k} are disjoint then their intersection number is equal to zero because we can approximate them sufficiently finely so that their approximations are also disjoint.

The intersection numbers satisfy the distributive law:

$$\mathcal{S}\left(A^k, B_1^{n-k} + B_2^{n-k}\right) = \mathcal{S}\left(A^k, B_1^{n-k}\right) + \mathcal{S}\left(A^k, B_2^{n-k}\right). \tag{8}$$

For when one approximates A^k in a cellular division $\mathfrak{M}_a$ and approximates B_1^{n-k} and B_2^{n-k} simultaneously in the dual cellular division $\mathfrak{M}_b$, the equation

$$\mathcal{S}\left(\overline{A}^k, \overline{B}_1^{n-k} + \overline{B}_2^{n-k}\right) = \mathcal{S}\left(\overline{A}^k, \overline{B}_1^{n-k}\right) + \mathcal{S}\left(\overline{A}^k, \overline{B}_2^{n-k}\right) \tag{$\overline{8}$}$$

holds for the approximation. When one appropriates with sufficiently fine cellular divisions $\mathfrak{M}_a$ and $\mathfrak{M}_b$, the intersection numbers of the chains will be the same as those of their approximations. Formula (8) then follows from $(\overline{8})$.

Just as in the case of (8) we can carry over the formulas

$$\mathcal{S}(A^k, B^{n-k}) = (-1)^{k(n-k)}\mathcal{S}(B^{n-k}, A^k) \tag{9}$$

and

$$\mathcal{S}(A^k, \mathcal{R}\partial B^{n-k}) = (-1)^k\mathcal{S}(\mathcal{R}\partial A^k, B^{n-k+1}) \tag{10}$$

to the case where A^k and B^{n-k+1} are any two singular chains whose boundaries do not intersect.

As earlier, it follows here that the intersection number of a closed k-chain and a division–null homologous $(n-k)$-chain is zero, and the intersection number of two closed chains does not change when one replaces either chain by a division-homologous chain. As a consequence of this fact, we can speak about the intersection number of two homology classes (with division) of respective dimension k and $n-k$ when they are given in a definite order. As a result, the assumption which we made implicitly in Theorems I and II of

§71 that the chains of dual Betti bases or dual torsion bases must lie on dual cellular divisions is no longer necessary. We can choose arbitrary singular chains.

If two orientable pseudomanifolds $\mathfrak{R}^k$ and $\mathfrak{R}^{n-k}$ which have been oriented in a particular manner are mapped by continuous transformations f and g, respectively, into an orientable manifold $\mathfrak{M}^n$, then the orienting chains B^k of $\mathfrak{R}^k$ and B^{n-k} of $\mathfrak{R}^{n-k}$ will transform to two singular closed chains $f(B^k)$ and $g(B^{n-k})$ (§27) which have a well-defined intersection number. If we choose other orienting chains $'B^k \sim B^k$ on $\mathfrak{R}^k$ and $'B^{n-k} \sim B^{n-k}$ on $\mathfrak{R}^{n-k}$ in place of B^k and B^{n-k}, then we clearly have

$$\left. \begin{array}{c} f(B^k) \sim f('B^k) \\ g(B^{n-k}) \sim g('B^{n-k}) \end{array} \right\} \quad \text{on} \quad \mathfrak{M}^n$$

and therefore

$$\mathcal{S}\big(f(B^k), g(B^{n-k})\big) = \mathcal{S}\big(f('B^k), g('B^{n-k})\big).$$

Singular images of the oriented pseudomanifolds $\mathfrak{R}^k$ and $\mathfrak{R}^{n-k}$ therefore have a well-defined intersection number, which depends only upon the orientation of $\mathfrak{R}^k$ and $\mathfrak{R}^{n-k}$ and upon the continuous mappings. The intersection number mod 2 exists even when $\mathfrak{R}^k$, $\mathfrak{R}^{n-k}$, or $\mathfrak{M}^n$ are nonorientable.

While dual cellular divisions are convenient for the purpose of presenting the general theory, it is now always practical to determine the intersection number in individual cases by a cellular approximation in dual cellular divisions. Before we turn to particular examples, we shall prove a lemma which will be of use in determining intersection numbers.

LEMMA. *Let $\mathfrak{M}^n$ be an orientable manifold having a coherently oriented simplicial decomposition and let E^n be an oriented n-simplex of the decomposition. Let X^k and X^{n-k} be two oriented geometric simplexes embedded in E^n which have no points in common except for a common midpoint P_k. If $\xi(P_0 P_1 \cdots P_k)$ is an oriented simplex of the normal subdivision of X^k and $\eta(P_k P_{k+1} \cdots P_n)$ is an oriented simplex of the normal subdivision of X^{n-k} and, furthermore, one has chosen the sign ζ so that the n-simplex $\zeta(P_0 P_1 \cdots P_k \cdots P_n)$ has the same orientation as E^n, then*

$$\mathcal{S}(X^k, X^{n-k}) = \xi \eta \zeta.$$

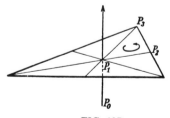

FIG. 127

Figure 127 shows the case $n - k = 2$, $k = 1$. The simplex E^3 is to be thought of as a large simplex containing X^1 and X^2, and has not been drawn in the figure.

Proof. In the proof we shall make use of the fact that the intersection numbers of two connected pairs of singular chains are the same. When we hold X^k fixed and deform X^{n-k} to a new simplex $'X^{n-k}$ in a way that the boundary of X^{n-k} never intersects X^k and the boundary of X^k never intersects X^{n-k}, then

$$\mathbb{S}(X^k, X^{n-k}) = \mathbb{S}(X^k, {'X^{n-k}})$$

since X^k, X^{n-k} and $X^k, {'X^{n-k}}$ form two connected pairs (cf. §31, Theorem II).

We first examine the case $k = 0$. Without loss of generality we can assume that X^0 is oriented with the sign $\xi = +1$ and X^n has the same orientation as E^n. That is, η and ζ are the same. We must then show that $\mathbb{S}(X^0, X^n) = 1$. To do this we transform X^n to the simplex E^n and transform X^0 to the midpoint of E^n in three steps. The first step consists of deforming X^n to a configuration X_1^n similar to E^n in such a way that X^0 becomes a center of similitude. Clearly, one can always manage to carry out this deformation so that the point X^0, which is held fixed, will never intersect the boundary of X^n. Thus

$$\mathbb{S}(X^0, X^n) = \mathbb{S}(X^0, X_1^n).$$

After this we project X_1^n from the center of similitude X^0 to the simplex E^n. Here X^0 will never intersect the boundary of X_1^n, so that

$$\mathbb{S}(X^0, X_1^n) = \mathbb{S}(X^0, E^n).$$

Finally, we transform X^0 to the midpoint X_1^0 of E^n and we orient it with the same sign, $\xi = 1$, as X^0. We then have

$$\mathbb{S}(X^0, E^n) = \mathbb{S}(X_1^0, E^n).$$

Putting these results together,

$$\mathbb{S}(X^0, X^n) = \mathbb{S}(X_1^0, E^n).$$

The simplicial decomposition of $\mathfrak{M}^n$ can be considered to be a cellular division in which E^n is an n-cell and X_1^0 is the 0-cell dual to E^n. We then have

$$\mathbb{S}(X_1^0, E^n) = 1.$$

Let us assume that the lemma has been proved for two simplexes of dimension $k - 1$ and $n - k + 1$, respectively. We shall now prove it for two simplexes X^k and X^{n-k}. The orientation of X^k is given by the subsimplex

$$\xi(P_0 P_1 \cdots P_k) \qquad\qquad (X^k)$$

and the orientation of X^{n-k} is given by the subsimplex

$$\eta(P_k P_{k+1} \cdots P_n).. \qquad (X^{n-k})$$

We can assume that P_{k-1} is the midpoint of a $(k-1)$-dimensional face X^{k-1} of X^k (and not, for example, the midpoint of a face of lower dimension). There then exists a simplex X^{n-k+1} which has X^{n-k} as a face and P_{k-1} as midpoint. The simplexes X^k and X^{n-k} have the interval $P_{k-1}P_k$ as their intersection.* Let X^{k-1} be oriented so that it appears with the sign $+1$ in the boundary of X^k. Its orientation is then given by the subsimplex

$$(-1)^k \xi(P_0 P_1 \cdots P_{k-1}). \qquad (X^{k-1})$$

In the same way, let X^{n-k+1} be oriented so that X^{n-k} appears in the boundary of X^{n-k+1} with the sign $+1$. Its orientation is then determined by the subsimplex

$$\eta(P_{k-1}P_k \cdots P_n). \qquad (X^{n-k+1})$$

By our induction assumption,

$$\mathbb{S}(X^{k-1}, X^{n-k+1}) = (-1)^k \xi \eta \zeta.$$

On the other hand, from $(10')$ we have

$$\mathbb{S}(X^k, X^{n-k}) = \mathbb{S}(X^k, \mathfrak{R} \partial X^{n-k+1}) = (-1)^k \mathbb{S}(\mathfrak{R} \partial X^k, X^{n-k+1})$$
$$= (-1)^k \mathbb{S}(X^{k-1}, X^{n-k+1}).$$

Therefore, as was to be proved,

$$\mathbb{S}(X^k, X^{n-k}) = \xi \eta \zeta.$$

To determine the intersection number of two singular chains A^k and B^{n-k} we can, first of all, omit all simplexes which are disjoint with the set theoretic intersection $\mathfrak{D}$ of the sets of points belonging respectively to A^k and B^{n-k}. For we can decompose the intersection number in accordance with the formula (8), and disjoint chains have intersection number 0. We wish to give special mention to the case where the intersection consists of finitely many points and has the following property: at each of these points of intersection there will always be exactly two simplexes which pass through the point in the manner described in the lemma. In this case we shall say that the two singular chains *pass smoothly through one another* at the common points. The intersection number $\mathbb{S}(A^k, B^{n-k})$ is then the sum of the intersection numbers

---

* If X^{n-k+1} protrudes out from E^n we can, in advance, proportionally diminish the simplexes X^k and X^{n-k} by a similarity transformation, with P_k the center of similitude. This can be accomplished so that neither the intersection number $\mathbb{S}(X^k, X^{n-k})$ nor the coefficients ξ, η, ζ change.

for the individual points of intersection, which in turn are determined by means of the lemma.

The methods and theorems of this section can be extended immediately to chains mod 2.

75. Examples

EXAMPLE 1. Let us look at the orientable surfaces. As a particular example we select the double torus, a surface of genus 2, because of the simplicity of the expressions involved. We assume that the double torus has been cut open to give its fundamental polygon. We know already that the closed chains a, b, c, d form a Betti basis (§41) and we wish to find a basis dual to this basis. The intersections can be followed more easily in Fig. 128 when we replace the chains a, b, c, d by homologous chains a', b', c', d' which run in the interior of the octagon. It is clear that upon making an appropriate simplicial decomposition the chains will pass smoothly through one another. We determine the intersection numbers by using such a decomposition. When the orientation of the surface is given by the boundary circle $aba^{-1}b^{-1}cdc^{-1}d^{-1}$ we get

$$\mathbb{S}(a', b') = 1 \qquad \text{and} \qquad \mathbb{S}(b', a') = -1$$

in agreement with formula (9) of §74. We have $\mathbb{S}(a', c') = 0$ since a' and c' are disjoint. Similarly, $\mathbb{S}(a', a') = 0$. For since $a' \sim a$, $\mathbb{S}(a', a') = \mathbb{S}(a', a)$ and the latter intersection number is zero because a' and a are disjoint. The matrix of intersection numbers* is then

$$
\begin{array}{c|cccc}
\mathbb{S} & a & b & c & d \\
\hline
a & & 1 & & \\
b & -1 & & & \\
c & & & & 1 \\
d & & & -1 &
\end{array}
\tag{1}
$$

where missing entries represent elements which are equal to zero.

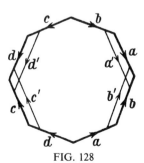

FIG. 128

---

*The element of the ith row and kth column is the intersection number of the ith chain with the kth chain, in that order.

By reversing orientations and rearranging the column entries we obtain the unit matrix:

$$
\begin{array}{c|ccccc}
\mathcal{S} & b & -a & d & -c \\
\hline
a & 1 & & & \\
b & & 1 & & \\
c & & & 1 & \\
d & & & & 1
\end{array}
\tag{2}
$$

The chains $b, -a, d, -c$ form the basis dual to a, b, c, d.

The basis a, b, c, d possesses the property that its chains can be arranged in pairs so that chains of the same pair have intersection number ± 1 while chains belonging to different pairs have intersection number 0. This is not an exclusive property of surfaces. It occurs for all dimensions of the form $2(2m + 1)$, $m = 0, 1, 2 \dots$. To show this we examine the chains of the "middle" dimension k in a manifold $\mathfrak{M}^{2k}$. From formula (9) of §74,

$$
\mathcal{S}(A^k, B^k) = (-1)^{kk}\mathcal{S}(B^k, A^k),
$$

so that $\mathcal{S}(A^k, B^k) = \pm\mathcal{S}(B^k, A^k)$ with the $+$ sign used when k is even and the $-$ sign when k is odd. The matrix of intersection numbers

$$
\mathcal{S}(B^k_\mu, B^k_\nu)
$$

of a Betti basis

$$
B^k_1, B^k_2, \dots, B^k_{\rho^k}
$$

is therefore skew symmetric when k is odd and is symmetric when k is even. In either case, its determinant is ± 1. For one can transform to dual bases (§71) by appropriate independent integer-valued unimodular transformations of the rows and columns. For these bases the matrix of intersection numbers is the unit matrix. For the case of odd k, which is the case of interest to us, there is a theorem of algebra which states that by means of cogredient, integer-valued unimodular transformations of the rows and columns one can bring a unimodular, skew symmetric matrix having integer-valued elements to the "box form" (1) in which principal minors of the form

$$
\begin{array}{cc}
0 & 1 \\
-1 & 0
\end{array}
$$

lie in sequence along the main diagonal, and all other elements are zero (see Hensel and Landsberg [1, p. 636 *et seq.*]). The new Betti basis obtained in this manner can be considered to be the $2(2m + 1)$-dimensional analog of the conjugate essential cuts on the closed orientable surfaces.[45]

It follows, further, that the middle Betti number (of dimension $k = 2m + 1$) of a $2(2m + 1)$-dimensional orientable manifold is even. In addition, the Euler characteristic N of such a manifold is even since, by the Poincaré duality theorem,

$$
N = -\sum_{\nu=0}^{2k} (-1)^\nu p^\nu = -2p^0 + 2p^1 - \cdots \pm 2p^{k-1} \mp p^k.
$$

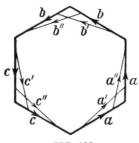

FIG. 129

EXAMPLE 2. In considering intersection numbers on nonorientable surfaces we must restrict ourselves to intersection numbers mod 2. Let us look at the nonorientable surface of genus 3. The nonoriented edges a, b, c of the fundamental polygon form a connectivity basis of the dimension 1 (§41). In Fig. 129 the edge a has been altered in two different ways to give a homologous chain mod 2, a' and a'', respectively. Since a' and a'' pass smoothly through one another, their intersection number mod 2 is

$$\mathfrak{S}(a, a) = \mathfrak{S}(a', a'') = \check{1},$$

while the intersection numbers of a' with the edges b and c are each $\check{0}$. The matrix of intersection numbers of mod 2 is

$\mathfrak{S}$	a	b	c
a	$\check{1}$		
b		$\check{1}$	
c			$\check{1}$

The connectivity basis is dual to itself.

EXAMPLE 3. Continuation of orientation along a path. Given a cellular division $\mathfrak{M}_a^n$ of an n-dimensional manifold $\mathfrak{M}^n$ we call the vertices of $\mathfrak{M}_a^n$ 0-cells and the edges of $\mathfrak{M}_a^n$ 1-cells. Let w be a closed edge path on $\mathfrak{M}_a^n$. Each vertex of w is in one-to-one correspondence with an n-cell and each edge of w is in one-to-one correspondence with an $(n-1)$-cell of the dual cellular division $\mathfrak{M}_b^n$. These n-cells and $(n-1)$-cells alternate with one another in a sequence determined by the path. They are now to be oriented. The n-cell corresponding to the initial point O is given an arbitrary orientation. The next n-cell is oriented so that opposite orientations are induced in the common $(n-1)$-cell, and further cells are oriented in this manner. In this way we can extend an orientation beginning at the initial cell along the whole path. After proceeding along the path, starting with the chosen initial orientation, one will return to the initial n-cell with either the orientation chosen initially or with the opposite orientation. According to which case occurs one says that the orientation is preserved or reversed, respectively, along the path. This definition only makes sense when one is given a particular cellular division and an edge path lying in it. The intersection numbers provide the means for us to make the definition independent of the choice of cellular division.

The manifold $\mathfrak{M}^n$ will be orientable in the sense of §24 if and only if the orientation is preserved along each and every closed edge path of $\mathfrak{M}_a^n$. Consequently we shall restrict ourselves to the nontrivial case of nonorientable manifolds in our discussion. In such manifolds there exists exactly one $(n-1)$-dimensional torsion coefficient having the value 2 (§24). There thus exists exactly one homology class for the dimension $n-1$ of order 2 and the (singular) chains in this class are characterized by the property that they themselves are not null homologous, but become so when doubled. We obtain an $(n-1)$-chain U^{n-1} having this property, as follows. We give the n-cells of $\mathfrak{M}_b^n$ an arbitrary but fixed orientation and form the boundary of the n-chain U^n which consists of the totality of n-cells oriented in this manner. An $(n-1)$-cell will then either not appear at all or will appear doubled in $\mathfrak{R}\partial U^n$ according to whether the two n-cells contiguous with it induce opposite or the same orientation, respectively, in this $(n-1)$-cell. Thus $\mathfrak{R}\partial U^n$ is the double of a chain U^{n-1}. We then have $2U^{n-1}\sim 0$, $U^{n-1}\not\sim 0$ (cf. §24; the difference between the present and earlier cases is that we are now using cells instead of simplexes).

We now see easily that orientation is preserved along the edge path w if and only if w intersects the chain U^{n-1} in an even number of points. For U^{n-1} contains exactly those $(n-1)$-cells of $\mathfrak{M}_b^n$ for which both contiguous n-cells induce the same orientation. We can then say: *Orientation is preserved along w if and only if the intersection number* mod 2 *of w with U^{n-1} is equal to $\overset{\circ}{0}$.*[*] This definition can be extended to arbitrary continuous paths which are not necessarily edge paths. The definition is independent of any simplicial decomposition of $\mathfrak{M}^n$ since, as we have already shown, U^{n-1} is determined by $\mathfrak{M}^n$ in a topologically invariant manner, up to the addition of null homologous chains.

It also follows that homologous paths behave in the same way, with respect to preservation of orientation, since the intersection number does not change when a chain is replaced by another chain homologous to it.

In addition, homotopic paths behave in the same way with respect to preservation of orientation, so that the elements of the fundamental group $\mathfrak{F}$ of $\mathfrak{M}^n$ can be decomposed into two classes according to how their paths behave with respect to preservation of orientation.[**] The path classes which preserve orientation form a subgroup $\mathfrak{H}$ of index 2 in $\mathfrak{F}$. For the product of two paths having reversal of orientation is a path on which orientation is preserved. Accordingly, the two-sheeted covering $\widetilde{\mathfrak{M}}^n$ belonging to the

---

[*] More exactly, we should speak of the intersection number of the chains mod 2 corresponding to w and U^{n-1}.

[**] In the discussion to follow we shall always mean paths w which depart from a fixed point O of $\mathfrak{M}^n$ and paths $\tilde{w}$ which depart from a fixed point $\tilde{O}$ of $\widetilde{\mathfrak{M}}^n$, where $\tilde{O}$ lies above O.

subgroup $\mathfrak{H}$ (§55) is related in a topologically invariant way to the manifold $\mathfrak{M}^n$.

The covering $\widetilde{\mathfrak{M}}^n$ is orientable. For one can lift the dual cellular divisions $\mathfrak{M}^n_a$ and $\mathfrak{M}^n_b$ to $\widetilde{\mathfrak{M}}^n$. If the orientation were to reverse along a closed edge path $\tilde{w}$ of $\widetilde{\mathfrak{M}}^n_a$, then orientation would also reverse along the corresponding base path w in $\mathfrak{M}^n_a$. The path w could not then belong to $\mathfrak{H}$ and $\tilde{w}$ would not be closed, which is contrary to our assumption.

$\widetilde{\mathfrak{M}}^n$ is the only orientable two-sheeted covering of $\mathfrak{M}^n$. For if $'\widetilde{\mathfrak{M}}^n$ were another orientable two-sheeted covering, then a closed edge path $'\tilde{w}$ of $'\widetilde{\mathfrak{M}}^n$ would project to a closed path of $\mathfrak{M}^n$ along which orientation is preserved and which would then belong to $\mathfrak{H}$. The subgroup $'\mathfrak{H}$ belonging to $'\widetilde{\mathfrak{M}}^n$ would thus be a subgroup of $\mathfrak{H}$ and since this subgroup is, like $\mathfrak{H}$, of index 2 in $\mathfrak{F}$, it would coincide with $\mathfrak{H}$.[46]

76. Orientability and Two-Sidedness

We have recognized that orientability is a property of a surface considered as a 2-dimensional manifold, without reference to any embedding in 3-space. In contrast to its property of being either orientable or nonorientable, the property of a surface being one-sided or two-sided depends upon the embedding of the surface in a 3-dimensional manifold. To intuitively understand the two-sided placement of a surface in a 3-dimensional manifold $\mathfrak{M}^3$, let us imagine that a small arrow has been stuck "perpendicular" to the surface and let us carry the arrow along a closed path on the suface until it returns to its point of departure. If the arrow never reverses direction after a traversal, then the surface lies two-sided in $\mathfrak{M}^3$, otherwise one-sided. At times, orientable surfaces are confused with two-sided surfaces. But the concepts of orientability and two-sidedness are not identical, because the following four mutually exclusive cases all occur.

There exist

1.   orientable two-sided surfaces, for example the 2-sphere and the torus, in Euclidean 3-space (closed to form the 3-sphere),

2.   nonorientable one-sided surfaces, for example the projective plane, in projective 3-space,

3.   orientable one-sided surfaces, and

4.   nonorientable two-sided surfaces.

An example illustrating the last two cases is given by the topological product of the projective plane with a circle. One can construct this product by identifying diametrically opposite points of the meridian circles on the boundary of a solid torus. Here a circular disk spanning a meridian circle closes to form the projective plane, a nonorientable surface which evidently lies two-sided in the 3-dimensional manifold. In Fig. 130 the intersection of the solid torus with a meridian plane has been drawn and the spanned circular disk is shown shaded.

In regard to the third case, the equatorial plane intersects the solid torus in an annulus, which closes to a torus when diametrically opposite points of the meridian circles are identified. This is

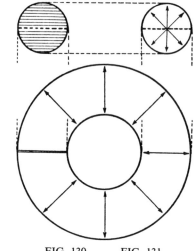

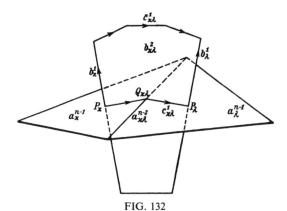

FIG. 130   FIG. 131

FIG. 132

an orientable surface which clearly lies one-sided in the manifold. Figure 131 shows the intersection with the equatorial plane.

The concepts introduced in this chapter give us the tools to define these intuitive ideas mathematically in a precise manner and extend them to spaces of arbitrary dimension.

Let $\mathfrak{M}^n$ be an n-dimensional manifold in which an $(n-1)$-dimensional manifold has been topologically embedded, that is, $\mathfrak{M}^{n-1}$ is a subset of $\mathfrak{M}^n$. We impose the additional requirement on the embedding that there exist a simplicial decomposition of $\mathfrak{M}^n$ for which $\mathfrak{M}^{n-1}$ is a subcomplex. We shall first define one-sidedness and two-sidedness of $\mathfrak{M}^{n-1}$ by choosing a particular decomposition which satisfies the above requirement. We regard the simplexes of the normal subdivision as cells of a cellular division $\mathfrak{M}_a^n$. This provides a cellular division $\mathfrak{M}_a^{n-1}$, of $\mathfrak{M}^{n-1}$ at the same time. Denote the

$(n - 1)$-cells of $\mathfrak{M}_a^{n-1}$ by a_κ^{n-1}. Referring to Fig. 132 let:

P_κ be the midpoint of a_κ^{n-1},

b_κ^1 be the 1-cell dual to a_κ^{n-1} in $\mathfrak{M}^n$,

$a_{\kappa\lambda}^{n-2}$ be the $(n - 2)$-cell of $\mathfrak{M}_a^{n-1}$ which is a common face of a_κ^{n-1} and a_λ^{n-1},

$Q_{\kappa\lambda}$ be the midpoint of $a_{\kappa\lambda}^{n-2}$,

$b_{\kappa\lambda}^2$ be the cell dual to $a_{\kappa\lambda}^{n-2}$ in $\mathfrak{M}^n$,

$c_{\kappa\lambda}^1$ be the cell dual to $a_{\kappa\lambda}^{n-2}$ in $\mathfrak{M}^{n-1}$ which has initial point P_κ and end point P_λ.

In the oriented cells b_κ^1 we have the mathematical realization of the arrow stuck through the surface. We call these the *transversal* 1-*cells* of $\mathfrak{M}^{n-1}$. Two transversal 1-cells b_κ^1 and b_λ^1 are said to be neighboring if the dual $(n - 1)$-cells a_κ^{n-1} and a_λ^{n-1} have a common face $a_{\kappa\lambda}^{n-2}$. Since incidence is preserved upon going over to dual cells, b_κ^1 and b_λ^1 appear in the boundary of $b_{\kappa\lambda}^2$:

$$\mathcal{R}\, \partial b_{\kappa\lambda}^2 = \varepsilon_\kappa b_\kappa^1 + \varepsilon_\lambda b_\lambda^1 + \cdots .$$

If the coefficients ε_κ and ε_λ are equal in magnitude and opposite in sign to one another, then we say that b_κ^1 and b_λ^1 *point in the same direction.*

The exact definition of one- and two-sidedness is now:

$\mathfrak{M}^{n-1}$ *lies two-sided in* $\mathfrak{M}^n$ *if the transversal* 1-*cells can be oriented so that each two neighboring* 1-*cells point in the same direction. If this is not possible, then* $\mathfrak{M}^{n-1}$ *lies one-sided in* $\mathfrak{M}^n$.

It will turn out that the dependence of this definition upon the particular choice of cellular division is only an apparent one.

We can regard $c_{\kappa\lambda}^1$ as the intersection of $\mathfrak{M}^{n-1}$ with $b_{\kappa\lambda}^2$. For $b_{\kappa\lambda}^2$ can have points in common with only those cells of $\mathfrak{M}_a^{n-1}$ which contain the midpoint $Q_{\kappa\lambda}$ of $b_{\kappa\lambda}^2$ (from Theorem VI, §66), that is, with the cells $a_{\kappa\lambda}^{n-2}, a_\kappa^{n-1}, a_\lambda^{n-1}$. The intersection of $b_{\kappa\lambda}^2$ with a_κ^{n-1} consists of the connecting interval, that is, the 1-simplex $P_\kappa Q_{\kappa\lambda}$. Let us call the 1-cells $c_{\kappa\lambda}^1$ the *edges* of $\mathfrak{M}^{n-1}$. If b_κ^1 and b_λ^1 point in the same direction, we define the *substitute arc* $\bar{c}_{\kappa\lambda}^1$ belonging to the edge $c_{\kappa\lambda}^1$ as the arc on the boundary circle of $b_{\kappa\lambda}^2$ which leads from the endpoint of b_κ^1 to the endpoint of b_λ^1 without meeting $\mathfrak{M}^{n-1}$. Now let C be a closed edge path on $\mathfrak{M}^{n-1}$, that is, a path which is composed only of edges $c_{\kappa\lambda}^1$. We shall assume that it passes through the points

$$P_1, P_2, \ldots, P_r, P_{r+1} = P_1.$$

We orient the transversal 1-cells $b_1^1, b_2^1, \ldots b_r^1, b_{r+1}^1$ so that each two consecutive cells point in the same direction. This can be done in only one way, given an arbitrary initial orientation of b_1^1. It is possible here that one and the same transversal cell may appear several times and can appear with opposite orientations. Depending upon whether $b_{r+1}^1 = +b_1^1$ or $b_{r+1}^1 = -b_1^1$, we say that *the arrow direction is preserved or reverses, respectively.* Obviously, when $\mathfrak{M}^{n-1}$ is two-sided the arrow direction is preserved along each edge path C, while for a one-sided embedding of $\mathfrak{M}^{n-1}$ there exists at least one edge path having reversal of the arrow direction. When one replaces each edge $c_{\kappa\lambda}^1$ of C by its corresponding substitute arc $\bar{c}_{\kappa\lambda}^1$, in the case of preservation of arrow direction there will exist a closed path $\bar{C}$ which does not meet $\mathfrak{M}^{n-1}$. On the other hand, when the arrow direction does reverse along C, the path will lead from the endpoint of b_1^1 to the initial point of b_1^1 and upon joining on the cell b_1^1 the path must become a closed path $\bar{C}$ having P_1 as initial point. The path C is clearly deformable to its substitute path. One needs only to transform continuously each edge $c_{\kappa\lambda}^1$, along half of the cell $b_{\kappa\lambda}^2$ to the arc $\bar{c}_{\kappa\lambda}^1$. Here the initial and endpoints of $c_{\kappa\lambda}^1$ will move along b_κ^1 and b_λ^1 to the endpoints of b_1^1 and b_λ^1, respectively. If C is a path which preserves arrow direction, then $\bar{C}$ and therefore the homotopic path C will have the residue class $\check{0}$ as intersection number mod 2 with $\mathfrak{M}^{n-1}$. This is so because $\mathfrak{M}^{n-1}$ and $\bar{C}$ are disjoint, considered as point sets. On the other hand, in the case of arrow direction reversal, the intersection number mod 2 is $\check{1}$ since $\bar{C}$ will pass smoothly through $\mathfrak{M}^{n-1}$ at the point P_1 and only at this point. Furthermore, since for each closed path of $\mathfrak{M}^{n-1}$ there exists a homologous edge path, because the totality of edges $c_{\kappa\lambda}^1$ form the edge complex of the cellular division dual to $\mathfrak{M}_a^{n-1}$, we have

THEOREM I. *If $\mathfrak{M}^{n-1}$ is two-sided, then the intersection number* mod 2 *for each and every path C on $\mathfrak{M}^{n-1}$ is* $\mathfrak{S}(\mathfrak{M}^{n-1}, C) = \check{0}$; *if $\mathfrak{M}^{n-1}$ is one-sided, then there exists a path such that* $\mathfrak{S}(\mathfrak{M}^{n-1}, C) = \check{1}$. *Here $\mathfrak{M}^{n-1}$ and C are to be regarded as chains* mod 2.

In this theorem we have characterized one-sidedness and two-sidedness in a way which no longer requires reference to a simplicial decomposition. The intersection numbers are independent of any such decomposition.

We shall give one more characterization of one- and two-sidedness. Since C is an edge path of the cellular division dual to $\mathfrak{M}_a^{n-1}$, we can continue the orientation of the $(n-1)$-dimensional "surface cells" a_κ^{n-1} along C by first orienting a_1^{n-1} and then orienting in succession each cell of the sequence

$$a_1^{n-1}, a_2^{n-1}, \ldots, a_r^{n-1} a_{r+1}^{n-1} = \pm a_1^{n-1}$$

so that each two successive cells induce opposite orientations in their common $(n-2)$-cell. According to whether $a_{r+1}^{n-1} = a_1^{n-1}$ or $a_{r+1}^{n-1} = -a_1^{n-1}$ we say that C is a path which preserves "surface orientation" or reverses "surface orientation," respectively. In a like manner we can continue the orientation of the "spatial cells" along the substitute path $\bar{C}$. For $\bar{C}$ consists of 1-cells of the cellular division dual to $\mathfrak{M}_a^n$ and it makes sense to say that $\bar{C}$ is a path which either preserves or reverses spatial orientation. Let a_κ^n denote the n-cell which is incident with a_κ^{n-1} and in which the endpoint of b_κ^1 lies. (We are again assuming that each two successive 1-cells in the sequence

$b_1^1, b_2^1, \ldots, b_r^1, b_{r+1}^1$ point in the same direction.) Let a_κ^n be oriented so that a_κ^{n-1} appears on its boundary with sign $+1$:

$$\Re \partial a_\kappa^n = a_\kappa^{n-1} + \cdots . \tag{1}$$

If we then move along the path $\bar{c}_\kappa^1$ from the midpoint of a_κ^n to the midpoint of a_λ^n ($\lambda = \kappa + 1$) and continue the orientation of a_κ^n along $\bar{c}_{\kappa\lambda}$, we shall arrive with the orientation $+a_\lambda^n$. (We can see this easily, since all n-cells through which we pass have the face a_κ^{n-2}.) *When we carry forward the spatial orientation along* $\bar{C}$, *we shall go from* $+a_1^n$ *to* $+a_{r+1}^n$.

We now distinguish between several cases.

Case I. The arrow direction is preserved along C.

(a) Surface orientation is preserved along C. Then $a_{r+1}^{n-1} = a_1^{n-1}$ and thus, from (1), $a_{r+1}^n = a_1^n$. That is, spatial orientation is also preserved along $\bar{C}$.

(b) Surface orientation reverses along C. Then $a_{r+1}^{n-1} = -a_1^{n-1}$ and hence $a_{r+1}^n = -a_1^n$. That is, spatial orientation reverses along $\bar{C}$.

Case II. The arrow direction reverses along C.

(a) Surface orientation is preserved along C. Then $a_{r+1}^{n-1} = a_1^{n-1}$. Now a_{r+1}^n and a_1^n lie on different sides of a_1^{n-1}. From (1), they induce the same orientations in a_1^{n-1}. That is, spatial orientation reverses along $\bar{C}$.

(b) Surface orientation reverses along C. Then $a_{r+1}^{n-1} = -a_1^{n-1}$. Cells a_{r+1}^n and a_1^n induce opposite orientations in a_1^{n-1}. Thus spatial orientation is preserved along $\bar{C}$.

When we also take into account the facts that the continuation of orientation along an arbitrary continuous path can be described in a topologically invariant manner (§75) and that homotopic paths, for example C and $\bar{C}$, will always both reverse or both preserve orientation simultaneously, we get

THEOREM II. *If* $\mathfrak{M}^{n-1}$ *lies two-sided in* $\mathfrak{M}^n$ (*case* I), *then surface orientation and spatial orientation along a path* C *of* $\mathfrak{M}^{n-1}$ *will either both reverse or will both be preserved simultaneously. If* $\mathfrak{M}^{n-1}$ *lies one-sided in* $\mathfrak{M}^n$ (*case* II) *then there will exist at least one path along which surface orientation alone or spatial orientation alone reverses.*

If $\mathfrak{M}^n$ is orientable then spatial orientation is preserved along each path. There thus follows

THEOREM III. *In an orientable manifold* $\mathfrak{M}^n$, *orientability of the embedded manifold* $\mathfrak{M}^{n-1}$ *is equivalent to two-sidedness and nonorientability is equivalent to one-sidedness.*

In particular, in Euclidean 3-space each orientable surface lies two-sided and each nonorientable surface lies one-sided.

77. Linking Numbers

Linking numbers are defined for two disjoint null homologous singular chains or, more generally, for two disjoint division–null homologous singular chains

$$A^{k-1} \quad \text{and} \quad B^{n-k}$$

in an orientable n-dimensional manifold $\mathfrak{M}^n$. Let us first consider the case where A^{k-1} is null homologous and B^{n-k} is division–null homologous (and thus possibly also null homologous). Then by definition A^{k-1} is the boundary of a singular chain A^k, and we define the *linking number*

$$\mathfrak{V}(A^{k-1}, B^{n-k})$$

to be the intersection number

$$\mathfrak{S}(A^k, B^{n-k}).$$

The linking number exists, since the closed chain B^{n-k} is disjoint with the boundary of A^k, by our assumption. In other words, the linking number $\mathfrak{V}(A^{k-1}, B^{n-k})$ is the intersection number of a k-chain spanning A^{k-1} with B^{n-k}. It is unimportant whether one chooses A^k or a different chain $'A^k$ spanning A^{k-1}. For since B^{n-k} is division–null homologous and $A^k - 'A^i$ is closed,

$$\mathfrak{S}(A^k - 'A^k, B^{n-k}) = 0,$$

that is,

$$\mathfrak{S}(A^k, B^{n-k}) = \mathfrak{S}('A^k, B^{n-k}).$$

As an example, we choose $\mathfrak{M}^n$ to be Euclidean 3-space closed by a point at infinity to form the 3-sphere; we choose A^{k-1} and B^{n-k} to be a meridian circle and the core of a solid torus, that is, two interlinked circles. We can then take A^k to be a circular disk spanned in the meridian circle. The core of the solid torus passes smoothly through this disk. The linking number of the meridian circle and the core is therefore ± 1.

If A^{k-1} is only division–null homologous as well, then there will exist an integer $\alpha \neq 0$ such that $\alpha A^{k-1} \sim 0$ (α does not have to be the smallest integer for which the homology $\alpha A^{k-1} \sim 0$ holds). If A^k is a chain spanning αA^{k-1}, that is,

$$\mathfrak{R}\partial A^k = \alpha A^{k-1},$$

then we define the *linking number*[47] by the equation

$$\mathfrak{V}(A^{k-1}, B^{n-k}) = (1/\alpha)\mathfrak{S}(A^k, B^{n-k}).$$

In general, $\mathfrak{V}$ is a fraction. If, instead of A^k, we use another chain $'A^k$ having boundary $'\alpha A^{k-1}$, then we will get

$$(1/'\alpha)\mathfrak{S}('A^k, B^{n-k})$$

as the intersection number. But $\alpha'A^k - '\alpha A^k$ is a closed chain, and has intersection number 0 with each division–null homologous chain. Thus we have

$$\alpha\mathfrak{S}('A^k, B^{n-k}) = '\alpha\mathfrak{S}(A^k, B^{n-k}),$$

or

$$(1/'\alpha)\mathfrak{S}('A^k, B^{n-k}) = (1/\alpha)\mathfrak{S}(A^k, B^{n-k}).$$

The definition is therefore independent of the choice of A^k.

As an alternative to letting B^{n-k} intersect with a chain A^k spanning αA^{k-1}, we can let A^{k-1} intersect with a chain B^{n-k+1} spanning βB^{n-k}, and we shall get the same linking number, except possibly for its sign. For

$$\mathfrak{V}(A^{k-1}, B^{n-k}) = (1/\alpha)\mathfrak{S}(A^k, B^{n-k}) = [1/(\alpha\beta)]\mathfrak{S}(A^k, \beta B^{n-k})$$

$$= (-1)^k[1/(\alpha\beta)]\mathfrak{S}(\alpha A^{k-1}, B^{n-k+1}) \qquad [\text{from (10), §74}]$$

$$= (-1)^k(1/\beta)\mathfrak{S}(A^{k-1}, B^{n-k+1})$$

Corresponding to the formulas (8) and (9) of §74, we have the formulas

$$\mathfrak{V}(A^{k-1}, B_1^{n-k} + B_2^{n-k}) = \mathfrak{V}(A^{k-1}, B_1^{n-k}) + \mathfrak{V}(A^{k-1}, B_2^{n-k})$$

and

$$\mathfrak{V}(A^{k-1}, B^{n-k}) = (-1)^{(k-1)(n-k)+1}\mathfrak{V}(B^{n-k}, A^{k-1})$$

for the linking numbers.

The linking number is not changed when we replace the chain A^{k-1} in the "complementary space" of B^{n-k} relative to $\mathfrak{M}^n$ by a homologous chain $'A^{k-1}$. For if U^k is disjoint with B^{n-k} and if

$$\mathfrak{R}\partial U^k = A^{k-1} - 'A^{k-1},$$

then we have

$$\mathfrak{V}(A^{k-1} - 'A^{k-1}, B^{n-k}) = \mathfrak{S}(U^k, B^{n-k}) = 0.$$

In a similar manner we can replace B^{n-k} by a homologous chain in the complementary space of A^{k-1}.

On the other hand, $\mathfrak{V}(A^{k-1}, B^{n-k})$ changes by an integer when one replaces A^{k-1} by any chain $'A^{k-1}$ which is disjoint from B^{n-k} and homologous to A^{k-1} in $\mathfrak{M}^n$. For since $A^{k-1} - 'A^{k-1}$ is an arbitrary null homologous chain in this case, then, as we saw originally, its linking number with B^{n-k} is an integer. Consequently, we can assign a linking number, determined up to an integer, to two homology classes H^{k-1} and H^{n-k} which, when considered as elements of the homology groups, are of finite order. This linking number is the linking number of any two nonintersecting representatives of the respective classes. In the particular case of a $(2m + 1)$-dimensional manifold there will exist a *self-linking number* corresponding to each m-dimensional homology class of finite order. This is the linking number of two disjoint homologous m-chains. The self-linking numbers are topological invariants of the manifold. In particular cases they may be used to show the distinctness of manifolds. We will demonstrate this, using the lens spaces as an example.

In order to find the self-linking number of the lens axis b in the lens space (p, q), we deform the axis by transforming the initial axis point P on the lower lens cap to a point P' on the equatorial circle of the lens. The equivalent endpoint Q will then automatically be moved on the upper lens cap to a particular point Q' of the equatorial circle. The other points of b will be deformed through the interior of the lens to the connecting arc $b' = P'Q'$ on the equatorial circle. This arc will subtend a fraction q/p of the periphery of the equatorial circle. The arcs b and b' are two homologous closed 1-chains. To find their linking number we choose the p-fold multiple of b', which is a curve running q times around the equatorial circle, and span a 2-chain in it, for example the circular disk K^2 which bisects the lens into two symmetrical halves, taken with multiplicity q. Since b and K^2 pass smoothly through one another, their intersection number is ± 1. Thus $\mathcal{S}(b, qK^2) = \pm q$ and the self-linking number which we are looking for is

$$\mathcal{V}(b, b') = \pm q/p.$$

We can find the self-linking numbers of the other homology classes by choosing the ν-fold multiple of the axis b ($\nu = 0, 1, \ldots, p - 1$). For the closed 1-chain b forms a homology basis of dimension 1. We have

$$\mathcal{V}(\nu b, \nu b') = \pm \nu^2 q/p,$$

where the upper or lower sign will respectively hold depending upon the orientation of the lens space.

In order for two lens spaces (p, q) and (p', q') to be homeomorphic it is first of all necessary that their fundamental groups be the same. That is, we require $p = p'$. In addition, the self-linking number q'/p of (p, q') must also appear as a self-linking number of (p, q). Therefore q'/p must be congruent mod 1 to one of the numbers $\pm \nu^2 q/p$. Thus there must exist an integer ν having the property that

$$q' \equiv \pm \nu^2 q \pmod{p}.$$

If we choose $(p, q) = (5, 1)$ as an example, then this congruence will only be satisfied for $q' = \pm 1$. Thus $q' = 2$ is not a solution. Hence the lens spaces $(5, 1)$ and $(5, 2)$ are not the same (Alexander [2], [10]).

The set of possible self-linking numbers of the 1-chains of a 3-dimensional manifold is, then, a topological invariant. On occasion it can be used to prove the distinctness of manifolds even when our strongest previous decision criterion, the fundamental group, fails.

For the lens spaces $(7, 1)$ and $(7, 2)$, however, not even the linking numbers will solve the homeomorphism problem, since the set of possible self-linking numbers is the same for both spaces, as is easily computed.

In the lens space $(3, 1)$ the possible self-linking numbers are

$$0, \quad \tfrac{1}{3}, \quad \tfrac{4}{3} \equiv \tfrac{1}{3} \pmod{1}$$

for one orientation and

$$0, \quad -\tfrac{1}{3}$$

for the other orientation of the space. Since the two sets of numbers are different it follows that it is impossible to map the lens space (3, 1) onto itself topologically with reversal of orientation. In such a case one also says that the lens space (3, 1) is an *asymmetric space*. This property appears for the first time in the 3-dimensional manifolds.[48]

Problems

1. One obtains two manifolds from two lens spaces (3, 1) by the process of connected sum formation (cf. Problem 3, §62). Show that these manifolds are not homeomorphic, even though they have the same fundamental group. (Find a homology basis for each composite manifold and determine the set of possible self-linking numbers.)

2. Show that in an orientable manifold $\mathfrak{M}^n$ there exists a k-dimensional torsion basis $A_1^k, A_2^k, \ldots, A_{\rho^k}^k$ and an $(n - k - 1)$-dimensional torsion basis $B_1^{n-k-1}, B_2^{n-k-1}, \ldots, B_{\rho^k}^{n-k-1}$ such that the linking number $\mathfrak{V}(A_\mu^k, B_\nu^{n-k-1}) = 0$ for $\mu \neq \nu$ and $= 1/c_\mu$ for $\mu = \nu$. Here c_μ denotes the torsion coefficient belonging to A_μ^k (cf. §71).

3. If a 3-dimensional orientable manifold $\mathfrak{M}^3$ has a single torsion coefficient, of dimension 1, then there exists at least one 1-chain having a self-linking number different than 1.

4. If an orientable 3-dimensional manifold has a prime integer of the form $4\kappa + 3$ as its only 1-dimensional torsion coefficient, then it is an asymmetric space.

We have already introduced the exterior space of a knot several times into our investigations. On one occasion the torus knots were used to give an illustration of the fundamental group (§52). On another occasion an arbitrary knot was used for the construction of 3-dimensional manifolds (§65) and in §58 we used the trefoil knot to demonstrate covering complexes. This last application is closely related to the linking numbers.

Let us embed a knot k provided with a particular orientation into the simplicially decomposed 3-sphere $\mathfrak{S}^3$. Let it be composed of edges of the simplicial decomposition. In addition, let us assume that the knot never passes through all three vertices of a 2-simplex and never contains a 1-simplex as a chord. If we regard the normally subdivided simplexes of $\mathfrak{S}^3$ as cells of a cellular division $\mathfrak{S}_a^3$, then the boring out of the knot, defined in §65, consists of removing all those 3-cells of the dual cellular division $\mathfrak{S}_b^3$ whose midpoints are vertices of k. The 2-cells dual to the edges of k are meridian cross sections of the bored-out solid torus $\mathfrak{B}$ and their boundary circles are meridian circles of $\mathfrak{B}$. After being oriented appropriately they are all homologous to one another on the torus $\mathfrak{T}$ which is the boundary surface of $\mathfrak{B}$. An arbitrarily chosen 1-cell chain u lying on $\mathfrak{S}_b^3$ of the exterior space $\mathfrak{A}$ will be null homologous in $\mathfrak{S}^3$ and will therefore be the boundary of a cell chain U^2. When one removes all cells from U^2 which are dual to edges of k, that is, all meridian cross sections of the solid torus $\mathfrak{B}$, then a chain $'U^2$ will result whose boundary will be formed by u and certain meridian circles of $\mathfrak{B}$. Thus, since all meridian circles are homologous on $\mathfrak{T}$ to a fixed meridian circle m_0,

$$u \sim \alpha m_0 \qquad \text{(on } \mathfrak{A}\text{)}.$$

Here α gives the linking number of u with k. For when $\mathfrak{S}^3$ is suitably oriented, m_0 will have the linking number 1 with k. Thus

$$\mathfrak{V}(k, u) = \mathfrak{V}(k, \alpha m_0) = \alpha.$$

If two cell chains u and v now have the same linking number α with k, then they will both be homologous to αm_0 and will therefore be homologous to one another on $\mathfrak{A}$. Furthermore, since each arbitrary singular closed k-chain on $\mathfrak{A}$ is homologous to a cell chain, we have the following theorem:

THEOREM. *Two singular 1-chains, u and v in the exterior space $\mathfrak{A}$ of a knot k will be homologous on $\mathfrak{A}$ if and only if they have the same linking number with k.*

As a consequence we have another proof that the homology group of $\mathfrak{A}$ is the free cyclic group (§65).

In the fundamental group $\mathfrak{F}$ of the exterior space $\mathfrak{A}$, the so-called *knot group*, those paths whose linking number with the knot k is divisible by a particular integer g form a subgroup $\mathfrak{H}$.

We wish to examine the covering $\tilde{\mathfrak{A}}$ of $\mathfrak{A}$ which belongs to the subgroup $\mathfrak{H}$.

A particular linking number will belong to each element of $\mathfrak{F}$, namely, the linking number of any representative path with the knot k. By assigning to each element of $\mathfrak{F}$ its linking number reduced mod g, we produce a homomorphic mapping of $\mathfrak{F}$ onto the cyclic group of order g. The subgroup $\mathfrak{H}$ will be mapped to the unit element. $\mathfrak{H}$ is then a normal subgroup and $\mathfrak{F}/\mathfrak{H}$ is cyclic of order g. *Consequently, the covering $\tilde{\mathfrak{A}}$ is regular and the group of covering transformations is cyclic. That is, $\tilde{\mathfrak{A}}$ is a cyclic covering in the sense of* §58. The covering $\tilde{\mathfrak{A}}$ is thereby characterized by the fact that a closed path of the base complex $\mathfrak{A}$ will be closed after being lifted into the covering complex if and only if its linking number with k is congruent to 0 (mod g).

The covering given here is, by the way, the only g-sheeted cyclic covering. For if $\mathfrak{H}$ is a normal subgroup of the knot group $\mathfrak{F}$ belonging to a given cyclic covering, then $\mathfrak{F}/\mathfrak{H}$ is cyclic of order g. In the homomorphism $\mathfrak{F} \to \mathfrak{F}/\mathfrak{H}$ each commutator $F_i F_k F_i^{-1} F_k^{-1}$ will map to the unit element of $\mathfrak{F}/\mathfrak{H}$. That is, each commutator belonging to $\mathfrak{F}$ will belong to $\mathfrak{H}$ and thus $\mathfrak{H}$ will contain the commutator group $\mathfrak{K}$ of $\mathfrak{F}$. Consequently, $\mathfrak{H}$ will consist of certain residue classes of the decomposition of $\mathfrak{F}$ with respect to $\mathfrak{K}$. These residue classes, regarded as elements of the factor group $\mathfrak{F}/\mathfrak{K}$, form a subgroup of index g in $\mathfrak{F}/\mathfrak{K}$. But $\mathfrak{F}/\mathfrak{K}$ is now the Abelianized knot group, that is, the free cyclic group. This group has only one subgroup of index g. Thus the uniqueness of $\mathfrak{H}$ has been proven and, at the same time, a purely group theoretical proof of the existence of the cyclic covering has been given.

It can be proved that an orientable surface, free of singularities, can be spanned in each knot so that the knot is the boundary of the surface. One can then obtain the g-sheeted cyclic covering by cutting apart the exterior space $\mathfrak{A}$ along this surface to form a "sheet" and then sewing g such sheets cyclically to one another. The resulting covering obviously admits a cyclic group of covering transformations of order g.

Cyclic coverings play a role in knot theory. Whereas the Abelianized knot group is always the free cyclic group, the finite sheeted cyclic coverings will in general possess 1-dimensional torsion coefficients. A necessary condition for two knots to be equivalent is that the torsion coefficients of their g-sheeted cyclic coverings be the same.*

Instead of defining the cyclic coverings for the exterior space $\mathfrak{A}$, that is, for the 3-sphere from which the knot has been bored, we could just as well have defined the cyclic coverings for the complementary space of the knot, which is the 3-sphere from which just the points of the knot have been removed. For the theorem of this section is also valid when we replace $\mathfrak{A}$ by $\mathfrak{S}^3 - k$. For if u and v are two singular 1-chains in $\mathfrak{S}^3 - k$ we can bore out k with a solid torus which is sufficiently slender that u and v lie entirely in the exterior space $\mathfrak{A}$. If u and v now have the same linking number $\mathfrak{V}(k, u) = \mathfrak{V}(k, v)$ with the knot, then $u \sim v$ in the exterior space $\mathfrak{A}$ and thus also in the complementary space $\mathfrak{S}^3 - k$. Let us note that $\mathfrak{A}$ is a finite complex, while $\mathfrak{S}^3 - k$ is an infinite complex.

---

*See Alexander [16] and Reidemeister [6], where one will also find further references to the literature.

CHAPTER ELEVEN

CONTINUOUS MAPPINGS

78. The Degree of a Mapping

The methods developed previously have extensive applications in the theory of continuous mappings of complexes and manifolds. We discuss two of these applications in this chapter: the degree of a mapping and the fixed point formula.

In §31 we divided the continuous mappings from a complex $\Re$ into another complex **K** into mapping classes, that is, classes of mappings homotopically deformable one to another. An important problem of topology is to find all possible mapping classes belonging to two given complexes $\Re$ and **K**. This general problem has been solved only for particular complexes, for example, the case where **K** is the n-sphere and $\Re$ is an arbitrary n-dimensional complex (cf. Hopf [18]). But we have already found necessary conditions for two mappings φ and ψ to belong to the same class: the homomorphic mappings from the homology groups of $\Re$ to the homology groups of **K** which are induced by φ must be the same as those induced by ψ; likewise, the homomorphic mapping (which is unique except for inner automorphisms) from the fundamental group of $\Re$ to that of **K** which is induced by φ must be the same as that induced by ψ.

In the case that $\Re$ and **K** are both closed orientable pseudomanifolds of dimension n and have been oriented, the nth homology group of each complex is a free cyclic group. The n-chain B^n which is generated by coherently orienting the simplexes of an arbitrary simplicial decomposition of $\Re$ will form a homology basis in $\Re$. Likewise, the n-chain $\mathbf{B}^n$ arising from a coherent orientation of the simplexes of a simplicial decomposition of **K** will form a homology basis in **K**. The homomorphic mapping of the nth homology group, induced by the continuous mapping φ of $\Re$ into **K**, is then specified by a single number γ; the image of the chain B^n is homologous to $\gamma \mathbf{B}^n$. We call γ the *degree of the mapping* φ.[49] *The degree of a mapping is, then, an invariant of the mapping class.* When we deform φ to a simplicial mapping ψ, having previously subdivided $\Re$ sufficiently finely, the simplexes of $\Re$ will transform under ψ to simplexes of **K** (in some cases these may be degenerate) and the image of the chain B^n will not only be homologous to $\gamma \mathbf{B}^n$ but will in fact be equal to the chain $\gamma \mathbf{B}^n$. Let us assume that α simplexes of $\Re$ map with preservation of orientation and β simplexes of $\Re$ map with reversal of orientation to a particular simplex Σ^n belonging to the simplicial decomposition of **K**. In this case, we have $\gamma = \alpha - \beta$. Intuitively, the degree of the mapping indicates how many times **K** is covered positively by the image of $\Re$.

EXAMPLES. The mapping of the boundary of an n-simplex onto itself in which two vertices are interchanged while all other vertices remain fixed (§31), has degree $\gamma = -1$.

We can find a mapping having an arbitrary degree as follows: On a 2-sphere $\mathfrak{S}^2$ let λ be the geographical longitude and let ϑ be the latitude. The formulas

$$\lambda' = \gamma \lambda, \qquad \vartheta' = \vartheta$$

produce a continuous mapping of $\mathfrak{S}^2$ into another 2-sphere $'\mathfrak{S}^2$ having the geographic coordinates λ', ϑ'. The degree of the mapping is γ when $\mathfrak{S}^2$ and $'\mathfrak{S}^2$ are suitably oriented. For if $\gamma \neq 0$, then a simplicial decomposition of $\mathfrak{S}^2$ produced by the equator and 3γ equidistant meridian circles will be transformed by the mapping to a simplicial decomposition of $'\mathfrak{S}^2$ produced by the equator and three equidistant meridian circles. Each triangle of $'\mathfrak{S}^2$ will then be covered in the same sense by $|\gamma|$ triangles of $\mathfrak{S}^2$. If, however, $\gamma = 0$, then all of $\mathfrak{S}^2$ will be mapped to a single meridian circle of $'\mathfrak{S}^2$. In that case the degree of the mapping will be 0, as a consequence of the following theorem:

THEOREM. *In a continuous mapping φ, if there is a point P of **K** which is not covered by the image of $\Re$, or if φ can be deformed to a mapping ψ which has this property, then the degree of φ is 0.*

Proof. Make P a vertex of a simplicial decomposition of **K** which is sufficiently fine so that the simplicial star $\mathfrak{St}^n$ about P is disjoint from $\psi(\Re)$. The singular chain $\psi(B^n)$ will then lie on the complex $\overline{\mathbf{K}}$, which is the complex which remains when one removes the n-simplexes of $\mathfrak{St}^n$ from **K**. By the approximation theorem (§28), there exists a homologous simplicial chain on $\overline{\mathbf{K}}$ which is equal to zero because it does not contain the simplexes of $\mathfrak{St}^n$. Thus $\psi(B^n) \sim 0$ on **K**, that is, ψ has degree 0.

If we let $\Re$ and **K** coincide we find: *A deformation* of a complex to itself has degree $+1$. This is because the identity map has degree $+1$.*

Let φ be a continuous mapping of $\Re$ into **K** and let φ_1 be a continuous

---

* *Editor's note*: The authors evidently intend "deformation" to mean any map which is deformable to the identity map.

mapping of $\mathbf{K}$ into $\mathbf{K}_1$. Then $\varphi_1\varphi$ is a continuous mapping of $\Re$ into $\mathbf{K}_1$. *If γ and γ_1 are the degrees of the mappings φ and φ_1, then the degree of the mapping $\varphi_1\varphi$ is $\gamma_1\gamma$.* For if the chain $\mathbf{B}_1^n$ on $\mathbf{K}_1$ is defined in the same way as B^n on $\Re$, then

$$\varphi(B^n) \sim \gamma\mathbf{B}^n \qquad \text{(on } \mathbf{K}\text{)}, \tag{1}$$

$$\varphi_1(\mathbf{B}^n) \sim \gamma_1\mathbf{B}_1^n \qquad \text{(on } \mathbf{K}_1\text{)}. \tag{2}$$

The homology (1) is preserved under the mapping φ_1 (§27, Theorem I). Thus

$$\varphi_1(\varphi(B^n)) \sim \gamma\varphi_1(\mathbf{B}^n) \sim \gamma\gamma_1\mathbf{B}_1^n \qquad \text{(on } \mathbf{K}_1\text{)},$$

which was to be proved.

In particular, if φ is a topological mapping of $\Re$ onto $\mathbf{K}$ and φ_1 is the reciprocal mapping, then $\varphi_1\varphi$ is the identity mapping, so that $\gamma\gamma_1 = 1$ and thus $\gamma = \gamma_1 = \pm 1$. *A topological mapping of $\Re$ onto $\mathbf{K}$ has the degree ± 1.* We already recognized this in §36. Based on this fact, we divided the topological mappings of $\Re$ onto $\mathbf{K}$ into mappings which preserve orientation and mappings which do not preserve orientation, that is mappings having degree $+1$ or -1, respectively.

If $\mathbf{K}$ is a manifold, we can also regard the degree of a mapping as the intersection number of a point P of $\mathbf{K}$ (where P is oriented with the $+$ sign), with the image chain $\varphi(B^n) \sim \gamma\mathbf{B}^n$. For $\mathcal{S}(P, \varphi(B^n)) = \mathcal{S}(P, \gamma\mathbf{B}^n) = \gamma\mathcal{S}(P, \mathbf{B}^n) = \gamma$ (§70). This definition can also be used in the case that $\Re$ is an orientable pseudomanifold with boundary and $\mathbf{K}$ is an orientable closed manifold, for example, when one deals with the mapping of a disk into the 2-sphere; the degree of the mapping is then of course defined only with respect to a particular point of $\mathbf{K}$ which must not belong to the image of the boundary of $\Re$.

Problem

Let an n-dimensional orientable pseudomanifold with boundary $\Re$ be mapped continuously into an n-dimensional orientable manifold $\mathbf{K}$.

(a) If two points P and Q on $\mathbf{K}$ are connected by a path which does not intersect the image of the boundary of $\Re$, show that the degree of the mapping is the same at P and at Q.

(b) If one deforms the mapping, and if the point P of $\mathbf{K}$ remains disjoint with the image of the boundary of $\Re$ during the whole course of the deformation, show that the degree of the mapping at P remains unchanged during the deformation.

79. A Trace Formula

We now turn our attention to theorems concerning the existence of fixed points of continuous mappings, and with this goal in mind we derive the fundamental "trace formula" of H. Hopf.

Let $\Re^n$ be a finite n-dimensional complex which has been given a particular simplicial decomposition. Let α^k be the number of k-simplexes and let

$$V_1^k, V_2^k, \ldots, V_{\alpha^k}^k$$

be a basis of the lattice $\mathfrak{T}^k$ of all k-chains. If we assign a chain

$$'V_\kappa^k = \sum_{\lambda=1}^{\alpha^k} \tau_{\kappa\lambda}^k V_\lambda^k \qquad (\kappa = 1, 2, \ldots, \alpha^k) \tag{1^k}$$

to each chain V_κ^k, this linear transformation will provide a homomorphic mapping $\mathbf{T}^k$ of the lattice $\mathfrak{T}^k$ into itself. Suppose that this has been done for all dimensions $k = 0, 1, \ldots, n$. The mappings $\mathbf{T}^k$ do not have to be completely independent of one another; rather, we require that they be *boundary faithful*. That is, if a chain

$$U^k = \sum_{\kappa=1}^{\alpha^k} u_\kappa V_\kappa^k$$

transforms to a chain

$$'U^k = \sum_{\kappa=1}^{\alpha^k} u_\kappa 'V_\kappa^k$$

as a consequence of the equations (1^k), then $\mathfrak{R}\partial U^k$ should transform to $\mathfrak{R}\partial 'U^k$ as a consequence of the equations (1^{k-1}). In the sections to follow, the boundary faithful mappings $\mathbf{T}^k$ will be given by a simplicial self-mapping of the complex $\mathfrak{R}^n$; here, however, we shall only make use of the assumed boundary faithfulness.

The requirement of boundary faithfulness implies that a closed chain transforms to a closed chain, a null homologous chain to a null homologous chain, and a division–null homologous chain to a division–null homologous chain. As an illustration of this, if $U^{k-1} \approx 0$ so that, for example, $\mathfrak{R}\partial U^k = cU^{k-1}$, then we have $\mathfrak{R}\partial 'U^k = c'U^{k-1}$ as a consequence of the boundary faithfulness, that is, $'U^{k-1}$ is also division–null homologous. Thus, to a class of k-chains which are division-homologous to one another there corresponds a particular image class of division-homologous chains. It follows from this that the Betti groups of dimensions $0, 1, \ldots, n$ are mapped homomorphically into themselves by the boundary faithful mappings $\mathbf{T}^k$.

If B_ρ^k $(\rho = 1, 2, \ldots, p^k)$ is a Betti basis in dimension k, then the homomorphic mapping $\mathbf{B}^k$ of the kth Betti group is given by the division homologies

$$'B_\rho^k \approx \sum_{\sigma=1}^{p^k} \beta_{\rho\sigma}^k B_\sigma^k \qquad (\rho = 1, 2, \ldots, p^k). \tag{2^k}$$

The formula which we wish to derive will give a relationship between the traces of the transformations $\mathbf{T}^k$ and $\mathbf{B}^k$. By the *trace* of a linear transformation we mean the sum of the coefficients of the entries on the principal diagonal of the transformation matrix. Thus the trace (German: Spur) of $\mathbf{T}^k$ is the number

$$\operatorname{Sp} \mathbf{T}^k = \sum_{\kappa=1}^{\alpha^k} \tau_{\kappa\kappa}^k$$

and the trace of $\mathbf{B}^k$ is

$$\operatorname{Sp} \mathbf{B}^k = \sum_{\rho=1}^{p^k} \beta_{\rho\rho}^k.$$

The trace is an invariant of the mapping and does not depend upon the choices of the basic elements V_κ^k and B_ρ^k, respectively (see, for example, Speiser [1, p. 148]).

The formula which we wish to find follows immediately, after we have exhibited the simultaneous normal form $\mathbf{K}^k$ $(k = 0, 1, \ldots, n - 1)$ of the incidence matrices (§21). Three types of chains A^k, B^k, and C^k appear in the entries of these matrices. The A^k are division–null homologous. The B^k form a Betti basis; we assume that the B^k appearing in formula (2^k) are identical with these chains. The C^k are chains which are not closed. We have

$$\mathcal{R}\partial C_\mu^k = c_\mu^{k-1} A_\mu^{k-1} \qquad (\mu = 1, 2, \ldots, \gamma^{k-1}). \tag{3}$$

The chains A^k, B^k, and C^k taken together form a basis for the lattice of all k-chains $\mathfrak{X}^k$, and they should be used as basis chains V^k in the linear transformations (1^k). If we do this, the square matrix $[\tau_{\kappa\lambda}^k] = \mathbf{T}^k$ of coefficients in the equations (1^k) will decompose into nine rectangular blocks:

$\mathbf{T}^k$	A^k	B^k	C^k	
$'A^k$	$\alpha_{\rho\sigma}^k$	(12)	(13)	γ^k
$'B^k$	(21)	$\beta_{\rho\sigma}^k$	(23)	p^k
$'C^k$	(31)	(32)	$\gamma_{\rho\sigma}^k$	γ^{k-1}

$(\mathbf{T}^k)$

We now make use of the boundary faithfulness of the mapping $(\mathbf{T}^k)$. From the fact that the chains A^k and B^k are closed it follows, first, that $'A^k$ and $'B^k$ are also closed. Thus none of the chains C^k can appear in the expressions for $'A^k$ and $'B^k$, and the coefficients in the rectangular blocks (13) and (23) must vanish. Thus the coefficients $\beta_{\rho\sigma}^k$ appearing in the submatrix above are the same as the coefficients $\beta_{\rho\sigma}^k$ appearing in the division homology (2^k), which were already given the same notation. It follows, second, that the chains A^k are division–null homologous and, therefore, so are the images $'A^k$. This can occur only if all of the coefficients (12) vanish. Third, Eq. (3) must transform to an equation which is still valid when one replaces the chains by their images:

$$\mathcal{R}\partial'C_\mu^k = c_\mu^{k-1} {}'A_\mu^{k-1} \qquad (\mu = 1, 2, \ldots, \gamma^{k-1}).$$

The linear mapping $\mathbf{T}^k$ of the k-chains results in a particular expression giving $'C^k_\mu$ in terms of the A^k, B^k, and C^k. In like manner, the linear mapping $\mathbf{T}^{k-1}$ of the $(k-1)$-chains results in a particular expression giving $'A^{k-1}_\mu$ in terms of the chains A^{k-1}. Upon introducing these expressions into the last equation and taking into account the fact that the chains A^k and B^k do not contribute to the boundary because they are closed, we get

$$\mathcal{R}\partial \sum_{\nu=1}^{\gamma^{k-1}} \gamma^k_{\mu\nu} C^k_\nu = c^{k-1}_\mu \sum_\nu \alpha^{k-1}_{\mu\nu} A^{k-1}_\nu.$$

Using (3),

$$\sum_\nu \gamma^k_{\mu\nu} c^{k-1}_\nu A^{k-1}_\nu = c^{k-1}_\mu \sum_\nu \alpha^{k-1}_{\mu\nu} A^{k-1}_\nu;$$

thus

$$\gamma^k_{\mu\nu} c^{k-1}_\nu = c^{k-1}_\mu \alpha^{k-1}_{\mu\nu}$$

and in particular, for $\mu = \nu$,

$$\gamma^k_{\mu\mu} = \alpha^{k-1}_{\mu\mu} \qquad (\mu = 1, 2, \ldots, \gamma^{k-1}).$$

The traces of the submatrices $\mathbf{\Gamma}^k = [\gamma^k_{\mu\nu}]$ and $\mathbf{A}^{k-1} = [\alpha^{k-1}_{\mu\nu}]$ are, therefore, the same:

$$\mathrm{Sp}\,\mathbf{\Gamma}^k = \mathrm{Sp}\,\mathbf{A}^{k-1}. \tag{4}$$

Making use of this equation, we form

$$\sum_{k=0}^{n} (-1)^k \mathrm{Sp}\,\mathbf{T}^k = \sum (-1)^k \mathrm{Sp}\,\mathbf{A}^k + \sum (-1)^k \mathrm{Sp}\,\mathbf{B}^k + \sum (-1)^k \mathrm{Sp}\,\mathbf{\Gamma}^k$$

$$= \sum (-1)^k \mathrm{Sp}\,\mathbf{B}^k + \mathrm{Sp}\,\mathbf{\Gamma}^0 + (-1)^n \mathrm{Sp}\,\mathbf{A}^n.$$

Since no division–null homologous n-chains exist, $\mathrm{Sp}\,\mathbf{A}^n = 0$, and we have $\mathrm{Sp}\,\mathbf{\Gamma}^0 = 0$ because all 0-chains are closed. Thus we have established *the trace formula of H. Hopf*:

$$\sum_{k=0}^{n} (-1)^k \mathrm{Sp}\,\mathbf{T}^k = \sum_{k=0}^{n} (-1)^k \mathrm{Sp}\,\mathbf{B}^k. \tag{H}$$

When one chooses $\mathbf{T}^k$ to be the identity mapping, then the matrix $\mathbf{T}^k$ is the unit matrix having α^k rows and $\mathbf{B}^k$ is the unit matrix having p^k rows; thus

$$\mathrm{Sp}\,\mathbf{T}^k = \alpha^k \qquad \text{and} \qquad \mathrm{Sp}\,\mathbf{B}^k = p^k$$

and the trace formula reduces to the Euler formula (§23)

$$\sum (-1)^k \alpha^k = \sum (-1)^k p^k,$$

of which it is a generalization.

80. A Fixed Point Formula

We shall now consider an arbitrary continuous mapping g_0 of a finite complex $\Re^n$ into itself. Let us construct two simplicial decompositions of $\Re^n$, one coarse and the other fine. Let the fine decomposition be derived from the coarse decomposition by means of an m-fold normal subdivision and let it be sufficiently fine so that the image of each simplicial star lies entirely in the interior of a simplicial star of the coarse decomposition. This is possible as a consequence of the theorem of uniform continuity.

From the deformation theorem, it is possible to deform g_0 to a simplicial mapping g_1. The mapping g_1 transforms each oriented k-simplex e_ν^k of the fine decomposition to an oriented simplex E_ν^k;

$$e_\nu^k \rightarrow E_\nu^k. \tag{1}$$

The simplex E_ν^k is either a k-simplex of the coarse decomposition or is a degenerate simplex equivalent to the k-chain 0, in the case where e_ν^k is transformed by the simplicial mapping to a simplex of dimension lower than k. Since the m-fold normal subdivision $\dot{E}_\nu^k$ of E_ν^k is a k-chain on the fine decomposition, then the simplicial mapping g_1 assigns a particular k-chain $\dot{E}_\nu^k$ of the fine decomposition to each k-simplex e_ν^k;

$$e_\nu^k \rightarrow \dot{E}_\nu^k. \tag{2}$$

The chain $\sum_\nu u_\nu \dot{E}_\nu^k$ will then correspond to an arbitrary chain $\sum_\nu u_\nu e_\nu^k$. This gives a homomorphic mapping $\mathbf{T}^k$ of the lattice $\mathfrak{T}^k$ of all k-chains of the fine decomposition into itself. This homomorphic mapping is boundary faithful since $\Re \partial e_\nu^k$ transforms to the m-fold normally subdivided boundary of E_ν^k, that is, to $\Re \partial \dot{E}_\nu^k$. For the boundary of the normal subdivision of a simplex is equal to the normal subdivision of the boundary of the simplex (§30). We can then apply the Hopf trace formula (H) of §79 to this homomorphic mapping.

We now make the additional assumption that the mapping g_0 has no fixed point. When we consider $\Re^n$ to be embedded in a Euclidean space, the distance of a point P from its image point $g_0(P)$ is a continuous function of P. The greatest lower bound δ of these distances will in fact belong to these distances, from Theorem III of §7, and since no point remains fixed in place under the mapping, δ is a positive number. If we have already chosen the coarse decomposition to be sufficiently fine so that the diameter of each simplex is smaller than $\delta/2$, then each point P and its image point $g_0(P)$ will belong to different simplexes. The same statement holds with respect to the approximating simplicial mapping g_1 because no point will leave a simplex to which it originally belonged during the course of the approximation (§31). The image chain $\dot{E}_\nu^k$ of e_ν^k will not, then, contain $e_\nu^{k'}$. Thus

$$\mathrm{Sp}\,\mathbf{T}^k = 0 \qquad (k = 0, 1, \ldots, n).$$

For the case of a mapping (2) with no fixed point, the trace formula (H) reduces to

$$\sum (-1)^k \text{ Sp } \mathbf{B}^k = 0.$$

Here we are making a statement about the mapping of the Betti groups induced by (2). This mapping is, however, the same as the mapping induced by (1). For the image of a closed chain under (2) is a subdivision of the image under (1) and the images are therefore homologous. The homomorphic mapping induced by (1) is the same as that induced by g_0 since the two mappings are deformable one to the other (Theorem III, §31).

We have won the following result, where approximations no longer appear:

THEOREM. *If a continuous mapping of a finite complex $\mathfrak{R}^n$ into itself is fixed point-free, then the alternating sum of the traces of the homomorphic mappings of the Betti groups is equal to 0:*

$$\sum_{k=0}^{n} (-1)^k \text{ Sp } \mathbf{B}^k = 0 \qquad\qquad (F)$$

(*fixed point formula*).[50]

81. Applications

In the following examples, three observations are of importance in applying formula (F):

Let the complex $\mathfrak{R}^n$ be connected. Then a Betti basis in dimension 0 will be formed by a point with $+$ orientation (§18). Under a continuous mapping it will transform to a point with $+$ orientation. Thus the trace of the mapping of the 0th Betti group is 1. In other words:

(I) $\text{Sp } \mathbf{B}^0 = 1$ follows from $p^0 = 1$.

In the case $p^k = 0$ the Betti basis consists of 0 k-chains. We have

(II) $\text{Sp } \mathbf{B}^k = 0$ follows from $p^k = 0$.

(III)   If $\mathfrak{R}^n$ is an orientable n-dimensional manifold, then the Betti number $p^n = 1$ and a Betti basis will be formed by the n-chain M^n which consists of the coherently oriented manifold. *If the image of M^n is homologous to γM^n, then* $\text{Sp } \mathbf{B}^n = \gamma$, whereby γ is the degree of the mapping.

EXAMPLE 1. *The closed n-ball.* From §19,

$$p^0 = 1, \qquad p^1 = p^2 = \cdots = p^n = 0,$$

and, therefore, from observations (I) and (II),

$$\sum_{k=0}^{n} (-1)^k \text{ Sp } \mathbf{B}^k = 1 - 0 + 0 - \cdots = 1 \neq 0.$$

A continuous mapping of the closed n-ball into itself will always possess at least one fixed point.

EXAMPLE 2. *The n-sphere.* From §19,

$$p^0 = 1, \qquad p^2 = \cdots = p^{n-1} = 0, \qquad p^n = 1.$$

If the mapping is to have no fixed points, then according to the fixed point formula (F),

$$(-1)^k \operatorname{Sp} \mathbf{B}^k = 1 + (-1)^n \gamma = 0.$$

Thus the degree of a fixed-point-free self-mapping of the n-sphere is $\gamma = (-1)^{n+1}$. This result can also be seen in a simple way. If P' is the image of a point P under a fixed point-free mapping g_0 of the unit n-sphere into itself, let P' move along the great circle determined by P and P' to the point diametrically opposite P. This is a homotopic deformation of the mapping g_0 to the mapping g_1 which interchanges diametrically opposite points. The mappings g_0 and g_1 have the same degree (§78). But the degree of the mapping g_1 is $(-1)^{n+1}$ since it is possible to regard the interchange of diametrically opposite points as a product of $n + 1$ mirror reflections each of which has degree -1. One can easily check this result for dimensions $n = 1, 2, 3$.

EXAMPLE 3. *Fixed-point-free deformations.* If the continuous mapping of a complex into itself is a deformation, then the homomorphic mapping of the Betti groups is the identity mapping (§31, Theorem IV) and therefore

$$\operatorname{Sp} \mathbf{B}^k = p^k.$$

Consequently, from formula (12) of §23,

$$\sum_{k=0}^{n} (-1)^k \operatorname{Sp} \mathbf{B}^k = \sum_{k=0}^{n} (-1)^k p^k = \sum_{k=0}^{n} (-1)^k \alpha^k = -N.$$

Thus, according to the fixed point formula, *the vanishing of the Euler characteristic is a necessary condition for the existence of fixed-point-free deformations.*[51]

In the particular case that the complex $\Re^n$ is an n-dimensional manifold $\mathfrak{M}^n$, this condition is always satisfied for odd n (§69, Theorem IV). For even n, on the other hand, it restricts the classes of manifolds which can admit fixed-point-free deformations. As an example, for closed surfaces ($n = 2$) we have

$$N = 2(h - 1) \qquad \text{or} \qquad N = k - 2, \qquad \text{respectively,}$$

where h and k are the numbers of handles or cross-caps, respectively, of the closed surface. The Euler characteristic will vanish only in the cases $h = 1$ and $k = 2$, that is, for the torus and for the nonorientable ring surface (Klein's bottle). Only these two closed surfaces can possibly allow deformations having no fixed points. Such deformations are easily found.

EXAMPLE 4. *The closed n-ball with holes.* Given a particular simplicial decomposition of the closed n-ball $\mathfrak{B}^n$, let us remove the interior points of l

n-simplexes $E_1^n, E_2^n, \ldots, E_l^n$ which are mutually disjoint from one another and are all disjoint with the boundary of $\mathfrak{B}^n$. Let $\overline{\mathfrak{B}}^n$ be the complex which remains. Then $p^0 = 1$ for this complex. For $k > 0$, a closed k-chain U^k of the simplicial decomposition will be null homologous in $\mathfrak{B}^n$ after again filling in the bored-out simplexes (§19). Thus U^k is the boundary of a simplicial chain U^{k+1}:

$$U^k = \Re \partial U^{k+1}. \tag{1}$$

When $k < n - 1$, the bored-out n-simplexes will not appear in U^{k+1}; thus U^{k+1} already lies in $\overline{\mathfrak{B}}^n$ and U^k is already null homologous in $\overline{\mathfrak{B}}^n$. Consequently, $\overline{\mathfrak{B}}^n$ has the Betti numbers

$$p^0 = 1, \qquad p^1 = p^2 = \cdots = p^{n-2} = 0. \tag{2}$$

If $k = n - 1$, on the other hand, then because of (1) the given chain U^{n-1} will be the boundary of a simplicial chain

$$U^n = \alpha_1 E_1^n + \cdots + \alpha_l E_l^n + {}'U^n \qquad ({}'U^n \text{ on } \overline{\mathfrak{B}}^n),$$

thus

$$U^{n-1} = \Re \partial U^n = \alpha_1 \Re \partial E_1^n + \cdots + \Re \partial E_l^n + \Re \partial \, {}'U^n$$

or

$$U^{n-1} \sim \alpha_1 \Re \partial E_1^n + \cdots + \alpha_l \Re \partial E_l^n \qquad (\text{on } \overline{\mathfrak{B}}^n).$$

That is, each $(n-1)$-chain of $\overline{\mathfrak{B}}^n$ is homologous to a linear combination of the l boundary chains

$$\Re \partial E_1^n, \ldots, \Re \partial E_l^n. \tag{3}$$

These boundary chains are homologously independent. For if a homology were to hold among them,

$$\beta_1 \Re \partial E_1^n + \cdots + \beta_l \Re \partial E_l^n \sim 0,$$

then there would exist an n-chain on $\mathfrak{B}^n$, $W^n \neq 0$, having the boundary

$$\Re \partial W^n = \beta_1 \Re \partial E_1^n + \cdots + \beta_l \Re \partial E_l^n.$$

After filling in the l cavities we would then have $W^n - \beta_1 E_1^n - \cdots - \beta_l E_l^n$ as a closed n-chain on $\mathfrak{B}^n$. But the only such chain is the chain 0. Since W^n does not contain any simplex E_i^n it follows that $\beta_1 = \cdots = \beta_l = 0$. Thus the l chains (3) form a Betti basis for the dimension $n - 1$ in $\overline{\mathfrak{B}}^n$, and $p^{n-1} = l$.

Let us assume that the external boundary of $\overline{\mathfrak{B}}^n$ is transformed to itself or, more generally, is transformed to a homologous $(n-1)$-chain; for example, it may shrink to a point. If, in addition, a of the l boundary spheres (3) transform individually to themselves or to homologous chains, while $b = l - a$ undergo a permutation, then there will appear a ones and b zeros in the principal diagonal of the transformation matrix $\mathbf{B}^{n-1}$. Thus

$$\mathrm{Sp}\, \mathbf{B}^{n-1} = a$$

and because of (2)

$$\mathrm{Sp}\,\mathbf{B}^0 = 1, \qquad \mathrm{Sp}\,\mathbf{B}^1 = \cdots = \mathrm{Sp}\,\mathbf{B}^{n-2} = 0.$$

Consequently, the fixed point formula (F) provides the following necessary condition for the mapping to have no fixed points:

$$1 - 0 + \cdots + (-1)^{n-1}a = 0 \qquad \text{or} \qquad a = (-1)^n.$$

Since a is an integer, there can exist no fixed-point-free mapping of $\overline{\mathfrak{B}}^n$ when n is odd. For even n, a fixed-point-free mapping can exist only when $a = 1$. An annulus, for example, will admit a fixed-point-free rotation about its center point. A spherical shell, on the other hand, will admit no fixed-point-free mapping into itself such that its two boundary spheres transform to homologous surfaces.

Problems

1. If, in a single-valued continuous mapping of the n-sphere into itself, at least one point of the n-sphere does not lie in the set of image points, then the mapping has a fixed point.

2. Show that a continuous mapping of projective n-space $\mathfrak{P}^n$ into itself will always possess a fixed point when n is even. If n is odd, the degree of the mapping must be 1 if a fixed-point-free mapping is to exist.

3. Show that a continuous mapping of the 2-sphere into itself will either possess a fixed point or a "diametrically opposite point," or both.

4. Prove that no continuous vector field can exist on the 2-sphere.

CHAPTER TWELVE

AUXILIARY THEOREMS FROM THE THEORY OF GROUPS

82. Generators and Relations

Topology is intimately associated with the theory of groups. In this chapter we shall present brief derivations and summarize those group theoretic theorems which were used in the course of our topological investigations. In contrast to the groups normally encountered in algebra or geometry, the groups which play a role in topology are usually given in terms of generators and relations. We shall therefore devote our initial discussion to this manner of determining a group.

Let $\mathfrak{F}$ be a finite or an infinite group* and let

$$A_1, A_2, \ldots, A_a \tag{1}$$

be a collection of (not necessarily distinct) elements of $\mathfrak{F}$. We call these elements *generators* of $\mathfrak{F}$ if each element of $\mathfrak{F}$ can be written as a product of elements $A_1, A_2, \ldots, A_a$ and their reciprocals

$$A_1^{-1}, A_2^{-1}, \ldots, A_a^{-1}. \tag{2}$$

Such a product is called a "*word*." As an example, $A_1 A_1^{-1}$ or $A_2^{-4} A_6 A_3^5$ is a word. On purely formal grounds, we also introduce the empty word, in which no generators appear. Each word will then represent a group element, but different words can represent the same group element. In particular, the empty word is the unit element of the group, which we also denote by 1. As an abbreviation, we shall also denote a word by $W(A_i)$ and we shall set

$$W_1(A_i) \equiv W_2(A_i)$$

(read: is identical to) if both words coincide element-for-element. On the

---

*Elementary examples and definitions can be found, for example, in van der Waerden [3, Chapter II], or in Reidemeister [7].

other hand, if $W_1(A_i)$ and $W_2(A_i)$ represent the same group element, without necessarily coinciding element-for-element, then we say that they are *equal*:

$$W_1(A_i) = W_2(A_i).$$

The word which arises when we reverse the order of all elements of a word $W(A_i)$ and simultaneously reverse the sign of all of the exponents appearing in $W(A_i)$ is called the reciprocal word $W^{-1}(A_i)$. Clearly, it will represent the group element which is reciprocal to the group element represented by $W(A_i)$.

If two words $W_1(A_i)$ and $W_2(A_i)$ are equal, then we say that the equation

$$W_1(A_i) = W_2(A_i)$$

is a *relation* of the group $\mathfrak{F}$ which holds among the generators $A_1, A_2, \ldots, A_a$. We usually write the equation so that the unit element of the group stands on the right-hand side:

$$W_1(A_i)W_2^{-1}(A_i) = 1.$$

The question of finding all relations is, accordingly, equivalent to the question of finding all representations of the unit element of the group. The *trivial relations*

$$A_iA_i^{-1} = 1 \qquad \text{and} \qquad A_i^{-1}A_i = 1 \tag{3}$$

always appear among the relations of the group.

If $R(A_i) = 1$ is a relation of $\mathfrak{F}$ and $W(A_i)$ is any word of $\mathfrak{F}$, then we can obtain a new word from $W(A_i)$ by "*application of the relation*" $R(A_i) = 1$ to $W(A_i)$. Application of the relation $R(A_i) = 1$ consists of either striking out R^{+1} when it appears in W, that is, by transforming $W \equiv W_1 R^{\pm 1} W_2$ to the new word $W_1 W_2$, or by inserting $R^{\pm 1}$ into W. The words obtained in these ways represent the same group element as the original word because $R^{\pm 1} = 1$.

Now let

$$R_1(A_i) = 1, \qquad R_2(A_i) = 1, \qquad \ldots \qquad R_r(A_i) = 1 \tag{4}$$

$r \geqq 0$, be finitely many relations of $\mathfrak{F}$. If it is possible to transform a word $W(A_i)$ to the empty word by means of a finite number of applications of the relations (4) and the trivial relations (3), then

$$W(A_i) = 1$$

is a relation of $\mathfrak{F}$ and is said to be a *consequence* of the relations (4). If each and every relation of $\mathfrak{F}$ is a consequence of (4), then (4) is said to be a *system of defining relations to the group* $\mathfrak{F}$. The system is characterized by the fact that one can transform an arbitrary representation of the group's unit element to the empty word by application of the relations (4) and the trivial relations (3). One can also transform any word to any other word equal to it by applying these relations.

We shall not require that the relations of a system of defining relations be independent. That is, we allow a relation to be a consequence of the other relations.

A system of defining relations (4) taken together with the generators (1) defines the group $\mathfrak{F}$ completely. For all other relations are determined by (4) and, as a result, so are all representations of the group's unit element. But this establishes whether two given words will represent the same group element. However, the question of deciding whether two words are equal by means of a computational procedure is, in general, an unsolved problem of group theory (the word problem).*

The restriction which we have made here, to finitely many generators and defining relations, is not justified by group theoretic considerations. There are in fact groups which can only be represented by infinitely many generators; for example, the rational numbers with the exception of zero form such a group, when algebraic multiplication is taken as the group multiplication, because there are infinitely many prime numbers. In the theory of finite complexes, which stands at the center of our investigations, we need only to deal with groups having finitely many generators and defining relations. We restrict our treatment to such groups for the sake of simplicity.

EXAMPLE 1. The integers, with algebraic addition as group multiplication, form an infinite group. We can choose the integer $+1$ as the generator A. The unit element of the group is the integer 0. The element A^k is the integer $+k$. A word $A^{\epsilon_1}A^{\epsilon_2}\cdots$, where each epsilon is equal to ± 1, can be transformed to A^k only with the help of the trivial relations. The word is equal to the unit element of the group only if $k = 0$. Each relation of the group is therefore a consequence of the trivial relations, and a system of defining relations is the empty system. Such a group, which has a single generator and no defining relations, is called a *free cyclic group*.

EXAMPLE 2. The residue classes of the integers mod g, with "elementwise" addition as group multiplication, form a group of order g. A generating element is the residue class A containing the integer 1. A system of defining relations is given by the single relation $A^g = 1$. In this relation 1 denotes the unit element of the group, which is the residue class containing the integer 0. Such a group is called a *cyclic group of order g*.

When we are given a group in terms of a system of generators and defining relations, we can derive other generators and defining relations by means of the following procedures:

1. Inclusion or removal of a consequence relation. If

$$R_1(A_i) = 1, \qquad R_2(A_i) = 1, \qquad \ldots, \qquad R_r(A_i) = 1 \qquad (5)$$

is a system of defining relations of $\mathfrak{F}$ and $R_{r+1}(A_i) = 1$ is a consequence of (5), then

$$R_1(A_i) = 1, \qquad R_2(A_i) = 1, \qquad \ldots, \qquad R_r(A_i) = 1, \qquad R_{r+1}(A_i) = 1 \qquad (6)$$

is also a system of defining relations. The converse holds also because the totality of consequence relations of the two systems (5) and (6) is the same.

2. Introduction or elimination of a generator. If

$$A_1, A_2, \ldots, A_a \qquad (7)$$

---

*Editor's note: There is no universal algorithmic procedure for solving the word problem in an arbitrary group defined by generators and relations (Rabin [1]).

are the generators of $\mathfrak{F}$ and (5) are its defining relations, then one can introduce an arbitrary product $W(A_i) = A_{a+1}$ as a new generator. Given the generators

$$A_1, A_2, \ldots, A_{a+1}, \tag{8}$$

the relations

$$R_1(A_i) = 1, \qquad R_2(A_i) = 1, \qquad \ldots, \qquad R_r(A_i) = 1, \qquad A_{a+1}^{-1} W(A_i) = 1 \tag{9}$$

form a system of defining relations. For if

$$R(A_1, A_2, \ldots, A_{a+1}) = 1$$

is an arbitrary relation in $\mathfrak{F}$, then one obtains the relation

$$R(A_1, A_2, \ldots, W(A_i)) = 1 \tag{10}$$

by application of the relation $A_{a+1} = W(A_i)$, where the new element A_{a+1} no longer appears in (10). The relation (10) is a consequence of (5) because (5) is a system of defining relations of $\mathfrak{F}$. The left side of (10) can also be transformed to the empty word by application of the relations (5) and the trivial relations.

Conversely, if we first assume that (9) is a system of defining relations among the generators (8), it follows that (7) are also generators of the group and (5) is a system of defining relations among them. For each product of the generators (8) can be expressed in terms of the generators (7) because $A_{a+1} = W(A_i)$. By a previous assumption, a relation $R(A_1, A_2, \ldots, A_a) = 1$ is a consequence of the relations (9). There thus exists a finite sequence of words

$$R(A_1, A_2, \ldots, A_a)$$
$$R'(A_1, A_2, \ldots, A_a, A_{a+1})$$
$$R''(A_1, A_2, \ldots, A_a, A_{a+1})$$
$$\vdots$$

such that each word comes from the preceding word by application of relation (9) or a trivial relation $A_i^\varepsilon A_i^{-\varepsilon} = 1$ ($\varepsilon = \pm 1$; $i = 1, 2, \ldots, a + 1$) and the last word of the sequence is the empty word. By substituting the word $W(A_i)$ for A_{a+1} everywhere, we form the sequence of words

$$R(A_1, A_2, \ldots, A_a)$$
$$R'(A_1, A_2, \ldots, A_a, W(A_i))$$
$$R''(A_1, A_2, \ldots, A_a, W(A_i))$$
$$\vdots$$

If we can show that each of these words can be transformed to the following word just by application of the relations (5) and the trivial relations

$A_i^\varepsilon A_i^{-\varepsilon} = 1$ ($i = 1, \ldots, a$), then we are finished. If the word

$$R^{(k+1)}(A_1, A_2, \ldots, A_a, A_{a+1})$$

follows from

$$R^{(k)}(A_1, A_2, \ldots, A_a, A_{a+1})$$

by application of one of the first r relations (9) or one of the trivial relations $A_i^\varepsilon A_i^{-\varepsilon} = 1$ ($i = 1, 2, \ldots, a$), then the word

$$R^{(k+1)}(A_1, A_2, \ldots, A_a, W(A_i))$$

will also follow from

$$R^{(k)}(A_1, A_2, \ldots, A_a, W(A_i)),$$

in a like manner, by application of the same relations. But if

$$R^{(k+1)}(A_1, A_2, \ldots, A_{a+1})$$

follows from

$$R^{(k)}(A_1, A_2, \ldots, A_{a+1})$$

by application of the relation $A_{a+1}^{-1} W(A_i) = 1$ or a trivial relation $A_{a+1}^\varepsilon A_{a+1}^{-\varepsilon} = 1$, then

$$R^{(k+1)}(A_1, A_2, \ldots, A_a, W(A_i))$$

follows from

$$R^{(k)}(A_1, A_2, \ldots, A_a, W(A_i))$$

by application of the relation $W^\varepsilon(A_i) W^{-\varepsilon}(A_i) = 1$, and is reduced using only trivial relations.

It is possible to prove that one can obtain any presentation of a group $\mathfrak{F}$ in terms of generators and defining relations, starting from any other such presentation, by means of a finite number of applications of the elementary operations which were just described. The proof is carried out by showing that, starting with two presentations of the group $\mathfrak{F}$, one may obtain a third presentation in which the generators of both the first and the second presentation appear, and that the third presentation may be obtained from either of the original presentations by introducing relations which are consequences of the given relations.

83. Homomorphic Mappings and Factor Groups

If, to each element F of a group $\mathfrak{F}$, an element $\overline{F}$ of a second group $\overline{\mathfrak{F}}$ has been assigned as the image element of a mapping and if, furthermore, each image $\overline{F}_3$ of the product of two elements $F_1 F_2 = F_3$ is equal to the product $\overline{F}_1 \overline{F}_2$ of their images, then we say that a *homomorphic mapping of $\mathfrak{F}$ into $\overline{\mathfrak{F}}$* has been given. If each element of $\mathfrak{F}$ is the image of an element of $\mathfrak{F}$, then we speak of a *homomorphic mapping of $\mathfrak{F}$ onto $\overline{\mathfrak{F}}$*. If the mapping of $\mathfrak{F}$ onto $\overline{\mathfrak{F}}$ is a one-to-one correspondence, then $\mathfrak{F}$ and $\overline{\mathfrak{F}}$ are said to be 1-isomorphic, or just *isomorphic*.

The following fundamental theorem holds with regard to homomorphic mappings of two groups. The theorem relates each homomorphic mapping of $\mathfrak{F}$ onto $\overline{\mathfrak{F}}$ to a factor group of $\mathfrak{F}$.

HOMOMORPHISM THEOREM. *If a group $\mathfrak{F}$ is mapped homomorphically onto a group $\overline{\mathfrak{F}}$, then $\overline{\mathfrak{F}}$ is isomorphic with the factor group $\mathfrak{F}/\mathfrak{N}$, where $\mathfrak{N}$ is the normal subgroup of $\mathfrak{F}$ whose elements are mapped to the unit element of $\overline{\mathfrak{F}}$. Conversely, $\mathfrak{F}$ maps homomorphically onto each factor group $\mathfrak{F}/\mathfrak{N}$ (where $\mathfrak{N}$ is a normal subgroup of $\mathfrak{F}$).*

The proof can be found, e.g., in van der Waerden [3] p. 35.

If $\mathfrak{F}$ is presented in terms of generators and defining relations [§82, (1) and (4)], it is valid to present a factor group $\overline{\mathfrak{F}} = \mathfrak{F}/\mathfrak{N}$ in terms of generators and relations. We shall assume that there exist finitely many elements $S_1(A_i)$, $S_2(A_i), \ldots, S_s(A_i)$ of $\mathfrak{N}$ such that $\mathfrak{N}$ is the smallest normal subgroup which contains these elements. (This assumption will always be satisfied if $\mathfrak{N}$ possesses finitely many generators $S_1(A_i), S_2(A_i), \ldots, S_s(A_i)$). Each element of $\mathfrak{N}$ will then be a product of finitely many elements of the form

$$F(A_i)S_\lambda^\varepsilon(A_i)F^{-1}(A_i) \qquad (\varepsilon = \pm 1). \tag{1}$$

Each such product will be an element of $\mathfrak{N}$ because $\mathfrak{N}$ is a normal subgroup; on the other hand, these products already form a normal subgroup thus they form the whole of $\mathfrak{N}$ since $\mathfrak{N}$ is to be the smallest normal subgroup containing each element $S_\lambda(A_i)$.

Let us denote the residue class in $\mathfrak{F}/\mathfrak{N}$ in which an element of $\mathfrak{F}$ lies by a bar written above the element. Then

$$\overline{A}_1, \overline{A}_2, \ldots, \overline{A}_a \tag{2}$$

will obviously form a system of generators of $\overline{\mathfrak{F}}$, and the relations

$$R_1(\overline{A}_i) = \overline{1}, \quad \ldots, \quad R_r(\overline{A}_i) = \overline{1}$$
$$S_1(\overline{A}_i) = \overline{1}, \quad \ldots, \quad S_s(\overline{A}_i) = \overline{1} \tag{3}$$

will hold. Here, the residue class 1, in which the unit element of $\mathfrak{F}$ lies, is the normal subgroup $\mathfrak{N}$.

We claim that (3) is a system of defining relations of $\overline{\mathfrak{F}}$, so that each relation $R(\overline{A}_i) = \overline{1}$ of $\overline{\mathfrak{F}}$ is a consequence of (3). The equation $R(\overline{A}_i) = \overline{1}$ denotes that $R(A_i)$ belongs to $\mathfrak{N}$. Now each element of $\mathfrak{N}$ is equal to a product of elements of the form (1). We can thus transform $R(A_i)$ to a product of elements of the form (1), with the help of the relations $R_1(A_i) = 1, \ldots, R_r(A_i) = 1$ and trivial relations. Likewise, by application of the relations $R_1(\overline{A}_i) = \overline{1}, \ldots, R_r(\overline{A}_i) = \overline{1}$ and trivial relations, we can transform $R(\overline{A}_i)$ to a product of transforms of $S_1^{\pm 1}(\overline{A}_i), \ldots, S_s^{\pm 1}(\overline{A}_i)$ and then transform this product, by again applying relations $S_1(\overline{A}_i) = \overline{1}, \ldots, S_s(\overline{A}_i) = \overline{1}$ and trivial relations, to the empty product.

We thus obtain the defining relations of the factor group by including certain additional relations with those of the group. Our result is stated in the following theorem:

THEOREM I. *If the group $\mathfrak{F}$ has the generators*

$$A_1, A_2, \ldots, A_a$$

and the relations

$$R_1(A_i) = 1, \qquad R_2(A_i) = 1, \qquad \ldots \qquad R_r(A_i) = 1,$$

and if $\mathfrak{N}$ is the smallest normal subgroup which contains the particular elements

$$S_1(A_i), S_2(A_i), \ldots, S_s(A_i),$$

then the factor group $\mathfrak{F}/\mathfrak{N} = \overline{\mathfrak{F}}$ has the generators

$$\overline{A}_1, \overline{A}_2, \ldots, \overline{A}_a$$

and the defining relations

$$R_1(\overline{A}_i) = \overline{1}, \qquad R_2(\overline{A}_i) = \overline{1}, \qquad \ldots, \qquad R_r(\overline{A}_i) = \overline{1},$$
$$S_1(\overline{A}_i) = \overline{1}, \qquad S_2(\overline{A}_i) = \overline{1}, \qquad \ldots, \qquad S_s(\overline{A}_i) = \overline{1};$$

where $\overline{A}_i$ denotes the residue class in which A_i lies.

In deriving Theorem I we started with a particular normal subgroup, $\mathfrak{N}$ and we tried to find the relations of $\mathfrak{F}/\mathfrak{N}$. But one can also start with the relations of $\mathfrak{F}$ and include arbitrary additional relations. In this way we can always obtain a factor group $\mathfrak{F}/\mathfrak{N}$ which is the smallest normal subgroup containing the left-hand sides of the additional relations.

We shall apply the homomorphism theorem to give a result which is of use in §20 Let three groups

$$\mathfrak{F}, \quad \mathfrak{D}, \quad \mathfrak{N} \tag{4}$$

be given such that $\mathfrak{D}$ and $\mathfrak{N}$ are normal subgroups of $\mathfrak{F}$ and let $\mathfrak{N}$ be contained in $\mathfrak{D}$. The subgroup $\mathfrak{N}$ determines a homomorphic mapping of $\mathfrak{F}$ onto the factor group $\mathfrak{F}/\mathfrak{N} = \overline{\mathfrak{F}}$. Under this mapping the groups (4) transform, respectively, to the groups

$$\overline{\mathfrak{F}} = \mathfrak{F}/\mathfrak{N}, \qquad \overline{\mathfrak{D}} = \mathfrak{D}/\mathfrak{N}, \qquad \overline{\mathfrak{N}} = \mathfrak{N}/\mathfrak{N} = \overline{1}$$

Since $\mathfrak{D}$ is a normal subgroup of $\mathfrak{F}$, then $\overline{\mathfrak{D}}$ is a normal subgroup of $\overline{\mathfrak{F}}$ and there will exist a homomorphic mapping of $\overline{\mathfrak{F}}$ onto $\overline{\mathfrak{F}}/\overline{\mathfrak{D}}$. The result of carrying out the two homomorphic mappings in succession is a homomorphic mapping of $\mathfrak{F}$ onto $\overline{\mathfrak{F}}/\overline{\mathfrak{D}}$ such that the elements which transform to the unit element of $\overline{\overline{\mathfrak{F}}}/\overline{\mathfrak{D}}$ are just the elements of $\mathfrak{D}$. Consequently, $\mathfrak{F}/\mathfrak{D}$ *and* $\overline{\overline{\mathfrak{F}}}/\overline{\mathfrak{D}} = (\mathfrak{F}/\mathfrak{N})/(\mathfrak{D}/\mathfrak{N})$ *are isomorphic.*

84. Abelianization of Groups

We shall apply Theorem I of the previous section to an important special case. Let $\mathfrak{F}$ be a group having the generators

$$A_1, A_2, \ldots, A_a \tag{1}$$

and the defining relations

$$(\mathfrak{F}) \qquad R_1(A_i) = 1, \qquad R_2(A_i) = 1, \qquad \ldots, \qquad R_r(A_i) = 1 \tag{2}$$

and let $\mathfrak{N}$ be the smallest normal subgroup which contains all of the commutators

$$A_i A_k A_i^{-1} A_k^{-1} \qquad (i, k = 1, 2, \ldots, a).$$

From Theorem I the factor group $\mathfrak{F}/\mathfrak{N} = \overline{\mathfrak{F}}$ has the generators

$$\overline{A}_1, \overline{A}_2, \ldots, \overline{A}_a \tag{$\overline{1}$}$$

and the defining relations

$$(\overline{\mathfrak{F}}) \qquad R_1(\overline{A}_i) = \overline{1}, \qquad R_2(\overline{A}_i) = \overline{1}, \qquad \ldots, \qquad R_r(\overline{A}_i) = \overline{1}, \tag{$\overline{2}$}$$

$$\overline{A}_i \overline{A}_k \overline{A}_i^{-1} \overline{A}_k^{-1} = \overline{1} \qquad (i, k, = 1, 2, \ldots, a). \tag{$\overline{3}$}$$

Here $\overline{A}_i$ denotes the residue class of the element A_i of $\mathfrak{F}$ with respect to the decomposition of $\mathfrak{N}$. For (3) we can also write

$$\overline{A}_i \overline{A}_k = \overline{A}_k \overline{A}_i.$$

The generating elements of $\overline{\mathfrak{F}}$ commute with one another. Consequently, all elements of $\overline{\mathfrak{F}}$ commute with one another; $\overline{\mathfrak{F}}$ is therefore Abelian and is called the *Abelianized group* $\mathfrak{F}$; it is obtained from $\mathfrak{F}$ by adding on the commutation relations $(\overline{3})$.

We must now show that $\overline{\mathfrak{F}}$ is determined by $\mathfrak{F}$ and does not depend upon the particular choice of generators and defining relations of $\mathfrak{F}$. We accomplish this by providing a second definition of $\overline{\mathfrak{F}}$ which clearly demonstrates its independence of the particular presentation of $\mathfrak{F}$. *The Abelianized group* $\mathfrak{F}$ *is the factor group of* $\mathfrak{F}$ *with respect to the commutator group* $\mathfrak{R}$. By the *commutator group* of a group $\mathfrak{F}$ we mean the normal subgroup which is generated by all of the commutators $F_\rho F_\sigma F_\rho^{-1} F_\sigma^{-1}$, where F_ρ and F_σ are arbitrary elements of $\mathfrak{F}$. We must show that $\mathfrak{N} = \mathfrak{R}$. From Theorem I of §83, $\mathfrak{N}$ is the smallest normal subgroup which contains the commutators $A_i A_k A_i^{-1} A_k^{-1}$, that is, particular commutators of $\mathfrak{F}$. Thus $\mathfrak{N}$ is contained in $\mathfrak{R}$. But $\mathfrak{R}$ is also contained in $\mathfrak{N}$, for an arbitrary $F_\rho F_\sigma F_\rho^{-1} F_\sigma^{-1}$ is transformed by the mapping $\mathfrak{F} \to \overline{\mathfrak{F}}$ to the unit element 1 since $\overline{\mathfrak{F}}$ is Abelian. Thus $\mathfrak{N}$ contains all commutators of $\mathfrak{F}$ and consequently contains $\mathfrak{R}$ as a subgroup. Thus $\mathfrak{N} = \mathfrak{R}$.

85. Free and Direct Products

A group $\mathfrak{F}$ is said to be the *free product* of the groups $\mathfrak{F}_1, \mathfrak{F}_2, \ldots, \mathfrak{F}_h$, expressed as a formula

$$\mathfrak{F} = \mathfrak{F}_1 \circ \mathfrak{F}_2 \circ \cdots \circ \mathfrak{F}_h, \tag{1}$$

if each element of $\mathfrak{F}$ other than the unit element can be represented uniquely as a product

$$F_{i\iota} F_{k\kappa} \cdots F_{z\zeta}. \tag{2}$$

Here $F_{i\iota}, F_{k\kappa}, \ldots$ are elements other than the unit element from the respective groups $\mathfrak{F}_i, \mathfrak{F}_k, \ldots$ and each two consecutive elements belong to different groups, that is, $i \neq k$ and so forth. The structure of the free product $\mathfrak{F}$ is determined by the structure of the groups $\mathfrak{F}_1, \mathfrak{F}_2, \ldots, \mathfrak{F}_h$. Two elements given in the form (2)

$$F_{i'\iota'} F_{k'\kappa'} \cdots F_{z'\zeta'} \quad \text{and} \quad F_{i''\iota''} F_{k''\kappa''} \cdots F_{z''\zeta''}$$

are multiplied by writing them alongside of one another. The element arising in this way

$$F_{i'\iota'} F_{k'\kappa'} \cdots F_{z'\zeta'} F_{i''\iota''} F_{k''\kappa''} \cdots F_{z''\zeta''} \tag{3}$$

is already in the form (2) if $z' \neq i''$. But if $F_{z'\zeta'}$ and $F_{i''\iota''}$ belong to the same group, then $F_{z'\zeta'} F_{i''\iota''}$ is a certain element $F_{z'}$ of the group $\mathfrak{F}_{z'}$. If $F_{z'} \neq 1$, then the product has the normal form (2). If $F_{z'} = 1$, we can simply strike out $F_{z'}$ and once again apply this procedure. By proceeding in this manner we eventually either obtain a unique normal form (2) or the unit element, for the product.

We shall not prove the theorem here that a free product of h arbitrary groups $\mathfrak{F}_1, \mathfrak{F}_2, \ldots, \mathfrak{F}_h$ exists. The existence proof can be found in Klein [1] p. 361.

If we are given the groups $\mathfrak{F}_1, \mathfrak{F}_2, \ldots, \mathfrak{F}_h$ in terms of their generators and defining relations, then we can derive the generators and defining relations of the free product from those of the individual groups. We shall demonstrate this for the case of two groups, $\mathfrak{F}_1$ and $\mathfrak{F}_2$, having the generators and defining relations

$$\left. \begin{array}{llll} A_1, A_2, \ldots, A_a \\ R_1(A_i) = 1, & R_2(A_i) = 1, & \ldots, & R_r(A_i) = 1 \end{array} \right\} \tag{$\mathfrak{F}_1$}$$

$$\left. \begin{array}{llll} B_1, B_2, \ldots, B_b \\ S_1(B_k) = 1, & S_2(B_k) = 1, & \ldots, & S_s(B_k) = 1 \end{array} \right\}. \tag{$\mathfrak{F}_2$}$$

The elements

$$A_1, A_2, \ldots, A_a, B_1, B_2, \ldots, B_b$$

obviously generate the free product $\mathfrak{F}_1 \circ \mathfrak{F}_2$ and the relations

$$R_1(A_i) = 1, \qquad \ldots, \qquad R_r(A_i) = 1,$$
$$S_1(B_k) = 1, \qquad \ldots, \qquad S_s(B_k) = 1 \qquad\qquad (\mathfrak{F})$$

are satisfied in the free product. These relations are defining relations of the free product, that is, each relation which is valid in the free product is a consequence of these relations. For given any product formed from the A's and B's which is equal to the unit element of $\mathfrak{F}$, one may decompose this product into maximally long subproducts each consisting of A's and B's alone in such a way that subproducts containing A's and B's alternate in the entire product. From (2), each of these subproducts must individually be equal to 1. They may therefore be transformed to the empty word by application of the relations $(\mathfrak{F})$. Each subproduct can in fact be transformed by using only the first r relations or only the last s relations, together with the trivial relations. The entire product can be transformed to the empty word in this way.

The proof is clearly also true for an arbitrary number of groups. As a result, we have

THEOREM I. *One obtains a system of generators and defining relations of the free product of a collection of group as the union of the generators and the union of the relations of the free factors, respectively.*

As an example let us consider the free product of h free cyclic groups $\mathfrak{F}_1, \mathfrak{F}_2, \ldots, \mathfrak{F}_h$. Each group $\mathfrak{F}_i$ has one generator A_i and no relations. The free product $\mathfrak{F}$ then has h generators and no relations. We call $\mathfrak{F}$ the *free group having h generators* $A_1, A_2, \ldots, A_h$.

It also follows that one can present an arbitrary number of generators and an arbitrary number of defining relations of arbitrary form, and there will always exist a group $\mathfrak{F}$ which is generated by these generators and has these relations as its defining relations. To construct $\mathfrak{F}$ we first construct the free group having the given generators. From Theorem I of §83, by including the relations we will obtain a factor group having the desired properties.

Problem

Prove that the free product of two finite groups has finite order if one factor consists of the unit element alone.

A group $\mathfrak{F}$ is said to be the *direct product* of the groups $\mathfrak{F}_1, \mathfrak{F}_2, \ldots, \mathfrak{F}_h$, expressed as a formula

$$\mathfrak{F} = \mathfrak{F}_1 \times \mathfrak{F}_2 \times \cdots \times \mathfrak{F}_h, \qquad\qquad (4)$$

if each element of $\mathfrak{F}$ can be represented uniquely as a product

$$F_{l_\iota} F_{2\kappa} \cdots F_{h\xi} \tag{5}$$

and each element of $\mathfrak{F}_\nu$ commutes with each element of $\mathfrak{F}_\mu$ if $\nu \neq \mu$. The structure of the direct product group is uniquely determined by the structures of the group factors; the equation

$$F_{l_{\iota'}} F_{2\kappa'} \cdots F_{h\xi'} \cdot F_{l_{\iota''}} F_{2\kappa''} \cdots F_{h\xi''} = (F_{l_{\iota'}} F_{l_{\iota''}})(F_{2\kappa'} F_{2\kappa''}) \cdots (F_{h\xi'} F_{h\xi''}) \tag{6}$$

reduces multiplication in the direct product to multiplications in the group factors.

The existence of the direct product of arbitrarily given factors is proved by construction. One introduces the symbolic product, that is, the sequence of letters (5), as elements and one defines the multiplication of elements by the equation (6).

If the groups $\mathfrak{F}_1, \mathfrak{F}_2, \ldots, \mathfrak{F}_h$ are presented in terms of their generators and defining relations, then the generators and defining relations of the direct product can also be presented. We shall again consider the case of two groups having the generators and defining relations $(\mathfrak{F}_1)$ and $(\mathfrak{F}_2)$. The generators.

$$A_1, A_2, \ldots, A_a, B_1, B_2, \ldots, B_b$$

obviously generate the direct product $\mathfrak{F}_1 \times \mathfrak{F}_2 = \mathfrak{F}$ and the relations

$$R_1(A_i) = 1, \qquad \ldots, \qquad R_r(A_i) = 1,$$
$$S_1(B_k) = 1, \qquad \ldots, \qquad S_s(B_k) = 1,$$

together with the commutativity relations of the generators

$$A_i B_k A_i^{-1} B_k^{-1} = 1 \qquad (i = 1, 2, \ldots, a, k = 1, 2, \ldots, b),$$

are clearly relations in $\mathfrak{F}$. They form a system of defining relations. For with the help of the commutativity relations any product of the A's and B's can be brought to the form

$$\prod_i A_i \cdot \prod_k B_k$$

and if this product is equal to 1, then individually $\prod_i A_i = 1$ and $\prod_k B_k = 1$ must hold. By definition, each element of the direct product must be uniquely representable in the form (5) and the unit element, in particular, can only be represented in the form $1 \circ 1$. In this case, the subproducts $\prod_i A_i$ and $\prod_k B_k$ transform individually to the empty word by use of the relations of the groups $\mathfrak{F}_1$ and $\mathfrak{F}_2$ and the trivial relations.

We then have

THEOREM II. *One obtains a system of generators and defining relations of the direct product* $\mathfrak{F} = \mathfrak{F}_1 \times \mathfrak{F}_2 \times \cdots \times \mathfrak{F}_h$ *by writing the generators and defining relations of the individual groups* $\mathfrak{F}_1, \mathfrak{F}_2, \ldots, \mathfrak{F}_h$ *together, and including the commutation relations of all pairs of groups* $\mathfrak{F}_i, \mathfrak{F}_k$ $(i \neq k)$.

Problem

Show that the order of the direct product of finite groups is equal to the product of the orders of its factors.

As an example, let us look at the direct product of h free cyclic groups $\mathfrak{F}_1, \mathfrak{F}_2, \ldots, \mathfrak{F}_h$ which are generated by $A_1, A_2, \ldots, A_h$, respectively. These A's are the generators of the direct product. The defining relations are just the commutation relations

$$A_i A_k A_i^{-1} A_k^{-1} = 1 \qquad (i, k, = 1, 2, \ldots, h).$$

Since the individual factors are Abelian groups in this case, their direct product is also Abelian. This direct product is called the *free Abelian group having h generators*. Since each element of $\mathfrak{F}_i$ can be uniquely represented in the form $A_i^{\alpha_i}$, it follows that each element of the free Abelian group having h generators can be uniquely represented in the form

$$F = A_1^{\alpha_1} A_2^{\alpha_2} \cdots A_h^{\alpha_h},$$

and the product of two elements

$$A_1^{\alpha_1'} A_2^{\alpha_2'} \cdots A_h^{\alpha_h'} \qquad \text{and} \qquad A_1^{\alpha_1''} A_2^{\alpha_2''} \cdots A_h^{\alpha_h''}$$

is the element

$$A_1^{\alpha_1' + \alpha_1''} A_2^{\alpha_2' + \alpha_2''} \cdots A_h^{\alpha_h' + \alpha_h''}.$$

If we associate an integer valued vector $(\alpha_1, \alpha_2, \ldots, \alpha_h)$ with each element F, we will have produced an isomorphism between the group $\mathfrak{F}$ and the group of all integer-valued h-dimensional vectors (with vector addition taken as the group multiplication). Each element of $\mathfrak{F}$ can therefore be represented by an h-dimensional integer-valued vector, that is, by a point in Euclidean h-space with has integer coordinates. These points form an h-dimensional point lattice. Thus at times we will use the term "h-dimensional lattice" instead of "free Abelian group with h generators." Instead of representing the element F by a vector, we can just as well represent F by the translation which this vector specifies for, as is well known, the group is isomorphic with the group of covering translations of the h-dimensional point lattice. We define the 0-dimensional lattice to be the group which consists only of the unit element.

We make one last observation! In the direct product group $\mathfrak{F} = \mathfrak{F}_1 \times \mathfrak{F}_2 \times \cdots \times \mathfrak{F}_h$, each of the groups $\mathfrak{F}_1, \mathfrak{F}_2, \ldots, \mathfrak{F}_h$ is a normal subgroup. When $\mathfrak{F}$ is decomposed with respect to $\mathfrak{F}_1$, a residue class in the decomposition will consist of the set of all elements which one obtains by holding the elements $F_{2\iota}, \ldots, F_{h\zeta}$ fixed in (5) and letting $F_{1\iota}$ run through all elements of $\mathfrak{F}_1$. The factor group $\mathfrak{F}/\mathfrak{F}_1$ is therefore isomorphic with the group $\mathfrak{F}_2 \times \cdots \times \mathfrak{F}_h$.

86. Abelian Groups

We have previously used the algebraic multiplication sign to indicate the process of group multiplication. When we deal with Abelian groups it is preferable to replace the multiplication sign with the addition sign and, correspondingly, to denote the unit element of the group by 0 instead of 1. In place of the product of two elements we then say the sum of two elements, in place of the reciprocal element the negative element, in place of the direct product the direct sum, and in place of the unit element the zero element. This renaming is done in order to make clear the close relationship of the theory of Abelian groups to the theory of integer-valued linear equations. Whenever we use *additive notation* and multiply group elements by using the + sign, it will be self-evident that the group multiplication is Abelian. We shall not expressly write out the commutation relations. In additive notation, the relation which defines a cyclic group of order p is no longer written $A^p = 1$ but, instead, as $pA = 0$. (One must be careful, in analogy with an algebraic congruence relation, not to conclude from this equation that $A = 0$. Group elements, not numbers, stand on the right-hand and left-hand sides of the equality sign.)

Let us first look at a free Abelian group $\mathfrak{F}_m$ having the generators

$$A_1, A_2, \ldots, A_m. \tag{1}$$

From §85, each element of $\mathfrak{F}_m$ can be represented uniquely in the form

$$\alpha_1 A_1 + \alpha_2 A_2 + \cdots + \alpha_m A_m. \tag{2}$$

A set of elements having the property that each element of $\mathfrak{F}_m$ can be represented uniquely as a linear combination of these elements is called a *basis* of $\mathfrak{F}_m$. The elements (1) thus form a basis of $\mathfrak{F}_m$. If

$$A'_1, A'_2, \ldots, A'_n$$

is another basis of $\mathfrak{F}_m$, then, first of all, $m = n$. If it were the case that $m < n$, for example, we could express the elements of the new basis in terms of the elements of the old basis:

$$A'_\nu = \sum_{\mu=1}^m \alpha_{\nu\mu} A_\mu \qquad (\nu = 1, 2, \ldots, n). \tag{4}$$

Since $m < n$, there would then exist a relation with rational coefficients between the rows of the matrix $[\alpha_{\nu\mu}]$, a relation with integer coefficients not all of which vanish. It would follow from this that such a relation holds between the A'_ν. This would contradict the basis property of the A'_ν. Consequently the matrix $[\alpha_{\nu\mu}]$ is square. This proves at the same time that the integer m (the dimension of the lattice) is a characteristic property of the lattice:

THEOREM I. *Two lattices are isomorphic if and only if they have the same dimension.*

Going in the opposite direction, the A_μ can be expressed uniquely in terms of the A'_μ:

$$A_\mu = \sum_{\lambda=1}^m \beta_{\mu\lambda} A'_\lambda \qquad (\mu = 1, 2, \ldots, m). \tag{5}$$

Setting (5) into (4),

$$A'_\nu = \sum_{\mu, \lambda = 1}^{m} \alpha_{\nu\mu} \beta_{\mu\lambda} A'_\lambda, \tag{6}$$

Because the A'_ν form a basis, we have

$$\sum_{\mu=1}^{m} \alpha_{\nu\mu} \beta_{\mu\lambda} = \delta_{\nu\lambda} \quad \begin{cases} = 1 & \text{for} \quad \nu = \lambda \\ = 0 & \text{for} \quad \nu \neq \lambda \end{cases} \tag{7}$$

or, in matrix notation,

$$[\alpha_{ik}][\beta_{ik}] = [\delta_{ik}].$$

This implies, for the determinants of the matrices,

$$|\alpha_{ik}||\beta_{ik}| = 1,$$

and because the matrix elements are integers,

$$|\alpha_{ik}| = \pm 1,$$

and (7) states that $[\beta_{ik}]$ is the reciprocal matrix of $[\alpha_{ik}]$. *Thus the transformation from one basis to another is unimodular and integer valued.*

THEOREM II. *A subgroup, $\mathfrak{G}$ of an m-dimensional lattice, $\mathfrak{F}_m$, is a lattice of dimension at most m.*

Proof. If $A_1, A_2, \ldots, A_m$ are the generators of $\mathfrak{F}_m$, then let $\mathfrak{F}_{m-1}$ be the sublattice generated by $A_1, A_2, \ldots, A_{m-1}$ and let $\mathfrak{H}$ be the subgroup of all elements of $\mathfrak{G}$ which belong to $\mathfrak{F}_{m-1}$. We will assume that the theorem has been proved for $\mathfrak{F}_{m-1}$. Then $\mathfrak{H}$ is a lattice of dimension at most $m - 1$. Let $B_1, B_2, \ldots, B_{k-1}$ be as basis of $\mathfrak{H}$ ($k \leq m$). Among all of the elements

$$G = g_1 A_1 + \cdots + g_m A_m$$

of $\mathfrak{G}$ we are attempting to find an element

$$B_k = g_1^* A_1 + \cdots + g_m^* + A_m$$

such that the last coefficient possesses the smallest possible positive value. (We can exclude the case when g_m is always equal to zero, since then $\mathfrak{G} = \mathfrak{H}$ and nothing more remainds to be proved.) The coefficient G_m is obviously a multiple of g_m^* and, therefore,

$$G - (g_m/g_m^*)HB_k$$

is an element of $\mathfrak{H}$ and is thus a linear combination of $B_1, B_2, \ldots, B_{k-1}$. Then $B_1, B_2, \ldots, B_k$ form a system of generators of $\mathfrak{G}$. Each element can be uniquely represented in terms of $B_1, B_2, \ldots, B_k$. Otherwise, there would exist a relation

$$x_1 B_1 + x_2 B_2 + \cdots + x_k B_k = 0$$

having coefficients which do not all vanish. In that case, we could express the

B's in terms of the A's and obtain a relation between the A's in which $x_k g_m^* \neq 0$ would appear as the coefficient of A_m. But no such relation holds between the A's. Thus $\mathfrak{G}$ is a k-dimensional lattice ($k \leqq m$) having the basis $B_1, B_2, \ldots, B_k$. Since the theorem is trivially true for for a 0-dimensional lattice, it is valid in general.

Because of Theorem II, we sometimes refer to a subgroup of a lattice as a *sublattice*.

Let $\overline{\overline{\mathfrak{F}}}$ now be an arbitrary Abelian group having the generators

$$\overline{A}_1, \overline{A}_2, \ldots, \overline{A}_m$$

and let $\mathfrak{F}_m$ be an m-dimensional lattice having the basis

$$A_1, A_2, \ldots, A_m.$$

If we assign the element

$$a_1\overline{A}_1 + a_2\overline{A}_2 + \cdots + a_m\overline{A}_m$$

of $\overline{\overline{\mathfrak{F}}}$ to each element

$$a_1A_1 + a_2A_2 + \cdots + a_mA_m$$

of $\mathfrak{F}_m$, then we obtain a homomorphic mapping of $\mathfrak{F}_m$ onto $\overline{\overline{\mathfrak{F}}}$. The elements of $\mathfrak{F}_m$ which map to the zero element of $\overline{\overline{\mathfrak{F}}}$ form a subgroup $\mathfrak{N}$ of $\mathfrak{F}_m$, that is a sublattice, and the elements of $\overline{\overline{\mathfrak{F}}}$ are in one-to-one correspondence with the residue classes in the decomposition of $\mathfrak{F}_m$ modulo $\mathfrak{N}$. The group $\overline{\overline{\mathfrak{F}}}$ is then the factor group $\mathfrak{F}_m/\mathfrak{N}$.

Let us now select any system of generators of $\mathfrak{N}$:

$$N_1, N_2, \ldots, N_n.$$

This is possible because $\mathfrak{N}$ is a lattice of dimension no larger than m. We do not require, here, that these elements form a basis of $\mathfrak{N}$. If

$$N_\nu = \sum_{\mu=1}^m a_{\nu\mu} A_\mu \qquad (\nu = 1, 2, \ldots, n),$$

then the N_ν, and therefore also $\mathfrak{N}$ and $\mathfrak{F}_m/\mathfrak{N} = \overline{\overline{\mathfrak{F}}}$, will be completely determined by the *matrix of coefficients*

$$[a_{\nu\mu}]. \tag{8}$$

However, the converse is obviously not valid, since the group $\overline{\overline{\mathfrak{F}}}$ can be described in arbitrarily many ways by means of a matrix $[a_{\nu\mu}]$. For example, we can obtain a new matrix by replacing the generators $A_1, A_2, \ldots, A_m$ and $N_1, N_2, \ldots, N_n$ of the lattices $\mathfrak{F}_m$ and $\mathfrak{N}$ by new generators $A_1', A_2', \ldots, A_m'$ and $N_1', N_2', \ldots, N_n'$ which are coupled to the former generators by unimodular integer-valued linear transformations. We shall now attempt to bring the matrix $[a_{\nu\mu}]$ to a normal form by means of such transformations. We do this step-by-step, by performing the following special unimodular transformations (elementary rearrangements) a finite number of times:

(a) Replacement of A_σ by $A_\sigma - A_\tau$ ($\sigma \neq \tau$).

In the matrix $[a_{\nu\mu}]$ this corresponds to addition of the σth column to the τth column. For

$$N_\nu = \sum a_{\nu\mu} A_\mu = \cdots + a_{\nu\sigma} A_\sigma + \cdots + a_{\nu\tau} A_\tau + \cdots$$
$$= \cdots + a_{\nu\sigma}(A_\sigma - A_\tau) + \cdots + (a_{\nu\tau} + a_{\nu\sigma})A_\tau + \cdots .$$

(b) Replacement of A_τ by $-A_\tau$.

This produces a change of sign of all elements in the τth column of $[a_{\nu\mu}]$. Corresponding replacements can be made in the generators N_1, $N_2, \ldots, N_n$:

(a') Replacement of N_κ by $N_\kappa + N_\lambda$.

This effects an addition of the λth row to the κth row.

(b') Replacement of N_λ by $-N_\lambda$.

This produces a change of sign of all elements of the λth row.

Out of these four operations, namely, row addition, column addition, change of sign of a row, and change of sign of a column, further operations can be derived, by appropriately combining them, for example, interchange of two rows or two columns. We shall prove in §87 that a coefficient matrix of rank γ can be reduced by means of such transformations to the following diagonal form:

$$n \left\downarrow \begin{array}{c} \xleftarrow{\hspace{1.2cm}} m \xrightarrow{\hspace{1.2cm}} \\ \begin{bmatrix} c_1 & & & 0 \\ & \ddots & & \\ & & c_\gamma & \\ 0 & & & 0 \end{bmatrix} \end{array} \right.$$

Each of the c's will be contained, as a factor in its predecessor. The first ρ diagonal elements $c_1, \ldots, c_\rho$ will differ from 1 and the last $\gamma - \rho$ diagonal elements will be equal to 1. We call the c's the *invariant factors* of the matrix $[a_{\nu\mu}]$.

If $A'_1, A'_2, \ldots, A'_m$ are the resulting new basis elements of $\mathfrak{F}_m$, then the elements

$$c_1 A'_1, c_2 A'_2, \ldots, c_\gamma A'_\gamma$$

and all linear combinations formed from them are just the elements of $\mathfrak{N}$. Two elements of $\mathfrak{F}_m$,

$$p_1 A'_1 + p_2 A'_2 + \cdots + p_m A'_m$$

and

$$q_1 A_1' + q_2 A_2' + \cdots + q_m A_m',$$

will belong to the same residue class of $\mathfrak{N}$ in $\mathfrak{F}_m$ if and only if

$$(p_1 - q_1)A_1' + (p_2 - q_2)A_2' + \cdots + (p_m - q_m)A_m'$$

belongs to $\mathfrak{N}$, that is,

$$p_1 \equiv q_1 \quad (\mathrm{mod}\ c_1), \qquad p_2 \equiv q_2 (\mathrm{mod}\ c_2), \qquad \ldots, \qquad p_\gamma \equiv q_\gamma \quad (\mathrm{mod}\ c_\gamma).$$

Thus each of the elements

$$\xi_1 A_1' + \xi_2 A_2' + \cdots + \xi_\gamma A_\gamma' + \eta_{\gamma+1} A_{\gamma+1}' + \cdots + \eta_m A_m' \tag{9}$$

$$(0 \leqq \xi_\mu < c_\mu; \qquad \eta_\mu \text{ arbitrary}) \tag{10}$$

will be a representative of exactly one of the residue classes. Expressed in another way: the elements of the factor group $\overline{\mathfrak{F}} = \mathfrak{F}_m / \mathfrak{N}$ (which we again denote by an overbar) can be written uniquely in the form

$$\xi_1 \overline{A}_1' + \xi_2 \overline{A}_2' + \cdots + \xi_\gamma \overline{A}_\gamma' + \eta_{\gamma+1} \overline{A}_{\gamma+1}' + \cdots + \eta_m \overline{A}_m' \tag{11}$$

if we require the normalization conditions (10).

Since the elements $\overline{A}_\mu'$ of the factor group $\overline{\mathfrak{F}}$ have the orders $c_1, c_2, \ldots, c_\gamma$, and ∞, respectively, $\overline{\mathfrak{F}}$ is the direct sum of the cyclic subgroups generated by $\overline{A}_1', \overline{A}_2', \ldots,$ and $\overline{A}_m'$.

Summarizing our results, we have

THEOREM III. *Each Abelian group $\overline{\mathfrak{F}}$ having a finite number of generators is the direct sum of ρ finite cyclic groups having orders $c_1, c_2, \ldots, c_\rho$ and p free cyclic groups. One can assume here that each c divides the preceeding c. The integers $c_1, c_2, \ldots, c_\rho$ which are different from 1 are the invariant factors of the matrix of coefficients $[a_{\nu\mu}]$ which determines $\overline{\mathfrak{F}}$. The integer p is the difference of the number of columns m and the rank γ of $[a_{\nu\mu}]$. The integers $c_1, c_2, \ldots, c_\rho$ and $p = m - \gamma$ are determined uniquely by the Abelian group, and are called the torsion coefficients and Betti number, respectively, of the group.*

This nomenclature is justified by the geometric meaning of these quantities when $\overline{\mathfrak{F}}$ is the homology group of a complex (§18 and §61).

We must still prove uniqueness. The subgroup $\overline{\mathfrak{U}}$ of order $c_1 c_2 \cdots c_\rho$ which is generated by $\overline{A}_1', \overline{A}_2', \ldots, \overline{A}_\rho'$ will obviously consist of the set of all elements of finite order of $\overline{\mathfrak{F}}$ and is therefore independent of the choice of generators. From §85 $\overline{\mathfrak{F}}/\overline{\mathfrak{U}}$ is isomorphic with the subgroup generated by $\overline{A}_{\gamma+1}', \ldots, \overline{A}_m'$ and is thus a p-dimensional lattice. Consequently, the integer p is an invariant of $\overline{\mathfrak{F}}$ and we need now to consider only the finite subgroup $\overline{\mathfrak{U}}$. When we multiply all elements of the cyclic group of order c_μ generated by $\overline{A}_\mu'$ by a positive integer x, these x-multiples form a cyclic group of order

$c_\mu/(c_\mu,x)$.* In a corresponding manner, when we multiply all elements of $\overline{\mathfrak{U}}$ by x, we obtain a subgroup of $\overline{\mathfrak{U}}$ of order

$$M(x) = \frac{c_1}{(c_1,x)} \frac{c_2}{(c_2,x)} \cdots \frac{c_\rho}{(c_\rho,x)}.$$

$M(x)$ will be equal to 1 only if every factor is equal to 1, that is, if x is a multiple of c_1. This characterizes c_1 in an invariant manner. After we have characterized c_μ ($\mu < \rho$) in an invariant manner, let us consider the integers x for which

$$M(x) = \frac{c_1}{(c_1,x)} \frac{c_2}{(c_2,x)} \cdots \frac{c_\mu}{(c_\mu,x)}.$$

This equation will hold if and only if the factors

$$\frac{c_{\mu+1}}{(c_{\mu+1},x)}, \ldots, \frac{c_\rho}{(c_\rho,x)}$$

are all equal to 1, that is, if x is a multiple of $c_{\mu+1}$. We have then characterized $c_{\mu+1}$ in an invariant manner.

We can use the invariants of an Abelian group to show the distinctness of non-Abelian groups. We can Abelianize any such group $\mathfrak{F}$ in a unique way to product an Abelian group $\overline{\mathfrak{F}}$. If we define the torsion coefficients and Betti number of the non-Abelian group $\mathfrak{F}$ to be the torsion coefficients and Betti number of $\overline{\mathfrak{F}}$, then it follows that a necessary condition for two groups $\mathfrak{F}_1$ and $\mathfrak{F}_2$ to be the same is that their torsion coefficients and Betti numbers be the same. If the group $\mathfrak{F}$ is given in terms of its generators

$$A_1, A_2, \ldots, A_m \tag{12}$$

and defining relations

$$R_1(A_i) = 1, \qquad R_2(A_i) = 1, \qquad \ldots, \qquad R_r(A_i) = 1, \tag{13}$$

then we can find its torsion coefficients and Betti numbers as follows: The defining relations of $\overline{\mathfrak{F}}$ are the relations (13) together with the commutation relations

$$A_i A_k = A_k A_i \qquad (i, k = 1, 2, \ldots, m). \tag{14}$$

Instead of regarding $\overline{\mathfrak{F}}$ as a factor group of $\mathfrak{F}$ with respect to the commutator group, we can also adopt the viewpoint that $\overline{\mathfrak{F}}$ arises from the m-dimensional lattice $\mathfrak{F}_m$, where $\mathfrak{F}_m$ is defined by the relations (14), when the additional relations (13) are included as well. The matrix describing the emplacement of $\mathfrak{N}$ in $\mathfrak{F}_m$ is simply the matrix of coefficients of the relations (13) when these relations are written in additive notation. The torsion coefficients and Betti number are determined from this matrix, in the manner just described.

---

* We shall use here the notation (a, b) to indicate the greatest common divisor of the integers a and b.

We shall demonstrate the procedure for the case of a group $\mathfrak{F}$ having three generators and the four defining relations

$$A^2 = B^3 = C^4 = ABC = 1.$$

This is the group of the octahedron. The element A is a rotation about the center of an edge, B is a rotation about the midpoint of a neighboring triangle, and C is a rotation about a vertex which leaves the diagonal through the vertex fixed.* The relations of the Abelianized group $\overline{\mathfrak{F}}$ are,

$$2\overline{A} = 3\overline{B} = 4\overline{C} = \overline{A} + \overline{B} + \overline{C} = 0.$$

The matrix of coefficients of this group is then

$$\begin{bmatrix} 2 & 0 & 0 \\ 0 & 3 & 0 \\ 0 & 0 & 4 \\ 1 & 1 & 1 \end{bmatrix}$$

and has the normal form

$$\begin{bmatrix} 2 & 0 & 0 \\ 0 & 1 & 0 \\ 0 & 0 & 1 \\ 0 & 0 & 0 \end{bmatrix}$$

One torsion coefficient exists, having the value 2. The Betti number is 0.

If a group $\mathfrak{F}$ possesses a Betti number other than zero, the Abelianized group $\overline{\mathfrak{F}}$, and thus $\mathfrak{F}$ itself, is infinite. On the other hand, if the Betti number is equal to zero, then we cannot decide by this criterion whether the group is infinite. As an example, the free product of two groups of order 2 and the direct product of these groups (the 4-group) both have two torsion coefficients of value 2 and Betti number 0. The free product is infinite, but the 4-group is finite.[52]

Problem:

Determine the torsion coefficients of the symmetric group of permutations of n digits.

87. The Normal Form of Integer Matrices

In §86 we made use of the theorem that each matrix **E** having integer elements can be brought to diagonal form by means of elementary transformations, that is, by addition or subtraction of a row (column) from another, by change of sign of all elements of a row (column), and by interchange of two rows (columns). We now carry out the proof.

Let the given matrix be

$$\mathbf{E} = [\varepsilon_{\iota\kappa}] \qquad (\iota = 1, 2, \ldots, n, \kappa = 1, 2, \ldots, m).$$

The integer $\varepsilon_{\iota\kappa}$ appears as the element in the ιth row and κth column. Let γ be the rank of **E**.

If $\gamma = 0$, then the matrix is already in normal form. If $\gamma > 0$, choose a nonzero element and bring it to the position $(\iota, \kappa) = (1, 1)$. We shall again

---

*The proof has been given by von Dyck [1].

denote the element in this position by ε_{11}. If all elements of the first row and first column are not divisible by ε_{11}, then let ε_{1k}, for example, be an element of the first row which leaves a remainder $\varepsilon_{11}' < \varepsilon_{11}$ when divided by ε_{11}. By repeatedly adding or subtracting the first column from the kth column we can eventually obtain the integer ε_{11}' at the position $(1, k)$. We can bring ε_{11}' to the position $(1, 1)$ be interchanging the first and kth columns. We proceed in the same manner if, in the first column, an element appears which is not divisible by ε_{11}. We repeat the procedure until all elements of the first row and column are divisible by the element in the position $(1, 1)$, and we again denote this element by ε_{11}. This case must occur ultimately, since the absolute value of ε_{11} decreases with each step. By repeatedly adding the first column (row) to the remaining columns (rows) sufficiently often, we shall obtain the result that all elements of the first row (column) except ε_{11} are equal to 0. In the resulting matrix, if there is an element ε_{ik} ($i \neq 1, k \neq 1$) which is not divisible by ε_{11}, we bring it to the first row by addition of rows and begin again with the procedure just applied. We continue in this way until all elements become divisible by the element in the position $(1, 1)$ and, except for ε_{11}, only zeros appear in the first row and first column.

We can carry out arbitrary elementary transformations of the submatrix $\mathbf{E}_1$ which arises from $\mathbf{E}$ by striking out the first row and column by means of elementary transformations of $\mathbf{E}$; these will not change the first row and column of $\mathbf{E}$ because of the zeros which stand there. By means of such transformations we can bring an element ε_{22} to the position $(2, 2)$ such that ε_{22} is contained as a factor of all remaining elements of $\mathbf{E}_1$, while all elements of the second row and second column are 0 except for ε_{22}. During the elementary transformations of $\mathbf{E}_1$, the property that all elements of $\mathbf{E}_1$ are divisible by ε_{11} is preserved. Thus ε_{22} is divisible by ε_{11}. By continuing the procedure, we eventually arrive at a matrix in which all elements are equal to 0, except for certain elements standing in the main diagonal. The number of nonzero elements must be γ since the rank of $\mathbf{E}$ does not change during elementary transformations. Each of the diagonal elements $\varepsilon_{11}, \varepsilon_{22}, \ldots, \varepsilon_{\gamma\gamma}$ is a divisor of the element following it. These diagonal elements, which are called the *invariant factors* of the matrix $\mathbf{E}$, differ from those mentioned in §86 only in their order, and this can be arbitrarily rearranged by means of row and column interchanges. Here, $\varepsilon_{11} = c_\gamma, \ldots, \varepsilon_{\gamma\gamma} = c_1$.

We have already proved in §86 that the normal form is unique, that is, independent of the details of the normalization procedure, and we have seen that the invariant factors are numbers characteristic of the Abelian group represented by means of the matrix. We can also see this independence directly as follows: If D_i denotes the greatest common divisor of all ι-rowed subdeterminants of $\mathbf{E}$, then D_i will not be changed by the transformations, according to familiar rules of determinant theory (Bocher [1]). When the normal form of the matrix is considered, D_i is the product of the first i

diagonal elements: $\varepsilon_{11} \cdots \varepsilon_{ii}$; thus

$$\varepsilon_{ii} = D_i / D_{i-1}$$

is already determined by **E**.

In practical cases it is not always advisable to employ the procedure just given. A different sequence of transformations may sometimes lead more quickly to the result. We shall demonstrate the procedure by applying it to an example encountered in §24, an incidence matrix for pseudomanifolds; this is a matrix **E** of n rows and m columns having the following properties:.

(a) Exactly two elements differ from 0 in each row. Their absolute value is 1.

(b) If we arbitrarily divide the columns into two classes, then there will exist at least one row whose ones stand in columns belonging to different classes.

The normalization is carried out in four steps:

Step 1

By interchanges of rows and columns **E** is transformed to the matrix

$$
\begin{bmatrix}
\pm 1 & \pm 1 & 0 & \cdots & 0 \\
* & * & \pm 1 & \cdots & 0 \\
\vdots & \vdots & \vdots & & 0 \\
* & * & * & \cdots & \pm 1 \\
\vdots & \vdots & \vdots & & \vdots \\
* & * & * & \cdots & *
\end{bmatrix}
\begin{matrix} \\ \\ m-1 \\ \\ \\ \end{matrix}
\Bigg\} n
\qquad\qquad (I)
$$

$$\longleftarrow \qquad m \qquad \longrightarrow$$

where the stars denote elements which are 0's or 1's and do not need to be specified in more detail. To get the form (I) and **E** we first bring the two ones of the first row to the first and second places by column interchanges. Because of (b) there exists a row, for example, the ith, having a 1 in the first or second column and having the other 1 in the kth column, $k > 2$. We then interchange the third column with the kth column and the ith row with the second row. The resulting matrix will have the same first two rows as (I). The rows are disposed of, in this way, row by row.

Step 2

By changes of sign in the columns, we can arrive at a configuration where exactly one $+1$ and one -1 stand in each of the first $m-1$ rows.

Step 3

Addition of columns 2 through m to the first column. The matrix then takes the configuration

$$\begin{bmatrix} 0 & 1 & 0 & \cdots & 0 \\ 0 & * & 1 & \cdots & 0 \\ \cdot & \cdot & \cdot & & \cdot \\ 0 & * & * & \cdots & 1 \\ \cdot & \cdot & \cdot & & \cdot \\ * & * & * & \cdots & * \end{bmatrix}. \tag{II}$$

The first column contains only 0's or 2's (which can appear only in the last $n - m + 1$ entries). When appropriate sign changes are made in the rows, the 1's appearing there will all have positive sign.

Step 4

By appropriate additions or subtractions of the first, second, . . . , $(m - 1)$ th row, respectively, to the remaining rows, we arrive at a configuration where only a single 1 stands in the second, third, . . . , mth column. The matrix will then have the form

$$\begin{bmatrix} 0 & 1 & 0 & \cdots & 0 \\ 0 & 0 & 1 & \cdots & 0 \\ \cdot & \cdot & \cdot & & \cdot \\ 0 & 0 & 0 & \cdots & 1 \\ \cdot & \cdot & \cdot & & \cdot \\ * & 0 & 0 & \cdots & 0 \end{bmatrix}. \tag{III}$$

The last $n - m + 1$ rows will contain only 0's, except in the first column. Consequently, the first column has not been altered. If no 2's now stand in the first column, that is, only 0's appear, the normal form will then result:

$$\begin{bmatrix} 0 & 1 & 0 & \cdots & 0 \\ 0 & 0 & 1 & \cdots & 0 \\ \cdot & \cdot & \cdot & & \cdot \\ 0 & 0 & 0 & \cdots & 1 \\ \cdot & \cdot & \cdot & & \cdot \\ 0 & 0 & 0 & \cdots & 0 \end{bmatrix}. \tag{IV}$$

If no more than one 2 appears in the first column then, by means of addition or subtraction of rows, we can produce just a single 2, which we can bring to the upper left-hand corner, by interchanging rows:

$$\begin{bmatrix} 2 & 0 & 0 & \cdots & 0 \\ 0 & 1 & 0 & \cdots & 0 \\ \cdot & \cdot & \cdot & \cdots & \cdot \\ 0 & 0 & 0 & \cdots & 1 \\ \cdot & \cdot & \cdot & & \cdot \\ 0 & 0 & 0 & \cdots & 0 \end{bmatrix}. \qquad\qquad \text{(V)}$$

We arrive at the normal form (IV) if the original matrix **E** has rank $m - 1$. We arrive at the normal form (V) if **E** has rank m.

COMMENTS

1. (§1) Knots in $\mathfrak{R}^4$. If we allow a knot to intersect itself, then it is possible to deform it in 3-dimensional (x, y, z)-Euclidean space to a circle. However, we can also achieve the same result without self-intersections if we regard the space (x, y, z) as a subspace of a 4-dimensional Euclidean space (x, y, z, t). For example, let us specify the two small pieces a and b of the knot which cross through one another as follows: Let a be the interval $-1 \leqq x \leqq +1$, $y = z = 0$; let b be the semicircle $y^2 + z^2 = 1$, $x = 0$, $z \geqq 0$. Our task, then, is to deform b to the semicircle b' given by $y^2 + z^2 = 1$, $x = 0$, $z \leqq 0$, without a self-crossing of the knot occurring, in the space (x, y, z, t). We accomplish this by means of a rigid Euclidean rotation of the semicircle b about the y-axis in the subspace $(x = 0, y, z, t)$, until it arrives at the position of b'.

2. (§2) The representation of the double torus is given a particularly clear description in Hilbert and Cohn-Vossen [1], p. 265. The same book gives pictures illustrating the intuitive content of the material.

3. (§2) Immersion of nonorientable surfaces. All possible ways of mapping the projective plane into Euclidean space $\mathfrak{R}^3$ so that its curves of self-intersection are free of double points have been ascertained by Boy [1]. Also see the end of the first volume of Schilling [1]: Topological realization of the projective plane by a singularity-free surface in space.

   The projective plane can be embedded into the Euclidean space $\mathfrak{R}^4$ without self-intersections by first projecting it to the boundary of a Möbius band which lies in a Euclidean 3-subspace $\mathfrak{R}^3$ from a point not in this subspace, and then sewing the projection cone to the Möbius band.

   In Appendix I of Hilbert and Cohn-Vossen [1] it is shown that the projective plane can even be embedded in $\mathfrak{R}^4$ as an algebraic surface. In this book one can also find interesting observations concerning the closing of the Möbius band in $\mathfrak{R}^3$ and the configurations of cross-caps.

4. (§2) Infinite surfaces have been treated systematically by B. von Kerékjártó [6, Section 5: open surfaces].

5. (§4) Closure of 3-space by groups. It seems obvious to ask the following question: Can one close Euclidean 3-space by means of groups other than the conformal and projective groups, and still satisfy the criterion that the group transformations be one-to-one correspondences? Seifert [1, p. 30], gives examples of other closings satisfying this criterion. The conformal and projective groups are, however, distinctive because they alone satisfy the Lie–Helmholtz conditions of motion (Weyl [1, p. 30]) in the resulting closed space.

   The closing of complex Euclidean space by groups of rational transformations is described in Behnke and Thullen [1] p. 3.

6. (§4) Position spaces in mechanics. The position space of a mechanical system has special significance not so much because its topology affects its dynamics as for the fact, pointed out by Jacobi [1, p. 44], that it can be provided with a metric such that the orbit curves are geodesics with respect to this metric. The total energy is constant with time and the potential energy can be specified as a function of position (see, for example, Whittaker [1, §104]).

As an example, let us consider the case of a plane double pendulum hanging in a gravitational field. If the total energy is sufficiently large so that each position can be achieved, then the position space is a torus. Since each closed curve on the torus can be deformed to a geodesic curve of the given metric (under certain conditions, which are fulfilled in this special case) (see, for example, Bieberbach [3, p. 195]) and since there exist infinitely many nonhomotopic closed curves on the torus, it follows that there are infinitely many distinct periodic motions of the plane double pendulum.

7. (§4) Mechanical phase spaces are discussed in Birkhoff [4].

8. (§5) Topological spaces. The concept of a neighborhood space, as it appears in the text, and which is determined through axioms A and B alone, has been constructed in such a way that the fundamental topological concepts, such as open subsets, closed subsets, continuous mappings, topological mappings, and identification, can be defined in terms of this concept. For our purposes, the neighborhood space is only an intermediate concept, leading us to the concept of a complex. It is too general a concept to allow one to derive essential geometric theorems from it per se. For this reason, set theoretic topology requires a more restrictive definition for a neighborhood space or a topological space. For example, the axioms $(\overline{A})$ and $(\overline{B})$ of Tietze–Vietoris [1, p. 156], are the same as our axioms A and B; but further axioms are added: $(\overline{B})_b$ The intersection of two neighborhoods of a point is again a neighborhood of a point; $(\overline{C})$ If the set $\mathfrak{U}$ is a neighborhood of a point P then so is the set of all "interior" points of $\mathfrak{U}$, that is, all points X for which $\mathfrak{U}$ is a neighborhood of X; $(\overline{D})$ If $P \neq Q$, then there exist neighborhoods of P and Q, respectively, which have no common point. The Tietze neighborhood space is therefore far more specialized than ours, and is equivalent to the "Hausdorff topological space" which more commonly appears in the literature. The axioms defining the latter space (Hausdorff [1, pp. 213 and 260]) are: (A) Axiom A of the text; (B) Given two neighborhoods of a point P, there exists a neighborhood of P which is a subset of both of these neighborhoods; (C) If the point Q lies in a neighborhood $\mathfrak{U}(P)$ of a point P, then it has a neighborhood $\mathfrak{U}(Q)$ which is a subset of $\mathfrak{U}(P)$ (the neighborhoods are therefore open sets); (D) is the same as axiom $(\overline{D})$ given above. These neighborhood axioms are satisfied, for example, by open ball neighborhoods and open cube neighborhoods in a Euclidean space. Since axiom $(\overline{B})$ of Tietze is not assumed, the neighborhoods of Hausdorff have a more restricted meaning than those of Tietze. For example, when Euclidean space is made into a topological space by choosing ball neighborhoods it must, at first, be distinguished from that in which the neighborhoods are open cubes. Later on, two systems of neighborhoods are declared to be equivalent if each neighborhood of the first system contains a neighborhood of the second system and each neighborhood of the second system contains one of the first system.

9. (§5) Homogeneity of numerical spaces. The points of the space of numerical n-tuples are not defined here to be mathematical objects which can be placed in one-to-one correspondence with n-tuples of real numbers; rather, we adopt the viewpoint that a point is identical with a numerical n-tuple. Regardless of this, we have introduced a "parallel" coordinate system in §9, in which the n-tuple $x_1, x_2, \ldots, x_n$ is assigned parallel coordinates $y_1, y_2, \ldots, y_n$; we must, therefore, distinguish the defining numerical n-tuple from the n-tuple of parallel coordinates, even through it is possible in a particular case that $x_1 = y_1, x_2 = y_2, \ldots, x_n = y_n$.

The space of numerical n-tuples is not "homogeneous" (Weyl [3, p. A7]) in the logical sense that two of its points are fundamentally indistinguishable from one another according to the axioms defining the space. This is not the case in a numerically defined space, because different numbers are distinct objects. On the other hand, the space of numerical n-tuples is homogeneous in the topological sense of §12, which is the only interpretation having mathematical significance.

10. (§12) Purely combinatorial topology. Kneser [4] and Tietze–Vietoris [1] have reviewed the state of development of the purely combinatorial methods. A complete, purely combinatorial presentation of surface topology has been given by Reidemeister. A summary of combinatorial surface topology has been given by Levi [1] and Chuard [1]. Reidemeister [6] has treated knot theory combinatorially. The encyclopedia article (Dehn–Heegaard [1]) is likewise presented combinatorially; however, the difficulties presented by a rigorous treatment, using purely combinatorial methods, were not fully brought out in this article. These difficulties, which are of a fundamental nature, and which are presently insurmountable in more than 3 dimensions, are discussed in Bilz [1], Kneser [4], Furch [2, 3, 5], and Weyl [2]. Attempts to remove these difficulties by appropriately defining combinatorial equivalence have been made by Newman [1], [4], Alexander [17], and Tucker [3], among others. See also the work of de Rham [1], which relates to Weyl [2]; also Mayer [1] and Bergmann [2].

11. (§14) Strictly speaking, we have only proved in this text that the n-sphere is generated when we identify the boundaries of two n-balls by mapping the boundaries congruently one onto the other. We can see that the same is true for an arbitrary topological mapping A of the boundaries $\mathfrak{B}_1$ and $\mathfrak{B}_2$ of the two n-balls, as follows: Let A' be a congruence mapping of the boundary of $\mathfrak{B}_1$ onto the boundary of a third ball $\mathfrak{B}_2'$. The boundaries of $\mathfrak{B}_2$ and $\mathfrak{B}_2'$ will then be mapped topologically onto one another by $A'A^{-1}$. This topological relation can be extended to give a topological mapping of the ball $\mathfrak{B}_2$ onto the ball $\mathfrak{B}_2'$, for example, by mapping corresponding radii of $\mathfrak{B}_2$ and $\mathfrak{B}_2'$ linearly one onto the other. From §8, we obtain a homeomorphic neighborhood space when we glue not $\mathfrak{B}_1$ and $\mathfrak{B}_2$ but, instead, $\mathfrak{B}_1$ and $\mathfrak{B}_2'$ to one another. This gives rise to the n-sphere, since the boundaries of $\mathfrak{B}_1$ and $\mathfrak{B}_2'$ are related congruently to one another.

12. (§14) Spaces of line elements. The problem gives two examples of spaces whose "points" are line elements of a closed surface. The question of finding all spaces consisting of oriented or nonoriented line elements of closed surfaces has been examined by three authors: Nielsen [6], who gives the fundamental groups of all spaces consisting of oriented line elements: Hotelling [1, 2]; and Threlfall [2]. One should also compare van der Waerden [5], problem 124 (nonoriented line elements of the projective plane), with Problem 4, §30 of this book.

13. (§20) On occasion, the entire homology group is referred to as the Betti group, and the group which we call the Betti group in this text is called the reduced Betti group; see, e.g., Pontrjagin [3].

14. (§21) By introducing the incidence matrices, Poincaré (especially in [4]) took the decisive step in the arithmetization of topology.

15. (§23) Chains mod m. Viewed algebraically, a chain is nothing other than a linear form whose indeterminates are the oriented simplexes. The coefficients of this form belong to a particular domain of coefficients. If this is the domain of integers, then we get the ordinary chains, as considered in §15–§22. If we choose the residue classes mod 2 as coefficients, then the chains mod 2 result. In general, one can use as coefficients the residue classes modulo an arbitrary integer $m \neq 1$. We then get homology groups and Betti numbers mod m whose topological invariance can be proved just as for the ordinary chains. The Betti numbers mod m are determined by the ordinary Betti numbers (which one also calls Betti numbers mod 0) and the torsion coefficients. In the reverse direction, the torsion coefficients can be calculated from the Betti numbers mod m ($m = 0, 2, 3, \dots$). Accordingly, the Betti numbers mod m offer no particular advantage over the ordinary homology groups in the study of homology questions. On the other hand, they have been useful in other investigations (duality and mapping theorems) (see, for example, Hopf [13, 18]; Pontrjagin [3]). Homology groups mod 2 were introduced by Tietze [1] and developed further by Veblen [4]. Homology groups mod m were first investigated by Alexander [15].

16. (§25) In more general neighborhood spaces, one defines the homology groups in an

appropriate manner so that as many theorems as possible from the theory of complexes remain valid. The following simple example may illustrate this.

Let us consider the point set in the Euclidean plane which is given, in a Cartesian coordinate system, by the equations

$$y = \sin(1/x) \quad \text{for} \quad x \neq 0,$$
$$-1 \leqq y \leqq \pm 1 \quad \text{for} \quad x = 0,$$

and add to this set the point at infinity which closes the Euclidean plane to form the 2-sphere $\mathfrak{S}^2$. We obtain a closed subset $\mathfrak{M}$ of $\mathfrak{S}^2$ on which each closed singular 1-chain is ~ 0, as one easily sees. On the other hand, $\mathfrak{M}$ decomposes $\mathfrak{S}^2$ into exactly two domains. According to Alexander's duality theorem (see Comment 47), the number of domains is one larger than the first Betti number of the embedded complex. Thus, if we use our definition of the singular homology groups but consider arbitrary closed sets instead of complexes, the Alexander duality theorem will no longer be valid. To learn how to extend the definition of the homology groups to spaces which are arbitrary closed subsets of Euclidean spaces see Alexandroff [7, 10], Cech [2], Lefschetz [12], and Vietoris [2, 4].

17. (§30) Proof of invariance of the homology groups. The idea of simplicially approximating singular chains to prove the topological invariance of the simplicial homology groups goes back to J. W. Alexander. In his first proof [1], he established the equality of the homology groups calculated from two different simplicial decompositions of a complex. This was accomplished without defining topologically invariant singular homology groups. Instead, the proof used the simpler concept of a singular chain (the continuous image of a simplicial chain); the additivity of the singular chains was not needed for an invariant definition of the homology groups.

A second proof by Alexander [15] did not use singular chains at all. Instead, he approximated the subdivision of each of the simplicial decompositions of the complex in the other decomposition.

In the text we proved the invariance of the connectivity numbers by expressing them in terms of the Betti numbers and torsion coefficients. We can see the invariance directly by introducing *singular chains* mod 2, that is, by using residue classes mod 2 in place of the integer chain coefficients. The proof of the approximation theorem, and consequently that of the connection groups for the new domain of coefficients, subsequently follows as in the text.

18. (§31) Complex projective spaces. Points of a complex projective space are $(n + 1)$-tuples of ratio numbers, which independently of one another range through all of the complex numbers, minus the $(n + 1)$-tuple consisting only of zeros. Neighborhoods are defined in a similar manner as for the case of a real projective space (§14); cf. van der Waerden [1].

Using the same methods as for the real projective spaces, one can prove that complex projective space is a $2n$-dimensional manifold, and that its homology groups of even dimension $0, 2, 4, \ldots, 2n$ are free cyclic groups, while those of odd dimension consist of the zero element alone. The homology bases are formed from the complex projective subspaces.

19. (p. 123) Invariance "in the large." In the special case of a closed pseudomanifold, the invariance of dimension follows from the results of Chapter IV, because the dimension n of a closed pseudomanifold can be characterized invariantly as the smallest integer such that the $(n + 1)$th connectivity number and all subsequent connectivity numbers are equal to 0.

20. (§32) Homology groups at a point. Instead of defining the homology groups at a point by using a particular simplicial decomposition and subsequently proving topological invariance, as we have done, we can also define these groups in a way which is topologically invariant from the start, as was done with the singular homology groups (§27). To do this, following van Kampen [3], we adopt the definition that two singular k-chains U^k and V^k of a complex $\Re$ are "equal at the point P" if $U^k - V^k$ is disjoint from P. A chain U^k is said to be "closed at P" or a cycle at P if $\Re \partial U^k$ does not pass through P, and one writes $U^k \sim V^k$ at P if $U^k - V^k$ is homologous to a chain disjoint with P. In other words, in all of the chains we

neglect the simplexes which are disjoint with P, that is, the simplexes belonging to $\Re - P$. We thus deal with "chains mod $(\Re - P)$." The k-cycles at P can be divided into homology classes. These classes form an Abelian group, which is topologically invariant because of the way it is defined, and when $k > 1$ this group is the same as the $(k - 1)$th homology group of the neighborhood complex at P. This can be proved by using a simplicial approximation.

Chains and homology groups mod $\mathfrak{L}$, where $\mathfrak{L}$ is an arbitrary closed subset of $\Re$, have been used extensively by Lefschetz [12], for example to assimilate the theory of manifolds with boundary into the theory of closed manifolds; cf. Comment 41.

21. (§34) By the "theorem of the invariance of dimension" one usually means the theorem of Problem 2, §32. The invariance of dimension was first proved by Brouwer [6]. Other proofs can be found in Alexandroff [12], Lefschetz [12], and Sperner [1].

22. (§38) In the purely combinatorial treatment of surface topology, in which the original objects are a finite or countable number of points, intervals, or surface elements, and only the question of elementary relatedness, not homeomorphism, makes sense, the theorem of the text is the principal theorem of surface topology. This theorem was first proved combinatorially by Dehn and Heegaard [1, p. 190]. The normal form used by these authors was not the fundamental polygon but consisted, instead, of three surface elements. Levi [1] and Reidemeister [7] have applied the fundamental polygon to the combinatorial proof of the fundamental theorem. For other normal forms see Threlfall [1]. See Chuard [1] for another combinatorial proof.

23. (§41) The fact that the homeomorphism problem for surfaces has been solved does not mean that all important problems of surface topology have been solved. Three-dimensional topology still gives rise to difficult questions of surface topology. An example is the construction and classification of all possible Heegaard diagrams (cf. Comment 34).

While the homeomorphism problem can be solved by making use of the homology properties of closed curves on surfaces, the other questions mentioned above are related to homotopy and isotopy properties. The question of when two surface curves are freely homotopic (§49) is a purely group theoretic question: one must find all classes of conjugate elements of the fundamental group of the surface. This problem has been solved completely by Dehn [2, 3]. It encompasses the question of the existence of *invertible curves*. These are curves which can be deformed homotopically, on the surface, to their reciprocals. That is, they are curves whose direction of traversal can be reversed by means of a homotopic deformation. See Baer [1, 2] regarding the isotopy of surface curves.

Another theorem which should be mentioned here is the following, proved by Nielsen [6]: Each automorphism of the fundamental group of an orientable surface can be induced by a topological mapping of the surface onto itself.

See Nielsen [2–6], Brouwer [1, 5, 14, 15], Kneser [7, 9], and Hopf [14] for mapping and fixed point theorems of surfaces.

Reidemeister [7] has presented a comprehensive description of purely combinatorial surface topology.

24. (§42) The definition of equality for paths is consequently different from the definition of equality for oriented singular 1-simplexes; we have required of the latter that their preimages can be mapped linearly, one to the other. If we were also to require this linear transformability for paths, then the associative law $ab(c) = a(bc)$ for the formation of product paths would not necessarily remain valid.

25. (§42) The concept of the fundamental group is due to Poincaré [2, 7]. The presentation of the generators and relations by means of auxiliary paths (§46) has been taken from Tietze [1].

26. (§43) THEOREM. *The Euclidean plane is the only 2-dimensional simply connected infinite homogeneous complex.*

To prove this theorem, we construct a mapping of a given complex $\Re$ of this type, onto the Euclidean plane. We start with a given simplicial decomposition of $\Re$. Each essential cut $\Re$ will cut a uniquely determined surface element from $\Re$. This is because $\Re$ is null homotopic and is thus $\sim 0 \bmod 2$ and, consequently, is the boundary of a 2-dimensional subcomplex $\mathfrak{E}$.

This subcomplex is a surface with boundary and is, in fact, a surface element. From §40, it will be sufficient to show that on $\mathfrak{E}$ an arbitrary 1-chain mod 2 $\mathfrak{U}$, will be null homologous mod 2. Certainly $\mathfrak{U} \sim 0$ mod 2 on $\mathfrak{R}$. Thus there will exist a finite subcomplex $\mathfrak{U}$ having $\mathfrak{U}$ as its boundary. $\mathfrak{U}$ cannot contain any triangle lying outside of $\mathfrak{E}$ because $\mathfrak{U}$ would then have to contain all triangles outside of $\mathfrak{E}$; for $\mathfrak{R} \partial \mathfrak{U}$ lies on $\mathfrak{E}$. But $\mathfrak{U}$ contains only finitely many triangles. Thus $\mathfrak{U}$ is already ~ 0 mod 2 on $\mathfrak{E}$. The subcomplex $\mathfrak{E}$ is unique, for if a second surface $\mathfrak{E}'$ could be spanned in $\mathfrak{R}$, then $\mathfrak{E}$ and $\mathfrak{E}'$ would together form a closed chain mod 2; but, aside from the chain 0, there exists no closed chain mod 2 on $\mathfrak{R}$ because if one triangle belonged to it then all others would have to belong to it.

If $\mathfrak{E}_1$ is now an arbitrary surface element of $\mathfrak{R}$ consisting of triangles and having the boundary $\mathfrak{R}_1$, then it is possible to extend $\mathfrak{E}_1$ in a unique way by adding on those triangles which have an edge or a vertex in common with triangles of $\mathfrak{E}_1$. We complete the resulting complex $\mathfrak{X}$ (which may conceivably show "holes") in a unique way by adding on all surface elements which have as their boundary essential cuts lying on the subcomplex.

We claim that the complex $\mathfrak{X}$ which arises thereby is a surface element. For it follows from the construction of $\overline{\mathfrak{X}}$ that the connectability condition (PM3) of §24 is satisfied. Let us then consider an essential cut $\mathfrak{R}_2$ consisting of edges of the boundary of $\overline{\mathfrak{X}}$. This cut will also lie on the boundary of $\mathfrak{X}$, since we have added no new boundary edges in extending $\mathfrak{X}$ to $\overline{\mathfrak{X}}$. Thus, by our construction, the surface element $\mathfrak{E}_2$ spanned in $\mathfrak{R}_2$ belongs to $\overline{\mathfrak{X}}$. On the other hand, because of (PM3) $\overline{\mathfrak{X}}$ consists of all triangles which can be connected with a triangle of $\overline{\mathfrak{X}}$ and thus, in particular, with a triangle of $\mathfrak{E}_2$. But these are just the triangles of $\mathfrak{E}_2$, for the boundary of $\mathfrak{E}_2$ also belongs to the boundary of $\overline{\mathfrak{X}}$. This implies that $\overline{\mathfrak{X}}$ also belongs to $\mathfrak{E}_2$. Thus $\overline{\mathfrak{X}} = \mathfrak{E}_2$ is a surface element.

The totality of triangles of $\mathfrak{E}_2$ which do not belong to $\mathfrak{E}_1$ form a punctured surface element, that is, an annulus $\mathfrak{Z}_1$ having $\mathfrak{R}_1$ as its inner boundary circle and $\mathfrak{R}_2$ as its outer boundary circle. Let us now map $\mathfrak{E}_1$ topologically onto the unit circle of the Euclidean plane $\mathfrak{Z}_1$ onto the annulus in the Euclidean plane having inner radius 1 and outer radius 2, so that the mappings are the same on the circle $\mathfrak{R}_1$ common to $\mathfrak{E}_1$ and $\mathfrak{Z}_1$. Let us next map $\mathfrak{Z}_2$ topologically onto the annulus having inner radius 2 and outer radius 3, and so forth. By this procedure, we get a topological mapping of $\mathfrak{R}$ onto the whole of the Euclidean plane.

27. (§44) This implies that two closed paths w_1 and w_2 departing from O which are continuously deformable one to the other are also combinatorially deformable one to the other. We first transform w_1 to $w_1 \cdot w_2^{-1} \cdot w_2$ and then to w_2 by means of combinatorial deformations. This is always possible because $w_1 w_2^{-1}$ is continuously null homotopic and therefore, from (II), is combinatorially null homotopic.

28. (§49) Among the closed surfaces, one should try to determine those on which invertible curves (cf. Comment 23) are to be found.

In projective 3-space, aside from the null homotopic curves, which are always invertible, the projective lines are invertible and, consequently, so are all curves. In contrast, only the null homotopic curves are invertible in the lens space (3, 1).

29. (§49) One can also achieve the transformation from a closed path w_1 to a homologous path by means of a deformation procedure, that of "deformation with tearing." By this we mean the following: We deform w_1 to a path $\overline{w}_1$ having a double point (that is, a point of self-intersection) P, and we assume that P is the initial point of $\overline{w}_1$. The path $\overline{w}_1$ is the product of two closed paths w_1' and w_1'', which can now be freely deformed independently of one another until they are finally again joined to give a single path w_2. We can apply the same procedure to w_2 that we applied to w_1, and so forth. After r steps we shall arrive at a path w_r which is obviously homologous to the initial path w_1. One can easily demonstrate that each path homologous to w_1 can in fact be obtained in this way. The three classes mentioned in §49, namely, path classes, classes of conjugate elements, and homology classes, consequently correspond, respectively, to the three deformation procedures: bound deformations, free deformations, and deformations with tearing.

30. (§52) The determination of the homology groups of a composite complex has been carried

out by Vietoris [6]; also see Mayer [1]. For the fundamental group of a composite complex see Seifert [1].

31. (§52) More generally, the group having the one relation $A^m = B^n$ will be isomorphic with the group $A^{m'} = B^{n'}$ only if the pairs of integers m, n and m', n' are the same, apart from the order of the integers. To prove this, form the factor group of the group $A^m = B^n$ with respect to the center generated by A^m. The factor group has the relations $\overline{A}^m = \overline{B}^n = 1$ and each of its elements of finite order can be transformed to a power of $\overline{A}$ or $\overline{B}$. The integers m and n are thereby characterized as invariants of the group (cf. Schrier [1]). A necessary condition for the equivalence (§65) of two torus knots m, n and m', n' is, then, the equality of the pairs of integers. In the case $m = m'$ and $n = n'$, the knots are not necessarily equivalent, however, since one must distinguish between a knot and its mirror image. There exist, for example, "right-handed" and "left-handed" trefoil knots (see Dehn [4]; Goeritz [2]; Seifert [4]).

32. (§58) See Reidemeister [4].

33. (§62) See Poincaré [7] regarding the discovery of the Poincaré spaces. The example of the spherical dodecahedron space is given there on p. 106 (see also Weber and Seifert [1]). The first procedure for constructing infinitely many different Poincaré spaces was given by M. Dehn (cf. §65). The theory of fibered spaces provides another procedure (cf. Comment 38 and Seifert [3, p. 207]). It turns out that each Poincaré space can be fibered in only one way, if it can be fibered at all. (The 3-sphere, which can be fibered in infinitely many ways, is not considered to be among the Poincaré spaces.) The fibered space is uniquely determined by the multiplicities $\alpha_1, \alpha_2, \ldots, \alpha_r$ of its exceptional fibers, and conversely. The multiplicities must satisfy the requirement that they are pairwise relatively prime.

Among the fibered Poincaré spaces, only the spherical dodecahedral space has a finite fundamental group. The fact that the manifold of §65 constructed from the trefoil knot by Dehn's procedure is homeomorphic with the spherical dodecahedron space follows from the fact that it can be fibered (cf. Threlfall and Seifert [1, II, p. 568]). Using the same procedure, it has been proven that the spherical dodecahedral space is not only a 5-fold cyclic covering space of the trefoil knot, but is also a 2-fold cyclic covering space of the torus knot 3, 5 and 3-fold cyclic covering space of the torus knot 5, 2 (Seifert [3, p. 222]).

34. (§63) See Seifert [1] and Goeritz [1]. For Heegaard diagrams of higher genus see Goeritz [4], Kreines [1], Reidemeister [8, 9], and Singer [1].

35. (§63) See Alexander [3, 8].

36. (§64) The proof in the text, given for the case of the projective plane, is clearly valid for any arbitrary nonorientable surface. It follows from the Alexander duality theorem (see Comment 47) that such a surface cannot be embedded in Euclidean 3-space as a subcomplex of a simplicial decomposition or, in fact, topologically in any way.

37. (§64) This result has been obtained in another way by Kneser [5].

38. (§65) The space form problem* and the theory of fibered spaces give two additional procedures for the construction of closed 3-dimensional manifolds. By a space form we mean an n-dimensional manifold which has the following two properties: first, it has been provided with either a spherical, Euclidean, or hyperbolic metric, that is, each point has a neighborhood which can be mapped congruently onto a ball of one of the three metric ground forms—n-dimensional spherical, Euclidean, or hyperbolic space (condition of metrical homogeneity); second, we postulate the "infiniteness" of the space form, that is, at each point it is possible to construct, in any direction, a geodesic ray which has no boundary (and in some cases may even return to the point). The second condition will exclude the case, for example, that each open subset of a space form is again a space form. The three ground forms then occupy a special place among the space forms, because they are the only ones which are simply connected. (If we also include, in addition to these three forms, the elliptic space which arises from the spherical metric space after identification of diametrically

---

* *Translator's note*: See Wolf [1] for a modern discussion of the space form problem. An older treatment is to be found in Klein [3].

opposite points, then these four space forms are the only ones which satisfy the Lie–Helmholtz conditions of motion.) It can be shown in 2 dimensions that each closed surface occurs as a region of discontinuity of a fixed point-free group of motions of one of the three metric ground forms: 2-sphere, Euclidean plane, and pseudosphere, that is, the hyperbolic plane (see Koebe [1] for example). The corresponding statement does not hold in 3 dimensions, as is demonstrated by the example of the topological product of the circle and the 2-sphere. But the regions of discontinuity in 3 dimensions do, at least, provide a wealth of illustrative material, from which we have taken all of the examples of Chapter IX. The groups of motions of the 3-sphere and Euclidean 3-space having a finite region of discontinuity can be completely classified. See Hopf [2], Threlfall and Seifert [1], and Hantsche and Wendt [1, Euclidean Space Forms]. Little is known about the regions of discontinuity of hyperbolic 3-space (see Löbell [4]; Weber and Seifert [1]).

The regions of discontinuity of groups of spherical motions, as well as the spaces of line elements of closed surfaces, provide motivation for the study of fibered spaces, since each is associated with a particular fibering. The fibers of the regions of discontinuity are the orbital curves of continuous groups of motions of the unit 3-sphere (also called the hypersphere). In the spaces of line elements, the fibers will be formed by line elements passing through a given point.

A fibered 3-space is, then, a 3-dimensional manifold whose points are distributed on a doubly infinite family of curves, the so-called fibers. Exactly one fiber passes through each point and each fiber H has a "fiber neighborhood," which is a subset of fibers containing H, such that it can be mapped fiber-faithfully onto a "fibered solid torus" and H transforms to the middle fiber of the solid torus. A fibered solid torus is a right circular cylinder of Euclidean 3-space which is fibered by means of lines parallel to its axis and whose floor and roof surfaces can be brought into coincidence by a rational screw rotation. We can now replace the unsolved homeomorphism problem of finding a complete system of invariants of 3-dimensional point manifolds with respect to topological mappings with the soluble problem of finding a complete system of invariants of fibered spaces with respect to fiber-faithful mappings (Seifert [3]). The fiber invariants so obtained, from which the fundamental group can be calculated, are not, of course, properties of a given manifold but, rather, of its fibering. Thus the question will sometimes remain open whether two differently fibered spaces, considered as point manifolds, cannot be mapped topologically one onto the other. There exist manifolds, moreover, which cannot be fibered in any way, since fiberability is related to a specific required property of the fundamental group. As examples, all closed hyperbolic space forms, as well as almost all topological sums (cf. §62), cannot be fibered. Nevertheless, fiber invariants are useful in the topology of manifolds because, in many cases, they allow one to decide whether manifolds are homeomorphic. Examples of this are the regions of discontinuity of groups of motions of the 3-sphere having fixed points and also the Poincaré spaces mentioned in Comment 33.

39. (§65) Equivalence of knots. We might attempt to consider two knots as equivalent if one knot can be deformed to the other by an isotopic deformation of the mapping of its preimage, as in §31. But we could then transform each knot to a circle. We would only have to pull the knot tight in a way so that the knotted portion pulls together to a point.

Another concept of the equivalence of two knots lying close to intuition, would be that two knots are equivalent if they can be transformed one to the other by means of an isotopic deformation (§31) of the whole space. We shall not prove that this concept is the same as that given in the text. Each isotopic deformation of 3-space obviously does, in fact, result in an orientation preserving homeomorphism of the space onto itself. But we leave open the converse question, whether each such topological mapping can be brought about by means of an isotopic deformation.

If one defines a knot as a rectilinear polygon, then one can work with combinatorial deformations which correspond to particular isotopic deformations of the mapping of the preimage and for which the pulling together of the knotted portion to a point is excluded. In this viewpoint, due to Reidemeister [6], one can regard equivalent knots as a class of spatial polygons which are combinatorially deformable one to the other.

40. (§68) This is a topologically invariant definition, since we have proved in §32 that the homology groups at a point are topological invariants.

The idea of basing the definition of manifolds on the homology properties at a point, instead of the homogeneity, was conceived independently by several authors: Alexander, L. Pontrjagin (unpublished), Vietoris [2], and Weyl [2]. The first complete treatment is due to van Kampen [3]. Pontrjagin calls these manifolds h-manifolds.

41. (§68) Manifolds with boundary. An n-dimensional manifold with boundary can be regarded as a pure complex with boundary whose doubling is a closed manifold, according to van Kampen [3, p. 37]. The star complex dual to a cellular division of the manifold with boundary is not necessarily again a cell complex. For this reason, the formal proof of the Poincaré duality theorem (§69) fails for manifolds with boundary, and the theorem loses validity. On the other hand, one again gets a cell complex if one removes from the dual star complex every star whose center point lies on the boundary. Thus if one wishes to extend the duality theorem to manifolds with boundary, this can be done, according to Lefschetz [12, p. 154], by replacing the chains with their residue classes modulo the boundary $\mathfrak{B}$ (cf. Comment 20). In that case, we get the following result: the (ordinary) Betti number p^k is equal to the $(n - k)$th Betti number mod $\mathfrak{B}$; similarly, the (ordinary) k-dimensional torsion coefficients are equal to the $(n - k - 1)$-dimensional torsion coefficients mod $\mathfrak{B}$.

In contrast, the theory of intersection numbers can be extended to manifolds with boundary without essential modifications. Aside from condition (R) of §73, the singular chains must satisfy the additional requirement that the intersection of the point sets which they cover must be disjoint with the boundary (cf. van Kampen [3]).

42. (§69) To a certain extent, the definition of the intersection number is somewhat arbitrary, and has not been standardized throughout the literature. Instead of setting $\mathfrak{S}(a^k, b^{n-k}) = \xi\eta\zeta$, one can also define the intersection number to be $\xi\eta\zeta\omega(k, n)$, where $\omega(k, n)$ denotes a function of k and n which takes the values ± 1. Thus, for example, van Kampen [3] uses $\omega = (-1)^k$.

A similar arbitrariness also occurs in the definitions of the boundary and the linking number.

43. (§71) The proof given here of Theorems I and II is due to Veblen [3].

44. (§74) We have made the transition from one cellular division to another using a method due to Lefschetz [6, 7, 12]. To show topological invariance, we applied mathematical induction to the intersection number $\mathfrak{S}(A^k, B^{n-k})$ to trace it back to an intersection number $\mathfrak{S}(A^0, B^n)$ whose invariance was easy to prove. Using the same method, Lefschetz proved that two closed singular chains A^r and B^s have an $(r + s - n)$-dimensional intersection chain C^{r+s-n}, when $r + s \geqq n$, whose homology class is determined uniquely by the homology classes of A^r and B^s. One calls the homology class of C^{r+s-n} the "product" of the homology classes of A^r and B^s. The homology classes, then, cannot only be added but also can be multiplied. They form a ring. The ring has been of significance in the theory of continuous mappings (Hopf [12, 14]). Hopf refers to this ring, together with the fundamental group, as the algebraic framework of the manifold, whose connectivity relations they describe without, admittedly, exhausting them.

45. (§75) Two essential cuts on an orientable closed surface are said to be conjugate if they pass smoothly through one another at exactly one point. As an example, in Fig. 128 a, b and c, d are pairs of conjugate essential cuts.

46. (§75) The intersection numbers play a role in the applications of topology to algebraic geometry. As an example, let us consider an algebraic curve C_m of order m in the complex projective plane, where the curve is specified by setting a ternary form of degree m equal to zero. It can be proved that the points of C_m form a subcomplex which, after simplicial decomposition and coherent orienting of its 2-simplexes, can be regarded as a singular 2-chain in a 4-dimensional manifold, the complex projective plane. It is possible to find an

appropriate projective line C_1 (which is a 2-sphere embedded in the complex projective plane) so that it has exactly m points of intersection with C_m. It is possible to show that the topological intersection number $\mathbb{S}(C_m, C_1)$ of the two 2-chains C_m and C_1 is also equal to m. This implies that the intersection number of C_m with an arbitrary projective line C_1' is also equal to m because $C_1 \sim C_1'$ (Comment 18), even though the actual number of points of intersection of C_m and C_1' could on occasion be smaller than m (for example, when C_1' is tangent with C_m). If we now define the intersection number of two arbitrary algebraic curves to be the topological intersection number of the singular chains which correspond to these curves, Bezout's theorem follows easily: The intersection number of a curve C_m of order m with a curve C_n of order n is equal to mn. Proof: Since the projective line represents a 2-dimensional homology basis, then $C_m \sim \mu C_1$. The null homologous chain $C_m - \mu C_1$ then has 0 as its intersection number with a projective line C_1'. Therefore $\mathbb{S}(C_m, C_1') = \mu \mathbb{S}(C_1, C_1')$. But, as determined above, $\mathbb{S}(C_m, C_1') = m$ and $\mathbb{S}(C_1, C_1') = 1$; thus $m = \mu$. That is, a curve of order m is homologous to a projective line taken m-fold. Accordingly, we have $\mathbb{S}(C_m, C_n) = \mathbb{S}(mC_1, nC_1') = mn\mathbb{S}(C_1, C_1') = mn$, which was to be proved.

This example illustrates only one of the ways in which topology can be applied to algebraic geometry. An extensive description of these topological methods can be found in van der Waerden [1, 4] and Lefschetz [2, 4, 12]; also see F. Severi.

47. (§77) Linking numbers were introduced by Brouwer [9].

One of the most beautiful and fruitful theorems of topology, the Alexander duality theorem (Alexander [5]), is closely related to the theory of linking numbers. Let a finite r-dimensional complex $\Re^r$ lie embedded in Euclidean n-space $\Re^n$ $(n > 1)$; by this we mean that $\Re^r$ is a subset of $\Re^n$. Let us denote the complement of the set $\Re^r$ in $\Re^n$ by $\Re^n - \Re^r$; this complement is an open set in a Euclidean space and it is also a complex; in fact, it is an infinite complex (§14). The same is true for the point set $\mathfrak{S}^n - \Re^r$ which arises when one closes $\Re^n$ with a single point at infinity to produce the n-sphere $\mathfrak{S}^n$. The Alexander duality theorem states a relation between the Betti groups of $\Re^r$ and those of $\mathfrak{S}^n - \Re^r$: If p^k is the kth Betti number of $\Re^r$ and $\bar{p}^k$ is the kth Betti number of $\mathfrak{S}^n - \Re^r$, then

$$p^k = \bar{p}^{n-k-1} \qquad (k \neq 0, k \neq n-1)$$

and in the two exceptional cases

$$p^0 = \bar{p}^{n-1} + 1, \qquad p^{n-1} = \bar{p}^0 - 1.$$

The deeper reason underlying the existence of such a relation involving the Betti numbers becomes evident only after one has chosen Betti bases for $\Re^r$ and $\mathfrak{S}^n - \Re^r$. It is possible to choose a k-dimensional Betti basis of $\Re^r$

$$B_1^k, \ldots, B_{p^k}^k$$

and an $(n - k - 1)$-dimensional Betti basis of $\mathfrak{S}^n - \Re^r$

$$\bar{B}_1^{n-k-1}, \ldots, \bar{B}_{\bar{p}^{n-k-1}}^{n-k-1}$$

(for $k \neq 0$ and $k \neq n - 1$) so that the matrix of linking numbers becomes

$$\mathbb{V}\left(B_i^k, \bar{B}_j^{n-k-1}\right) = \delta_{ij} = \begin{cases} 1 & \text{for} \quad i = j \\ 0 & \text{for} \quad i \neq j. \end{cases}$$

For the case $k = 0$, the Betti basis consists of points whose number is equal to the number of isolated connected subcomplexes contained in the complex; in this case, instead of using the Betti basis, one uses a complete system of homologously independent 0-chains having 0 as its sum of coefficients (§18), and this system contains one fewer point. A corresponding remark holds with regard to the basis B_j^0 in the case $k = n - 1$.

The same theorem is also valid mod 2. One merely replaces the Betti bases and Betti numbers by connectivity bases and connectivity numbers (Alexander [5]; Pontrjagin [3]). The theorem has also been extended from the case of the n-sphere to that of arbitrary manifolds (van Kampen [3]; Pontrjagin [3]).

Some consequences of the Alexander duality theorem:

1. The number of domains into which the n-sphere $\mathfrak{S}^n$ is decomposed by an embedded complex $\mathfrak{K}^r$ is determined by the complex alone, and is independent of the manner of its embedding. It is equal to the $(n-1)$th Betti number plus 1, and is also equal to the $(n-1)$th connectivity number plus 1, of $\mathfrak{K}^r$. It follows from this that an $(n-1)$-dimensional nonorientable pseudomanifold cannot be embedded in $\mathfrak{S}^n$ (Brouwer [10]). For in this case $p^{n-1} \doteq 0$ and the connectivity number $q^{n-1} = 1$, so that the number of domains into which $\mathfrak{S}^n$ is decomposed would be 1 in one computation and 2 in the other. We have already proved this theorem for surfaces in $\mathfrak{S}^3$, admittedly not for the case of an arbitrary topological embedding, but for a simplicial embedding of the surfaces (§64 and Comment 36). An orientable $(n-1)$-dimensional manifold $\mathfrak{M}^{n-1}$ embedded in $\mathfrak{S}^n$ will, on the other hand, decompose $\mathfrak{S}^n$ into exactly two domains, because $p^{n-1} = 1$. The domain containing the "point at infinity" of $\mathfrak{S}^n$ is called the outside of $\mathfrak{M}^{n-1}$; the other domain is called the inside of $\mathfrak{M}^{n-1}$. The Jordan curve theorem is a special case, and states that the Euclidean plane is decomposed into two path connected open subsets by the topological image of a circle.

2. The theorem of invariance of domain: If one topologically maps a domain $\mathfrak{G}$ of Euclidean n-space $\mathfrak{R}^n$ (that is, an open subset of $\mathfrak{R}^n$) onto a subset $\mathfrak{G}'$ of another Euclidean n-space $'\mathfrak{R}^n$, then $\mathfrak{G}'$ is also a domain (Brouwer [12]). Let us first prove the following lemma: Let $\mathfrak{S}^n$ be the n-sphere which arises from $\mathfrak{R}^n$ when $\mathfrak{R}^n$ is closed with a point P at infinity. Let $\mathfrak{E}^n$ be a topological n-simplex lying in $\mathfrak{R}^n$ and having the $(n-1)$-sphere $\mathfrak{S}^{n-1}$ as its boundary. If $\mathfrak{I}$ and $\mathfrak{A}$ are, respectively, the inside and outside domains into which $\mathfrak{S}^n$ is decomposed by $\mathfrak{S}^{n-1}$ [from (1) above], then $\mathfrak{E}^n$ consists of just the point set $\mathfrak{I} + \mathfrak{S}^{n-1}$. Proof: The intersection number mod 2 of P with $\mathfrak{E}^n$ is 0, since P is disjoint with $\mathfrak{E}^n$. The intersection number mod 2 of $\mathfrak{E}^n$ with each arbitrary point of $\mathfrak{A}$ then vanishes. For one can connect any two points of $\mathfrak{A}$ with a path which does not intersect $\mathfrak{S}^{n-1}$. If there were to exist a point in $\mathfrak{I}$ which was disjoint with $\mathfrak{E}^n$, then, for the same reason, the intersection number mod 2 of each point of $\mathfrak{I}$ would have to vanish. But in that case, each arbitrary 0-chain mod 2 $\mathfrak{U}^0$ lying in $\mathfrak{S}^n - \mathfrak{S}^{n-1}$ which was null homologous in $\mathfrak{S}^n$ would have vanishing linking number, that is, $\mathcal{V}(\mathfrak{S}^{n-1}, \mathfrak{U}^0) = \mathfrak{S}(\mathfrak{E}^n, \mathfrak{U}^0)$, in contradiction to the Alexander duality theorem. Thus all points of $\mathfrak{I}$ belong to $\mathfrak{E}^n$. A point of $\mathfrak{A}$ cannot, then, belong to $\mathfrak{E}^n$, since one can connect any two inner points of a simplex with an interval which does not intersect the boundary of the simplex.

It follows, as a special case of the lemma, that the midpoint of $\mathfrak{E}^n$ is an interior point of $\mathfrak{E}^n$ with respect to $\mathfrak{R}^n$. The theorem of invariance of domain follows immediately when one constructs a small geometric simplex having the midpoint X about an arbitrary point X of $\mathfrak{G}$. The image of the n-simplex is then a topological simplex belonging to $\mathfrak{G}'$ and X' is an interior point.

48. (§77) The asymmetry of the lens space (3, 1) was noticed by Kneser [8]. Further details on demonstrating the distinctness of manifolds with the help of linking numbers are to be found in Alexander [10], de Rham [1], Reidemeister [9], and Seifert [4].

49. (§78) The concept of the degree of a mapping was introduced by Brouwer [8].

The problem of constructing a continuous mapping of $\mathfrak{R}^n$ to $\mathbf{K}^n$ having a given degree does not always have a solution. For example, if $\mathfrak{R}^n$ is an orientable surface of genus $p > 0$ and $\mathbf{K}^n$ is an orientable surface of genus q, the degree of the mapping γ is subject to the restrictions $|\gamma|(q-1) \leqq p - 1$. In particular, γ can take only the values $\gamma = 0, +1, -1$ when a surface of genus $p > 1$ is mapped to itself (Kneser [9]).

It seems obvious to ask the question: Can a mapping of degree γ always be deformed in a way so that a (small) domain of $\mathbf{K}^n$ can be covered exactly $|\gamma|$ times? This question has been answered in a positive sense by Hopf [8, Part II] and by Kneser [9]. The theory of the degree of a mapping is only one of the algebraic group theoretic methods which are useful in dealing with continuous mappings. This algebra of mappings is described in Hopf [12, 14, 18]. Additional references to the literature are also to be found there.

50. (§80) The fixed point formula. The text derivation of the fixed point formula follows from the work of Hopf [9]. This formula is a special case of the general Lefschetz–Hopf fixed point formula: *Given an arbitrary continuous mapping f of a pure complex $\mathfrak{N}^n$ into itself, then $\sum(-1)^k \operatorname{Sp} \mathbf{B}^k$ is equal to the negative of the algebraic sum of the indexes of all fixed points.* The index of an isolated fixed point P is defined as follows, when P is an inner point of an n-simplex. Construct a small $(n-1)$-sphere $\mathfrak{S}^{n-1}$ about P. From each point Q of this sphere draw the vector $\mathfrak{V}_Q$ to the image point $f(Q)$. The ray through P drawn parallel to $\mathfrak{V}_Q$ cuts $\mathfrak{S}^{n-1}$ at a point $\varphi(Q)$. The mapping φ is a continuous mapping of $\mathfrak{S}^{n-1}$ into itself. *The index of the fixed point* is defined to be the degree of this mapping.

The concept of the index of a fixed point is due essentially to Poincaré [1, Parts 3 and 4]. Hopf [7, 11] proved the general fixed point formula by means of a reduction to the special case whose derivation appears in the text. The original proof by Lefschetz [6, 7] made use of the "*topological product method,*" which we shall briefly sketch here.

One can describe a continuous mapping f of a complex $\mathfrak{N}^n$ into a complex $\mathbf{K}^m$ by forming the topological product space $\mathfrak{N}^n \times \mathbf{K}^m$ and assigning the point $P \times f(P)$ in this space to the point P of $\mathfrak{N}^n$. The points assigned in this manner form a subset of the topological product which is homeomorphic to $\mathfrak{N}^n$. We shall call this subset the *characteristic subcomplex* of the mapping f in the topological product space. If we are now given, besides f, an additional continuous mapping g of $\mathfrak{N}^n$ into $\mathbf{K}^m$, then we define a coincidence point of the mappings f and g to be a point P such that $f(P) = g(P)$. The coincidence points correspond to the common points of the characteristic subcomplexes of f and g. In particular, if $\mathfrak{N}^n$ and $\mathbf{K}^m$ are manifolds having the same dimension, then the characteristic subcomplexes of f and g have an intersection number which is defined to be the algebraic number of coincidence points. Lefschetz has given a formula which allows one to compute this number, when one knows the homeomorphic mappings, of the homology groups induced by f and g. The general fixed point formula then results when one examines the special case that $\mathfrak{N}^n$ and $\mathbf{K}^m$ are the same and g is the identity mapping. Lefschetz' proof will consequently hold only for manifolds (with or without boundary); on the other hand, it also includes the case of multivalued mappings. The topological product method has proved useful in other mapping problems (see, for example, Hopf [12, 14]).

The fixed point formula is closely related to the theory of *continuous vector fields*. A "small transformation" f, which is a self-mapping which displaces points only slightly, will determine a vector field. The vectors of the field run from the original points to their respective images. The field is continuous except at the singular positions, which correspond to the fixed points of f and which we assume to be finite in number. We define the index of a singular position of the vector field to be the index at the fixed point in question, of the small transformation. Since, for a small transformation, we have $\operatorname{Sp} \mathbf{B}^i = p^i$ and the alternating sum of the Betti numbers is equal to the negative of the Euler characteristic, there then follows Hopf's theorem [5]: *The sum of the indexes of the singular positions of a vector field is equal to the negative of the Euler characteristic of the manifold.*

On occasion, the fixed point formula can be used to ascertain the existence of a fixed point (§81). One must distinguish this question from another question, which is worth consideration: *What is the smallest number of fixed points of a mapping class?* That is, what is the smallest number of fixed points which can be obtained by means of a deformation of a given mapping? J. Nielsen has succeeded in evaluating this minimal number by dividing the fixed points into *fixed point classes*. Two fixed points are assigned to the same fixed point class if it is possible to connect them with a path w such that w together with its image path form a null homotopic closed path. An equivalent definition is obtained if one lifts the mapping of the base complex into the universal covering complex, which in general can be done in several ways. The fixed points of such a covering mapping will, after projection into the base complex, then belong to the same class. See Nielsen [4], [6] and Hopf [6], where one can find references to, in part, earlier work by Alexander, Brouwer, Birkhoff, Feigl, and others.

51. (§81) The theorem that a $(2k + 1)$-dimensional manifold admits fixed point-free deformations is valid in general (see Hopf [5]).

52. (§86) In §58 we have already encountered in regard to the covering complexes, an example where topological considerations yield group theoretic results which are not obtained easily (if at all) by arithmetic procedures. Among the other examples of this type is the question of determining the structure of a group from its generators and relations and, in particular, of solving the isomorphism problem: Are two groups, which are presented in terms of their generators and relations, isomorphic?

A particularly fine example is to be found in Artin's theory of braids [2]. He shows that the symmetric permutatiom group on n symbols is given by two generators a and σ and by the relations $a^n = (a\sigma)^{n-1}$, $\sigma^2 = 1$, and these commutation relations: σ commutes with $a^i \sigma a^{-i}$ $(2 \leqq i \leqq n/2)$. Another example is given in Threlfall and Seifert [1, p. 577].

A procedure which allows insight into the properties of the group (theoretically in all cases, and practically in many cases) is the construction of a group diagram. This is a topological edge or surface complex, which can be obtained from the group generators and relations (Dehn [3]; Threlfall [1]). As an example, when one cuts a manifold in some manner to form a polyhedron (as in §60) and constructs the universal covering complex, the polyhedron will then lift to a cellular division. The edge complex of the dual cellular division is a Dehn group diagram of the fundamental group. Thus, for example, by cutting apart the spherical dodecahedron space, we find that the group having the relations $A^5 = B^2 = C^3 = ABC$ is the binary icosahedral group of order 120. We can get an idea of the difficulty of the isomorphism problem when we select two ways of cutting apart the manifold from among the unforseeably many possibilities, and attempt to convert the systems of generators and relations, to be obtained as in §62, to one another.

BIBLIOGRAPHY

It is not claimed that the present bibliography is complete. One can find bibliographic listings of the literature in M. Dehn and P. Heegard [1] to 1907; B. von Kerékjártó [6] to 1923; B. L. van der Waerden [2] to 1930; S. Lefschetz [12] to 1930; H. Tietze and L. Vietoris [1] to 1930 (complete to 1926); K. Reidemeister [6], knot theory to 1932.

ALEXANDER, J. W.

[1] A proof of the invariance of certain constants in Analysis Situs. *Trans. Amer. Math. Soc.* **16** (1915), 148–154.

[2] Note on two three-dimensional manifolds with the same group. *Trans Amer. Math Soc.* **20** (1919), 339–342.

[3] Note on Riemann spaces. *Bull. Amer. Math. Soc.* **26** (1919), 370–372.

[4] On transformations with invariant points. *Trans. Amer. Math. Soc.* **23** (1922), 89–95.

[5] A proof and extension of the Jordan-Brouwer separation theorem. *Trans. Amer. Math. Soc.* **23** (1922), 333–349.

[6] On the deformation of an *n*-cell. *Proc. Nat. Acad. Sci. U.S.A.* **9** (1923), 406–407.

[7] Invariant points of a surface transformation of given class. *Trans. Amer. Math. Soc.* **25** (1923), 173–184.

[8] A lemma on a system of knotted curves. *Proc. Nat. Acad. Sci. U.S.A.* **9** (1923), 93–95.

[9] On the subdivision of 3-space by a polyhedron. *Proc. Nat. Acad. Sci. U.S.A.* **10** (1924), 6–8.

[10] New topological invariants expressible as tensors. *Proc. Nat. Acad. Sci. U.S.A.* **10** (1924), 99–101.

[11] On certain new topological invariants of a manifold. *Proc. Nat. Acad. Sci. U.S.A.* **10** (1924), 101–103.

[12] Topological invariants of a manifold. *Proc. Nat. Acad. Sci. U.S.A.* **10** (1924), 493–494.

[13] On the intersection invariants of a manifold. *Proc. Nat. Acad. Sci. U.S.A.* **11** (1925), 143–146.

[14] Note on a theorem of H. Kneser. *Proc. Nat. Acad. Sci. U.S.A.* **11** (1925), 250–251.

[15] Combinatorial Analysis Situs. *Trans. Amer. Math. Soc.* **28** (1926), 301–329.

[16] Topological invariants of knots and links. *Trans. Amer. Math. Soc.* **30** (1928), 275–306.

[17] The combinatorial theory of complexes. *Ann. Math.* (2) **31** (1930), 294–322.

[18] Some problems in topology. *Verh. Int. Math.-Kongr.* **1** (1932), 249–257.

[19] A matrix knot invariant. *Proc. Nat. Acad. Sci. U.S.A.* **19** (1933), 272–275.

ALEXANDER, J. W., and G. B. BRIGGS

[1] On types of knotted curves. *Ann. Math.* (2) **28** (1927), 562–586.

ALEXANDER, J. W., and L. W. COHEN
  [1] A classification of the homology groups of compact spaces. *Ann. Math.* (2) **33** (1932), 538–566.
ALEXANDROFF, P.
  [1] Zur Begründing der *n*-dimensionalen mengentheoretischen Topologie. *Math. Ann.* **94** (1925), 296–308.
       (On the foundations of set-theoretical topology in *n* dimensions.)
  [2] Simpliziale Approximationen in der allgemeinen Topologie. *Math. Ann.* **96** (1927), 489–511.
       (Simplicial approximations in general topology.)
  [3] Über kombinatorische Eigenschaften allgemeiner Kurven. *Math. Ann.* **96** (1927), 512–554.
       (On combinatorial properties of general curves.)
  [4] Über stetige Abbildungen kompakter Räume. *Math. Ann.* **96** (1927), 555–571.
       (On continuous mappings of compact spaces.)
  [5] Über die Dualität zwischen den Zusammenhangszahlen einer abgeschlossenen Menge und des zu ihr komplementären Raumes, *Nachr. Ges. Wiss. Göttingen* (1927), 323–329.
       (On the duality between the connectivity numbers of a closed set and those of its complementary space.)
  [6] Une définition des nombres de Betti pour un ensemble fermé quelconque. *C. R. Acad. Sci. Paris* **184** (1927), 317–329.
       (A definition of Betti's numbers for an arbitrary closed set.)
  [7] Sur le décomposition de l'espace par des ensembles fermés. *C. R. Acad. Sci. Paris* **184** (1927), 425–428.
       (On the decomposition of space by closet sets.)
  [8] Über den allgemeinen Dimensionsbegriff und seine Beziehung zur elementaren geometrischen Anschauung. *Math. Ann.* **98** (1928), 617–636.
       (On the general concept of dimension and its relation to elementary geometric intuition.)
  [9] Zum allgemeinen Dimensionsproblem. *Nachr. Ges. Wiss. Göttingen* (1928), 25–44.
       (On the general problem of dimension.)
  [10] Untersuchungen über Gestalt und Lage beliebiger abgeschlossener Mengen. *Ann. Math.* (2) **30** (1928), 101–187.
       (Studies of the form and position of arbitrary closed sets.)
  [11] Dimensionstheorie, ein Beitrag zur Theorie der abgeschlossenen Mengen. *Math. Ann.* **106** (1932), 161–238.
       (Dimension theory, a contribution to the theory of closed sets.)
  [12] "Einfachste Grundbegriffe der Topologie." Berlin, 1932.
       ("Basic Fundamentals of Topology.")
  [13] Sur la notion de dimension des ensembles fermés. *J. Math. Pures Appl.* (9) **11** (1932), 283–298.
       (On the concept of dimension of closed sets.)
  [14] Über einen Satz von Herrn Borsuk. *Monatschr. Math. Phys.* **40** (1933), 127–128.
       (On a theorem of Borsuk.)
  [15] Über die Urysohnschen Konstanten. *Fund. Math.* **20** (1933), 140–150.
       (On Urysohn's constants.)
ALEXANDROFF, P., and P. URYSOHN
  [1] Mémoire sur les espaces topologiques compacts. *Verh. Akad. Wetensch. Amsterdam* (1) **14**, No. 1 (1929).
       (A note on compact topological spaces.)
ANTOINE, L.
  [1] Sur l'homéomorphie de deux figures et de leurs voisinages. *J. Math. Pures Appl.* (8) **4** (1921), 221–325.
       (On the homeomorphism of two figures and of their neighborhoods.)

[2] Sur les ensembles parfaits partout discontinus. *C. R. Acad. Sci. Paris* **173** (1921), 284–285.
(On perfectly discontinuous sets.)

ARTIN, E.
[1] Zur Isotopie zweidimensionaler Flächen im R_4. *Abh. Math. Sem. Univ. Hamburg* **4** (1925), 174–177.
(On the isotopy of two-dimensional surfaces in R_4.)
[2] Theorie der Zöpfe. *Abh. Math. Sem. Univ. Hamburg* **4** (1925), 47–72.
(The theory of braids.)
[3] See also Klein [1] p. 346.

BAER, R.
[1] Kurventypen auf Flächen. *J. Reine Angew. Math.* **156** (1927), 231–246.
(Types of curves on surfaces.)
[2] Isotopie von Kurven auf orientierbaren geschlossenen Flächen. *J. Reine Angew. Math.* **159** (1928), 101–116.
(The isotopy of curves on closed orientable surfaces.)

BANKWITZ, C.
[1] Über die Torsionszahlen der zyklischen Überlagerungs räume des Knotenausenraumes. *Ann. Math.* (2) **31** (1930), 131.
(On the torsion numbers of the cyclic covering spaces of the exterior space of a knot.)
[2] Über die Fundamentalgruppe des inversen Knotens und des gerichteten Knotens. *Ann. Math.* (2) **31** (1930), 129.
(On the fundamental group of inverse and directed knots.)
[3] Über die Torsionszahlen der alternierenden Knoten. *Math. Ann.* **103** (1930), 145–162.
(On the torsion numbers of alternating knots.)

BEHNKE, H., and P. THULLEN
[1] "Theorie der Funktionen mehrerer komplexer Veränderlichen." Berlin, 1934.

BERGMANN, G.
[1] Zwei Bemerkungen zur abstrakten und kombinatorischen Topologie. *Monatschr. Math.* **38** (1931), 245–256.
(Two comments on abstract and combinatorial topology.)
[2] Zur algebraischen-axiomatischen Begründung der Topologie. *Math. Z.* **35** (1932), 502–511.
(On the algebraic-axiomatic foundation of topology.)

BETTI, E.
[1] Sopra gli spazi un numero qualunque di dimensioni. *Ann. Mat. Pura Appl.* (2) **4** (1871), 140–158.
(On spaces of arbitrary dimensionality.)

BIEBERBACH, L.
[1] "Projektive Geometrie." Leipzig, 1931.
[2] "Höhere Geometrie." Leipzig, 1933.
[3] "Differentialgleichungen, " 3rd ed.

BILZ, E.
[1] Beitrag zu den Grundlagen der kombinatorische Analysis Situs. *Math. Z.* **18** (1923), 1–41.
(A contribution to the foundations of combinatorial Analysis Situs.)

BIRKHOFF, G. D.
[1] Proof of Poincaré geometric theorem. *Trans. Amer. Math. Soc.* **14** (1913), 14–22.
[2] Dynamical systems with two degrees of freedom. *Trans. Amer. Math. Soc.* **18** (1917), 199–300.
[3] Une generalisation à *n* dimensions du dernier théorème de géométrie de Poincaré. *C. R. Acad. Sci. Paris* **192** (1921), 196–198.
(A generalization on Poincaré's last geometric theorem to *n* dimensions.)
[4] Dynamical systems. *Amer. Math. Soc. Colloq. Publ.* **9** (1927).

[5]   Einige Probleme der Dynamik. *Jahresber. Deutsch. Math.-Verein.* **38** (1929), 1–16.
      (Some problems in dynamics.)
BIRKHOFF, G. D., and O. D. KELLOG
[1]   Invariant points in function space. *Trans. Amer. Math. Soc.* **23** (1922), 96–115.
BIRKHOFF, G. D., and P. A. SMITH
[1]   Structure analysis of surface transformations. *J. Math. Pures Appl.* (9) **7** (1928), 345–379.
BLASCHKE, W.
[1]   *Differentialgeometrie* **1** (1921).
BOCHER, M.
[1]   "Einführung in die höhere Algebra."
BORSUK, K.
[1]   Über Schnitte der *n*-dimensionalen euklidischen Räume. *Math. Ann.* **106** (1932), 239–248.
      (On intersections of *n*-dimensional Euclidean spaces.)
[2]   Über die Abbildungen der metrischen kompakten Räume auf die Kreislinie. *Fund. Math.*
      20 (1933), 224–231.
      (On mappings of compact metric spaces onto the circle.)
[3]   Drei Sätze uber die *n*-dimensionale euklidische Sphäre. *Fund. Math.* **20** (1933), 177–190.
      (Three theorems on the *n*-dimensional Euclidean sphere.)
BOY, W.
[1]   Über die Curvatura integra und die Topologie geschlossener Flächen. *Math. Ann.* **57**
      (1903), 151–184.
      (On the integral curvature and the topology of closed surfaces.)
BRAUNER, K.
[1]   Zur Geometrie der Funktionen zweir komplexen Veränderlichen. II. Das Verhalten der
      Funktionen in der Umgebung ihrer Verzweigungsstellen. *Abh. Math. Sem. Univ.*
      Hamburg **6** (1928), 1–55.
      (On the geometry of functions of two complex variables. II. The behavior of the
      functions in neighborhoods of their branching locations.)
BRODY, E. J.
[1]   The topological classification of lens spaces. *Ann. Math.* **71** (1960), 163–184.
BROUWER, L. E. J.
[1]   On one-one continuous transformations of surfaces into themselves. *Proc. K. Ned. Akad.*
      Wetensch. **11** (1908), 788–798; **12** (1909), 286–297; **13** (1910), 767–777; **14** (1911),
      300–310; **15** (1912), 352–360; **22** (1920), 811–814; **23** (1920), 232–234.
[2]   On continuous vector distributions on surfaces. *Proc. K. Ned. Akad. Wetensch.* **11** (1908),
      850–858; **12** (1909), 716–734; **13** (1910), 171–186.
[3]   Zur Analysis situs. *Math. Ann.* **68** (1910), 422–434.
[4]   Beweis des Jordanschen Kurvensatzes. *Math. Ann.* **69** (1910), 169–175.
      (A proof of Jordan's curve theorem.)
[5]   Über eindeutige stetige Transformationen von Flächen in sich. *Math. Ann.* **69** (1910),
      176–180.
      (On single-valued continuous transformations of surfaces into themselves.)
[6]   Beweis der Invarianz der Dimensionenszahl. *Math. Ann.* **70** (1911), 161–165.
      (A proof of the invariance of the number of dimensions.)
[7]   Sur le théorème de M. Jordan dans l'espace à *n* dimensions. *C. R. Acad. Sci. Paris* **153**
      (1911), 542–544.
      (On Jordan's theorem in a space of *n* dimensions.)
[8]   Über Abbildung von Mannigfaltigkeiten. *Math. Ann.* **71** (1912), 97–115 and 598.
      (On the mapping of manifolds.)
[9]   On looping coefficients. *Proc. K. Ned. Akad. Wetensch.* **15** (1912), 113–122.
[10] Über Jordansche Mannigfaltigkeiten. *Math. Ann.* **71** (1912), 320–327.
      (On Jordan manifolds.)

[11] Beweis des Jordanschen Satzes für den *n*-dimensionalen Raum. *Math. Ann.* **71** (1912), 314–319.
(A proof of Jordan's theorem for *n*-dimensional space.)

[12] Beweis der Invarianz des *n*-dimensionalen Gebietes. *Math. Ann.* **71** (1912), 305–313; **72** (1912), 55–56.
(A proof of the invariance of *n*-dimensional domains.)

[13] Über den natürlichen Dimensionsbegriff. *Journ. f. Math.* **142** (1913), 146–152; *Proc. K. Ned. Akad. Wetensch.* **26** (1923), 795–800; *Math. Z.* **21** (1924), 312–314.
(On the natural concept of dimension.)

[14] Aufzählung der Abbildungsklassen endlichfach zusammenhängender Flächen. *Math. Ann.* **82** (1921), 280–286.
(Enumeration of the mapping classes of surfaces with finite connectivity.)

[15] Über die Minimalzahl der Fixpunkte bei den Klassen von eindeutigen stetigen Transformationen der Ringflächen. *Math. Ann.* **82** (1921), 94–96.
(On the minimal number of fixed points of the classes of single-valued continuous transformations of the toroidal surfaces.)

[16] On transformations of projective spaces. *Proc. K. Ned. Akad. Wetensch.* **29** (1926), 864–865.

BROWN, A. B.

[1] An extension of the Alexander duality theorem. *Proc. Nat. Acad. Sci. U.S.A.* **16** (1930), 407–408.

[2] On the join of two complexes. *Bull. Amer. Math. Soc.* **37** (1931), 417–420.

[3] Group invariants and torsion coefficients. *Ann. Math.* (2) **33** (1932), 373–376.

[4] Topological invariance of subcomplexes of singularities. *Amer. J. Math.* **54** (1932), 117–122.

BURAU, W.

[1] Über Zopfinvarianten. *Abh. Math. Sem. Univ. Hamburg* **9** (1932), 117–124.
(On braid invariants.)

[2] Kennzeichnung der Schlauchknoten. *Abh. Math. Sem. Univ. Hamburg* **9** (1932), 125–133.
(On the classification of tubular knots.)

CARATHÉODORY, C.

[1] Über die Begrenzung einfach zusammenhängender Gebiete. *Math. Ann.* **73** (1913), 323–370.
(On the boundaries of simply-connected domains.)

[2] "Reele Funktionen," 2nd ed. Chelsea, New York, 1947.

ČECH, E.

[1] Trois théorèmes sur l'homologie. *Publ. Fac. Sci. Univ. Masaryk* **144** (1931), 1–21.
(Three homology theorems.)

[2] Théorie générale de l'homologie dans un espace quelconque. *Fund. Math.* **19** (1932), 149–183.
(General homology theory for an arbitrary space.)

[3] Eine Verallgemeinerung des Jordan-Brouwerschen Satzes. *Ergebn. Math. Kolloq.* H. 5 (1933), 29–31.
(A generalization of the Jordan-Brouwer theorem.)

[4] Théorie générale des variétés et de leurs théorèmes de dualité. *Ann. Math.* (2) **34** (1933), 621–730.
(A general theory of varieties and their duality theorems.)

[5] Introduction à la théorie de l'homologie. *Publ. Fac. Sci. Univ. Masaryk* **184** (1933), 1–34 (in Czech., French summary, pp. 35–36).
(Introduction to homology theory.)

CHUARD, J.

[1] Questions d'analysis situs. *Rend. Circ. Mat. Palermo* (2) **46** (1922), 185–224.

DEHN, M.
[1] Über die Topologie des dreidimensionalen Raumes. *Math. Ann.* **69** (1910), 137–168. (on the topology of three-dimensional space.)
[2] Transformation der Kurven auf zweiseitigen Flächen. *Math. Ann.* **72** (1912), 413–421. (Transformations of curves on two-sided surfaces.)
[3] Über unendliche diskontinuierliche Gruppen. *Math. Ann.* **71** (1912), 116–144. (On infinite discontinuous groups.)
[4] Die beiden Kleeblattschlingen. *Math. Ann.* **75** (1914), 402–413. (The two trefoil knots.)
DEHN, M., and P. HEEGARD
[1] Analysis Situs. *Enzymol. Math. Wiss.* (III) AB 3 (1907).
DE RHAM, G.
[1] Sur l'analysis situs des variétés à *n* dimensions. *J. Math. Pures Appl.* (9) **10** (1931), 115–120.
(On the Analysis Situs of *n*-dimensional varieties.)
[2] Sur la théorie des intersections et les imtegrales multiples. *Comment. Math. Helv.* **4** (1932), 151–154.
(On the theory of intersections and multiple integrals.)
EHRESMANN, C.
[1] Sur la topologie de certaines variétés algebriques. *C. R. Acad. Sci. Paris* **196** (1933), 152–154.
(On the topology of certain algebraic varieties.)
EPHRAIMOWITSCH, W.
[1] Zur Theorie der nichtorientierbaren Mannigfaltigkeiten. *Math. Z.* **29** (1928), 55–59. (On the theory of non-orientable manifolds.)
ERRERA, A.
[1] L'origine et les problémes de l'analysis situs. *Rev. Univ. Brux.* **7–8** (1921), 1–15. (The origin and problems of Analysis Situs.)
[2] Sur les polyèdres réguliers de l'analysis situs. *Mém. Acad. Belg.* (2) **7** (1922). (On the regular polygon polyhedra of Analysis Situs.)
FEIGL, G.
[1] Fixpunktsätze für spezielle *n*-dimensionale Mannigfaltigkeiten. *Math. Ann.* **98** (1927), 355–398.
(Fixed point theorems for special *n*-dimensional manifolds.)
[2] Geschichtliche Entwicklung der Topologie. *Jahresber. Deutsch. Math.-Verein.* **37** (1928), 273–286.
(The historical development of topology.)
FENCHEL, W.
[1] Elementare Beweise und Anwendungen einiger Fixpunktsätze. *Mat. Tidskr.* B.H. **3–4** (1932), 66–87.
(Elementary proofs and applications of some fixed point theorems.)
FISCHER, A.
[1] Gruppen und Verkettungen. *Comment. Math. Helv.* **2** (1930), 253–268. (Groups and links.)
FLEXNER, W.W.
[1] The Poincaré duality theorem for topological manifolds. *Ann. Math.* (2) **32** (1931), 539–548.
[2] On topological manifolds. *Ann. Math.* (2) **32** (1931), 393–406.
FLEXNER, W.W., and S. LEFSCHETZ
[1] On the duality theorem for the Betti numbers of topological manifolds. *Proc. Nat. Acad. Sci. U.S.A.* **16** (1930), 530–533.
FRANKL, F.
[1] Topologische Beziehungen in sich kompakter Teilmengen euklidischer Räume zu ihren Komplementen usw. *Sitzungsber. Akad. Wiss. Wien* **136** (1927), 689–699.

(Topological relations between relatively compact subsets of Euclidean spaces and their complements.)

[2] Zur Topologie des dreidimensionalen Raumes. *Monatschr. Math. Phys.* **38** (1931), 357–364.
(On the topology of three-dimensional space.)

FRANKL, F., and L. PONTRJAGIN
[1] Ein Knotensatz mit Anwendung auf die Dimensionstheorie. *Math. Ann.* **102** (1930), 785–789.
(A knot theorem having application to dimension theory.)

FURCH, R.
[1] Orientierung von Hyperflächen im projektiven Raum. *Abh. Math. Sem. Univ. Hamburg* **1** (1922), 210–212.
(On the orientation of hypersurfaces in projective space.)
[2] Zur Grundlegung der kombinatorischen Topologie. *Abh. Math. Sem. Univ. Hamburg* **3** (1924), 69–88.
(On the foundations of combinatorial topology.)
[3] Zur kombinatorischen Topologie des dreidimensionalen Raumes. *Abh. Math. Sem. Univ. Hamburg* **3** (1924), 237–245.
(On the combinatorial topology of three-dimensional space.)
[4] Über den Schnitt zweier Sphären in R^3. *Math. Z.* **28** (1928), 556–566.
(On the intersection of two spheres in R^3.)
[5] Polyedrale Gebilde verschiedener Metric. *Math. Z.* **32** (1930), 512–544.
(Polyhedral figures with different metrics.)

GAWEHN, J.
[1] Über unberandete zweidimensionale Mannigfaltigkeiten. *Math. Ann.* **98** (1927), 321–354.
(On two-dimensional manifolds without boundary.)

GIESEKING, H.
[1] "Analytische Untersuchungen über topologische Gruppen." Hilchenbach, 1912.
(Analytic investigations of topological groups.)

GOERITZ, L.
[1] Die Heegaard-Diagramme des Torus. *Abh. Math. Sem. Univ. Hamburg* **9** (1932), 187–188.
(The Heegaard diagrams of the torus.)
[2] Knoten und quadratische Formen. *Math. Z.* **36** (1933), 647–654.
(Knots and quadratic forms.)
[3] Normalformen der Systeme einfacher Kurven auf orientierbaren Flächen. *Abh. Math. Sem. Univ. Hamburg* **9** (1933), 223–243.
(Normal forms of the systems of simple curves on orientable surfaces.)
[4] Die Abbildungen der Brezelfläche und der Vollbrezel vom Geschlecht 2. *Abh. Math. Sem. Univ. Hamburg* **9** (1933), 244–259.
(The mappings of the pretzel surface and the solid pretzel of genus 2.)

HADAMARD, J.
[1] La géometrie de situation et sone rôle en mathématiques. *Rev. Mois* **8** (1909), 38–60.
(The geometry of place and its role in mathematics.)
[2] Sur quelques applications de l'indice de Kronecker. Appendix to J. Tannery, "Théorie des Fonctions." Paris, 1930.
(Some applications of Kronecker's index.)

HAUSDORFF, F.
[1] "Grundzüge der Mengenlehre," 1st ed. Leipzig, 1914.
(Foundations of the theory of sets.)
[2] "Grundzüge der Mengenlehre," 2nd ed. Leipzig, 19.

HEEGAARD, P.
[1] Sur l'Analysis Situs. *Bull. Soc. Math. France* **44** (1916), 161–242.

HEESCH, H.
[1] Über topologisch reguläre Teilungen geschlossener Flächen. *Nachr. Ges. Wiss. Göttingen* (1932), 268–273.
(On topologically regular divisions of closed surfaces.)

HENSEL, K., and G. LANDSBERG
[1] "Theorie der algebraischen Funktionen einer Variable," Leipzig, 1902.

HILBERT, D., and S. COHN-VOSSEN
[1] "Anschauliche Geometric." Berlin, 1932.

HILTON, P. J., and S. WYLIE.
[1] "Homology Theory." Cambridge Univ. Press, London and New York, 1960.

HOPF, H.
[1] Die Curvatura integra Clifford-Kleinscher Raumformen. *Nachr. Ges. Wiss. Göttingen* (1925), 131–141.
(The integral curvature of Clifford-Klein space forms.)
[2] Zum Clifford-Kleinschen Raumproblem. *Math. Ann.* **95** (1925), 313–339.
(On the Clifford-Klein space problem.)
[3] Über die Curvatura integra geschlossener Hyperflächen. *Math. Ann.* **95** (1925), 340–367.
(On the integral curvature of closed hypersurfaces.)
[4] Abbildungsklassen *n*-dimensionaler Mannigfaltigkeiten. *Math. Ann.* **96** (1927), 209–224.
(Mapping classes of *n*-dimensional manifolds.)
[5] Vektorfelder in *n*-dimensionalen Mannigfaltigkeiten. *Math. Ann.* **96** (1927), 225–250.
(Vector fields in *n*-dimensional manifolds.)
[6] Über Mindestzahlen von Fixpunkten. *Math. Z.* **26** (1927), 726–774.
(On the minimal numbers of fixed points.)
[7] A new proof of the Lefschetz formula on invariant points. *Proc. Nat. Acad. Sci. U.S.A.* **14** (1928), 149–153.
[8] On some properties of one-valued transformations. *Proc. Nat. Acad. Sci. U.S.A.* **14** (1928), 206–214.
[9] Eine Verallgemeinerung der Euler-Poincaréschen Formel. *Nachr. Ges. Wiss. Göttingen* (1928), 127–136.
(A generalization of the Euler-Poincaré formula.)
[10] Zur Topologie der Abbildungen von Mannifaltigkeiten. I. Neue Darstellung der Theorie des Abbildungsgrades für topologische Mannigfaltigkeiten. *Math. Ann.* **100** (1928), 579–608. II. Klasseninvarianten von Abbildungen. *ibid.* **102** (1929), 562–623.
(On the topology of the mappings of manifolds. I. A new description of the theory of the degree of a mapping for topological manifolds. II. Class invariants of mappings.)
[11] Über die algebraische Anzahl von Fixpunkten. *Math. Z.* **29** (1929), 493–524.
(On the algebraic number of fixed points.)
[12] Zur Algebra der Abbildungen von Mannigfaltigkeiten. *J. Reine Angew. Math.* **163** (1930), 71–88.
(On the algebra of the mappings of manifolds.)
[13] Über wesentliche und unwesentliche Abbildungen von Komplexen. *Moskau Math. Samml.* (1930), 53–62.
(On essential and inessential mappings of complexes.)
[14] Beiträge zur Klassifizierung der Flächenabbildungen. *J. Reine Angew. Math.* **165** (1931), 225–236.
(Contributions to the classification of mappings of surfaces.)
[15] Geometrie infinitesmale et Topologie. *Enseign. Math.* (2) **30** (1931), 233–240.
[16] Über die Abbildungen der dreidimensionalen Sphäre auf die Kugelfläche. *Math. Ann.* **104** (1931), 637–665.
(On the mappings of the three-sphere onto the two-sphere.)
[17] Differentialgeometrie und topologische Gestalt. *Jahresber. Deutsch. Math.-Verein.* **41** (1932), 209–229.
(Differential geometry and topological configuration.)

[18]   Die Klassen der Abbildungen der n-dimensionalen Polyeder auf die n-dimensionale Sphäre. *Comment. Math. Helv.* **5** (1933), 39–54.
(The classes of mappings of n-dimensional polyhedra onto the n-sphere.)

HOPF, H., and E. PANNWITZ

[1]   Über stetige Deformationen von Komplexen in sich. *Math. Ann.* **108** (1933), 433–465.
(On continuous deformations of a complex into itself.)

HOPF, H., and W. RINOW

[1]   Über den Begriff der vollständigen differentialgeometrischen Fläche. *Comment. Math. Helv.* **3** (1931), 209–225.
(On the concept of a complete differential-geometric surface.)

HOTELLING, H.

[1]   Three dimensional manifolds of states of motions. *Trans. Amer. Math. Soc.* **27** (1925), 329–344.

[2]   Multiple-sheeted spaces and manifolds of states of motions. *Trans. Amer. Math. Soc.* **28** (1926), 479–490.

JACOBI, C. G. J.

[1]   "Gesammelte Werke," Suppl. Vol. Berlin, 1884.

JOHANSSON, J.

[1]   Topologische Untersuchungen über unverzweigte Uberlagerungsflächen. *Norske Vid.-Akad. Oslo Math.-Natur. Kl. Skr.* No. 1 (1931), 1–69.
(Topological investigations of unbranched covering surfaces.)

[2]   Zu den zweidimensionalen Homotopiegruppen. *Norsk Mat. Foren. Skr.* (2) No. 1/12 (1933), 55–59.
(On the two-dimensional homotopy groups.)

KÄHLER, E.

[1]   Über die Verzweigung einer algebraischen Funktion zwier Veränderlichen in der Umgebung einer singulären Stelle. *Math. Z.* **29** (1929), 188–204.
(On the branching of an algebraic function of two variables in the neighborhood of a singularity.)

KIANG, TSAI-HAN

[1]   On the groups of orientable two-manifolds. *Proc. Nat. Acad. Sci. U.S.A.* **17** (1931), 142–144.

KIRBY, and SIEBENMANN

[1]   On the triangulation of manifolds and the Hauptvermutung. *Bull. Amer. Math. Soc.* **75** (1969), 742–749.

KLEIN, F.

[1]   "Höhere Geometrie," 3rd ed. Berlin, 1926.

[2]   "Nichteuklidische Geometrie." Berlin, 1928.

[3]   "Vorlesungen über nicht-Euklidische Geometrie." Chelsea, New York.

KNESER, H.

[1]   Kurvenscharen auf geschlossenen Flächen. *Jahresber. Deutsch. Math.-Verein.* **30** (1921), 83–85.
(Families of curves on closed surfaces.)

[2]   Reguläre Kurvenscharen auf den Ringflächen. *Math. Ann.* **91** (1924), 135–154.
(Regular families of curves on the ring surfaces.)

[3]   Ein topologischer Zerlegungssatz. *Proc. K. Ned. Akad. Wetensch.* **27** (1924), 601–616.
(A topological decomposition theorem.)

[4]   Die Topologie der Mannigfaltigkeiten. *Jahresber. Deutsch. Math.-Verein.* **34** (1925), 1–14.
(The topology of manifolds.)

[5]   Eine Bemerkung über dreidimensionale Mannigfaltigkeiten. *Nachr. Ges. Wiss. Göttingen* (1925), 128–130.
(A comment on three-dimensional manifolds.)

[6]   Die Deformationssätze der einfachzusammenhängenden Flächen. *Math. Z.* **25** (1926), 362–372.
(Deformation theorems for simply-connected surfaces.)

[7] Glättung von Flächenabbildungen. *Math. Ann.* **100** (1928), 609–617.
(The smoothing of surface mappings.)

[8] Geschlossene Flächen in dreidimensionale Mannigfaltigkeiten. *Jahresber. Deutsch. Math.-Verein.* **38** (1929), 248–260.
(Closed surfaces in three-dimensional manifolds.)

[9] Die kleinste Bedeckungszahl innerhalb einer Klasse von Flächenabbildungen. *Math. Ann.* **103** (1930), 347–358.
(The smallest covering number within a class of surface mappings.)

KNOPP, K.
[1] "Functionentheorie." Göschen, 1926.

KOEBE, P.
[1] Riemannsche Mannigfaltigkeiten und nichteuklidische Raumformem. I. *Sitzungsber. Preuss. Akad. Wiss.* (1927), 164–196; II. (1928), 345–348; III. (1928), 385–442; IV. (1929), 414–457; V. (1930), 304–364; VI. (1930), 505–541; VII. (1931), 506–534; VIII. (1932), 249–284.
(Riemann manifolds and non-Euclidean space forms.)

KOOPMAN, B. O., and A. B. BROWN
[1] On the covering of analytic loci by complexes. *Trans. Amer. Math. Soc.* **34** (1932), 231–251.

KOWALEWSKI, G.
[1] "Analytische Geometrie." Leipzig, 1923.

KREINES, M.
[1] Zur Konstruktion der Poincaré-Räume. *Rend. Circ. Mat. Palermo* **56** (1932), 277–280.
(On the construction of Poincaré spaces.)

KRONECKER, L.
[1] Über Systeme von Funktionen mehrerer Variabeln. *Monatsber. Berlin. Akad.* (1869), 159–193 and 688–698.
(On systems of functions of several variables.)

KÜNNETH, H.
[1] Zur Bestimmung der Fundamentalgruppe einer Produktmannigfaltigkeit. *Sitzungsber. Phys.-Med. Soz. Erlangen* **54–55** (1922–1923), 190–196.
(On the determination of the fundamental group of a product manifold.)

[2] Über die Bettischen Zahlen einer Produktmannigfaltigkeit. *Math. Ann.* **90** (1923), 65–85.
(On the Betti numbers of a product manifold.)

[3] Über die Torsionszahlen von Produktmannigfaltigkeiten. *Math. Ann.* **91** (1924), 125–134.
(On the torsion coefficients of product manifolds.)

KURATOWSKI, C.
[1] "Topologie. I. Espaces métrisables, espaces complets." Warsaw, 1933.
("Topology. I. Metrizable Spaces, Complete Spaces.")

LANDSBERG, G.
[1] Beiträge zur Topologie geschlossener Kurven mit Knotenpunkten und zur Kroneckerschen Charakteristikentheorie. *Math. Ann.* **70** (1911), 563–579.
(Contributions to the topology of closed curves having modal points and on Kronecker's theory of characteristics.)

LEFSCHETZ, S.
[1] Algebraic surfaces, their cycles and integrals. *Ann. Math.* (2) **21** (1920), 225–258.
[2] On certain numerical invariants of algebraic varieties, with application to Abelian varieties. *Trans. Amer. Math. Soc.* **22** (1921), 327–482.
[3] Continuous transformations of manifolds. *Proc. Nat. Acad. Sci. U.S.A.* **9** (1923), 90–93.
[4] "L'Analysis Situs et la Géométrie algébrique." Paris, 1924.
[5] Intersections of complexes and manifolds. *Proc. Nat. Acad. Sci. U.S.A.* **11** (1925), 287–289.
[6] Intersections and transformations of complexes and manifolds. *Trans. Amer. Math. Soc.* **28** (1926), 1–49.

[7]   Manifolds with a boundary and their transformations. *Trans. Amer. Math. Soc.* **29** (1927), 429–462.

[8]   The residual set of a complex on a manifold and related questions. *Proc. Nat. Acad. Sci. U.S.A.* **13** (1927), 614–622 and 805–807.

[9]   Closed point sets on a manifold. *Ann. Math.* (2) **29** (1928), 232–254.

[10]   Duality relations in topology. *Proc. Nat. Acad. Sci. U.S.A.* **15** (1929), 367–369.

[11]   On Transformations of closed sets. *Ann. Math.* (2) **32** (1930), 273–282.

[12]   "Topology." New York, 1930.

[13]   On compact spaces. *Ann. Math.* (2) **32** (1931), 521–538.

[14]   On singular chains and cycles. *Bull. Amer. Math. Soc.* **39** (1933), 124–129.

[15]   On generalized manifolds. *Amer. J. Math.* **55** (1933), 469–504.

LENNES, N. J.

[1]   Theorems on the simple finite polygon and polyhedron. *Amer. J. Math.* **33** (1911), 37–62.

[2]   Curves and surfaces in analysis situs. *Bull. Amer. Math. Soc.* (2) **17** (1911), 525.

LENSE, J.

[1]   Über die Indikatrix der projektiven Räume. *Jahresber. Deutsch. Math.-Verein.* **34** (1926), 243–244.
   (On the indicatrix of the projective spaces.)

LEVI, F.

[1]   "Geometrische Konfigurationen." Leipzig, 1929.

LISTING, J. B.

[1]   Vorstudien zur Topologie. *Gött. Studien* (1847), 811–875.
   (Preliminaries to topology.)

[2]   Der Census räumlicher Komplexe. *Abh. Ges. Wiss. Göttingen* **10** (1861), 97–180; *Nachr. Ges.Wiss. Göttingen* (1861), 352–358.
   (The enumeration of spatial complexes.)

LÖBELL, F.

[1]   Über die geodätischen Linien der Clifford-Kleinschen Flächen. *Math. Z.* **30** (1929), 572–607.
   (On the geodesic curves of Clifford-Klein surfaces.)

[2]   Ein Satz über die eindeutigen Bewgungen Clifford-Kleinscher Flächen in sich. *J. Reine Angew. Math.* **162** (1930), 114–125.
   (A theorem on single-valued transformations of Clifford-Klein surfaces.)

[3]   Zur Frage der Struktur der geschlossenen geodätischen Linien in den offenen Clifford-Kleinschen Flächen mit positive Charakteristik. *J. Reine Angew. Math.* **162** (1930), 125–131.
   (On the question of the structure of the closed geodesic curves on open Clifford-Klein surfaces of positive characteristic.)

[4]   Beispiele geschlossener dreidimensionaler Clifford-Kleinscher Räume neativer Krümmung. *Ber. Sächs Akad. Wiss.* **83** (1931), 168–174.
   (Some examples of closed three-dimensional Clifford-Klein spaces having negative curvature.)

MARKOV, A. A.

[1]   *Proc. Int. Congr. Math., 1958.*

MAYER, W.

[1]   Über abstrakte Topologie. *Monatschr. Math. Phys.* **36** (1929), 1–42 and 219–258.
   (On abstract topology.)

MENGER, K.

[1]   "Dimensionstheorie." Leipzig, 1928.

MOISE, E.

[1]   Affine structures in 3 manifold. V. The triangulation theorem and the Hauptvermutung. *Ann. Math.* (2) **56** (1952), 96–114.

MORSE, H. M.

[1] Recurrent geodesics on a surface of negative curvature. *Trans. Amer. Math. Soc.* **22** (1921), 84–100.

[2] A fundamental class of geodesics on any closed surface of genus greater than one. *Trans. Amer. Math. Soc.* **26** (1924), 25–60.

[3] Singular points of vector fields under general boundary conditions. *Amer. J. Math.* **51** (1929), 165–178.

NEWMAN, M. H. A.

[1] On the foundations of Combinatory Analysis Situs. *Proc. K. Ned. Akad. Wetensch.* **29** (1926), 611–641; **30** (1927), 670–673.

[2] A property of 2-dimensional elements. *Proc. K. Ned. Akad. Wetensch.* **29** (1927), 1401–1405.

[3] On the superposition of n-dimensional manifolds. *J. London Math. Soc.* (2) **2** (1927), 56–64.

[4] Topological equivalence of complexes. *Math. Ann.* **99** (1928), 399–412.

[5] Combinatory topology of convex regions. *Proc. Nat. Acad. Sci. U.S.A.* **16** (1930), 240–242.

[6] Combinatory topology and Euclidean n-space. *Proc. London Math. Soc.* (2) **30** (1930), 339–346.

[7] Intersection-complexes. I. Combinatory theory. *Proc. Cambridge Philos. Soc.* **27** (1931), 491–501.

[8] A theorem in combinatory topology. *J. London Math. Soc.* **6** (1931), 186–192.

[9] A theorem on periodic transformations of spaces. *Quart. J. Math. Oxford Ser.* (2) (1931), 1–8.

[10] On the products $C_h C_k$ and $C_h \times C_k$ in topology. *J. London Math. Soc.* **7** (1932), 143–147.

NIELSEN, J.

[1] Die Isomorphismen der allgemeinen unendlichen Gruppe mit zwei Erzeugenden. *Math. Ann.* **78** (1918), 385–397.
(The isomorphisms of the general infinite group having two generators.)

[2] Über fixpunktfreie topologisch Abbildungen geschlossener Flächen. *Math. Ann.* **81** (1919), 94–96.
(On fixed-point-free topological mappings of closed surfaces.)

[3] Über die Minimalzahl der Fixpunkte bei den Abbildungstypen der Ringflächen. *Math. Ann.* **82** (1920), 83–93.
(On the minimal number of fixed points of the mapping classes of the ring surfaces.)

[4] Über topologischen Abbildungen geschlossener Flächen. *Abh. Math. Sem. Univ. Hamburg* **3** (1924), 246–260.
(On topological mappings of closed surfaces.)

[5] Zur Topologie der geschlossenen Flächen. *Proc. Scand. Math. Congr., 6th*, (1925).
(On the topology of closed surfaces.)

[6] Untersuchungen zur Topologie der geschlossenen zweiseitigen Flächen. I. *Acta Math.* **50** (1927), 189–358; II. **53** (1929), 1–76; III. **58** (1932), 86–167.
(Investigations on the topology of closed two-sided surfaces.)

[7] Über reguläre Riemannsche Flächen. (In Danish.) *Mat. Tidsskr.* BH **1** (1932), 1–18.
(On regular Riemann surfaces.)

NÖBELING, G.

[1] Die neusten Ergebnisse der Dimensionstheorie. *Jahresber. Deutsch. Math.-Verein.* **41** (1931), 1–17.
(Recent advances in dimension theory.)

PANNWITZ, E.

[1] Eine elementargeometrische Eigenschaft von Verschlingungen und Knoten. *Math. Ann.* **108** (1933), 629–672.
(An elementary-geometric property of links and knots.)

POINCARÉ, H.

[1] Sur les courbes définies par une équation différentielle. *J. Math. Pures Appl.* (3) **7** (1881), 375–424; (3) **8** (1882), 251–296; (4) **1** (1885), 167–244; (4) **2** (1886), 151–217. (On the curves defined by a differential equation.)

[2] Analysis Situs. *J. École Polytech., Paris* (2) **1** (1895), 1–121.

[3] (First) Complément (to Analysis Situs). *Rend. Circ. Mat. Palermo* **13** (1899), 285–343.

[4] (Second) Complément. *Proc. London Math. Soc.* **32** (1900), 277–308.

[5] (Third) Complément. *Bull. Soc. Math. France* **30** (1902), 49–70.

[6] (Fourth) Complément. *J. Math. Pures Appl.* (5) **8** (1902), 169–214.

[7] (Fifth) Complément. *Rend. Circ. Mat. Palermo* **18** (1904), 45–110.

[8] Sur un théorème de Géométrie. *Rend. Circ. Mat. Palermo* **33** (1912), 375–407.

PONTRJAGIN, L.

[1] Zum Alexanderschen Dualitätssatz. *Nachr. Ges. Wiss. Göttingen* (1927), 315–322 and 446–456. (On Alexander's duality theorem.)

[2] Sur une hypothèse fondamentale de la théorie de la dimension. *C. R. Acad. Sci. Paris* **190** (1930), 1105–1107. (On a fundamental hypothesis of dimension theory.)

[3] Über den algebraischen Inhalt topologischer Dualitätssätze. *Math. Ann.* **105** (1931), 165–205. (On the algebraic content of topological duality theorems.)

RABIN, M. A.

[1] *Ann. Math.* **67** (1958), 172.

RADO, T.

[1] Über den Begriff der Riemannschen Fläche. *Acta Univ. Szeged.* **2** (1925), 101–121. (On the concept of the Riemann surface.)

REIDEMEISTER, K.

[1] Knoten und Gruppen. *Abh. Math. Sem. Univ. Hamburg* **5** (1926), 7–23. (Knots and Groups.)

[2] Elementare Begründung der Knotentheorie. *Abh. Math. Sem. Univ. Hamburg* **5** (1926), 24–32. (Basic elements of knot theory.)

[3] Über Knotengruppen. *Abh. Mat. Sem. Univ. Hamburg* **6** (1928), 56–64. (On knot groups.)

[4] Fundamentalgruppe und Überlagerungsräume. *Nachr. Ges. Wiss. Göttingen* (1928), 69–76. (The fundamental group and covering spaces.)

[5] Knoten und Verkettungen. *Math. Z.* **29** (1929), 713–729. (Knots and links.)

[6] "Knotentheorie." Berlin, 1932.

[7] "Einführung in die kombinatorische Topologie." Braunschweig, 1932. (An introduction to combinatorial topology.)

[8] Zur dreidimensionalen Topologie. *Abh. Math. Sem. Univ. Hamburg* **9** (1933), 189–194. (On three-dimensional topology.)

[9] Heegaarddiagramme und Invarianten von Mannigfaltigkeiten. *Abh. Math. Sem. Univ. Hamburg* **10** (1934). (Heegaard diagrams and invariants of manifolds.)

[10] Homotopiegruppen von Komplexen. *Abh. Math. Sem. Univ. Hamburg* **10** (1934). (Homotopy groups of complexes.)

REY PASTOR, J.

[1] Sulla topologia dei domini di uno spazio ad *n* dimensioni. *Atti Accad. Naz. Lincei, Rend. Cl. Sci. Fis. Mat. Natur.* (6) **15** (1932), 524–527. (On the topology of domains of an *n*-dimensional space.)

RIEMANN, B.
   [1]   Fragment aus der Analysis Situs. "Werke, " 2nd ed., p. 474.
         (Excerpt from Analysis Situs. "Collected Works, " 2nd ed., p. 474.)
   [2]   Abelschen Funktionen. *Werke* §19, 84.
         (Abelian functions.)
RYBARZ, J.
   [1]   Über drei Fragen der abstrakten Topologie. *Monatschr. Math. Phys.* **38** (1931), 215–244.
         (On three questions of abstract topology.)
SCHERRER, W.
   [1]   Geometrische Deutung des Gaußchen Verschlingungsintegrals. *Comment. Math. Helv.* **5**
         (1933), 25–27.
         (On the geometric interpretation of Gauss' linking integral.)
SCHILLING, F.
   [1]   "Projektive und nichteuklidische Geometrie." Leipzig, 1931.
SCHRIER, O.
   [1]   Über die Gruppen $A^a B^b = 1$. *Abh. Math. Sem. Univ. Hamburg* **3** (1924), 167–169.
         (On the groups $A^a B^b = 1$.)
   [2]   Die Verwandschaft stetiger Gruppen im Großen. *Abh. Math. Sem. Univ. Hamburg* **5**
         (1927), 233–244.
         (The relationship of continuous groups in the large.)
SCHRIER, O., and E. SPERNER
   [1]   "Analytische Geometrie." Leipzig, 1931.
SEIFERT, H.
   [1]   Konstruktion dreidimensionaler geschlossener Räume. *Ber. Sächs. Akad. Wiss.* **83** (1931),
         26–66.
         (The construction of three-dimensional closed spaces.)
   [2]   Homologiegruppen berandeter dreidimensionaler Mannifgaltigkeiten. *Math. Z.* **35**
         (1932), 609–611.
         (Homology groups of three-dimensional manifolds with boundary.)
   [3]   Topologie dreidimensionaler gefaserte Räume. *Acta Math.* **60** (1932), 147–238.
         (The topology of three-dimensional fibered spaces.)
   [4]   Verschlingungsinvarianten. *Sitzungsber. Preuss. Akad. Wiss.* **16** (1933), 811–828.
         (Linking invariants.)
SEVERI, F.
   [1]   Sulla topologia e sui fondamenti dell'analisi generale. *Rend. Sem. Mat. Roma* (2) **7**
         (1931), 5–37.
         (On the topological foundations of general analysis.)
   [2]   Über die Grundlagen der algebraischen Geometrie. *Abh. Math. Sem. Univ. Hamburg* **9**
         (1933), 335–364.
         (On the foundations of algebraic geometry.)
SIEBENMANN, L.
   [1]   Are nontriangulable manifolds triangulable? *In* "Top of Manifolds" (J. C. Cantrell and
         C. H. Edwards, eds.), pp. 77–84. Markham Press, 1970.
SINGER, C.
   [1]   Three-dimensional manifolds and their Heegaard diagrams. *Trans. Amer. Math. Soc.* **35**
         (1933), 88–111.
SMALE
   [1]   Generalized Poincaré conjecture in dimensions greater than 4. *Ann. Math.* (2) **74** (1961),
         391–406.
SPEISER, A.
   [1]   "Theorie der Gruppen von endlicher Ordnung." Berlin, 1927.

SPERNER, E.
[1] Neuer Beweis für die Invarianz der Dimensionszahl und des Gebiets. *Abh. Math. Sem. Univ. Hamburg* **6** (1928), 265–272.
(A new proof of the invariance of the dimension number and of domain.)
[2] "Über die fixpunktfreien Abbildungen der Ebene," Hamburg Math. Monogr. H 14. Leipzig, 1933.
("On the Fixed-point-free Mappings of the Plane.")

STEINITZ, E.
[1] Beiträge zur Analysis Situs. *Sitzungs ber. Berlin Math. Ges.* **7** (1908), 29–49.
(Contributions to Analysis Situs.)
[2] Polyeder und Raumeinteilungen. *Encykl. Math. Wiss.* (III) **AB 12** (1916), 139.
(Polyhedra and divisions of space.)

STUDY, E.
[1] Geometrie der Kreise und Kugeln. *Math. Ann.* **86** (1922).

TAIT, P. G.
[1] On knots. *Trans. Roy. Soc. Edinburgh* **28** (1879), 145–190; **32** (1887), 327–339 and 493–506.

THRELFALL, W.
[1] Gruppenbilder. *Abh. Sächs Akad. Wiss. Leipzig Math.-Phys. Kl.* **41**, No. 6 (1932), 1–59.
(Group diagrams.)
[2] Räume aus Linienelemente. *Jahresber. Deutsch. Math.-Verein.* **42** (1932), 88–110.
(Line element spaces.)

THRELFALL, W. and H. SEIFERT
[1] Topologische Untersuchung der Discontinuitätsbereiche endlicher Bewegungsgruppen des dreidimensionalen sphärischen Raumes. I. *Math. Ann.* **104** (1930), 1–70; II. **107** (1932), 543–586.
(Topological investigations of regions of discontinuity of finite groups of motions of the 3-sphere.)

TIETZE, H.
[1] Über die topologischen Invarianten mehrdimensionaler Mannigfaltigkeiten. *Monatschr. Math. Phys.* **19** (1908), 1–118.
(On the topological invariants of higher-dimensional manifolds.)
[2] Sur les représentations continues des surfaces sur elles-mêmes. *C. R. Acad. Sci. Paris* **157** (1913), 509–512.
(On continuous mappings of surfaces onto themselves.)
[3] Über stetige Abbildungen einer Quadratfläche auf sich selbst. *Rend. Circ. Mat. Palermo* **38** (1914), 247–304.
(On the continuous mappings of a quadratic surface onto itself.)
[4] Über den Richtungssinn und seine Verallgemeinerung. *Jahresber. Deutsch. Math.-Verein.* **29** (1920), 95–123.
(On the sense of direction and its generalization.)
[5] Über Analysis Situs. *Abh. Math. Sem. Univ. Hamburg* **2** (1923), 37–68.
[6] Beiträge zur allgemeinen Topologie. I. Axiome für verschiedene Fassungen des Umgebungsbegriffs. *Math. Ann.* **88** (1923), 290–312. II. Über die Einfuhrung uneigentlicher Elemente. *ibid.* **91** (1924), 210–224. III. Über die Komponenten offener Mengen. *Monatschr. Math. Phys.* **33** (1923), 15–17.
(Contributions to general topology. I. Axioms for different versions of the neighborhood concept. II. On the introduction of improper elements. III. On the components of open sets.)
[7] Zur Topologie berandeter Mannigfaltigkeiten. *Monatschr. Math. Phys.* **35** (1928), 25–44.
(On the topology of manifolds with boundary.)

TIETZE, H., and L. VIETORIS
[1] Beziehungen zwischen den verschiedenen Zweigen der Topologie. *Encykl. Math. Wiss.* (III) **AB 13** (1930).
(Relations between the different branches of topology.)

TUCKER, A. W.
[1] On combinatorial topology. *Proc. Nat. Acad. Sci. U.S.A.* **18** (1932), 86–89.
[2] Modular homology characters. *Proc. Nat. Acad. Sci. U.S.A.* **18** (1932), 467–471.
[3] An abstract approach to manifolds. *Ann. Math.* (2) **34** (1933), 191–243.

URSELL, H. D.
[1] Intersections of complexes. *J. London Math. Soc.* **3** (1928), 37–48.

VAN DER WAERDEN, B. L.
[1] Topologische Begründung des Kalkuls der abzählenden Geometrie. *Math. Ann.* **102** (1929), 337–362.
(Topological foundations of the calculus of enumerative geometry.)
[2] Kombinatorische Topologie. *Jahresber. Deutsch. Math.-Verein.* **39** (1930), 121–139.
[3] "Moderne Algebra I." Berlin, 1930.
[4] Zur algebraischen Geometrie. IV. Die Homologeizahlen der Quadriken und die Formeln von Halphen der Liniengeometrie. *Math. Ann.* **109** (1933), 7–12.
(On algebraic geometry. IV. The homology numbers of the quadrics and Halphen's line-geometric formulas.)
[5] *Jahresber. Deutsch. Math.-Verein.* **42** (1933), 112.

VAN KAMPEN, E. R.
[1] Eine Verallgemeinerung des Alexanderschen Dualitätssatzes. *Proc. K. Ned. Akad. Wetensch.* **31** (1928), 899–905.
(A generalization of Alexander's duality theorem.)
[2] Zur Isotopie zweidimensionaler Flächen im R_4. *Abh. Math. Sem. Univ. Hamburg* **6** (1928), 216.
(On the isotopy of two-dimensional surfaces in R_4.)
[3] "Die kombinatorische Topologie und die Dualitätssätze." The Hague, 1929.
(Combinatorial Topology and its Duality Theorems.)
[4] Komplexe in euklidischen Räumen. *Abh. Math. Sem. Univ. Hamburg* **9** (1932), 72–78 and 152–153.
(Complexes in Euclidean spaces.)
[5] Some remarks on the join of two complexes and on invariant subsets of a complex. *Amer. J. Math.* **54** (1932), 543–550.
[6] On the fundamental group of an algebraic curve. *Amer. J. Math.* **55** (1933), 255–260.
[7] On the connection between the fundamental groups of some related spaces. *Amer. J. Math.* **55** (1933), 261–267.
[8] On some lemmas in the theory of groups. *Amer. J. Math.* **55** (1933), 268–273.

VEBLEN, O.
[1] Theory of plane curves in non-metrical analysis situs. *Trans. Amer. Math. Soc.* **6** (1905), 83–98.
[2] On the deformation of an n-cell. *Proc. Nat. Acad. Sci. U.S.A.* **3** (1917), 654–656.
[3] The intersection numbers. *Trans. Amer. Math. Soc.* **25** (1923), 540–550.
[4] "Analysis Situs," 2nd ed., Amer. Math. Soc. Colloq. Publ. No. 5, Part II. Amer. Math. Soc., New York, 1931.

VEBLEN, O., and J. W. ALEXANDER
[1] Manifolds of n dimensions. *Ann. Math.* (2) **14** (1913), 163–178.

VIETORIS, L.
[1] Über stetige Abbildungen einer Kugelfläche. *Proc. K. Ned. Akad. Wetensch.* **29** (1926), 443–453.
(On continuous mappings of a spherical surface.)

[2] Über den höheren Zusammenhang kompakter Räume und eine Klasse von zusammenhängstreuen Abbildungen. *Math. Ann.* **97** (1927), 454–472.
(On the higher connectivity of compact spaces and a class of connectivity-faithful mappings.)

[3] Über die Symmetrie der Zusammenhangszahlen kombinatorischer Mannigfaltigkeiten. *Monatschr. Math.* **35** (1928), 165–174.
(On the symmetry of the connectivity numbers of combinatorial manifolds.)

[4] Zum höheren Zusammenhang der kompakten Räume. *Math. Ann.* **101** (1929), 219–225.
(On the higher connectivity of compact spaces.)

[5] Erzeugung der regulären Unterteilung von simplizialen Komplexen durch wiederholte Zweiteilung. *Monatschr. Math. Phys.* **37** (1930), 97–102.
(Production of the normal subdivision of simplicial complexes by means of repeated bisection.)

[6] Über die Homologiegruppen der Vereinigung zweier Komplexe. *Monatschr. Math. Phys.* **37** (1930), 159–162.
(On the homology groups of the join of two complexes.)

[7] Über den höheren Zusammenhang von Vereinigungsmengen und Durchschnitten. *Fund. Math.* **19** (1932), 265–273.
(On the higher connectivity of set unions and intersections.)

VON DYCK, W.
[1] *Math. Ann.* **20** (1882), 35.
[2] Beiträge zur Analysis Situs. *Ber. Sächs Akad. Wiss.* **37** (1885), 314–325 (I); **38** (1886), 53–69 (II); **39** (1887), 40–52 (III).
[3] Beiträge zur Analysis Situs. I. *Math. Ann.* **32** (1888), 457–512; II. **37** (1890), 273–316.
(Contributions to Analysis Situs.)

VON KERÉKJÁRTÓ, B.
[1] Beweis des Jordanschen Kurvensatzes. *Trans. Hung. Acad. Sci.* **38** (1919), 194–198.
(A proof of Jordan's curve theorem.)
[2] Über die Brouwerschen Fixpunktsätze. *Math. Ann.* **80** (1919), 29–32.
(On Brouwer's fixed point theorems.)
[3] Über Transformationen des ebenen Kreisringes. *Math. Ann.* **80** (1919), 33–35.
(On transformations of the planar annulus.)
[4] Zur Gebietsinvarianz. *Trans. Hung. Acad. Sci.* **39** (1921), 220–221.
(On the invariance of domain.)
[5] Kurvenscharen auf Flächen. *Nachr. Ges. Wiss. Göttingen* (1922), 71–79.
(Families of curves on surfaces.)
[6] "Vorlesungen über Topologie." Berlin, 1923.

WEBER, C., and H. SEIFERT
[1] Die beiden Dodekaederräume. *Math. Z.* **37** (1933), 237–253.
(The two dodecahedron spaces.)

WEYL, H.
[1] "Mathematische Analyse des Raumproblems." Berlin, 1923.
[2] Analysis Situs Combinatorio. *Rev. Mat. Hisp.-Amer.* (4) **5** (1923), 209–218, 241–248, and 273–279; **6** (1924), 33–41.
[3] "Philosphie der Mathematik und Naturwissenschaflen" (Spec. reprint from "The Handbuch der Philosophie"). Munich, 1927.
[4] "Über die Idee der Riemannschen Fläche." Chelsea, New York, 1947 (reprint).
(The concept of the Riemann surface.)

WHITTAKER, E. T.
[1] "Analytische Dynamik." Berlin, 1914.

WILDER, R. L.

[1] Point sets in three and higher dimensions and their investigation by means of a unified analysis situs. *Bull. Amer. Math. Soc.* **38** (1932), 649–692.

WILSON, W. A.

[1] Representation of a simplicial manifold on a locally simplicial manifold. *Proc. K. Ned. Akad. Wetensch.* **29** (1926), 1129–1133.

[2] Representation of manifolds. *Math. Ann.* **100** (1928), 552–578.

WIRTINGER, W.

[1] Über die Verzweigungen bei Funktionen von zwei Veränderlichen. *Jahresber. Deutsch. Math.-Verein.* **14** (1905), 517.

(On the branchings of functions of two variables.)

WOLF, J. A.

[1] "Spaces of Constant Curvature." McGraw-Hill, 1967.

ZARISKI, O.

[1] On the topology of algebroid singularities. *Amer. J. Math.* **54** (1932), 453–465.

SEIFERT:
TOPOLOGY OF 3-DIMENSIONAL
FIBERED SPACES

TOPOLOGY OF 3-DIMENSIONAL FIBERED SPACES*

The subject of this paper is related to the homeomorphism problem for 3-dimensional closed manifolds. The fundamental theorem for 2-manifolds tells us how many topologically distinct 2-manifolds there are. The methods for its proof cannot yet be applied to 3 or more dimensions. There are two ways to approach the 3-dimensional problem. The first one is to examine fundamental regions (Diskontinuitätsbereiche) of groups acting on a 3-dimensional metric space (Bewegungsgruppen). In the 2-dimensional case, every closed surface is a fundamental region of a fixed-point-free action; however, there are 3-manifolds for which this is not true. The fundamental regions of 3-dimensional spherical actions are endowed with a certain fibration: the fibers are trace curves (Bahnkurven) of a continuous action on the hypersphere; examples will be given in §3 and can also be found in DB II.[1] This leads us to the second way: instead of investigating a complete system of *topological* invariants of 3-dimensional manifolds, we search for a system of invariants for *fiber preserving* maps of *fibered* 3-manifolds. This task is completely solved in this paper. These invariants refer of course to the fibering of the manifold, not to the manifold itself, so that so far the question remains whether two spaces with different fibrations can be homeomorphic. Furthermore there are 3-manifolds that do not admit a fibration (§15). Even so, in many cases the fiber invariants can be used to decide whether 3-manifolds are homeomorphic. Examples for this are given in §12–§14 and in DB II.

A knowledge of the topology of surfaces, the fundamental group, and the

---

*Reprinted from H. Seifert, *Acta Mathematica* **60** (1933), 147–288 (translated by Wolfgang Heil).

[1] Cf. W. Threlfall and H. Seifert, Topologische Untersuchungen der Diskontinuitätsbereiche endlicher Bewegungsgruppen des dreidimensionalen sphärischen Raumes. *Math. Ann.* **107**. This will be referred to as DB II; the first part in *Math. Ann.* **104** will be cited as DB I.

homology group is assumed. The spaces of line elements[2] (Linienelemente) provide introductory examples. Other examples are given in this paper.

1. Fibered Spaces

We define a manifold[3] to be a set of points such that for each point there is a system of subsets, called *neighborhoods*, which satisfy the axioms (1)–(4) below.

(1)   Hausdorff axioms:

   (a)   Each point P has at least one neighborhood $U(P)$; each neighborhood of P contains P.

   (b)   If $U(P)$ and $V(P)$ are neighborhoods of P, then there exists a neighborhood $W(P) \subset U(P) \cap V(P)$.

   (c)   If Q lies in $U(P)$, then there exists a neighborhood $U(Q)$ of Q which is contained in $U(P)$.

   (d)   For two distinct points there exist disjoint neighborhoods.

A system of neighborhoods satisfying these axioms is called a *topological space*. Two equivalent systems of the same point set determine the same topological space. Here systems are equivalent if each neighborhood $U(P)$ of one system contains a neighborhood $U'(P)$ of the other system, and vice versa. A subset of a topological space is *open* if it contains for each of its points a neighborhood of this point. The system of all open subsets of a topological space is a system of neighborhoods, which is equivalent to all other systems of neighborhoods of this space. From now on we always choose this system of neighborhoods.

(2)   Each point of M has a neighborhood homeomorphic to an open 3-ball in 3-dimensional Euclidean space.

(3)   If an arbitrary neighborhood is assigned to each point, then countably many of these cover the manifold. If already finitely many suffice to cover the manifold, it is called *closed*, otherwise *open*.[4]

(4)   The manifold is *connected*, i.e., any two points can be connected by an arc, or equivalently, the manifold is not the union of two disjoint open sets.

---

[2] W. Threlfall, Räume aus Linienelementen. *Jahresber. Deutsch. Math.-Verein.* **42** (1932), 88–110.

[3] Cf. H. Kneser, Topologie der Mannigfaltigkeiten. *Jahresber. Deutsch. Math.-Verein.* **34** (1926), 1.

[4] Instead of (3) we could require the second Hausdorff countability axiom in addition to (1) and (2): There exists an equivalent system of neighborhoods that consists of countably many distinct point sets. The following axiom would do just as well: The manifold can be covered with countably many subsets, each of which is homeomorphic to an open 3-dimensional Euclidean ball.

In combinatorial topology manifolds are required to admit a *triangulation*. This requirement is redundant for our purpose, since fibered spaces can be triangulated, as will be shown in §4. One could say a manifold is fibered if it is decomposed into curves, called *fibers*, such that each point lies on exactly one fiber and a neighborhood of each point can be mapped homeomorphically onto a neighborhood of a point in a Euclidean space in such a way that fibers are mapped to line segments of a bundle of parallel lines. This requirement is a local one. But even if we postulated this for all points of the manifold, we would still find this definition of a fibered manifold to be too general.

In the present paper we consider only those fibered manifolds which satisfy in addition to the four manifold axioms the three following axioms which relate to properties of the fibering in the large. (We call these manifolds *fibered spaces*.)

(5) The manifold can be decomposed into fibers, where each fiber is a simple closed curve.

(6) Each point lies on exactly one fiber.

(7) For each fiber H there exists a *fiber neighborhood*, that is, a subset consisting of fibers and containing H, which can be mapped under a *fiber preserving* map onto a *fibered solid torus*, where H is mapped onto the "middle fiber."

A *fibered solid torus* is obtained from a fibered cylinder $D^2 \times I$ where the fibers are the lines $x \times I$, $x \in D^2$, by rotating $D^2 \times 1$ (but keeping $D^2 \times 0$ fixed) through an angle of

$$2\pi(\nu/\mu)$$

and then identifying $D^2 \times 0$ and $D^2 \times 1$ (i.e., $x \times 0$ is identified with $\rho(x) \times 1$, where ρ is the rotation). Here ν, μ are coprime integers. Without loss of generality we can assume that

$$\mu > 0 \qquad \text{and} \qquad 0 \leqslant \nu \leqslant \tfrac{1}{2}\mu.$$

For if ν is replaced by $\nu + k\mu$ or by $-\nu$, then the new solid torus can be mapped onto the old one by a fiber preserving map.

A map is *fiber preserving* if it (1) is a homeomorphism and (2) maps fibers to fibers. Two solid tori which are homeomorphic under a fiber preserving map will not be distinguished.

When identifying the cylinder $D^2 \times I$ with the solid torus the lines (fibers) of $D^2 \times I$ are decomposed into classes such that each class contains exactly μ lines, which match together to give one fiber of the solid torus, except that the class containing the axis of $D^2 \times I$ consists of the axis alone, which also makes up a fiber. If $\mu = 1$, we call the solid torus an *ordinary solid torus*.

The fiber neighborhoods are (in contrast to point neighborhoods) closed sets: each fiber neighborhood contains its boundary torus.

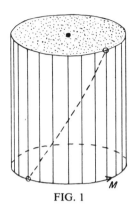

FIG. 1

A *meridian* M of a solid torus V is a simple closed oriented curve on the boundary torus T which is not contractible on T but contractible in V. A homeomorphism of V onto itself maps a meridian to a meridian. If we forget about orientation, we can map a meridian onto any other meridian under a continuous deformation of T. In Fig. 1, e.g., the oriented boundary curve of the bottom surface $D^2 \times 0$ is a meridian. A *longitude* B of the solid torus is a simple closed curve on T which intersects M in exactly one point.

B is determined (modulo deformations of T) up to its orientation and multiples of M. Any pair of meridian and longitude can be mapped onto another such pair by a topological map of the solid torus onto itself; however, even though any meridian can be mapped onto any other by a deformation of T, this is not necessarily true for longitudes. The topological map of the solid torus, which sends a longitude to another which is not homologous (on T), cannot be obtained by a deformation of the identity.

We now orient a fiber H of a solid torus. Thus, if we have chosen a fiber H, a meridian M, and a longitude B on the boundary T of a given fibered solid torus V, we can just as well choose instead of H, M, B any other system H', M', B' which is related to the first system as follows:

$$H \sim \varepsilon_1 H', \tag{1}$$

$$M \sim \varepsilon_2 M', \tag{2}$$

$$B \sim \varepsilon_3 B' + xM'. \tag{3}$$

Here $\varepsilon_i = \pm 1$; x is an integer. Instead of the equal sign we have chosen the homology sign, which denotes *homology on* T. For homology is all that matters to us and we allow, for example, that H' be a fiber disjoint to H and M' be a meridian obtained from M by a deformation of T.

Throughout, we write relations of the homology group additively and relations of the fundamental group multiplicatively.[5]

The numbers μ and ν not only determine the fibered solid torus V, but

[5] Cf. B. L. Van der Waerden, "Moderne Algebra I." p. 19. Berlin, 1930.

conversely V determines μ and ν uniquely, i.e., two fibered solid tori can be mapped onto each other by a fiber preserving map iff they have the same defining numbers μ, ν. For, choosing the longitude B suitably (shortest path on $\partial D^2 \times I$ from a point $x \in \partial D^2 \times 0$ to its equivalent point on $\partial D^2 \times 1$, the dotted line in Fig. 1) and orienting M and H suitably, we have on T the homology

$$H \sim \nu M + \mu B, \tag{H}$$

which means precisely that μ and ν are the defining numbers of the fibered solid torus. If we were to choose instead of H, M, B an arbitrary system H', M', B' of the fibered solid torus, then we would get

$$H' \sim nM' + mB' \tag{H'}$$

since M' and B' are a fundamental system[6] of curves on T which is a basis for the homology. Here m and n are coprime integers since the fiber is a simple closed curve, and $m \neq 0$ since it is not homologous to the meridian. On the other hand, we can express the homology (H) in terms of H', M', B' via the formulas (1), (2), (3):

$$\varepsilon_1 H' \sim (\varepsilon_2 \nu + x\mu) M' + \varepsilon_3 \mu B'.$$

Therefore

$$\varepsilon_1 \big[(\varepsilon_2 \nu + x\mu) M' + \varepsilon_3 \mu B' \big] \sim nM' + mB'.$$

Comparing the coefficients, we see that μ and ν are determined by m and n. To see this, note that $|\mu| = |m|$, also $\mu > 0$, so $\mu = |m|$; also ν is equal to $|n|$, reduced modulo m to a number in the interval $[-\frac{1}{2}m, \frac{1}{2}m]$. Thus the numbers μ and ν are characteristic for the given fibered solid torus.

Meridian and longitude are already defined on a nonfibered solid torus. We need to define still another curve, the *crossing curve* Q (Querkreis), presuming the fibering. It is a simple closed curve on T that intersects each fiber of T in exactly one point. It is therefore (except for its orientation and multiples of the fiber) determined by the fibering of T, i.e., if Q and Q' are two crossing curves, we have the formula

$$Q \sim \varepsilon_4 Q' + yH' \tag{4}$$

in addition to the transformation formulas (1)–(3). The fiber H and crossing curve Q are a fundamental system of curves on T similar to meridian and longitude, i.e., any other closed curve on T is homologous to a linear combination of H and Q.

The boundary of an arbitrary fibered solid torus is a fibered torus. Therefore the boundaries of any two fibered solid tori can be mapped onto each other under a fiber preserving homeomorphism. The fibered solid torus

---

[6] Meridian and longitude are also called a canonical system of curves or a pair of conjugate Rückkehrschnitte.

is determined by the fibering of its boundary torus only if on this torus a closed curve M is distinguished as meridian. Of course, M must satisfy the conditions to be a simple closed curve not homologous to zero (on T) and not homologous to a fiber. If on a fibered torus the fiber H is oriented and a crossing curve Q is chosen, M can be expressed [with coprime integers α ($\neq 0$) and β] as follows:

$$M \sim \alpha Q + \beta H.$$

We claim that the fibered solid torus is uniquely determined by the fibering of its boundary and by M, hence by α, β. We show this by computing the characteristic numbers μ, ν. If

$$B \sim \rho Q + \sigma H$$

is a longitude on the fibered torus, we can assume (choosing orientation of B suitably) that

$$\begin{vmatrix} \alpha & \beta \\ \rho & \sigma \end{vmatrix} = 1 \tag{5}$$

since both of Q, H and M, B are a fundamental system of curves on the torus. Then

$$H \sim \alpha B - \rho M$$

ρ is determined by α and β up to multiples of α by (5). As before from (H′), the last equation gives us now the characteristic numbers μ and ν uniquely: $\mu = |\alpha|$, $\nu =$ the absolute value of the number ρ, reduced mod α to $[-\frac{1}{2}\alpha, \frac{1}{2}\alpha]$. In particular if the meridian is a crossing curve we have an ordinary fibered solid torus.

The simplest example of a fibered space is $S^1 \times S^2$. It is obtained from $S^2 \times I$ by identifying the points $x \times 0$ and $x \times 1$. Figure 2 shows a cross section through the center point of $S^2 \times I \subset R^3$. The fibers correspond to the radii of the hollow ball. We have a fibered space, since each fiber has a fiber neighborhood which can be mapped onto a fibered solid torus with the numbers $\mu = 1$, $\nu = 0$.

2. Orbit Surface

The most important concept in the study of fibered spaces is that of the *orbit surface* (Zerlegungsfläche). Every fibered space F has an orbit surface f. Now f is not a subset of the space F and can in general not be embedded in F,[7] but is defined as follows: there is a one-to-one correspondence between the fibers of F and the points of f.[8] Since each point of F lies on exactly one

---

[7] Our definition of Zerlegungsfläche is not related to G. D. Birkhoff's surface of section, Dynamical systems with two degrees of freedom [*Trans. Amer. Math. Soc.* **18** (1917), 268; cf. also L. Bieberbach, "Differentialgleichungen," p. 136. Berlin, 1923].

[8] The orbit surface thus indicates how the manifold is "decomposed" into fibers [cf. H. Tietze and L. Vietoris, *Encykl. Math. Wiss.* (III) **AB 13** (1930), 178].

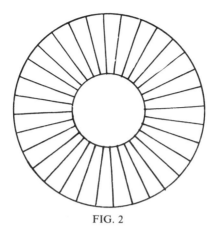

FIG. 2

fiber, it follows that each point of F has exactly one image on f. The neighborhoods of f are defined as images of the neighborhoods in F (i.e., of the open subsets of F). The following can be proved:

(1) f is a Hausdorff space.

(2) Each point of f has a neighborhood homeomorphic to an open 2-cell. (For the proof use the fact that each fiber neighborhood can be mapped topologically onto a solid torus.)

(3) Any covering of f by neighborhoods has a countable subcovering. f is an open or closed manifold if F is open or closed, respectively.

(4) f is connected.

(1)–(4) imply that f is triangulable, by a theorem of T. Radó.[9] Therefore we can apply all the theorems of the theory of 2-manifolds. If F is closed, then f is an orientable surface of genus p (number of handles) or a nonorientable surface of genus k (number of cross-caps). In the example $S^1 \times S^2$, the orbit surface is a 2-sphere which can be embedded into $S^1 \times S^2$ so that each fiber meets it in exactly one point.

Any closed or open, orientable or nonorientable surface f is the orbit surface of some fibered space, for example of the product $f \times S^1$ (the fibers are $x \times S^1$, $x \in f$). Here the orbit surface can again be embedded into the fibered space, as above. In §3 we shall give an example where this is no longer possible.

We use throughout the following notation. Passing from the fibered space F to the orbit surface f we pass from capital letters to small letters. Thus to the fiber H of the space F corresponds the point h of the orbit surface f.

If Ω_H is a fiber neighborhood of the fiber H, we call its image ω_h an *orbit neighborhood* (Zerlegungsumgebung) of the image point h of H. The *orbit neighborhood* is obtained from the meridian disk of the fiber neighborhood,

[9]T. Radó, Über den Begriff der Riemannschen Fläche, *Acta Univ. Szeged.* **2** (1925), 101.

i.e., from the bottom disk of the cylinder of Fig. 1, by identifying points which belong to the same fiber. Therefore, the orbit neighborhood is a circle sector of an angle $2\pi/\mu$ whose boundary radii have been identified, or in other words: it is the orbit surface of a cyclic rotation group of order μ of the disk about its center point. Hence the orbit neighborhood can be mapped homeomorphically onto a disk with boundary; hence it is a 2-cell. The orbit neighborhoods are just like the fiber neighborhoods *closed* point sets. They satisfy the neighborhood axioms only after removing their boundary curves.

The orbit neighborhoods satisfy the following:

LEMMA 1. *If ω_h is an orbit neighborhood of the point h and if e is a 2-cell contained in ω_h such that h is not on the boundary of e, then e is also an orbit neighborhood (a) of h, if h is an interior point of e, (b) of each interior point of e, if h does not belong to e. The fiber neighborhoods E (resp. Ω_H) which map onto e (resp. ω_h) are in case (a) homeomorphic under a fiber preserving map; in case (b) E is an ordinary fibered solid torus.*

Proof. (a) The fibers that map to the points of e constitute a fibered subset E of Ω_H which contains the fiber H in its interior. If we think of Ω_H as a fibered cylinder with boundary disks identified under a rotation, we obtain the orbit neighborhood ω_h (Fig. 3) from the meridian disk $\tilde{\omega}_{\tilde{h}}$ of Ω_H (Fig. 4) if we identify those points of $\tilde{\omega}_{\tilde{h}}$ which are equivalent under the cyclic rotation group of order μ acting on $\tilde{\omega}_{\tilde{h}}$.

The points of $\tilde{\omega}_{\tilde{h}}$ which map to points of e constitute a 2-cell $\tilde{e}$ (shaded in Fig. 4) which contains the center point $\tilde{h}$ of $\tilde{\omega}_{\tilde{h}}$ in its interior and which is mapped to itself under the cyclic rotation group. The subspace E of Ω_H consists of the lines parallel to the axis of the cylinder Ω_H which pass through the points of $\tilde{e}$. We shall show that we can map $\tilde{e}$ onto $\tilde{\omega}_{\tilde{h}}$ under an

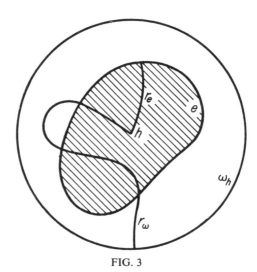

FIG. 3

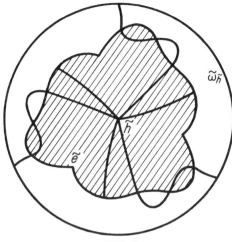

FIG. 4

orientation preserving homeomorphism $\tilde{a}$ keeping $\tilde{h}$ fixed and such that any μ points which are equivalent under the cyclic rotation group are again mapped onto μ such points. Taking the corresponding map on the lines of E and Ω_H, we obtain a topological map of E onto Ω_H which maps fibers to fibers and keeps the middle fiber H fixed, as claimed in the lemma.

The map $\tilde{a}$ is obtained as follows: Let a be an orientation preserving map that maps e onto ω_h and keeps h fixed, let r_e be a simple arc from h to the boundary of e, and let r_ω be the image of r_e which is a simple arc from h to the boundary of ω_h. Now $\tilde{e}$ (resp. $\tilde{\omega}_{\tilde{h}}$) is decomposed by the μ (pre-)images of r_e (resp. r_ω) into μ consecutive sectors

$$\tilde{e}^1, \tilde{e}^2, \ldots, \tilde{e}^\mu \qquad (\text{resp.} \quad \tilde{\omega}^1, \tilde{\omega}^2, \ldots, \tilde{\omega}^\mu)$$

which are cyclically interchanged by the rotation group. The map a determines a map of the sector $\tilde{e}^i$ onto the sector $\tilde{\omega}^i$ and hence a map $\tilde{a}$ of $\tilde{e}$ onto $\tilde{\omega}_{\tilde{h}}$, as required.

(b) In this case, to the 2-cell in ω_h there correspond in $\tilde{\omega}_{\tilde{h}}$ now μ disjoint 2-cells $\tilde{e}^1, \tilde{e}^2, \ldots, \tilde{e}^\mu$ which are interchanged under the cyclic rotation group. The fiber set E corresponding to e is in the cylinder Ω_H made up of μ congruently fibered cylinders which lie over $\tilde{e}^1$ to $\tilde{e}^\mu$. Now E is obtained from these pieces by pasting them together (one after the other) and finally identifying top and bottom disks under the identity map. Therefore E is an ordinary fibered solid torus, in which we can take each inner fiber as the middle fiber.

From Lemma 1 we obtain

LEMMA 2. *If Ω_H^1 and Ω_H^2 are two fiber neighborhoods of the fiber H, they are homeomorphic under a fiber preserving map which keeps H fixed.*

Proof. On the orbit surface there exists a 2-cell e containing h and lying in the interior of the intersection of the orbit neighborhoods ω_h^1 and ω_h^2. By Lemma 1, e is the image of a fiber neighborhood E of the fiber H, and E can be mapped under a fiber preserving map (keeping H fixed) to each of Ω_H^1 and Ω_H^2, respectively.

This lemma implies that for a given fiber H the numbers μ, ν are the same for all fiber neighborhoods of H; hence they are an invariant of H. If $\mu > 1$, we call H an *exceptional fiber of order* μ of the space; if $\mu = 1$, an *ordinary* fiber. If a fiber in the neighborhood of an exceptional fiber H of order μ approaches H, its limit runs μ times around H. In a fibered solid torus all the fibers are ordinary fibers, except possibly for the middle fiber. In a fiber neighborhood of an exceptional fiber H of order μ we have that $\mu \cdot H$ is homologous to an ordinary fiber. The points of the orbit surface that are images of exceptional fibers are *exceptional points*; as points of the orbit surface, they cannot be distinguished from ordinary points.

THEOREM 1. *A closed fibered space contains at most finitely many exceptional fibers.*

For otherwise there would exist a point of the space such that any neighborhood of it meets infinitely many exceptional fibers. The fiber through this point would not have a fiber neighborhood.

3. Fiberings of S^3

Before studying fiberings in general, we construct examples of fiberings of S^3 with exceptional fibers. We think of S^3 as lying in R^4, where it is a hypersurface with the equation

$$x_1^2 + x_2^2 + x_3^2 + x_4^2 = 1,$$

where x_1, x_2, x_3, x_4 are Cartesian coordinates. The *fibers* are the trace curves of certain groups of rigid motions in a single variable (eingliedrigen) of the hypersphere into itself. As hypersphere curves of R^4 they are given by the equations

$$
\begin{aligned}
x_1' &= x_1 \cos mt + x_2 \sin mt, \\
x_2' &= - x_1 \sin mt + x_2 \cos mt, \\
x_3' &= x_3 \cos nt + x_4 \sin nt, \\
x_4' &= - x_3 \sin nt + x_4 \cos nt.
\end{aligned}
$$

Here m and n are coprime positive integers; t is a continuous parameter. The trace curves are closed curves which are traversed once if t runs from 0 to 2π.

We visualize the sphere by projecting it stereographically from the north pole $(0, 0, 0, 1)$ into the equator plane $x_4 = 0$. The equator plane is a

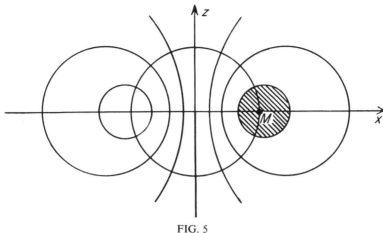

FIG. 5

3-dimensional Euclidean space with the Cartesian coordinates x, y, z which we close to the conformal space by adjoining one single point of infinity, the image of the north pole.[10] Each point (x_1, x_2, x_3, x_4) distinct from the north pole has a unique image with coordinates x, y, z; the x-, y-, and z-axes are identified with the x_1-, x_2-, and x_3-axes of R^4. The Euclidean space has now in addition to the Euclidean metric (from R^4) a spherical metric which comes from the stereographic projection of the hypersphere. The projection transforms the rigid motions of the hypersphere into conformal (or spherical-rigid) motions which permute diametrical balls of the unit sphere $x^2 + y^2 + z^2 = 1$. In particular, the above described continuous group is mapped into a group which sends the z-axis and the unit circle $x^2 + y^2 = 1$, $z = 0$, to itself.

Then the ∞^1 tori, which have the z-axis as axis of rotation and which intersect each of the spheres through the unit circle orthogonally, are all mapped into themselves. Figure 5 shows a section of the torus with the x, z-plane. Each of the tori bounds a solid torus which contains the unit circle in its interior and is fibered by the trace curves of the group of motions. For a half-plane bounded by the z-axis is under a motion of the group rotated about the z-axis. The circular section of the half-plane with a solid torus (shaded in the figure) is spherical-rigidly rotated about its spherical center M about the angle $2\pi n/m$ during the time that the half-plane is rotated once about the z-axis. The characteristic numbers μ and ν of the fibered solid torus are therefore $\mu = m$ and $\nu = $ absolute value of n, reduced $\mathrm{mod}\, m$ to $[-\frac{1}{2} m, \frac{1}{2} m]$.

The part of the hypersphere lying outside the torus considered is also a solid torus fibered by trace curves which has the z-axis as middle fiber. For under the rigid motion $x_1' = x_3$, $x_2' = x_4$, $x_3' = x_1$, $x_4' = x_2$, that is, under the

---

[10] Cf. DB II §7, §1, and §2.

corresponding spherical motion of the conformal space, the unit circle and z-axis are interchanged. The characteristic numbers of this solid torus are $\mu = n$ and $\nu = m$ (reduced mod n to the interval $[-\frac{1}{2}n, \frac{1}{2}n]$). The unit circle is therefore an exceptional fiber of multiplicity m, and the z-axis an exceptional fiber of multiplicity n. Each other trace curve is an ordinary fiber since it is contained in a fibered solid torus neighborhood of the unit circle. Each ordinary fiber wraps m times about the z-axis and n times about the unit circle; hence it is knotted, namely, a torus knot[11] if m and n are different from 1.

The orbit surface of a hypersphere fibration is always the 2-sphere. For each closed curve in S^3 can be deformed into a point; therefore the same holds for the orbit surface. Since S^3 is closed, so is the orbit surface (§2); hence it can only be the 2-sphere. Here is a direct verification in the case that $m = n = 1$, in which case there are no exceptional fibers. In this case the trace curves are circles, which include the z-axis and the unit circle. Each circle intersects the interior of the unit circle exactly once, except for the unit circle. If a point in the interior of the unit disk approaches the boundary, then the trace curve through this point approaches the unit circle. Thus one has to close the interior of the unit circle with one single point, the image point of the unit circle, to obtain the orbit surface. This completion gives us the 2-sphere.

The orbit surface cannot be embedded into the hypersphere so that each fiber intersects it in its image point, because a 2-sphere in S^3 intersects any closed curve in an even number of points.[12]

In §11 we shall show that the fiberings described above are the only possible fiberings of the hypersphere; i.e., any fibering of S^3 can be mapped to one of these under a fiber preserving map.

4. Triangulations of Fibered Spaces

The fibered spaces are defined as topological spaces via point sets, but it is well known that there are also other, purely combinatorial, definitions of manifolds which use different things for their construction, namely, cells of dimensions 0 to 3. A combinatorial manifold determines a topological manifold if we fill in the cells (which can be chosen to be simplexes) with points. In 2 dimensions, any topological space satisfying the corresponding axioms (1)–(4) of §1 can be triangulated (see Footnote 9), and therefore one can base theorems about 2-manifolds on the topological or the combinatorial definition, whichever is more convenient. In three and more dimensions,

---

[11] K. Reidemeister, Knoten und Gruppen. *Abh. Math. Sem Univ. Hamburg* **5** (1927), 19.

[12] Since each point of the hypersphere is mapped to a point of the orbit surface, we have a map of S^3 onto S^2. It is the same map which H. Hopf investigates in "Über die Abbildungen der 3-dimensionalen Sphäre auf die Kugelfläche" [*Math. Ann.* **104** (1931), 637–665].

however, it is not yet proved that a manifold satisfying axioms (1)–(4) of §1 can be triangulated.* Therefore it is important to know that fibered spaces can be triangulated, so that we can use both the methods of point set and combinatorial topology. We now present a lemma which is useful but not necessary for the proof of the triangulation of fibered spaces.

LEMMA 3. *If ω is a (closed) 2-cell on the orbit surface f which contains no exceptional points, then ω is an orbit neighborhood of each of its interior points. If ω contains exactly one exceptional point in its interior, then ω is an orbit neighborhood of this exceptional point.*

Proof. Let h be the exceptional point, or if ω has no exceptional points, let h be an arbitrary interior point of ω. Take a triangulation of ω which is so small that each 2-simplex is covered by an orbit neighborhood. Furthermore we require that h lie in the interior of a 2-simplex. Such a triangulation exists, for mapping ω onto a disk of R^2, we find a positive radius ε such that a disk of radius ε about an arbitrary point p of ω is covered by an orbit neighborhood (which is not necessarily the orbit neighborhood of p). If the ε-disk is not contained in the disk, we consider only the part belonging to ω. If there did not exist such an ε, there would exist a sequence of disks whose radii and center points converge to 0 and a point p_0, respectively, and each of which could not be covered with an orbit neighborhood. Then we could take a disk of radius $\rho > 0$ about p_0 which is covered by the orbit neighborhood of p_0. This disk contains almost all disks of the sequence, almost all of which can therefore be covered by one orbit neighborhood. This contradiction assures the existence of an ε as above. We now triangulate ω so small that each 2-simplex can be covered by a disk of radius ε. Then we apply Lemma 1 to the ε-disks and find that all 2-simplexes are orbit neighborhoods. The corresponding fiber neighborhoods are ordinary fibered solid tori, except possibly for the orbit neighborhood Δ_H of the fiber H which is mapped into the 2-simplex δ_H containing h. Now, as is well known, there is a sequence of 2-cells $\omega_1 = \delta_h, \omega_2, \ldots, \omega_\sigma = \omega$, which all are 2-simplexes of the triangulation of ω and such that each is obtained from its predecessor by adjoining an adjacent 2-simplex along one or two edges, a fact which, by the way, may not be true in 3 dimensions. The corresponding fiber sets $\Omega_1 = \Delta_H, \Omega_2, \ldots, \Omega_\sigma = \Omega$ are fiber neighborhoods of H. For as ω_i is obtained from ω_{i-1} by pasting on a 2-simplex δ along a single 1-cell s (which may consist of one or two edges of δ), we obtain Ω_i from Ω_{i-1} by pasting an ordinary fibered solid torus Δ fiber preservingly to Ω_{i-1} along a fibered annulus S. It is easy to see that this gives us again a fibered solid torus.

THEOREM 2. *Every fibered space can be triangulated.*

Proof. We take a triangulation of the orbit surface such that the exceptional points are contained in the interior of the 2-simplexes and such

* *Translator's note*: This paper was printed December 14, 1932.

that no 2-simplex contains more than one exceptional point. By Lemma 3 each 2-simplex is an orbit neighborhood. The fibered space is therefore decomposed into a finite or countable number of fibered solid tori. Two adjacent such solid tori have a fibered annulus in common, which is mapped onto a 1-simplex of the orbit surface and which can be mapped onto a rectangle of R^2 after removing a spanning arc. We can therefore speak about straight lines in such an annulus. These are lines which map to straight lines of the rectangle. Now we triangulate each of the fibered solid tori so that the triangulation of the three annuli which make up the boundary of the solid torus is "linear." On each of these annuli there are now two triangulations which come from the triangulations of the two adjacent solid tori and which can be replaced by a common subdivision since they are linear. This gives us a decomposition of the fibered space into cells. From this we can deduce a simplicial triangulation by barycentric subdivision.

5. Drilling and Filling (Surgery)

An essential aid for the classification of fibered spaces will be the method of *drilling out* exceptional fibers and replacing the *drill hole* by ordinary fibered solid tori. To drill out a fiber H from a fibered space F means to remove from F the interior points of a fiber neighborhood Ω_H of H. This results in a fibered space $\overline{F}$ *with boundary*. The boundary is a fibered torus. The orbit surface $\bar{f}$ of $\overline{F}$ is obtained from the orbit surface f of F by removing the interior points of the orbit neighborhood ω_h into which the fiber neighborhood Ω_H is mapped.

We first show that the space $\overline{F}$ is independent of the choice of the fiber neighborhood of the fiber H and second that $\overline{F}$ is independent of the choice of H if H is an ordinary fiber. Then we get back fibered spaces F by *closing* an arbitrary fibered space $\overline{F}$ with boundary with suitable fibered *torus seals* (Verschluss ring).

LEMMA 4. *If Ω and Ω' are two fiber neighborhoods of a fiber H in a fibered space F, there exists a fiber preserving deformation of F which sends Ω to Ω' and leaves H fixed.*

Proof. Between Ω and Ω' we put a fiber neighborhood Ω_1 of H which lies in the interior of Ω and Ω' and show that there exists a fiber preserving deformation of F that keeps H fixed and sends Ω to Ω_1. Then there is also a deformation which sends Ω' to Ω since Ω' is not distinguished from Ω. The required deformation is the first deformation followed by the inverse of the second. The existence of such a fiber neighborhood Ω_1 follows from Lemma 1 since for any two orbit neighborhoods ω and ω' of h there exists an orbit neighborhood ω_1 of h which lies in the interior of ω and ω'.

We now take another orbit neighborhood ω_a of h which contains ω in its interior. This is possible; one can choose for ω_a a 2-cell which contains ω in

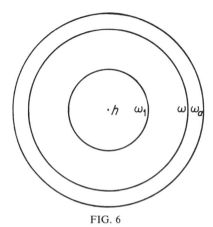

FIG. 6

its interior and contains no exceptional points except h. This 2-cell exists since the orbit neighborhoods are closed and exceptional points have no accumulation points, and it is an orbit neighborhood by Lemma 3.

To get a model, we map ω_a onto a disk of R^2, with the image of h as center point, and such that ω and ω_1 are mapped to concentric circles (Fig. 6). Now we perform on the disk a deformation which sends ω_1 to ω (for example, by radially blowing up ω_1). This deformation of the orbit neighborhood of h corresponds to a fiber preserving deformation of the fiber neighborhood Ω_a of H which keeps H and the boundary of Ω_a pointwise fixed. We obtain this deformation of Ω_a by cutting Ω into a Euclidean cylinder and transferring the deformation of ω_a to all meridian disks which are μ-fold branched covering surfaces of ω_a.

Lemma 4 implies that the fibered space $\overline{F}$, which is obtained from F by drilling out a fiber H, is independent of the choice of the (infinitely many) fiber neighborhoods of H.

LEMMA 5. *The fibered space with boundary $\overline{F}$, which is obtained from F by drilling out an ordinary fiber H, is independent of the choice of the ordinary fiber H.*

Proof. If H and H' are two ordinary fibers of F, h and h' their image points on the orbit surface f, there exists a 2-cell ω which contains h and h' in its interior and contains no points which are images of exceptional fibers. Then there exists a deformation of ω which sends h to h' and keeps the boundary of ω fixed. By Lemma 3, ω is an orbit neighborhood of each of its interior points and hence the image of an ordinary fibered solid torus Ω. The deformation of ω corresponds to a fiber preserving deformation of Ω which sends H to H' and leaves the boundary torus of Ω pointwise fixed.

The same arguments apply to the drilled-out space $\overline{F}$ and show that the space obtained from F by drilling out an arbitrary number of ordinary fibers is independent from the choice of the ordinary fibers which are drilled out. The only requirement is that the drilled-out fiber neighborhoods be mutually disjoint.

From the fibered space with boundary $\overline{F}$ that is obtained from F by drilling out a fiber we can construct new (closed) fibered spaces by closing the boundary torus $\overline{\Pi}$ of $\overline{F}$ with a fibered solid torus, the *torus seal V*. This is achieved by a fiber preserving pasting of the boundary torus Π of V to the torus $\overline{\Pi}$. Given the torus seal V, this closing can be made in infinitely many essentially different ways. But the closing is completely determined if one knows the image $\overline{M}$ of a meridian curve M of V on the torus $\overline{\Pi}$. Obviously, $\overline{M}$ can neither be null homologous nor homologous to a fiber on $\overline{\Pi}$ since otherwise this would be true for M on Π; furthermore, $\overline{M}$ is without singular points. These are all requirements for $\overline{M}$. For we have

LEMMA 6. *If on the boundary torus $\overline{\Pi}$ of a fibered space with boundary $\overline{F}$ we have a simple closed curve $\overline{M}$ on $\overline{\Pi}$ which is neither homologous to 0 nor to a fiber, then there exists exactly one fibered solid torus V whose boundary torus Π can be mapped under a fiber preserving map onto $\overline{\Pi}$ such that $\overline{M}$ is homotopic to 0 in V. The thus resulting (closed) fibered space F_1 is uniquely determined by $\overline{F}$ and the homology class of $\overline{M}$ on $\overline{\Pi}$.*

Proof. (a)   First we show that there exists one and only one fibered solid torus V that satisfies the requirements of the theorem. If $\overline{Q}$ is a crossing curve, $\overline{H}$ an oriented fiber on $\overline{\Pi}$, we have

$$\overline{M} \sim \alpha\overline{Q} + \beta\overline{H} \qquad (\alpha = 0, (\alpha, \beta) = 1).$$

In §1 it was shown that there exists exactly one fibered solid torus V with meridian M, fiber H, and suitable chosen crossing curve Q such that on the boundary Π of V we have

$$M \sim \alpha Q + \beta H.$$

We can map Π onto $\overline{\Pi}$ under a fiber preserving map such that Q goes to $\overline{Q}$ and H to $\overline{H}$. For we can cut Π, $\overline{\Pi}$ along Q and H, $\overline{Q}$ and $\overline{H}$, respectively, into two rectangles which are hatched by the fibers and we can map these rectangles onto each other under a fiber preserving map. Then $\overline{M}$ is mapped to M, and thus $\overline{M}$ becomes a meridian of V.

(b)   We now show that the fibered space F_1 is uniquely determined by $\overline{F}$ and the homology class of $\overline{M}$ (on $\overline{\Pi}$). All possible fiber preserving maps of $\overline{\Pi}$ onto Π under which $\overline{M}$ becomes homotopic to 0 in V are obtained from a single such map followed by a fiber preserving map A_{Π} from Π onto Π which maps the meridian M, or more precisely its homology class, to itself or its negative. We shall have proved the independence of the resulting fibering F_1 from the choice of the above maps once we have shown that we can extend A_{Π} to a fiber preserving map A_V of V onto V whose restriction to Π is A_{Π}.

We first check how the homology classes of Π are transformed under A_Π. Let H, Q, and M be fiber, crossing curve, and meridian curve on Π, respectively, with an arbitrary but fixed orientation, and let

$$M \sim \alpha Q + \beta H.$$

Because of the transformations (4) in §1 we can choose Q a priori such that $\alpha > 0$ and $0 \leqslant \beta < \alpha$; of course, since M is a simple closed curve, α and β are coprime. Let H', Q', M' be the images of these curves under A_Π. Since A_Π is fiber preserving, we have from §1

$$H' \sim \varepsilon_1 H, \qquad Q' \sim \varepsilon_2 Q + \lambda H \qquad (\varepsilon_1, \varepsilon_2 \pm 1). \tag{1}$$

The meridian curve M is mapped under A_Π into

$$M' \sim \alpha Q' + \beta H' \sim \varepsilon_2 \alpha Q + (\varepsilon_1 \beta + \alpha\lambda)H.$$

Now we must have that $M' \sim \varepsilon_3 M$, hence

$$\varepsilon_2 \alpha Q + (\varepsilon_1 \beta + \alpha\lambda)H \sim \varepsilon_3(\alpha Q + \beta H).$$

Comparing coefficients, we get $\varepsilon_2 = \varepsilon_3$ and

$$\alpha\lambda + \varepsilon_1 \beta = \varepsilon_2 \beta. \tag{2}$$

If $\alpha > 2$, this implies $\lambda = 0$ and for (1) there are only the two possibilities

$$\alpha > 2 \quad \begin{cases} (1) & H' \sim H, & Q' \sim Q \\ (2) & H' \sim -H, & Q' \sim -Q. \end{cases}$$

For $\alpha = 2$ we must have $\lambda = +1, -1$, or 0, since $0 < \beta < \alpha$. Thus there are 4 possibilities

$$\alpha = 2 \quad \begin{cases} (1) & H' \sim H, & Q' \sim Q \\ (2) & H' \sim -H, & Q' \sim -Q \\ (3) & H' \sim -H, & Q' \sim Q + H \\ (4) & H' \sim H, & Q' \sim -Q - H. \end{cases}$$

For $\alpha = 1$ we again get $\lambda = 0$ and we obtain the four possibilities

$$\alpha = 1 \quad \{ H' \sim \pm H, \quad Q' \sim \pm Q$$

with all four combinations of the signs.

The map A_V which we have to construct will be the composition of two fiber preserving maps $A_V = J_V \cdot B_V$.[13] B_V is an arbitrary fiber preserving map which transforms the homology classes on Π in the same way as A_Π does. J_V maps each class to itself. We cut V into a right circular cylinder. In case that $H' \sim -H$, $Q' \sim -Q$ we let B_V be a rotation of Π about a line orthogonal to the cylinder axis. Then B_V is fiber preserving and sends each homology class on Π to its negative. In the case $\alpha = 1$ we obtain the desired map B_V by the

---

[13] $J_\nu \cdot B_\nu$ is the map obtained by first applying B_ν, then J_ν.

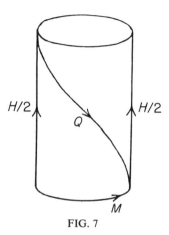

$H/2$ $H/2$

Q

M

FIG. 7

rotation as in the previous case or by a reflection on a plane which is orthogonal to or passes through the cylinder axis. In the case $\alpha = 2$, the fiber is made up of two lines lying diametrical to the middle fiber. Since $M \sim 2Q + H$, the crossing curve appears as in Fig. 7. A transformation (3) is obtained by reflecting the cylinder at the plane orthogonal to the cylinder axis and going through its center point; a transformation (4) is obtained by reflecting at a plane which goes through the axis.

It remains to be shown that for an arbitrary fiber preserving map J_Π of Π onto itself which maps each homology class of Π to itself, there exists a fiber preserving map J_V of V onto itself whose restriction to Π is J_Π. We show first that J_Π can be deformed to the identity by a fiber preserving deformation. We can show this, e.g., by first taking a rigid translation of the fiber into itself such that the image Q' of Q is mapped onto Q. Such a deformation is possible since by hypothesis Q' is homologous to Q on the boundary torus. This is followed by a fiber preserving deformation which interchanges the fibers and such that the composition keeps Q pointwise fixed. The map J_Π so deformed appears in the fibered rectangle, which is obtained from Π by cutting along a fiber H and Q, as a fiber preserving map C which leaves the two parallel edges Q pointwise fixed and which translates the inner points only along their fibers. To transform this map of the rectangle into the identity by a fiber preserving map, we proceed as in the proof of the Tietze deformation theorem by Alexander. We complete the rectangle to a strip by the region which is shaded in Fig. 8 and define a map C' of this strip which coincides with C in the rectangle and is the identity in the shaded region. Let $T(t)$ be a stretching of the band upward which leaves the lower boundary Q of the band fixed: the ordinate ξ of a point of the band should go over to $t\xi$. Then $T(t)^{-1} C' T(t) = C'(t)$ is a topological map of the strip, which maps the rectangle fiber preservingly into itself for $t \geqslant 1$. For $t = 1$ this map coincides with C in the rectangle. As $t \to \infty$, $C'(t)$ continuously approaches the identity.

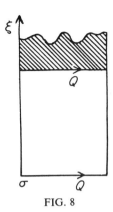

FIG. 8

Thus C and therefore J_Π is deformed into the identity by a fiber preserving deformation.

We now describe this deformation by a parameter τ which decreases from 1 to $\frac{1}{2}$. Let the map corresponding to τ be $J_\Pi(\tau)$. To extend J_Π to the desired map J_V, we cut V to a cylinder (of radius 1) and introduce cylindrical coordinates z, φ, ρ. Then $\rho = \text{const}$ gives a concentric torus of radius ρ. We map each of the tori onto itself under a fiber preserving map. The boundary torus is mapped under $J_\Pi = J_\Pi(1)$. If the map $J_\Pi(\tau)$ in the coordinates z, φ is given by

$$\left. \begin{array}{l} z' = z'(z, \varphi, \tau) \\ \varphi' = \varphi'(z, \varphi, \tau) \end{array} \right\}, \qquad (J_\Pi(\tau))$$

the map J_V for $1 \geqslant \rho \geqslant \frac{1}{2}$ is defined by

$$\left. \begin{array}{l} z' = z'(z, \varphi, \rho) \\ \varphi' = \varphi'(z, \varphi, \rho) \\ \rho' = \rho \end{array} \right\}, \qquad (J_V)$$

whereas for $\frac{1}{2} \geqslant \rho \geqslant 0$ it is the identity. This construction of the map A_V completes the proof of Lemma 6.

Instead of constructing A_V as above, we could have described this map directly in terms of cylindrical coordinates. For if

$$\left. \begin{array}{ll} \bar{z} = \bar{z}(z, \varphi) & [= z'(z, \varphi, 1)] \\ \bar{\varphi} = \bar{\varphi}(z, \varphi) & [= \varphi'(z, \varphi, 1)] \end{array} \right\}, \qquad (J_\Pi)$$

describes the map J_Π of the torus Π in terms of cylindrical coordinates, then the desired map A_V is given in the range $1 \geqslant \rho \geqslant \frac{1}{2}$ by

$$\left. \begin{array}{l} z' = 2(\rho - \frac{1}{2})\bar{z} - 2(\rho - 1)z \\ \varphi' = 2(\rho - \frac{1}{2})\bar{\varphi} - 2(\rho - 1)\varphi \\ \rho' = \rho \end{array} \right\}, \qquad (A_V)$$

and for $\frac{1}{2} \geqslant \rho \geqslant 0$ it is the identity. However, since it is not quite easy to demonstrate that this map A_V is a homeomorphism, we have chosen the method above.

6. Classes of Fibered Spaces

If w is a path on the orbit surface f from a point h_1 to a point h_2, we can in the fibered space deform the fiber H_1 into the fiber H_2 over fibers so that the image on f runs along w. The path w does not determine the mapping of H_1 to H_2 pointwise, but during the deformation the fiber can be translated in itself. But the map of H_1 to H_2 is determined up to orientation preserving autohomeomorphisms of H_2. Therefore, if H_1 is oriented, then the orientation is translated uniquely to H_2 along the path w. We shall take up this point more closely at the end of this section.

If w' is another path of h_1 to h_2, the translation of a fixed orientation of h_1 along w' can lead to a different result as translation along w. However, the end orientations agree if w can be deformed to w' on the orbit surface. In particular, if w is a closed curve on f, it is possible that running along w the orientation of the fiber is preserved or changed. Depending on whether we have the first or second case, we associate the value $+1$ or -1 to the curve w. Since this value is invariant under deformations of the curve, to each element of the fundamental group there corresponds a unique value. To the product $a \cdot b$ of two elements of the fundamental group corresponds the product of the two corresponding values; the inverse of a has the same value as a. This implies that the value of a curve is determined already by its homology class. For each null homologous curve has value $+1$ since it represents an element of the commutator subgroup of the fundamental group, and is therefore a product of commutators, and each commutator $aba^{-1}b^{-1}$ has value $+1$. Therefore the values of all curves are known if the values of a fundamental system of curves of the fundamental group, or even the homology group, are known.

We say that two fibered spaces F and F' belong to the same *class* if their orbit surfaces f and f' can be mapped onto each other under a homeomorphism such that each curve is mapped to one with the same value. The class of a fibered space is therefore determined by its "*valuated orbit surface*." Two fibered spaces belong certainly to different classes if their orbit surfaces are not homeomorphic. However, spaces belonging to different classes may have the same orbit surface. We shall give a complete enumeration of the classes in §7 and §8. For example, for the projective plane there are two classes, depending on whether the orientation of the fiber is preserved or reversed along the projective line. For a simply connected surface there is only one class since each closed curve on it is null homologous, hence has value $+1$.

If we drill out a fiber of the space and replace the drill hole by a new torus seal as in §5 , the class of the fibered space is not changed. For the class is already determined if we know the value of one curve in each homology class. The representatives of the homology classes can then be chosen so that they are not affected by the drilling and filling, i.e., this process of changing the space does not affect the valuation of the curves, as it does not affect the orbit surface.

If we drill out all the exceptional fibers from a fibered space F and fill in the drill holes with ordinary torus seals, we obtain from F by this process (but not in a unique way) another space F_0 which has no exceptional fibers and belongs to the same class as F. Conversely, we can get back F from F_0. Therefore we first would like to characterize all spaces without exceptional fibers belonging to the same class. To this end, we cut the orbit surface f of a space F_0 into the fundamental polygon v, where we have to require that f be closed, hence F be a closed space. We adopt this restriction from now on. We change the fundamental polygon to a polygon $\bar{v}$ by cutting off the vertices, which means that we change the surface f to a punctured surface $\bar{f}$ by cutting out a 2-cell which contains the vertex h of v. Figure 9 shows the punctured fundamental polygon of the orientable surface of genus $p = 2$. We can think of $\bar{f}$ as the orbit surface of a space $\bar{F}_0$ which is obtained from F_0 by drilling out a fiber H. Then $\bar{F}_0$ is uniquely determined by F_0 since $\bar{F}_0$ does not depend on the choice of the drilled out ordinary fiber, by Lemma 5 (§5). Now we triangulate $\bar{f}$ using the edges of the polygon $\bar{v}$ (dotted lines of Fig. 9). The fibers of $\bar{F}_0$ which map to points of a 2-simplex of the triangulation constitute an ordinary fibered solid torus, by Lemma 3 (§4). As in the proof of Lemma 3 we can build up the polygon $\bar{v}$ step by step from 2-simplexes so that after each step we obtain a 2-cell. This construction corresponds to a construction of $\bar{F}_0$ from ordinary fibered solid tori, which gives us an ordinary fibered solid torus $\bar{V}$. The edges of $\bar{v}$ correspond in $\bar{V}$ to fibered annuli. If two edges a' and a'' in $\bar{v}$ are identified with an arc a of $\bar{f}$, we have to identify the corresponding

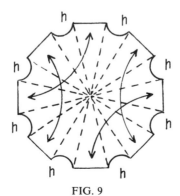

FIG. 9

annuli A' and A'' in $\overline{V}$ with a fibered annulus A of $\overline{F}_0$ under a fiber preserving map. If we identify in this way all the corresponding annuli of $\overline{V}$, we get $\overline{F}_0$. If we know how two edges a' and a'' of $\overline{v}$ are identified (under an orientation preserving or orientation reversing map) and whether the fiber orientation is preserved or reversed along a closed curve of $\overline{f}$ which intersects the edges of $\overline{v}$ only in one point of the edge a, then the identification of the annuli A' and A'' is uniquely determined up to an orientation preserving and fiber preserving map of one of the annuli onto itself, say A'. This map of A' can be induced by a fiber preserving map of the solid torus $\overline{V}$ which keeps all the other annuli (which correspond in pairs) fixed. The map of A' to A' has therefore no effect on the closing of $\overline{V}$ to $\overline{F}_0$. All fibered spaces with boundary obtained in this way can be mapped onto $\overline{F}_0$ under a fiber preserving map.

This shows that all closed fibered spaces F_0 without exceptional fibers which belong to the same class give the same fibered space (with boundary) $\overline{F}_0$ after drilling out an arbitrary fiber. If we drill out $r + 1$ fibers instead of just one, we again obtain the same fibered space (bounded by $r + 1$ tori), namely, the sapce obtained from $\overline{F}_0$ by drilling out r fibers. As the proof of Lemma 5 (§5) shows, it does not matter which fibers of $\overline{F}_0$ are drilled out. We sum up:

THEOREM 3. *Each class of closed fibered spaces determines (and is determined by) a unique fibered space with boundary, the* **classifying space** $\overline{F}_0$. *The classifying space is the only fibered space with boundary and without exceptional fibers which has as orbit surface the punctured valuated orbit surface which characterizes the class. From $\overline{F}_0$ we obtain all spaces of the class by drilling out a finite number r of fibers and closing the $r + 1$ boundary tori with arbitrary torus seals. The enumeration of all classes will be given in Theorem 7, §8.*

So far, we started with a given fibered space F and defined its class, i.e., its valuated orbit surface. Now we start with an arbitrary valuated closed surface and show that it is the valuated orbit surface of a class. We cut the given surface f into the fundamental polygon v as above and puncture it by cutting off the vertices of v to get $\overline{v}$. The ordinary fibered solid torus $\overline{V}$ which has $\overline{v}$ as meridian disk can be made into a fibered space (with boundary) $\overline{F}_0$ by identifying under a fiber preserving map any two annuli A' and A'' on the boundary on $\overline{V}$ which map to corresponding edges a' and a'' of $\overline{v}$ such that a fiber of A' is identified with a fiber of A'' if the point of a' is identified with the corresponding point of a''. Then there exist essentially two distinct maps of A' to A''. For if we orient the the fibers of $\overline{V}$ simultaneously so that any two oriented fibers on $\overline{V}$ are homologous, we can map A' to A'' under a map which preserves and under a map which reverses the fiber orientation. In the first case the orientation of the fiber is preserved along a curve which goes from a point of A' through the interior of $\overline{V}$ to the equivalent point of A''; in the second case it is reversed. If we identify in this way any two annuli of $\overline{V}$ which correspond to two equivalent edges of $\overline{v}$ under one of the two maps, we get a space with a boundary Π_0 which consists of fibers. These boundary fibers correspond to the boundary curve $\overline{v}$. Therefore Π_0 is a torus or a Klein bottle. To show that Π_0 is a torus, we observe that if we run along the boundary curve of $\overline{f}$, we cross each edge of the polygon v exactly twice. In both cases the fiber orientation is either preserved or reversed so that if we run once along the boundary curve the fiber orientation is preserved; but this is the case only for the torus. The space obtained from $\overline{V}$ under the identifications is therefore a fibered space (with boundary) without exceptional fibers. Its orbit surface is the punctured surface f, whose valuation was obtained from an arbitrary valuation of the edges of a fundamental polygon (namely the fundamental polygon dual to v). This proves

THEOREM 4. *For an arbitrary valuated closed surface there is a corresponding class of fibered spaces. A valuation of the surface is obtained by an arbitrary valuation of a canonical system of fundamental curves, i.e., the edges of a Poincaré fundamental polygon of the surface.*

We proved the last remark by constructing for any arbitrarily given valuation of the fundamental curves a space $\overline{F}_0$ whose orbit surface is the given punctured surface; the valuation of the orbit surface determined by $\overline{F}_0$ agrees for the fundamental curves with the arbitrarily given valuation. One could easily have shown directly that an arbitrary valuation of the fundamental curves, i.e., of the generators of the fundamental group, leads to a well-defined valuation of the whole group since each generator appears exactly twice in the single relation of the fundamental group, and therefore an arbitrary valuation of the generators gives a well-defined valuation of the single defining relation and hence of each relation between elements of the fundamemtal group.

Theorems 3 and 4 give us the tools to determine complete invariants of fibered spaces under fiber preserving maps. We now describe in detail the translation of the fiber orientation along a path which was used in the definition of class. If w is a path on the orbit surface from a point h_1 to a point h_2 and if s is a continuous parameter from 0 to 1 on w, we have for each value s of the parameter a point $h(s)$ of f and hence a fiber $H(s)$. Orient each fiber $H(s)$ arbitrarily. If the same fiber H belongs to different values s, which happens if w has multiple points, we give H the same number of mutually independent orientations. A fiber neighborhood of $H(s)$ or, more precisely, the corresponding orbit neighborhood on f cuts out from w a neighborhood of the point $h(s)$. If for each value of s all the fibers corresponding to the path near $h(s)$ are homologous in the fiber neighborhood of $H(s)$, where a μ-fold exceptional fiber counts μ times, we say that *the fibers are oriented simultaneously along* w. It is clear that there exists such a simultaneous orientation of the fibers along w if w is covered by one orbit neighborhood ω; because then we need only orient all the fibers which map to points of w so that they are homologous in Ω.[14] In the general case we decompose w into finitely many pieces so that each piece lies in the interior of an orbit neighborhood. The fibers of the individual pieces can be oriented simultaneously so that each fiber at the intersection of two pieces gets the same orientation from the two pieces. Then all the fibers are oriented simultaneously along w. The fibers can be oriented simultaneously along w apparently only in two opposite ways; the orientation along w is determined by the orientation of a single fiber, e.g., the initial fiber. Under a simultaneous orientation of the fibers of w, the orientation of the first fiber is translated along w to the last fiber.

If w and w' are homotopic curves of the orbit surface which both go from h_1 to h_2, and if the fiber H_1 is oriented, then the translation of the orientation along w and w' to H_2 gives the same result; i.e., the fiber orientation is preserved under translation along the closed path ww'^{-1}. For ww'^{-1} bounds a singular 2-cell on f, i.e., the continuous image of a 2-cell e. We triangulate e so small that the image of each 2-simplex is contained in an orbit neighborhood on f. Since the path ww'^{-1} can be built up from boundary paths of 2-simplexes by canceling out edges which are traveled in opposite directions, and since the fiber orientation is preserved along a closed path which lies in an orbit neighborhood, the fiber orientation is preserved along ww'^{-1}.

We now want to solve the problem whether and in how many different ways the orbit surface $\bar{f}_0$ can be embedded in the classifying space $\bar{F}_0$ so that each fiber intersects it exactly in its image point. To this end, we cut $\bar{f}_0$ into a fundamental polygon $\bar{u}$ which, in contrast to the fundamental polygon $\bar{v}$ above, contains the hole of $\bar{f}_0$ in its interior, i.e., $\bar{u}$ is a punctured 2-cell. This corresponds to a cutting of $\bar{F}_0$ into a fibered hollow torus $\bar{U}$. The "inner" boundary surface Π_0 of $\bar{U}$ is mapped onto the boundary of the hole of $\bar{u}$, whereas the "outer" boundary Σ is decomposed into an even number $2j$ of pairwise equivalent fibered annuli which map onto edges of the polygon $\bar{u}$. Suppose we have succeeded in embedding $\bar{f}_0$ into $\bar{F}_0$; then $\bar{f}_0$ appears in $\bar{U}$ necessarily as an annulus which meets Σ in a crossing curve Q and Π_0 in a crossing curve Q_0. If $Q'_1, \ldots, Q''_j, Q''_1, \ldots, Q''_j$ are the $2j$ oriented edges which make up Q and which correspond to the $2j$ lateral surfaces of Σ, then if two such lateral surfaces (annuli) A'_i and A''_i are identified, the two edges Q'_i and Q''_i which they contain have to be identified under an orientation preserving or reversing map. (Conversely,) a crossing curve Q with this property can always be found on Σ by choosing the crossing lines $Q'_1, Q'_2, \ldots, Q'_j$ arbitrarily, but such that their end points go to the same

---

[14] In this case we say that the fibers of Ω are oriented simultaneously. More generally we talk about a simultaneous orientation of all fibers of a fibered space if in each fiber neighborhood any two fibers are homologous, where a μ-fold execptional fiber counts μ times. Not every fibered space admits a simultaneous orientation of fibers but only the spaces of the classes Oo and Nn I of p. 391.

point under the identification of the lateral surfaces. Then $\bar{f}_0$ can be embedded into $\bar{F}_0$; for example, we can cut $\bar{U}$ into a hollow cylinder (annulus $\times$ I) and draw from the points of Q radii which lie orthogonal to the cylinder axis. These radii in $\bar{U}$ make up the required orbit surface.

Suppose now we have $\bar{f}_0$ embedded into $\bar{F}_0$ in a different way, with crossing curves Q^* and Q_0^* instead of Q and Q_0. The lines Q_i' and $Q_i'^*$ of Q (resp. Q^*), which lie in the same lateral side A_i' of Σ, have (after choosing an orientation of Σ) a certain intersection number[15] γ_i'; here we assume that Q_i' and $Q_i'^*$ have no common endpoints, which can be achieved by a small deformation of the embedded orbit surfaces. Since under the identification of the corresponding lateral sides A_i' and A_i'' the lines Q_i' and $Q_i'^*$ are identified with the lines Q_i'' and $Q_i''^*$ (resp. with $-Q_i''$ and $Q_i''^*$), the intersection number is $\gamma_i'' = -\gamma_i'$ or $= +\gamma_i'$, depending on whether A_i' and A_i'' form an association of type one or two.[16] $\gamma = \sum_{i=1}^{j} \gamma_i' + \gamma_i''$, i.e., the intersection number of Q and Q^* is 0 if all the lateral sides of Σ are identified in the first way, i.e., if $\bar{F}_0$ is orientable. Otherwise we can choose Q^* such that γ is a given even number. Therefore, if $\bar{F}_0$ is orientable, Q can be deformed into Q^* and hence Q_0 into Q_0^*, i.e., on the boundary surface Π_0 of $\bar{F}_0$ there exists a crossing curve Q_0 which is determined up to orientation and deformations, such that Q_0 is the intersection of Π_0 and the orbit surface $\bar{f}_0$ is embedded in $\bar{F}_0$. If $\bar{F}_0$ is nonorientable, there are besides Q_0 infinitely many crossing curves Q_0^* which can be the intersection of $\bar{f}_0$ and Π_0. They all differ from Q_0 by an even multiple of the fiber. If we cut the fibered torus $\times$ I, $\bar{U}$ along the embedded orbit surface $\bar{f}_0$, we obtain a drilled-out fibered prism in which bottom and top surface are equivalent and the lateral surfaces are pairwise equivalent. We shall use this representation of the classifying space in §10 to determine the fundamental group.

7. The Orientable Fibered Spaces

Our task to determine all fibered spaces and to characterize them by invariants splits into two parts: first, to determine all the classes; second to list all spaces of a given class. We first solve this problem for orientable spaces.

First suppose the orbit surface is orientable of genus p. Since the space is orientable, the fiber orientation is preserved along any curve of the surface. For if w is a closed curve of value -1 on the orbit surface (which misses exceptional points), there is a fiber preserving deformation of the space which traces the fiber H along the curve w. This is so because w can be covered with finitely many orbit neighborhoods without exceptional points (Heine–Borel).

[15] O. Veblen, "Analysis Situs," 2nd ed., Amer. Math. Soc. Colloq. Publ. No. 5, Part 2. Amer. Math. Soc., New York, 1931.

[16] H. Tietze, Topologische Invarianten, *Monatsh. Math. Phys.* **19** (1907). [See Seifert and Threlfall, this Lehrbuch p. 220.]

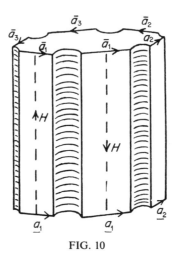

FIG. 10

Inside each orbit neighborhood one can apply the fiber preserving deformation of the proof of Lemma 5 and thus deform the fiber step by step along w into its initial position. In particular we can choose the deformation such that an orbit neighborhood ω of the point h comes back to itself, since along w the orientation of the surface is not changed because it is orientable. The corresponding fibered solid torus Ω is then mapped onto itself under an orientation reversing map. But, by a well-known theorem, the orientation of an orientable space is not reversed under a deformation. Therefore, all curves have value $+1$, and there is for each orientable orbit surface a single class of orientable fibered spaces. Now the fibered topological product of a punctured surface of genus p and S^1 is an orientable fibered space whose orbit surface is the punctured surface of genus p and all of whose curves are of value $+1$. Since this space has no exceptional fibers, it is the classifying space $\bar{F}_0$.

Even if the orbit surface is nonorientable, there is only one corresponding class of orientable spaces. As in the above case we first observe that the fiber orientation is preserved along an orientation preserving curve of the orbit surface. But if w is an orientation reversing curve of the orbit surface, then the space is orientable only if the fiber orientation is reversed along w. Therefore the valuation is determined by the surface. The classifying space is in this case not the topological product of the punctured surface of genus k and S^1, but has to be constructed by the method of §6. Figure 10 shows it for $k = 3$. In the prism we have to identify bottom and top disks under a translation. The two lateral surfaces in which we have drawn the fiber H are to be identified so that the edge a_1 of one surface is identified with the edge $\overline{a_1}$ of the other surface. Similarly we have to identify the other four unshaded lateral sides of the prism. The six shaded sides become the boundary torus of the classifying space and the bottom surface becomes the orbit surface.

This finishes off the determination of the class and we now proceed to determine the invariants of an orientable fiber space F. We orient the space and the invariants depend on the orientation. We shall obtain the invariants by drilling out the exceptional fibers of F and replacing them by ordinary solid tori whose meridians are uniquely determined by the fibered space F up to orientation. In this way we get from the oriented space F a unique oriented space F_0 without exceptional fibers. Let C_1 be an exceptional fiber of F and Ω_1 a fiber neighborhood of C_1. The solid torus Ω_1 gets a certain orientation from F, which induces on the boundary torus Π_1 of Ω_1 a certain orientation o. On Π_1 we choose an oriented crossing curve Q and an oriented fiber H. These two curves determine an orientation o′ on Π_1. For, cutting Π_1 along Q and H into a rectangle, a certain orientation of it is determined by the sequence $QHQ^{-1}H^{-1}$. By reversing the orientation of one of the curves Q and H, we reverse the orientation o′. But o′ is not changed by reversing the orientation of both curves simultaneously. We now orient Q and H so that o′ agrees with o. This can be expressed by saying that using the orientation o the curves Q and H shall have intersection number $+1$. Another pair of curves Q_1 and H_1 which determines the same orientation o′ = o on Π_1 is related to Q and H (on Π_1) as follows:

$$H \sim \varepsilon H_1, \qquad Q \sim \varepsilon Q_1 + y H_1 \qquad (\varepsilon = \pm 1, \quad y \text{ arbitrary integer}). \qquad (1)$$

For if Q_1, H_1 determine the same orientation as Q, H, the determinant of the transformation must have value $+1$. This implies that in the transformation formulas (1) and (4) of §1, $\varepsilon_1 = \varepsilon_4 (= \varepsilon)$. The meridian curve M_1 of the solid torus Ω_1 can now be expressed in terms of Q and H as

$$M_1 \sim \alpha Q + \beta H \sim \varepsilon \alpha Q_1 + (\alpha y + \varepsilon \beta) H_1 = \alpha_1 Q_1 + \beta_1 H_1. \qquad (2)$$

We can choose Q_1 and H_1 such that

$$\alpha_1 > 1 \quad \text{and} \quad 0 < \beta_1 < \alpha_1, \qquad (3)$$

which determines ε and y. If instead of M_1 we choose the meridian curve with opposite orientation, we only have to reverse both the orientations of Q_1 and H_1 to obtain the same homology $M_1 \sim \alpha_1 Q_1 + \beta_1 H_1$. Hence the numbers α_1, β_1 are determined uniquely by the nonoriented meridian of Ω_1 and the crossing curve Q_1 is determined up to its orientation. We now drill out Ω_1 and replace the drill hole with a new torus seal V_1 which has Q_1 as meridian curve. Then V_1 is an ordinary fibered solid torus since the meridian is a crossing curve. Thus we have derived an orientable fibered space F_1 from F which is uniquely determined by F, the orientation of F, and the drilled-out exceptional fiber. For F_1 is independent of which fiber neighborhood Ω_1 of C_1 is drilled out because by Lemma 4 (§5) we can deform an arbitrary fiber neighborhood of the fiber C_1 onto another under a fiber preserving deformation of F.

We now apply this construction to F_1, i.e., we drill out an exceptional fiber C_2 and obtain the pair α_2, β_2 as additional invariants of the oriented space F.

Continuing in this way, we finally obtain an oriented space $\overline{F}_0$ without exceptional fibers which is determined by F and its orientation. F_0 is independent of the order in which we have drilled out the exceptional fibers of F because we can drill them all out at the same time by choosing the fiber neighborhoods sufficiently small.

From F_0 we drill out an arbitrary fiber neighborhood V_0 and obtain the class space $\overline{F}_0$ of F. It inherits the orientation from F. Since $\overline{F}_0$ is orientable there is a distinguished crossing curve Q_0 on the boundary torus Π_0 of $\overline{F}_0$ which is determined up to orientation and deformations as the boundary of the orbit surface $\bar{f}_0$ embedded in $\overline{F}_0$ (see §6). We orient Q_0 and a fiber H_0 of Π_0 so that they give on Π_0 the same orientation as that induced by V_0. The meridian curve M_0 of V_0, which is a crossing curve, is in the system Q_0, H_0 of the form

$$M_0 \sim Q_0 + bH_0. \tag{4}$$

The integer b is determined by the oriented space F_0, hence by F and its orientation.

This gives us a complete system of invariants of F, by the following:

THEOREM 5. *An orientable fibered space F together with its orientation is determined by a one-to-one correspondence by a system of invariants*

$$(O, o; p \mid b; \alpha_1, \beta_1; \alpha_2, \beta_2; \ldots; \alpha_r, \beta_r)$$

or

$$(O, n; k \mid b; \alpha_1, \beta_1; \alpha_2, \beta_2; \ldots; \alpha_r, \beta_r).$$

Here O means that F is orientable; o (resp. n) means that the orbit surface is orientable (resp. nonorientable). p and k are the genus [number of handles (resp. cross-caps)] of the orientable (resp. nonorientable) orbit surface. The three symbols to the left of the bar determine therefore the class of F. The number b determines uniquely the construction of the space without exceptional fibers F_0 from the class space $\overline{F}_0$. The numbers α_i, β_i determine uniquely (one-to-one) the exceptional fibers in F.

The theorem tells us when two orientable fibered spaces with given orientations are homeomorphic under an orientation and fiber preserving map. Theorem 6 shows how the invariants change if the orientation is reversed.

We have seen how to find the system of invariants for a given oriented space F. To show that this system is complete, we construct conversely to a given system of invariants a unique oriented space F. The numbers p (resp. k) determine the class (see p. 384) and hence by Theorem 3 (§6) the class space $\overline{F}_0$. We can orient $\overline{F}_0$ arbitrarily since there exists a fiber preserving and orientation reversing map of $\overline{F}_0$ onto itself (reflection of the solid torus $\overline{V}$ of §6 on a meridian disk). This determines the crossing curve Q_0 of the boundary torus Π_0 of $\overline{F}_0$ and a fiber H_0 up to simultaneous reversion of their orientation. b determines $M_0 \sim Q_0 + bH_0$ up to orientation and therefore the closing of $\overline{F}_0$ to F_0 uniquely. From F_0 we have to drill out r arbitrary fibers; the resulting space which is bounded by r tori is independent of the choice of the drilled-out fibers by Lemma 5. On each of the boundary tori there is a distinguished (up to orientation) crossing curve Q_i, namely, the meridian of the drilled-out solid torus, and the orientation of F_0 therefore determines a pair of curves Q_i, H_i up to simultaneous reversion of orientation. This determines uniquely the meridian $M_i \sim \alpha_i Q_i + \beta_i H_i$ of the new torus seal (up to orientation) and therefore uniquely the closing of F_0 to F.

We now describe for an orientable fibered space F a useful "diagram" $\overline{V}_0$ which together with $\overline{F}_0$ determines the space. Choose in F disjoint fiber neighborhoods Ω_i of the exceptional fibers. Then the ordinary torus seals V_i which replace the drill holes in F_0 are disjoint. We can choose the fiber neighborhood V_0, which we removed from F_0 to obtain the class space $\overline{F}_0$, in such a way that it contains all torus seals V_i in its interior by Lemma 3. The fibered space with boundary $\overline{V}_0$ that is obtained from V_0 after removing the V_i, and which is the topological product of S^1 and a disk punctured r times, is the *diagram* of the fibered space F if the distinguished crossing curve Q_0 of $\overline{F}_0$ is drawn on the boundary torus Π_0 of $\overline{V}_0$, and the meridian curves M_i of the drill holes Ω_i are drawn on the remaining r boundary tori Π_i. Obviously Q_0 determines how one has to glue on the class space $\overline{F}_0$ [which is determined by p (resp. k)] to the boundary torus Π_0. By Lemma 6, M_i determines the filling in of the drill hole Ω_i. Furthermore, if we orient $\overline{V}_0$, we get an orientation of F.

To obtain the invariants $b; \alpha_1, \beta_1; \ldots; \alpha_r, \beta_r$ of F from the diagram $\overline{V}_0$, we orient the fibers of $\overline{V}_0$ simultaneously, i.e., so that they are homologous in $\overline{V}_0$. Then the orientation of the fibers $H_0, H_1, \ldots, H_r$ on the boundary tori $\Pi_0, \Pi_1, \ldots, \Pi_r$ is determined. Hence the crossing curves $Q_1, \ldots, Q_r$ on the boundary tori are determined together with their orientation by requiring that the orientation on Π_i which is induced by Q_i and H_i shall be opposite to the orientation induced by $\overline{V}_0$, and by requiring that the numbers α_i, β_i in

$$M_i \sim \alpha_i Q_i + \beta_i H_i \qquad \text{(on } \Pi_i) \tag{5}$$

satisfy $\alpha_i > 1, 0 < \beta_i < \alpha_i$. The Q_i are meridians of the torus seals V_i. Closing $\overline{V}_0$ with the V_i, we obtain an ordinary solid torus V_0 with the meridian

$$M_0 \sim Q_0 + bH_0 \qquad \text{(on } \Pi_0) \tag{6}$$

and it is easily proved that

$$M_0 \sim Q_1 + Q_2 + \cdots + Q_r \qquad \left(\text{in } \overline{V}_0 \right)$$

and hence

$$-Q_0 + Q_1 + Q_2 + \cdots + Q_r \sim bH_0 \qquad \left(\text{in } \overline{V}_0 \right). \tag{7}$$

Figure 11 shows $\overline{V}_0$ with $r = 3, b = 4$.

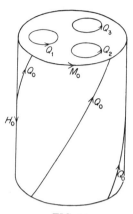

FIG. 11

We now want to find out how the invariants are changed if the orientation of F is reversed. In the diagram $\overline{V}_0$ only the orientation is reversed, but not the curves M_i and Q_0. It is useful to reverse the orientations of the fibers of $\overline{V}_0$ simultaneously; let $H'_0, H'_1, \ldots, H'_r$ be the fibers $H_0, H_1, \ldots, H_r$, but with opposite orientation:

$$H'_i \sim -H_i \qquad (\text{on } \Pi_i, i = 0, 1, \ldots, r). \tag{8}$$

We have to replace the $Q_1, Q_2, \ldots, Q_r$ by the crossing curves $Q'_1, Q'_2, \ldots, Q'_r$. Then

$$Q'_i \sim Q_i + y_i H_i \qquad (\text{on } \Pi_i, i = 1, 2, \ldots, r). \tag{9}$$

The sign of Q_i is $+1$ since the determinant of the transformation of the pair (8) and (9) has value -1, so that the orientation on $\overline{V}_0$ is reversed and hence the orientation of Π_i. For the same reason

$$Q'_0 \sim Q_0 \qquad (\text{on } \Pi_0). \tag{10}$$

Then we have for the meridian M_i

$$M_i \sim \alpha_i Q_i + \beta_i H_i \sim \alpha_i Q'_i + (\alpha_i y_i - \beta_i) H'_i = \alpha'_i Q'_i + \beta'_i H'_i.$$

The requirement $\alpha'_i > 1$ and $0 < \beta'_i < \alpha'_i$ gives us $\alpha'_i = \alpha_i$ and $\beta'_i = \alpha_i - \beta_i$, i.e., $y_i = 1$. b' is [as b from (7)] now determined by

$$-Q'_0 + Q'_1 + \cdots + Q'_r \sim b' H'_0. \tag{11}$$

Using (7)–(10), we get $b' = -r - b$.

THEOREM 6. *The oriented fibered space F with invariants*

$$(O, o; p \mid b; \alpha_1, \beta_1; \ldots; \alpha_r, \beta_r)$$

[*resp.*

$$(O, n; k \mid b; \alpha_1, \beta_1; \ldots; \alpha_r, \beta_r)]$$

has after reversing its orientation the invariants

$$(O, o; p| - r - b; \alpha_1, \alpha_1 - \beta_1; \ldots; \alpha_r, \alpha_r - \beta_r)$$

[*resp.*

$$(O, n; k \mid - r - b; \alpha_1, \alpha_1 - \beta_1; \ldots; \alpha_r, \alpha_r - \beta_r)].$$

If we had normed the numbers β_i to the interval

$$-\tfrac{1}{2}\alpha_i < \beta_i \leqslant \tfrac{1}{2}\alpha_i$$

instead of norming to $0 < \beta_i < \alpha_i$ by (3), the invariants $b, \beta_1, \ldots, \beta_r$ would only change their signs if the orientation of F were reversed, in the case that no exceptional fibers of order 2 were present, i.e., all $\alpha_i > 2$. But if $\alpha_1 = 2, \ldots, \alpha_s = 2$, only the last $r - s$ invariants β would change their signs if the orientation were reversed, but b would have to be replaced by $-s - b$, so that choosing the new normalization would not lead to an essential simplification for the purpose of reorientation.

8. The Nonorientable Fibered Spaces

As in the orientable case we first determine the classes. First assume the orbit surface f is orientable. Then the genus of f is > 0, since otherwise F is orientable (see §6 and §7). We show: *For each orientable orbit surface of genus $p > 0$ there is exactly one class of nonorientable spaces.* The claim is true for $p = 1$. For if a and b are two conjugate simple closed curves on a torus, then a, say, has value -1. We can assume that then b has value -1; otherwise we replace b by ab. Now suppose the claim is true for genus $p - 1$ ($\geqslant 1$). We prove it for p by showing that on a surface of genus $p > 1$ there is a handle on which all curves have value $+1$. Cutting off this handle we get a punctured surface of genus $p - 1$ having some curves of value -1 which is unique by the induction hypothesis. To show the existence of such a handle choose a system of curves which cuts the surface into a fundamental polygon with boundary $a_1 b_1 a_1^{-1} b_1^{-1} \cdots a_p b_p a_p^{-1} b_p^{-1}$. If there is a pair a_i, b_i of value $+1$, we are done. Otherwise a_1, say, has value -1. Assume b_1 has value $+1$ (otherwise replace b_1 by $a_1 b_1$). There is a curve a_j or b_j ($j > 1$) of value -1; thus one of the curves $a_1 a_j$ or $a_1 b_j^{-1}$ has value $+1$ and spans together with b_1 a handle with each curve of value $+1$.

Since the class is unique we can choose (on a surface of genus $p \geqslant 1$) a canonical system all whose curves have value -1.

If the orbit surface f is nonorientable of genus k we represent it as a sphere with k cross-caps $x_1, \ldots, x_k$ (see Fig. 12). Then a_i is a curve which intersects the cross-cap in one point; i.e., a_i is orientation reversing. Then $H_1(f) = \{a_1, \ldots, a_k : 2a_1 + \cdots + 2a_k \sim 0\}$. The valuation of f is therefore determined by the valuation of $a_1, a_2, \ldots, a_k$. If all the a_i have value -1, F is orientable. Thus at least one a_i has value $+1$.

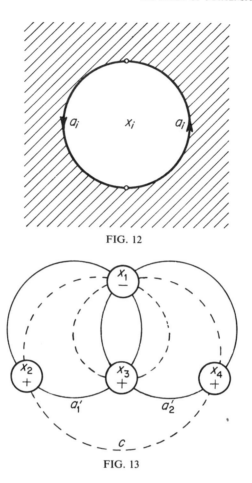

FIG. 12

FIG. 13

This leads to the following cases:

Case (a) a_i has value $+1$ for each i. Then $F \approx f \times S^1$ and $\overline{F}_0 \approx$ (punctured f) $\times S^1$.

Case (b) k_1 of the a_i have value $+1$, $k_2 = k - k_1$ have value -1 ($k_1 > 0$, $k_2 > 0$). Suppose $f \neq P^2$ ($k = 1$) and $f \neq$ Klein bottle ($k = 2$). We claim that we can always assume that $k_1 = 1$ or $= 2$. This is clear for $k = 3$. Suppose $k > 3$ and $k_1 \neq 1$, $k_1 \neq 2$. There exist at least three a_i, say a_2, a_3, a_4 of value $+1$ and one, say a_1, of value -1. Let l be a curve which separates the cross-caps x_1, x_2, x_3, x_4 from the others. l separates f into φ and ψ, where φ is a sphere with the cross-caps $x_1, \ldots, x_4$ and one boundary l. On φ there are two disjoint simple closed curves $a_1' \sim a_1 + a_2 + a_3$ and $a_2' \sim a_1 + a_3 + a_4$ of value -1. There is a simple closed orientation reversing curve c, disjoint to $a_1' \cup a_2'$, (see Fig. 13), such that the surface $\overline{\varphi}$, obtained from φ by cutting along a_1' and a_2', is nonorientable. We can represent $\overline{\varphi}$ as a sphere with two cross caps and three boundary curves l, $a_1'^2, a_2'^2$. φ is obtained from $\bar{\varphi}$ by identifying diametrical points of $a_1'^2$ and $a_2'^2$. Gluing back ψ and l, we have a new representation of f as sphere with k cross-caps. Since a_1' and a_2' now have value -1, the number of negative cross-caps has increased at least by 1. Continuing, we get $k_1 = 1$ or $k_1 = 2$.

We now show that the latter two valuations are distinct. Let d be a curve on f such that

$$d \sim \sum \gamma_i a_i \not\sim 0 \qquad \text{and} \qquad 2d \sim \sum 2\gamma_i a_i \sim 0. \tag{1}$$

(For example, d can be chosen to be a simple closed curve that intersects each cross-cap exactly once. In this case, cutting f along d we obtain an orientable surface with one or two holes, depending on whether k is odd or even. d is called an *orientation producing* simple closed curve.) Since $2d \sim 0$ is a consequence of $2a_1 + \cdots + 2a_k \sim 0$, $\sum 2\gamma_i a_i$ differs from $\sum 2a_i$ only by a factor and all the γ_i are equal and odd since otherwise $d \sim 0$. Hence d has value $(-1)^{k_2}$. Thus the valuations of f with even k_2 are different from those with odd k_2; in particular the valuations $k_1 = 1$ and $k_1 = 2$ yield different valuations of f. The investigation of all classes of fibered spaces is complete.

THEOREM 7. *For each orientable orbit surface f of genus p there is exactly one class of orientable fibered spaces, and if $p > 0$, exactly one class of nonorientable fibered spaces. For each nonorientable orbit surface f of genus k there is exactly one class of orientable fibered spaces, and if $k > 2$, exactly three classes of nonorientable fibered spaces; for $k = 1$ there is one class, for $k = 2$ there are two classes.*

The following table lists the different classes. O, N refer to orientability and nonorientability of F, and o, n to the orbit surface f, whose genus must be given in order for the class to be determined. Recall that a closed curve w of f is given the value $+1$ if the fiber orientation is preserved along w; otherwise w gets the value -1, and note that the class and therefore the classifying space $\bar{F}_0$ is uniquely determined by the valuation of all the curves of f.

Oo	All curves have value $+1$; $\bar{F}_0 \approx$ (punctured f) $\times S^1$;
On	All one-sided curves have value -1;
No	There are curves of value -1;
Nn I	All curves have value $+1$; $\bar{F}_0 \approx$ (punctured f) $\times S^1$;
Nn II	There are one-sided curves of value -1 and of value $+1$; each orientation producing simple closed curve has value -1;
Nn III	There are one-sided curves of value -1 and of value $+1$; each orientation producing simple closed curve has value $+1$.

For $p = 0$ there is only the class Oo, for $k = 1$ only On and Nn I, for $k = 2$ only On, Nn I, Nn II. $\bar{F}_0$ can now be constructed as in §6.

We now characterize the nonorientable fibered spaces F by invariants. Let C_1 be an exceptional fiber in F, Ω_1 a fiber neighborhood of C_1, Π_1 the boundary of Ω_1, M_1 a meridian on Π_1, Q an arbitrary crossing curve on Π_1, and H a fiber; then

$$M_1 \sim \alpha Q + \beta H \qquad \text{(on } \Pi_1\text{)}. \tag{2}$$

Using formulas (1) and (4) of §1,

$$H \sim \varepsilon_1 H_1, \qquad Q \sim \varepsilon_4 Q_1 + y H_1. \tag{3}$$

we can choose a new crossing curve Q_1 and fiber H_1 such that

$$M_1 \sim \alpha_1 Q_1 + \beta_1 H_1 \tag{4}$$

$$\text{with } \alpha_1 > 1, \qquad 0 < \beta_1 \leqslant \tfrac{1}{2}\alpha_1.$$

For the first requirement determines ε_4. Choosing y suitably we reduce β_1 to $[-\tfrac{1}{2}\alpha_1, \tfrac{1}{2}\alpha_1]$ and finally we choose ε_1. There is no orientation on Π_1 determined by F since F is nonorientable; hence ε_1 and ε_4 can be chosen independently (cf. §7 in the orientable case). α_1, β_1 are uniquely determined by Ω_1 and hence by C_1. The same holds, if $\alpha_1 > 2$, for Q_1 and H_1, up to simultaneously changing their orientation, which is permitted since the orientation of M_1 is not given by Ω_1. But for $\alpha_1 = 2$ there is besides Q_1, H_1 another system

$$Q_1' \sim Q_1 + H_1, \qquad H_1' \sim -H_1 \tag{5}$$

in which M_1 also appears in normal form (4):

$$M_1 \sim 2Q_1' + H_1'. \tag{6}$$

If $\alpha_1 > 2$, we drill out Ω_1 and replace it by an ordinary torus seal V_1 having Q_1 as meridian and do the same for all exceptional fibers of multiplicity > 2. This determines uniquely a nonorientable fibered space F_s, which has only $s \geqslant 0$ exceptional fibers of multiplicity 2. To investigate F_s further, we need

LEMMA 7. *A nonorientable fibered space $\overline{F}$ with boundary which is obtained from a (closed) fibered space by drilling out finitely many exceptional fibers admits a fiber preserving autohomeomorphism keeping the boundary tori pointwise fixed except for one, $\overline{\Pi}$. On $\overline{\Pi}$ a given crossing curve Q is mapped to a crossing curve of the form*

$$Q' \sim +(Q + 2zH) \qquad \text{or} \qquad Q' \sim -(Q + 2zH), \tag{7}$$

where z is an arbitrary integer and H is an oriented fiber on $\overline{\Pi}$. Furthermore, one can choose the homeomorphism orientation preserving or reversing on Π.[17]

Proof. (a) Let $z = 0$. To find an orientation reversing homeomorphism we glue on $\overline{\Pi}$ an ordinary fibered solid torus V having Q as meridian and get a space $\overline{F} + V$. The required map will be the end result of a fiber preserving deformation of $\overline{F} + V$. Choose on $\overline{F} + V$ a simple closed curve W from an interior point P of V and disjoint to the exceptional fibers which is

---

[17] The theorem does not claim that we can choose the sign in (7) arbitrarily.

orientation reversing. Deform $\overline{F} + V$ (fiber preservingly) so that P runs along W and at the end V is mapped to itself (see §7). Then V and hence $\overline{\Pi}$ is mapped to itself under an orientation reversing homeomorphism which maps Q to $Q' \sim +Q$ or $Q' \sim -Q$, depending on whether the fiber orientation is changed along the curve W. Finally, remove V to get the desired map of $\overline{F}$.

(b) By (a) there is a fiber preserving map of $\overline{F}$ mapping $Q + zH$ to $\pm(Q + zH)$ and orientation reversing on $\overline{\Pi}$. Here Q is mapped to $Q' \sim \pm(Q + 2zH)$. To get such an orientation preserving map, follow this map by a homeomorphism of $\overline{F}$ sending Q' to $\pm Q'$ and reversing orientation on $\overline{\Pi}$.

We use the lemma to show that F_s is uniquely determined by the class and s, if $s > 0$. Drill out the s exceptional fibers. The resulting $\overline{F}_s$ is determined by the class of F_s (= class of F) and by s, because $\overline{F}_s = \overline{F}_0$ (for $s = 1$) or $\overline{F}_s = \overline{F}_0$ (drilled out $(s-1)$ times) (see §6). From $\overline{F}_s$ we obtain F_s by closing with s solid tori of multiplicity $\mu = 2$. This closing is independent of how the torus seal Ω is sewn (fiber preservingly) onto the boundary Π of $\overline{F}_s$. For if Q is a crossing curve, H a fiber of $\overline{\Pi}$, and M a meridian of Ω, then

$$M \sim 2Q + yH \qquad (\text{on } \overline{\Pi}).$$

We show that the result is independent of y. Since M is a simple closed curve, y is odd. If $y \equiv 1$ (mod 4), there is a fiber preserving map of $\overline{F}_s$ keeping all boundary components fixed except for $\overline{\Pi}$ and such that $\overline{\Pi}$ is mapped orientation preservingly and Q is mapped to

$$Q' \sim \pm\{Q + 2[(1 - y)/4]H\};$$

hence M is mapped to

$$M' \sim 2Q' + yH' \sim \pm(2Q + H)$$

(Lemma 7). If $y \equiv -1$ (mod 4) we choose a fiber preserving map of $\overline{F}_s$ which is orientation reversing on $\overline{\Pi}$ and which sends Q to

$$Q' \sim \pm\{Q + 2[(1 + y)/4]H\},$$

hence M to

$$M' \sim \pm(2Q + H).$$

Thus instead of

$$M \sim 2Q + yH$$

we can choose $M' \sim \pm(2Q + H)$ as meridian of the torus seal. Therefore F_s depends only on $\overline{F}_0$ and on s.

If $s = 0$, we obtain F_0 from $\overline{F}_0$ by closing with an ordinary solid torus having a crossing curve Q on Π_0 as meridian. On Π_0 there are exactly two essentially distinct crossing curves. For by Lemma 7, Q can be mapped to $Q' \sim \pm(Q + 2zH)$ by a fiber preserving map of $\overline{F}_0$.

Therefore, if Q_0 is a crossing curve of Π_0, for example $Q_0 = \bar{f}_0 \cap \Pi_0$, where $\bar{f}_0$ is the orbit surface embedded in $\bar{F}_0$, we have only the two cases: $Q \sim Q_0$ or $Q \sim Q_0 + H$. If $\bar{f}_0$ can be embedded into $\bar{F}_0$ so that $\bar{f}_0 \cap \Pi_0 = Q_0$, then $\bar{f}_0$ cannot be embedded into $\bar{F}_0$ so that $\bar{f}_0 \cap \Pi_0 = Q_0 + H$, and vice versa (see §6). Therefore the two cross curves Q_0 and $Q_0 + H$ are essentially different, i.e., there is no fiber preserving map of $\bar{F}_0$ to itself which sends Q_0 to $\pm(Q_0 + H)$. Therefore the fibered spaces F_0 and F_0' obtained from $\bar{F}_0$ by taking Q_0 and $Q_0 + H$, respectively, as meridian Q of the torus seal are different. For a fiber preserving map $F_0 \to F_0'$ could be so deformed that the torus seals and hence the meridians of F_0 and F_0' correspond; hence there would be a fiber preserving map of $\bar{F}_0$ sending Q_0 to $\pm(Q_0 + H)$. The two distinct spaces F_0 and F_0' are therefore determined by $\bar{F}_0$ and by the number $b = 0$ or $= 1$.

Now suppose we know F_s $(s \geqslant 0)$. Then F is uniquely determined by

$$\alpha_i, \beta_i \qquad (\alpha_i > 2, 0 < \beta_i < \tfrac{1}{2}\alpha_i), \quad i = s + 1, \ldots, r.$$

For, drilling out $r - s$ arbitrary fibers from F_s, there is a unique (unoriented) crossing curve Q_i on each boundary torus Π_i, namely, the meridian of the drilled-out solid torus. Choosing an oriented fiber H_i on Π_i, the meridian M_i of the new torus seal is determined by

$$M_i \sim \alpha_i Q_i + \beta_i H_i,$$

by Eq. (4). But, since the orientation of Q_i and H_i is arbitrary, we obtain besides M_i another possible meridian

$$M_i' \sim \alpha_i Q_i - \beta_i H_i.$$

By Lemma 7 there is a fiber preserving map of the bounded space which keeps Π_j pointwise fixed $(j \neq i)$ and maps Π_i under an orientation reversing map to itself such that $Q_i \to \pm Q_i$. Then $M_i \to \pm M_i' \sim \pm(\alpha_i Q_i - \beta_i H_i)$. Hence it does not matter which of M_i or M_i' is chosen as meridian of the torus seal. Thus F is uniquely determined by its class and the numbers α_i, β_i, s, and b. Analogously to Theorem 5 we formulate the result in:

THEOREM 8. *A nonorientable fibered space F is uniquely determined by a system of invariants*

$$(\mathrm{No}; p \mid b; \alpha_1, \beta_1; \ldots; \alpha_s, \beta_s; \alpha_{s+1}, \beta_{s+1}; \ldots; \alpha_r, \beta_r)$$

or

$$(\mathrm{Nn\ I}; k \mid b; \alpha_1, \beta_1; \ldots; \alpha_s, \beta_s; \alpha_{s+1}, \beta_{s+1}; \ldots; \alpha_r, \beta_r)$$

or

$$(\mathrm{Nn\ II}; k \mid b; \alpha_1, \beta_1; \ldots; \alpha_s, \beta_s; \alpha_{s+1}, \beta_{s+1}; \ldots; \alpha_r, \beta_r)$$

or

$$(\mathrm{Nn\ III}; k \mid b; \alpha_1, \beta_1; \ldots; \alpha_s, \beta_s; a_{s+1}, \beta_{s+1}; \ldots; \alpha_r, \beta_r).$$

Here N *means that* F *is nonorientable;* o (*resp.* n) *means that the orbit surface is orientable* (*resp. nonorientable). The numbers* α_i, β_i *determine the exceptional fibers.* $\alpha_i = 2$, $\beta_i = 1$ *for* $i \leqslant s$ *and* $\alpha_i > 2$, $0 < \beta_i < \frac{1}{2}\alpha_i$ *for* $i > s$. b *is of any significance only if* $s = 0$. *In this case* $b = 0$ *or* $= 1$ *and determines the closing of the classifying space* $\bar{F}_0$ *to* F_0. *If* $s > 0$, *then* F *is already uniquely determined without specifying* b, *and* b *is omitted.*

EXAMPLE. Let F be a nonorientable fibered space with one exceptional fiber of multiplicity 3, with $\bar{F}_0$ determined by Nn I; k. Here $\bar{F}_0 \approx$ (punctured nonorientable surface of genus k) $\times S^1$. We obtain the two different fibered spaces:

$$(\text{Nn I}; k \mid 0; 3, 1) \qquad \text{and} \qquad (\text{Nn I}; k \mid 1; 3, 1).$$

But adding an exceptional fiber of multiplicity 2, both spaces go over into the same space

$$(\text{Nn I}; k \mid -; 2, 1; 3, 1).$$

9. Covering Spaces

1. Let $\tilde{F}$ be a (unbranched) covering of F (i.e., there is a covering map p of $\tilde{F}$ onto F such that for each point P of F and each P_i of $p^{-1}(P)$ there exist neighborhoods $U(P)$, $U(P_i)$ such that $p \mid U(P_i) : U(P_i) \to U(P)$ is a homeomorphism).

Let F be a fibered space, H a fiber. Let $\tilde{H}$ be a component of $p^{-1}(H)$. Then $\tilde{H} \approx S^1$ or R^1. Let S be the collection of all the curves $\tilde{H}$, for all fibers H of F. When is S a fibering of $\tilde{F}$?

2. Let Ω_C be a fiber neighborhood of a fiber C of F and let $\tilde{\Omega}_{\tilde{C}}$ be a component of $p^{-1}(\Omega_C)$. Then $\tilde{\Omega}_{\tilde{C}}$ consists of curves of S and contains the fiber $\tilde{C}$ [which is a component of $p^{-1}(C)$] in its interior. $\tilde{\Omega}_{\tilde{C}}$ is determined by Ω_C and an integer σ (including ∞) which denotes the multiplicity of the covering $\tilde{\Omega}_{\tilde{C}} \to \Omega_C$. Thus $\tilde{C} \to C$ is a σ-fold covering.

3. If $\sigma < \infty$, then all the curves of $\tilde{\Omega}_{\tilde{C}}$ are closed; if $\sigma = \infty$, they are all open. Thus each curve of S has a neighborhood which consists entirely of closed or of open curves of S. Hence $\tilde{F}$ is the union of two disjoint open sets, the sets of closed and open curves of S. Since $\tilde{F}$ is connected one of these is the empty set. Hence, S *cannot contain closed and open curves at the same time. If* (all) *the curves of* S *are closed, then* S *is a fibering of* $\tilde{F}$, since a finite covering of a fiber neighborhood, Ω_C is again a fibered solid torus.

4. From now on we assume that S is a fibering of $\tilde{F}$. Since the covering $\tilde{\Omega}_{\tilde{C}} \to \Omega_C$ is completely determined by the integer σ, we can compute the invariants $\tilde{\mu}, \tilde{\nu}$ of $\tilde{\Omega}_{\tilde{C}}$ from the invariants μ, ν of Ω_C and from σ. Cutting $\tilde{\Omega}_{\tilde{C}}$ into a fibered cylinder, we have to identify the top and bottom disks under a

rotation through

$$2\pi\,\frac{\tilde{\nu}}{\tilde{\mu}} = 2\pi\,\frac{\nu}{\mu}\,\sigma = 2\pi\nu\,\frac{\sigma/(\mu,\sigma)}{\mu/(\mu,\sigma)}\,;$$

(μ,σ) = gcd of μ and σ. Therefore, by definition of the characteristic numbers (§1),

$$\tilde{\mu} = \frac{\mu}{(\mu,\sigma)}\,,\qquad \tilde{\nu} \equiv \pm\nu\,\frac{\sigma}{(\mu,\sigma)}\quad (\mathrm{mod}\ \tilde{\mu}). \tag{1}$$

Thus in the cylinder $\tilde{\Omega}_{\tilde{C}}$ there are $\tilde{\mu} = \mu/(\mu,\sigma)$ parallel lines, which form one ordinary fiber of $\tilde{\Omega}_{\tilde{C}}$. Thus each ordinary fiber of Ω_C is covered by (μ,σ) ordinary fibers of $\tilde{\Omega}_{\tilde{C}}$, but the middle fiber C is covered only by one fiber $\tilde{C}$ of $\tilde{\Omega}_{\tilde{C}}$. Therefore $p:\tilde{F}\to F$ induces a continuous map $\bar{p}$ of the orbit surface $\tilde{f}$ onto f. If c and $\tilde{c}$ are the points corresponding to the fibers C and $\tilde{C}$, respectively, then if $(\mu,\sigma) > 1$, the covering of f by $\tilde{f}$ is branched over c of branch index (μ,σ). The index of the branching always divides the multiplicity of the exceptional fiber C. Hence only exceptional points can occur as branch points.

5. Since $\tilde{\mu} \leqslant \mu$ by (1), the covering $\tilde{C}$ of C is always an ordinary fiber if C is ordinary. But if C is an exceptional fiber ($\mu > 1$), then $\tilde{C}$ may or may not be exceptional. For example, identify two congruently fibered solid tori with an α-fold exceptional fiber along their boundary so that congruent points are identified. The result is a fibered space F with invariants $(O,\mathrm{o};\ 0\mid -1;\ \alpha,\beta;\alpha,\alpha-\beta)$ which is homeomorphic to $S^2\times S^1$. Taking the α-fold covering of each of the solid tori and identifying equivalent points, we get an α-fold covering $\tilde{F}\to F$ without exceptional fibers. For the invariants in (1) are $\mu = \alpha,\ \sigma = \alpha$; hence $\tilde{\mu} = 1$ for both (exceptional) fibers.

If $\tilde{H}$ and $\tilde{H}'$ are two fibers of $\tilde{F}$ which cover two *ordinary* fibers H and H', ρ and ρ' times, respectively, then $\rho = \rho'$. For, join $\tilde{H}$ and $\tilde{H}'$ in $\tilde{F}$ by a path whose projection in F does not meet exceptional fibers. Since in a (sufficiently small) neighborhood of an ordinary fiber the multiplicity of the covering is not changed, it remains constant along the entire path.

6. The universal covering space $\hat{F}$ of F is a fibered space if and only if for a fiber H of F a component $\hat{H}$ of $p^{-1}(H)$ is closed (by 3). Then H is covered ρ times by $\hat{H}$, $\rho < \infty$. Since $\hat{H}\simeq 0$ in $\hat{F}$ (simply connected), $H^\rho\simeq 0$ in F. Therefore, $\hat{F}$ *is a fibered space if and only if a finite multiple of the fiber of F is homotopic to 0 in F.* Clearly, if this holds for a single fiber H, it holds for all fibers of F.

7. Let F be a nonorientable fibered space and $\tilde{F}$ the 2-fold orientable covering of F. Since any fiber H of F is orientation preserving, H lifts to two closed curves $\tilde{H}$ and $\tilde{H}'$. Hence $\tilde{H}$ is closed and $\tilde{F}$ is a fibered space, and $\sigma = 1$. Therefore $p\mid\tilde{\Omega}_{\tilde{H}}:\tilde{\Omega}_{\tilde{H}}\to\Omega_H$ is a fiber preserving homeomorphism. Let $T:\tilde{F}\to\tilde{F}$ be the fiber preserving involution (without fixed

points) which is the nontrivial covering transformation. T reverses the orientation of $\tilde{F}$ and induces a fixed point-free involution of $\tilde{f}$.

For example, let F be the space

$$(\text{No}; p \mid b; \alpha_1, \beta_1; \ldots; \alpha_r, \beta_r). \tag{2}$$

$\tilde{F}$ has $2r$ exceptional fibers; if H is an exceptional fiber with invariants α_1, β_1, then H is covered by two exceptional fibers $\tilde{H}$ and $\tilde{H}'$ with invariants α_1, β_1 and $\alpha_1, \alpha_1 - \beta_1$, respectively (by Theorem 6). For the fiber preserving involution of $\tilde{F}$ maps $\tilde{H}$ to $\tilde{H}'$ and reverses the orientation of $\tilde{F}$. Since furthermore $\tilde{f}$ is an (unbranched) 2-fold covering of f, $\tilde{f}$ is orientable of genus $2p - 1$; hence $\tilde{F}$ is the space

$$(O, o; 2p - 1 \mid \tilde{b}; \alpha_1, \beta_1; \ldots; \alpha_r, \beta_r; \alpha_1, \alpha_1 - \beta_1; \ldots; \alpha_r, \alpha_r - \beta_r). \tag{3}$$

Since $\tilde{F}$ admits an orientation reversing fiber preserving homeomorphism, the invariants are the same if the orientation of $\tilde{F}$ is reversed. By Theorem 6, $\tilde{F}$ has the invariants

$$(O, o; 2p - 1 \mid -2r - \tilde{b}; \alpha_1, \beta_1; \ldots; \alpha_r, \beta_r; \alpha_1, \alpha_1 - \beta_1; \ldots; \alpha_r, \alpha_r - \beta_r). \tag{4}$$

For (3) and (4) to be equal we must have that $\tilde{b} = -2r - \tilde{b}$, hence $\tilde{b} = -r$, independent of b. Similarly for the other cases. Result:

Let $\tilde{F}$ be the orientable 2-sheeted covering of F.

$$\begin{cases} F(\text{No}; p \mid b; \alpha_1, \beta_1; \ldots; \alpha_r, \beta_r) \\ \tilde{F}(\text{Oo}; 2p - 1 \mid -r; \alpha_1, \beta_1; \ldots; \alpha_r, \beta_r; \alpha_1, \alpha_1 - \beta_1; \ldots; \alpha_r, \alpha_r, \alpha_r - \beta_r), \end{cases}$$

$$\begin{cases} F(\text{Nn I}; k \mid b; \alpha_1, \beta_1; \ldots; \alpha_r, \beta_r) \\ \tilde{F}(\text{Oo}; k - 1 \mid -r; \alpha_1, \beta_1; \ldots; \alpha_r, \beta_r; \alpha_1, \alpha_1 - \beta_1; \ldots; \alpha_r, \alpha_r - \beta_r), \end{cases}$$

$$\begin{cases} F(\text{Nn II}; k \mid b; \alpha_1, \beta_1; \ldots; \alpha_r, \beta_r) \\ \tilde{F}(\text{On}; 2k - 2 \mid -r; \alpha_1, \beta_1; \ldots; \alpha_r, \beta_r; \alpha_r, \beta_r; \alpha_1, \alpha_1 - \beta_1; \ldots; \alpha_r, \alpha_r - \beta_r), \end{cases}$$

$$\begin{cases} F(\text{Nn III}; k \mid b; \alpha_1, \beta_1; \ldots; \alpha_r, \beta_r) \\ \tilde{F}(\text{On}; 2k - 2 \mid -r; \alpha_1, \beta_1; \ldots; \alpha_r, \beta_r; \alpha_1, \alpha_1 - \beta_1; \ldots; \alpha_r, \alpha_r - \beta_r). \end{cases}$$

In the two latter cases the orbit surface $\tilde{f}$ is nonorientable since there are one-sided curves on f along which the fiber orientation is reversed, i.e., which are orientation preserving in F.

8. Let F be a fibered space with orbit surface f. Let $\tilde{f}$ be an (unbranched) covering of f, $\tilde{p}$ a point over a point p of f, and P a point of F which maps to p. Let $\tilde{F} = \{(P, \tilde{p})\}$. A neighborhood of a point $(P_0, \tilde{p}_0)$ consists of all points $(P, \tilde{p})$ where P lies in a neighborhood of P_0 (in F) and $\tilde{p}$ in a neighborhood of $\tilde{p}_0$. Defining $g(P, \tilde{p}) = P$, we see that $g: \tilde{F} \to F$ is a covering of F. The

multiplicity of this covering is the multiplicity of the covering $\tilde{f} \to f$. If a point P of F runs along a fiber H, then $(P, \tilde{p})$ for fixed $\tilde{p}$ runs along a curve $\tilde{H}$ which lies one-to-one over H. Hence $\tilde{F}$ is a fibered space by 3 above and a fiber neighborhood $\tilde{\Omega}_{\tilde{H}}$ of $\tilde{F}$ is mapped onto Ω_H under a fiber preserving homeomorphism.

For example, let F be the orientable space $(On; 1 \mid b; \alpha_1, \beta_1; \dots; \alpha_r, \beta_r)$ with orbit surface the projective plane. Let f be the 2-sphere. Then $\tilde{F}$ is orientable, hence of class $(Oo; 0)$. Orienting $\tilde{F}$ so that $g : \tilde{F} \to F$ is orientation preserving, the fiber neighborhoods $\tilde{\Omega}_{\tilde{H}}$ and $\tilde{\Omega}_{\tilde{H}}'$ map to the same Ω_H preserving orientations, and therefore to the exceptional fiber with invariants α, β correspond in $\tilde{F}$ two exceptional fibers both with invariants α, β. Drilling out the exceptional fibers of F and filling in ordinary solid tori and doing the same thing in $\tilde{F}$, we obtain F_0 and $\tilde{F}_0$ without exceptional fibers and $\tilde{F}_0$ is a 2-fold covering of F_0. We find that $\tilde{b} = 2b$; hence $\tilde{F}$ is

$$(Oo; 0 \mid 2b; \alpha_1, \beta_1; \dots; \alpha_r, \beta_r; \alpha_1, \beta_1; \dots; \alpha_r, \beta_r).$$

10. Fundamental Groups of Fibered Spaces

We cut the classifying space $\overline{F}_0$ of a fibered space F into a fibered prism with a drill hole, as in §6 but so that the drill hole touches the prism along an edge H. Similarly we drill out the r ordinary tori $V_1, \dots, V_r$ (which have to be replaced by exceptional tori) so that they touch H. Then the $r + 1$ boundary tori $\Pi_0, \Pi_1, \dots, \Pi_r$ intersect the bottom surface in the cross curves $Q_0, Q_1, \dots, Q_r$. (See Fig. 14 for $p = 2$ and $r = 2$).

We obtain the fundamental group of this space $\overline{\overline{F}}_0 = \overline{F}_0 - \text{int}(V_1 \cup \cdots \cup V_r)$ by running around the 2-cells. Then for an orientable orbit surface of genus $p \geqslant 0$ we have[18]

$$\pi_1\left(\overline{\overline{F}}_0\right) = \{A_1, B_1, \dots, A_p, B_p, Q_0, Q_1, \dots, Q_r, H :$$

$$A_i H A_i^{-1} = H^{\varepsilon_i}, \; B_i H B_i^{-1} = H^{\varepsilon_i'} \; (i = 1, \dots, p; \varepsilon_i, \varepsilon_i' = \pm 1), \quad (1)$$

$$Q_0 Q_1 \cdots Q_r = A_1 B_1 A_1^{-1} B_1^{-1} \cdots A_p B_p A_p^{-1} B_p^{-1},$$

$$Q_j H Q_j^{-1} = H \; (j = 0, 1, \dots, r)\}.$$

Here ε_i $(\varepsilon_i') = \pm 1$ or -1 depending on whether the fiber orientation is preserved or reversed along A_i (B_i).

For $p = 0$ we get the relations

$$Q_0 Q_1 \cdots Q_r = 1,$$
$$Q_j H Q_j^{-1} = H \quad (j = 0, 1, \dots, r). \quad (2)$$

---

[18] Cf. H. Seifert, Konstruction dreidimensionaler geschl. Räume, *Ber. Sächs. Akad. Wiss.* **83** (1931), 33. The auxiliary paths and therefore the relations of the first type are redundant, since $\overline{\overline{F}}_0$ contains only one vertex.

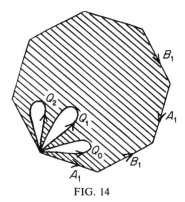

FIG. 14

For a nonorientable orbit surface of genus k we get

$$\pi_1\left(\overline{\overline{F}}_0\right) = \{A_1, \ldots, A_k, Q_0, Q_1, \ldots, Q_k, H :$$

$$A_i H A_i^{-1} = H^{\varepsilon_i} \ (i = 1, 2, \ldots, k; \varepsilon_i = \pm 1), \tag{3}$$

$$Q_0 Q_1 \cdots Q_r = A_1^2 \cdots A_k^2,$$

$$Q_j H Q_j^{-1} = H, \ (j = 0, 1, \ldots, r)\}.$$

$\pi_1(F)$ is obtained from $\pi_1(\overline{\overline{F}}_0)$ by adding $r + 1$ relations which correspond to the $r + 1$ torus seals of the boundary tori $\Pi_0, \Pi_1, \ldots, \Pi_r$. They are

$$Q_0 H^b = Q_1^{\alpha_1} H^{\beta_1} = \cdots = Q_r^{\alpha_r} H^{\beta_r} = 1. \tag{4}$$

For example, $Q_1^{\alpha_1} H^{\beta_1} = 1$ means that the meridian $M_1 \sim \alpha_1 Q_1 + \beta_1 H_1$ of the torus seal belonging to Π_1 is null homotopic in the torus seal. For example, the fundamental group of the space $(Oo; 0 \mid b; \alpha_1, \beta_1, \ldots, \alpha_r, \beta_r)$ has the relations

$$Q_o H^b = Q_1^{\alpha_1} H^{\beta_1} = \cdots = Q_r^{\alpha_r} H^{\beta_r} = Q_0 Q_1 \cdots Q_r = 1,$$

$$Q_j H Q_j^{-1} = H \qquad (j = 0, 1, \ldots, r). \tag{5}$$

Adding the relations $Q_0 = Q_1 = \cdots = Q_r = H = 1$ we obtain from $\pi_1(F)$ the fundamental group $\pi_1(f)$ of the orbit surface f. Geometrically this can be seen as follows: The mapping of $F \to f$ induces a homomorphism[19] of $\pi_1(F)$ onto $\pi_1(f)$ and therefore $\pi_1(f)$ is a quotient group of $\pi_1(F)$. Similarly $H_1(f)$ is a quotient group of $H_1(F)$, and this is also true for open fibered spaces. (We shall use this fact in §14.)

Among the closed 3-dimensional manifolds the ones which occur as fundamental regions (Diskontinuitätsbereiche) of 3-dimensional spherical groups of motions, and thus have finite fundamental groups, have been thoroughly investigated. Therefore we are interested in the question whether

---

[19] A homomorphism is sometimes called a "one- or multiple-to-one isomorphism".

the fibered spaces give us new manifolds of finite fundamental group, or if they are already included among the fundamental regions. In DB II (see footnote 1) we shall show that the fibered spaces with finite fundamental group coincide with the fundamental regions of fixedpoint-free spherical groups of motions. A necessary condition for the finiteness of the fundamental group of F is that the fundamental group of the orbit space f be finite since the latter is a quotient of the former. Hence f is a 2-sphere or projective plane.

If f is a 2-sphere, then (5) are the relations of the fundamental group of F. Adding $H = 1$, we obtain the factor group

$$\{ \overline{Q}_0, \overline{Q}_1, \ldots, \overline{Q}_r : \overline{Q}_1^{\alpha_1} = \cdots = \overline{Q}_r^{\alpha_r} = \overline{Q}_1 \cdots \overline{Q}_r = 1 \}. \tag{6}$$

For $r \geqslant 3$ this is a polygon net group. Taking an r-gon with angles $\pi/\alpha_1, \ldots, \pi/\alpha_r$ on the 2-sphere, the Euclidean plane, or the hyperbolic plane, depending on whether

$$\sum_{i=1}^{r} \frac{1}{\alpha_i} >, =, \text{ or } < r - 2, \tag{7}$$

and reflecting it successively on its sides, we obtain a polygon net which covers the sphere, or the Euclidean or hyperbolic plane, with alternating congruent and mirror imaged (black and white) r-gons. It admits a group of orientation preserving covering translations which has as fundamental region a double polygon, i.e., a white and adjacent black r-gon. This group is the above factor group (6).[20] Since for $r > 3$ this polygon cannot lie on the 2-sphere so as to cover it, it follows that (6) and hence (5) is infinite. For $r = 3$ the group (6) is finite only if it is a Platonian group, i.e., if $\alpha_1, \alpha_2, \alpha_3$ is one of the triples $(2, 2, n), (2, 3, 3), (2, 3, 4), (2, 3, 5)$ $(n \geqslant 2)$. It can be shown (DB II, §7) that for these triples the group (5) is finite. If $r \leqslant 2$, then (5) is cyclic (finite or infinite).

If f is the projective plane, then F is the space

$$(\text{On}; 1 \mid b; \alpha_1, \beta_1; \ldots; \alpha_r, \beta_r) \tag{8}$$

since a nonorientable (closed) 3-manifold has infinite fundamental group. This follows also since the first Betti number of the fundamental groups of fibered spaces is > 0.[21] The space (8) has a 2-fold orientable covering (§9), namely,

$$(\text{Oo}; 0 \mid 2b; \alpha_1, \beta_1; \alpha_1, \beta_1; \ldots; \alpha_r, \beta_r; \alpha_r, \beta_r).$$

This space has infinite fundamental group unless $r = 1$. Therefore follows

THEOREM 9. *A fibered space F with finite fundamental group has the projective plane or the 2-sphere as orbit surface. In the first case F has at most*

---

[20] Cf. W. Threlfall, Gruppenbilder *Abh. Sächs. Akad. Wiss.* **41** No. 6 (1932).
[21] Poincaré has introduced $P_1 = p_1 + 1$ as Betti number. We follow H. Weyl.

one exceptional fiber, in the latter case F has at most three exceptional fibers. If F has three exceptional fibers, they have to be of multiplicity $(2,2,n)$, $(2,3,3)$, $(2,3,4)$, or $(2,3,5)$.

When are two fibered spaces homeomorphic but not homeomorphic under a fiber preserving map?

THEOREM 10. *Suppose F and F' have the 2-sphere as orbit surface and have at least three exceptional fibers of multiplicities $\alpha_1, \ldots, \alpha_r$ and $\alpha'_1, \ldots, \alpha'_{r'}$, respectively. If F is homeomorphic to F' (not necessarily under a fiber preserving map) then the tuples $\alpha_1, \ldots, \alpha_r$ and $\alpha'_1, \ldots, \alpha'_{r'}$ must be equal (up to order).*

Proof. For $r = 3$, the center of (5) is the subgroup $\{H\}$ generated by H. For if the center were bigger than $\{H\}$, then (6) would have a nontrivial center. This is not the case if (6) is a group of the Euclidean or hyperbolic plane. If (6) is a Platonian group it has a nontrivial center only if it is a dihedral group whose order is a multiple of 4. It can be shown that in this case the center of (5) is not bigger than $\{H\}$ (DB II, §6). Hence (6) is the quotient of (5) by its center. If $F \approx F'$, then $\pi_1(F) \cong \pi_1(F')$ and $\pi_1(F)/\{H\} \simeq \pi_1(F')/\{H'\}$. But two polygon net groups (6) are isomorphic if and only if the polygons have the same number of vertices and the same angles, which proves the theorem. To see this, we can assume that none of the polygon net groups is a Platonian group, for such a group has necessarily the vertex number 3 and the triples of Theorem 9. The elements $\overline{Q}_1, \ldots, \overline{Q}_r$ of (6) are rotations about the r vertices of a polygon Π through $2\pi/\alpha_1, \ldots, 2\pi/\alpha_r$. Since an element of finite order of (6) is (as a transformation of a metric plane) necessarily a rotation about a fixed point, i.e., about a vertex of the polygon net, it follows that each nontrivial element of finite order of (6) is conjugate to a rotation about a vertex of Π, i.e., to a power $Q_i^{\gamma_i}$ ($\gamma_i = 1, \ldots, \alpha_i - 1$). But two such powers $\overline{Q}_i^{\gamma_i}$ and $\overline{Q}_j^{\gamma_j}$ are never conjugates (as can be seen from the geometry). Therefore the numbers $\alpha_1, \ldots, \alpha_r$ determine uniquely the number of conjugate classes of elements of finite order and conversely one can easily verify that the numbers $\alpha_1, \ldots, \alpha_r$ are determined by the number of conjugate classes of elements of given finite order.

11. Fiberings of the 3-Sphere (Complete List)

In §3 we described fiberings of S^3 with two exceptional fibers of orders m, n where $(m, n) = 1$. We now show that these are the only fiberings of S^3. More generally, we look at all simply connected (closed) fibered spaces.

Let F be a fibered space with $\pi_1(F) = 1$. Then $f \approx S^2$ and F is

$$(\text{Oo}; 0 \mid b; \alpha_1, \beta_1; \ldots; \alpha_r, \beta_r).$$

A necessary condition for $\pi_1(F)$ to be finite is that $r \leqslant 3$ (by Theorem 9). For

$r = 3$ the quotient group (6) of $\pi_1(F)$, where F is as in Theorem 9, is a Platonian group, and hence not trivial. Therefore if $\pi_1(F) = 1$, then $r \leqslant 2$.

For $r = 0$, $\pi_1(F) = \{Q_0, H : Q_0 H^b = 1 = Q_0\} = \{H : H^b = 1\}$. Hence $b = \pm 1$. Therefore $(Oo; 1 \mid 1)$ or $(Oo; 0 \mid -1)$ are the only simply connected fibered spaces without exceptional fibers. They differ only in their orientation (by Theorem 6) and are the fibering of S^3 by circles since this is free from exceptional fibers.

For $r = 1$, $b\alpha_1 + \beta_1 = \pm 1$ is necessary and sufficient for $\pi_1(F) = 1$. Now α_1 ($\geqslant 2$) is arbitrary. For b and β_1 there are then two solutions, $b = 0$, $\beta_1 = 1$ and $b = -1$, $\beta_1 = \alpha_1 - 1$. The two spaces $(Oo; 0 \mid 0; \alpha_1, 1)$ and $(Oo; 0 \mid -1; \alpha_1, \alpha_1 - 1)$ differ only in their orientation (Theorem 6), and therefore there is a unique simply connected fibered space (up to orientation) having a single exceptional fiber of order α_1. This space is therefore the trace curve fibering of S^3 with the values $m = 1$, $n = \alpha_1$.

For $r = 2$, $\pi_1(F)$ is cyclic of order $|b\alpha_1\alpha_2 + \beta_1\alpha_2 + \beta_2\alpha_1|$. The equation

$$b\alpha_1\alpha_2 + \beta_1\alpha_2 + \beta_2\alpha_1 = \pm 1$$

has a solution only if $(\alpha_1, \alpha_2) = 1$. But for any given coprime α_1, α_2 ($\geqslant 2$) there are exactly two solutions for b, β_1, β_2, for which $0 < \beta_1 < \alpha_1$ and $0 < \beta_2 < \alpha_2$. The corresponding spaces differ only in their orientation. This will be proved in §12 for an arbitrary r. Therefore there is only one fibering (up to orientation) for any two given coprime exceptional fibers, which therefore has to agree with that of §3. This proves

THEOREM 11. *A closed simply connected fibered space is S^3. Any fibering of S^3 is uniquely determimed by two positive coprime integers m and n. For $m = n = 1$ there are no exceptional fibers; if only one of m (or n) is 1 there is one exceptional fiber of order n (or m). If m and n are different from 1, they are the orders of the two exceptional fibers. All fiberings of S^3 agree with those of §3.*

The ordinary fibers for $m \neq 1$, $n \neq 1$ are torus knots which wind m times around the z-axis and n times around the unit circle in the conformal space. For $m = 2$, $n = 3$ they are trefoil knots.

12. The Fibered Poincaré Spaces

We now determine which fibered spaces are Poincaré spaces, that is, which have trivial first homology group[22] and which are not homeomorphic to S^3. By §10 if $H_1(F) = 1$, then $H_1(f) = 1$; hence $f \approx S^2$ and F is $(Oo; 0 \mid b; \alpha_1, \beta_1; \ldots; \alpha_r, \beta_r)$. $H_1(F)$ is the Abelianized $\pi_1(F)$ and has the $r + 2$ generators

$$Q_0, Q_1, \ldots, Q_r, H$$

[22] Cf. DB I, p. 51.

and in addition to being commutative has the relations

$$Q_0 H^b = Q_1^{\alpha_1} H^{\beta_1} = \cdots = Q_r^{\alpha_r} H^{\beta_r} = Q_0 Q_1 \cdots Q_r = 1.$$

In additive notation,

$$
\begin{aligned}
Q_0 \qquad\qquad\qquad + bH &= 0 \\
\alpha_1 Q_1 \qquad\qquad + \beta_1 H &= 0 \\
\vdots \qquad\qquad\qquad& \\
\alpha_r Q_r + \beta_r H &= 0 \\
Q_0 + Q_1 + \cdots + Q_r \qquad\qquad &= 0
\end{aligned}
\tag{1}
$$

We obtain equivalent relations and generators for $H_1(F)$ by transforming the generators and relations by unimodular substitutions. In this way we can transform the coefficient matrix into a normal form which has all entires 0 except possibly in the main diagonal, where the entries are the invariant factors of the original matrix. If $H_1(F) = 1$, then in the normal form all the elements in the main diagonal are 1 (otherwise we would have a nontrivial relation $k_i Q_i = 1$). That is, the Betti number $= 0$ and there are no torsion coefficients. Since the given matrix is square the two conditions are equivalent to

$$
\Delta = \begin{vmatrix}
1 & 0 & \cdots & 0 & b \\
0 & \alpha_1 & \cdots & 0 & \beta_1 \\
\vdots & \vdots & & \vdots & \vdots \\
0 & 0 & \cdots & \alpha_r & \beta_r \\
1 & 1 & \cdots & 1 & 0
\end{vmatrix} = \pm 1.
\tag{2}
$$

Computing Δ we get the equation

$$
\begin{aligned}
\Delta &= b\alpha_1 \cdots \alpha_r + \beta_1 \alpha_2 \cdots \alpha_r + \alpha_1 \beta_2 \alpha_3 \cdots \alpha_r + \cdots + \alpha_1 \alpha_2 \cdots \alpha_{r-1} \beta_r \\
&= \varepsilon \qquad (\varepsilon = \pm 1).
\end{aligned}
\tag{3}
$$

If we reverse the orientation of F, i.e., if we consider $(O, o; 0 \mid -r - b; \alpha_1, \alpha_1 - \beta_1; \ldots; \alpha_r, \alpha_r - \beta_r)$, we would get a determinant $\Delta' = -\Delta$. Therefore we can assume that $\varepsilon = \pm 1$. This determines the orientation of F. To solve (3) with $\varepsilon = +1$, we let $\alpha_1, \ldots, \alpha_r$ be given $(\alpha_i \geqslant 2)$ and try to solve for b, $\beta_1, \ldots, \beta_r$. For $r = 0$ and $r = 1$ we get $b = 1$ and $b\alpha_1 + \beta_1 = 1$, which was discussed in §11. Thus assume $r \geqslant 2$. There exists no solution of (3) if two of the α_i have a common divisor. Hence assume the α_i are pairwise coprime. Then

$$\gcd(\alpha_1 \cdots \alpha_r, \alpha_2 \cdots \alpha_r, \alpha_1 \alpha_3 \cdots \alpha_r, \cdots, \alpha_1 \alpha_2 \cdots \alpha_{r-1}) = 1.$$

Hence there exists a solution $b, \beta_1, \ldots, \beta_r$ and $(\beta_i, \alpha_i) = 1$; otherwise the left-hand side of (3) would have a common factor $\neq 1$. But β_i need not satisfy

$$0 < \beta_i < \alpha_i.$$

But this condition can be satisfied by replacing β_i by $\beta_i + x_i\alpha_i$ and at the same time b by $b - x_i$, which also satisfies (3). This normalized solution is unique for if b', $\beta'_1, \ldots, \beta'_r$ is any other normalized solution of (3), then

$$(b - b')\alpha_1 \cdots \alpha_r + (\beta_1 - \beta'_1)\alpha_2 \cdots \alpha_r + \cdots = 0.$$

This implies $\beta_i - \beta'_i \equiv 0 \pmod{\alpha_i}$, hence $\beta_i = \beta'_i$.

This solution of (3) completes the proof of Theorem 11. Hence for $r = 2$ the fibered spaces with trivial first homology group are homeomorphic to S^3. For $r > 2$ they are Poincaré spaces since by Theorem 11 a fibration of S^3 has at most two exceptional fibers. Thus follows

THEOREM 12. *A fibered Poincaré space ($\neq S^3$) has at least three exceptional fibers; their multiplicities $\alpha_1, \ldots, \alpha_r$ are pairwise coprime. Conversely, for any $r \geqslant 3$ pairwise coprime integers $\geqslant 2$, there exists a unique fibered Poincaré space having r exceptional fibers with the given multiplicities. Two fibered Poincaré spaces are homeomorphic if and only if they are homeomorphic under a fiber preserving map; i.e., a Poincaré space admits at most one fibering. The only fibered Poincaré space with finite fundamental group is the dodecahedral space.*[23]

It remains to prove the two latter claims. If two fibered Poincaré spaces are homeomorphic, they must have the same multiplicities for the exceptional fibers, by Theorem 10. But these determine already the fibering of a Poincaré space.

By Theorem 9, a fibered Poincaré space with finite fundamental group can have only three exceptional fibers with the multiplicities $2, 3, 5$ because this is the only triple in Theorem 9 with pairwise coprime integers. This space has by (3) the invariants

$$(Oo; 0 \mid -1; 5, 1; 2, 1; 3, 1)$$

and its fundamental group has relations

$$Q_0 H^{-1} = Q_1^5 H = Q_2^2 H = Q_3^3 H = Q_0 Q_1 Q_2 Q_3 = 1.$$

(These relations imply that H commutes with the Q_i). Eliminating H, we obtain the presentation of the binary icosahedral group[24] of order 120:

$$Q_1^5 = Q_2^2 = Q_3^3 = Q_1 Q_2 Q_3.$$

In DB II, §7, it is shown that this is the dodecahedral space by exhibiting a fibering of the dodecahedral space.

[23] Cf. DB I, §12.
[24] Cf. Aufgabe 84 in *Jahresber. Deutsch Math.-Verein.* **41** (1936), 6.

13. Constructing Poincaré Spaces from Torus Knots

M. Dehn[25] described a method for constructing Poincaré spaces as follows: Let A be the complement of a regular neighborhood of a knot C in S^3, and let $\Pi = \partial A$. Then $H_1(M)$ is the free cyclic group generated by a meridian M on Π. If B is a simple closed curve on Π intersecting M in one point, $B \sim xM$ (in A) and we can assume that $x = 0$ by replacing (if necessary) B by $B - xM$. Then B is uniquely determined by requiring that $M \cap B$ be a point and $B \sim 0$ in A (up to orientation and deformation on Π). Closing A with a torus seal V' having as meridian

$$M' \sim M + qB \qquad (\text{on } \Pi; \, q \neq 0), \tag{1}$$

we get a closed space R with $H_1(R) = 0$.

Now suppose C is a torus knot. Such knots are ordinary fibers of the fiberings of S^3, given in §3, which are characterized by two coprime integers m and n ($\geqslant 2$). Drill out an ordinary fiber C. Then a fiber H of Π can be deformed in A into n times the z-axis, and since the z-axis is $\sim mM$ in A (with suitable orientation of M), we have that $H \sim mnM$ (in A). Hence $h - mnM \sim 0$ in A, i.e., $H = B$. By (1), $M' \sim M + qB \sim (1 - qmn)M + qH$ on Π. Since M is a crossing curve on Π, the torus seal has an exceptional fiber of multiplicity $|qmn - 1|$, for since m and $n > 1$ (otherwise C would be unknotted and we would not get a Poincaré space), $|qmn - 1| > \max(m,n) > 1$. Thus R is the unique Poincaré space (by Theorem 12) with three exceptional fibers of multiplicities m, n, $|qmn - 1|$. Furthermore, since $|q_1 mn - 1| \neq |q_2 mn - 1|$, if $q_1 \neq q_2$, two Poincaré spaces obtained from the same torus knot with different q's are not homeomorphic by Theorem 12. Finally, two Poincaré spaces obtained from different torus knots are never homeomorphic. For if a Poincaré space with exceptional fibers $\alpha_1 < \alpha_2 < \alpha_3$ is obtained from a torus knot, then it can only be the knot $m = \alpha_1$, $n = \alpha_2$, since $|qmn - 1| > \max(m,n)$. This implies by the way that two torus knots $m < n$ and $m' < n'$ are topologically equivalent only if $m = m'$, $n = n'$, since only in this case are the Poincare spaces which can be constructed from them the same.

THEOREM 13. *A Poincaré space can be constructed from a torus knot if and only if it can be fibered and the fibering has exactly three exceptional fibers of multiplicites* $\alpha_1 < \alpha_2 < \alpha_3$, *where* $\alpha_1, \alpha_2, \alpha_3$ *are pairwise coprime integers* (> 1) *and* $\alpha_3 = |q\alpha_1\alpha_2 - 1|$ *(q an arbitrary integer). Such a Poincaré space can only be constructed from a unique torus knot in a unique way.*

For example, the Dehn trefoil space constructed from a trefoil knot $m = 2$,

[25] M. Dehn, Über die Topologie des dreidimensionalen Raumes, *Math. Ann.* **69** (1910), 137–168.

$n = 3$, $q = 1$ is homeomorphic with the unique fibered Poincaré space with three exceptional fibers of multiplicities $2, 3, 5$. Its fiber invariants are listed in §12.

14. Translation Groups of Fibered Spaces

A *translation group* $\mathfrak{G}$ of a fibered space F is a finite group of homeomorphisms $F \to F$ such that each map of $\mathfrak{G}$ maps each fiber H onto itself and preserves orientation of H. For an arbitrary fiber H of F let $\mathfrak{H} = \{\varphi \mid H, \varphi \in \mathfrak{G}\}$. We claim that $\mathfrak{H}$ is a finite cyclic group of rotations of a circle. For if P is a point of H and $P', P'', \ldots, P^{(i)} = P$ are its equivalent points such that P' is next to P with respect to the given orientation of H, the points $P, P', P'', \ldots$ and the arcs between them are cyclically permuted under a map of $\mathfrak{G}$. In particular, if P is a fixed point, then the arc $\overline{PP'}$ is mapped onto itself keeping P, P' fixed, and since the map has finite order it must be the identity. There is a map in $\mathfrak{H}$ which sends P to $P^{(k)}$ (k arbitrary). Therefore $\mathfrak{H}$ consists of the powers of the map which sends P to P'.

Claim. Every translation group $\mathfrak{G}$ is cyclic. It suffices to show that a map S of $\mathfrak{G}$ which leaves an ordinary fiber H fixed is the identity, for then $\mathfrak{G}$ is isomorphic to $\mathfrak{H}$, which we know to by cyclic. The maximum of the translations of the points of a fiber H' under S converges to 0 as H' converges to H. But this maximal translation cannot be arbitrarily small since S is of finite order. Therefore S is the identity on a fiber neighborhood of H. The set of all ordinary fibers which are fixed under S is therefore open. The set of all ordinary fibers which are not pointwise fixed is also open, hence empty since F is connected. But then clearly all the exceptional fibers are also left pointwise fixed under S.

The following theorem deals with the existence of translation groups:

THEOREM 14. *A closed fibered space of class* (Oo; p) *or* (Nn I; k) *admits a translation group of arbitrary order g.*

Proof. We first show that a fibered solid torus with invariants μ, ν admits such a group. Cut the solid torus into a Euclidean cylinder of height 1, and let z be the height of a point P; then there is a continuous transformation group of the solid torus such that each point runs along its fiber and the z-coordinate changes continuously, $z' = z + t$. Here z' is the coordinate of the image point and t the continuous parameter of the group. z has to be considered mod 1. If t increases continuously from 0, then $t = 1$ is the first value for which the middle fiber is mapped to itself, $t = \mu$ is the first value for which the map is the identity. The cyclic translation group g consists of the transformations belonging to $t = 0, \mu/g, \ldots, \mu(g-1)/g$.

Let F be a fibered space with simultaneously oriented fibers; triangulate f

so that each exceptional point lies in the interior of a 2-simplex and each 2-simplex contains at most one exceptional point and so that any two 2-simplexes with exceptional points do not intersect. This corresponds to a decomposition of F into solid tori. We define a cyclic translation group of order g in each of the solid tori with exceptional fibers and on the remaining fibers of F which map to vertices of f. As generator Z of $\mathfrak{G}$ we take the translation which rotates the ordinary fibers by as little as possible in positive direction. Let K be a fibered annulus that maps to an edge of the triangulation of f. If K lies on an exceptional torus, then $\mathfrak{G}$ is already defined on K. If K lies on an ordinary torus, then $\mathfrak{G}$ is already defined on the boundary curves a and b of K. It is clear that $\mathfrak{G}$ can be defined on all of K (Z is a rotation of K about $2\pi/g$), since $a \sim b$ in F, since the fibers are oriented simultaneously. Now $\mathfrak{G}$ is defined on the boundary Π of each ordinary fibered solid torus V.

We think of V as being embedded in Euclidean space, symmetric with respect to an axis of rotation and such that each fiber of V is mapped to itself under a rotation about this axis. We choose a fiber preserving auto-homeomorphism A of the boundary torus Π of V such that $AZA^{-1}:\Pi\to\Pi$ is a rigid rotation about the axis of rotation through an angle of $2\pi/g$. This is always possible since the translation Z restricted to each of the three fibered annuli which form Π (and which map to the three edges of a 2-simplex of the triangulation of the orbit surface) is conjugate to a rigid rotation of a Euclidean annulus through an angle of $2\pi/g$. We can choose A such that each class of curves on Π is mapped to itself. As shown in §5 we can extend A to a fiber preserving autohomeomorphism of V. Therefore V can be mapped homeomorphically to a rotation symmetric solid torus V' in Euclidean space (which has the property that a rotation about the axis of rotation rotates each fiber in itself) such that $Z \mid \Pi$ is then conjugate to a rigid rotation of the boundary torus Π' of V' through an angle of $2\pi/g$. This rotation Π' can be extended to a rigid rotation of V' through the same angle. This defines a translation Z of order g on the sapce F, and proves Theorem 14.

We now show that the orbit space of F under $\mathfrak{G}$ is a fibered space F'. First, let $\mathfrak{G}$ act on a solid torus V. If V is an ordinary fibered solid torus, then clearly the orbit space of V is again an ordinary solid torus. Suppose V is a torus with invariants μ, ν. Suppose $\mathfrak{U}$ is a nontrivial subgroup of $\mathfrak{G}$ keeping the exceptional fiber pointwise fixed. $\mathfrak{U}$ is cyclic or order u. We claim that there exists a meridian disk of V which is mapped to itself under $\mathfrak{U}$. Cut V into a Euclidean cylinder of height 1 and let E_0 be the meridian disk of height $\frac{1}{2}$. Let $E_1, E_2, \ldots, E_{u-1}$ be the images of E_0 under $\mathfrak{U}$. We can assume that no E_i intersects the top and bottom disk of the cylinder by choosing V sufficiently small. Each fiber of the cylinder intersects $E_0, E_1, \ldots, E_{u-1}$ in u (not necessarily distinct) points. Choosing the highest such point on each fiber we obtain a meridian disk E of V which is mapped to itself under $\mathfrak{U}$.[26] Therefore we can cut V along E into a cylinder on which $\mathfrak{U}$ acts as a group of rigid rotations about the axis and translations of the fibers in themselves. The orbit space is a cylinder sector of an angle $2\pi/u$, where the two vertical faces have to be identified such that we get a fibered cylinder. In this cylinder, top and bottom disks are identified under a rotation of $2\pi\nu/\mu'$, where $\mu' = \mu/u$, hence $(\mu', \nu) = 1$. Therefore the orbit space of $\mathfrak{U}$ is a fibered solid torus V' with a (μ/u)-fold exceptional fiber.

The translation group $\mathfrak{G}$ of V maps to a translation group $\mathfrak{G}'$ of V', where $\mathfrak{G}'$ has order $v = g/u$ and does not contain a translation $\neq 1$ which keeps the exceptional fiber of V' pointwise fixed. The cylinder corresponding to V' is then divided by the $v - 1$ images of the bottom disk into v equivalent parts. In each part, bottom and top disks correspond under a rotation of $2\pi\nu''/\mu''$, $(\mu'', \nu'') = 1$. The orbit space D of $\mathfrak{G}'$ (on V'), which is also the orbit space of $\mathfrak{G}$ (on V), is a fibered solid torus which is covered by V' (unbranched) v times. Since the fibers of D correspond one-to-one to those of V', the orbit surface of V' covers (unbranched) that of D. From §9 we have $(\mu'', v) = 1$ and hence by (1) in §9 $\mu' = \mu''$, i.e., D has a μ'-fold exceptional fiber. Now $(g, \mu) = (uv, u\mu') = u(v, \mu') = u$ and $v = g/u = g/(g, \mu)$. The numbers u, v are therefore determined by the order g of $\mathfrak{G}$ and the multiplicity μ of the exceptional fiber of V.

Result. The orbit space D of a translation group $\mathfrak{G}$ of order g on a fibered solid torus V with a μ-fold exceptional fiber is a fibered torus with exceptional fiber of multiplicity $\mu/(\mu, g)$. For $(\mu, g) > 1$, the covering $V \to D$ is branched, where the exceptional fiber of V is a branch curve of order (μ, g). This implies that the orbit space of F under $\mathfrak{G}$ is a fibered space F', and $F \to F'$ is a branched covering.

We now compute the invariants of F'. Let F be the space $(Oo; p \mid b;$ $\alpha_1, \beta_1; \ldots; \alpha_r, \beta_r)$. Drilling out the exceptional fibers and an ordinary fiber we get $\bar{F} \approx \bar{f} \times S^1$, where $\bar{f}$ is an $(r + 1)$ times punctured surface of genus p. On the boundary tori $\Pi_0, \Pi_1, \ldots, \Pi_r$ we have the crossing curves $Q_0, Q_1, \ldots, Q_r$. The Q_i and H_i ($H_0, H_1, \ldots, H_r$ simultaneously oriented) determine on Π_i orientations opposite to that induced by F, and

$$Q_0 + Q_1 + \cdots + Q_r \sim 0 \qquad (\text{in } \bar{F}).$$

---

[26] To see that $\mathfrak{U}$ maps E to itself, suppose there is a map B in $\mathfrak{U}$ which sends a point P of E to a point P' not on E. Then the line segment parallel to the axis of the cylinder V intersects E in a point $Q' \neq P'$. The line segment $P'Q'$ is mapped under B^{-1} to a line segment PQ, where Q lies on one of the disks E_i. But P is the highest of the n intersections of the line segment through P and the disks $E_1, \ldots, E_{u-1}$ and therefore PQ contains a point R of the top disk of V, whose image under B is a point R' on the line segment $P'Q'$. Now if P approaches continuously the axis of the cylinder, P', Q', R' move continuously, and since at last P' and Q' coincide, R' must at some time coincide with P' or Q', i.e., there is a map B^{-1} of $\mathfrak{U}$ that maps a point of a certain E_i into a point of the top disk. This contradicts the choice of the disks E_i.

We get F by taking $Q_0 + bH_0, \alpha_1 Q_1 + \beta_1 H_1, \ldots, \alpha_r Q_r + \beta_r H_r$ as meridians of the torus seals V_i. The orbit space $\bar{F}'$ of $\mathfrak{G} \mid \bar{F}$ is the product of an $(r + 1)$ times punctured surface of genus p and S^1. The orientation (and fiber orientation) of $\bar{F}$ carries over to $\bar{F}'$. Let $\check{Q}_0, \check{Q}_1, \ldots, \check{Q}_r$ and $\check{H}_0, \check{H}_1, \ldots, \check{H}_r$ be the images of $Q_0, Q_1, \ldots, Q_r, H_0, H_1, \ldots, H_r$ in $\bar{F}'$. Then $\check{Q}_0, \check{Q}_1, \ldots, \check{Q}_r$ are crossing curves on the boundary tori $\Pi_0', \Pi_1', \ldots, \Pi_r'$ of $\bar{F}'$, whereas $\check{H}_i$ covers a fiber H_i' of Π_i' g times: $\check{H}_i = gH_i'$. We have

$$\check{Q}_0 + \check{Q}_1 + \cdots + \check{Q}_r \sim 0 \qquad (\text{in } \bar{F}')$$

and the orientation determined by $\check{Q}_i$ and H_i' on Π_i' is opposite to that induced by $\bar{F}'$. The orbit space F' is determined by the meridians $M_i' \sim \check{\alpha}_i \check{Q}_i + \check{\beta}_i H_i'$ and $M_0' \sim \check{Q}_0 + \check{b} H_0'$ of the torus seals V_i'. $M_i \simeq 0$ in V_i, hence $\check{M}_i \simeq 0$ in V_i'. Therefore

$$M_i \sim \alpha_i Q_i + \beta_i H_i \qquad (\text{on } \Pi_i)$$

implies

$$\check{M}_i \sim \alpha_i \check{Q}_i + \beta_i \check{H}_i \quad \sim \alpha_i \check{Q}_i + \beta_i g H_i' \qquad (\text{on } \Pi_i')$$
$$\sim 0 \qquad (\text{in } V_i').$$

Therefore

$$M_i' \sim \frac{\alpha_i}{(\alpha_i, g)} \check{Q}_i + \frac{\beta_i g}{(\alpha_i, g)} H_i' \qquad (\text{in } V_i')$$

$$= \check{\alpha}_i \check{Q}_i + \check{\beta}_i H_i' \sim 0$$

and since $\check{\alpha}_i$ and $\check{\beta}_i$ are coprime, M_i' is a meridian on V_i'.

Similarly $M_0' \sim \check{Q}_0 + bg H_0' \sim \check{Q}_0 + \check{b} H_0'$ is a meridian on V_0'. But $\check{b}, \check{\alpha}_i, \check{\beta}_i$ are not yet the sought after fiber invariants of F' since $\check{\beta}_i$ need not satisfy $0 \leqslant \check{\beta}_i < \check{\alpha}_i$. But taking instead of $\check{Q}_1, \ldots, \check{Q}_r$ the crossing curves $Q_1' \sim \check{Q}_1 + x_1 H_1', \ldots, Q_r' \sim \check{Q}_r + x_r H_r'$, and instead of $\check{Q}_0$ the crossing curve $Q_0' \sim \check{Q}_0 - (x_1 + \cdots + x_r) H_0'$, we have the correct homology

$$Q_0' + Q_1' + \cdots + Q_r' \sim 0 \qquad (\text{in } \bar{F}')$$

and the orientation induced by Q_i' and H_i' on Π_i' is the same as that from $\check{Q}_i$ and H_i'. Now in the new basis curves the meridians M_i' are as follows:

$$M_i' \sim \check{\alpha}_i Q_i' + (\check{\beta}_i - \check{\alpha}_i x_i) H_i' = \alpha_i' Q_i' + \beta_i' H_i' \qquad (i = 1, \ldots, r),$$

$$M_0' \sim Q_0' + (\check{b} + x_1 + \cdots + x_r) H_0' = Q_0' + b' H_0'.$$

Choosing x_i such that $0 \leqslant \beta_i' < \alpha_i'$ and omitting those α_i', β_i' for which $\alpha_i' = 1$ ($\beta_i' = 0$), we obtain the fiber invariants of F'.

If F is $(\text{Nn I}; k \mid b; \alpha_1, \beta_1; \ldots; \alpha_r, \beta_r)$ we get a similar result.

EXAMPLE. The trefoil space of Dehn $(\text{Oo}; 0 \mid -1; 2, 1; 3, 1; 5, 1)$ with translation group of order $g = 5$. Now

$$(\alpha_1, g) = 1, \qquad\qquad (\alpha_2, g) = 1, \quad (\alpha_3, g) = 5,$$

$$\check{\alpha}_1 = \frac{\alpha_1}{(\alpha_1, g)} = 2, \quad \check{\alpha}_2 = 3 \qquad \check{\alpha}_3 = 1, \qquad \check{b} = bg = -5,$$

$$\check{\beta}_1 = \frac{\beta_1 g}{(\alpha_1, g)} = 5, \quad \check{\beta}_2 = 5, \qquad \check{\beta}_3 = 1,$$

hence $x_1 = 2$, $x_2 = 1$, $x_3 = 1$. Therefore the orbit space F' is the space

$$(\text{Oo}; 0 \mid b'; \alpha_1', \beta_1'; \alpha_2', \beta_2') = (\text{Oo}; 0 \mid -1; 2, 1; 3, 2).$$

$\pi_1(F')$ is of order $\Delta' = b'\alpha_1'\alpha_2' + \beta_1'\alpha_2' + \alpha_1'\beta_2' = 1$. Hence $F' \approx S^3$ and the fibers are trefoil knots. In particular, the 5-fold exceptional fiber of F is mapped to an ordinary fiber of F', a trefoil knot. Therefore, F is a 5-sheeted branched covering of S^3 with a trefoil as branch curve.

This result can be generalized. Let F be a Poincaré space $(\text{Oo}; 0 \mid b; \alpha_1, \beta_1; \ldots; \alpha_r, \beta_r)$. Necessary and sufficient for F to be a Poincaré space is that the determinant

$$\Delta = \begin{vmatrix} 1 & 0 & \cdots & 0 & b \\ 0 & \alpha_1 & \cdots & 0 & \beta_1 \\ \vdots & \vdots & & \vdots & \vdots \\ 0 & 0 & \cdots & \alpha_r & \beta_r \\ 1 & 1 & \cdots & 1 & 0 \end{vmatrix} = \pm 1.$$

Now $\bar{F}' \approx \bar{f} \times S^1$, where $\bar{f}$ is a $(r + 1)$ times punctured 2-sphere. The generators of $H_1(\bar{F}')$ are $\check{Q}_0, \check{Q}_1, \ldots, \check{Q}_r$ and an arbitrary fiber H' and we have the single relation $\check{Q}_0 + \check{Q}_1 + \cdots + \check{Q}_r \sim 0$. Closing $\bar{F}'$ to F' we get the additional relations

$$\check{Q}_0 + \check{b}H' = \check{\alpha}_1\check{Q}_1 + \check{\beta}_1 H' = \cdots = \check{\alpha}_r\check{Q}_r + \check{\beta}_r H' \sim 0.$$

Here

$$\check{b} = bg, \qquad \check{\alpha}_i = \alpha_i/(\alpha_i, g), \qquad \check{\beta}_i = \beta_i g/(\alpha_i, g).$$

The relation matrix of $H_1(F')$ is therefore

$$\begin{bmatrix} 1 & 0 & \cdots & 0 & \check{b} \\ 0 & \check{\alpha}_1 & \cdots & 0 & \check{\beta}_1 \\ \vdots & \vdots & & \vdots & \vdots \\ 0 & 0 & \cdots & \check{\alpha}_r & \check{\beta}_r \\ 1 & 1 & \cdots & 1 & 0 \end{bmatrix}$$

and its determinant Δ' is

$$
\Delta' = \begin{vmatrix}
1 & 0 & \cdots & 0 & bg \\
0 & \dfrac{\alpha_1}{(\alpha_1,\, g)} & \cdots & 0 & \dfrac{\beta_1 g}{(\alpha_1,\, g)} \\
\vdots & \vdots & & \vdots & \vdots \\
0 & 0 & \cdots & \dfrac{\alpha_r}{(\alpha_r,\, g)} & \dfrac{\beta_r g}{(\alpha_r,\, g)} \\
1 & 1 & \cdots & 1 & 0
\end{vmatrix}
$$

$$
= \frac{g}{(\alpha_j,\, g)(\alpha_2,\, g) \cdots (\alpha_r,\, g)}\, \Delta
$$

$$
= \pm \frac{g}{(\alpha_1,\, g)(\alpha_2,\, g) \cdots (\alpha_r,\, g)}\,.
$$

F' is a Poincaré space or S^3 only if $\Delta' = \pm 1$. Since $\alpha_1, \alpha_2, \ldots, \alpha_r$ are relatively coprime we have

$$
(\alpha_1,\, g)(\alpha_2,\, g) \cdots (\alpha_r,\, g) = (\alpha_1 \alpha_2 \cdots \alpha_r,\, g)
$$

and $\Delta' = \pm 1$ if and only if g divides $\alpha_1 \alpha_2 \cdots \alpha_r$. The multiplicities of the excpetional fibers of F' are the $\check{\alpha}_1, \check{\alpha}_2, \ldots, \check{\alpha}_r$ different from 1. By Theorem 12, the $\check{\alpha}_i$ characterize F'. Hence follows

THEOREM 15. *The orbit space F' of a translation group of a fibered space F with invariants*

$$
(\mathrm{Oo};\, p \mid b; \alpha_1,\, \beta_1;\, \ldots;\, \alpha_r,\, \beta_r)
$$

or

$$
(\mathrm{Nn\ I};\, k \mid b; \alpha_1,\, \beta_1;\, \ldots;\, \alpha_r,\, \beta_r)
$$

is a fibered space of the same class, whose invariants are determined by those of F and the order g or $\mathfrak{G}$. If F is the Poincaré space with r exceptional fibers of multiplicites $\alpha_1, \alpha_2, \ldots, \alpha_r$, then F' is a Poincaré space or S^3 if and only if $g \mid \alpha_1 \cdots \alpha_r$. In this case F' is the Poincaré space whose exceptional fibers have as multiplicities the following of the numbers which are $\neq 1$:

$$
\frac{\alpha_1}{(\alpha_1,\, g)},\ \frac{\alpha_2}{(\alpha_2,\, g)},\ \ldots,\ \frac{\alpha_r}{(\alpha_r,\, g)}\,.
$$

The covering of F' by F is branched over the exceptional fibers of F for which $(\alpha_i,\, g) > 1$ of branching index $(\alpha_i,\, g)$.

Specializing, we get

THEOREM 16. *The orbit space F' of a Poincaré space F with r exceptional fibers of orders $\alpha_1, \alpha_2, \ldots, \alpha_r$ under a translation group of order $g = \alpha_1 \alpha_2 \cdots \alpha_i$ is a Poincaré space or S^3 with exceptional fibers of orders $\alpha_{i+1}, \alpha_{i+2}, \ldots, \alpha_r$.*

THEOREM 17. *Let $\alpha_1, \alpha_2, \ldots, \alpha_r$ be $r \geqslant 3$ pairwise coprime integers $\geqslant 2$, and let $k_1, k_2, \ldots, k_{r-2}$ be $r - 2$ torus knots in S^3 (which are ordinary fibers of a fibering of S^3) of type m, n, where m and n are any two of the numbers $\alpha_1, \ldots, \alpha_r$. Delete these two numbers from the sequence $\alpha_1, \ldots, \alpha_r$ and take a one-to-one correspondence between the remaining α_i and the knots k_i. Construct the branched covering of S^3 having the knots $k_1, \ldots, k_{r-2}$ as branch curves and having the following property* E: *A curve $\tilde{w}$ of the covering space which lies over a closed curve w of $S^3 - (k_1 \cup \cdots \cup k_{r-2})$ is closed if and only if the linking number $x_i(w, k_i)$ is divisible by the number α_j which corresponds to the knot k_i $(i = 1, \ldots, r - 2)$. This covering is $(\alpha_1 \alpha_2 \cdots \alpha_r / mn)$-sheeted and is a Poincaré space which is the same regardless of how one picks out the numbers m, n from $\alpha_1, \alpha_2, \ldots, \alpha_r$.*

Proof. Assume $m = \alpha_{r-1}$, $n = \alpha_r$, and α_i corresponds to k_i $(i = 1, \ldots, r - 2)$. Letting a translation group of order $g = \alpha_1 \cdots \alpha_{r-2}$ act on the Poincaré space F with r exceptional fibers of multiplicities $\alpha_1, \ldots, \alpha_r$, we obtain as orbit space F' a fibered space with two exceptional fibers α_{r-1} and α_r by the previous theorem. Since a Poincaré space has at least three exceptional fibers (Theorem 12), $F' \approx S^3$ with a fibering having torus knots of type $m = \alpha_{r-1}$, $n = \alpha_r$ as ordinary fibers (§3). By Theorem 15, F is a branched covering of F'; the branch curves are the exceptional fibers of orders $\alpha_1, \ldots, \alpha_{r-2}$ which map to ordinary fibers in F', hence to $r - 2$ torus knots $k_1, \ldots, k_{r-2}$ or type m, n. The branching index is $(\alpha_i, g) = \alpha_i$, i.e., a curve in F winding once around the ith branch curve maps to a curve in F' winding α_i times around k_i. The covering $F \to F'$ is regular and the covering transformation group is cyclic of order $g = \alpha_1 \cdots \alpha_{r-2}$. Therefore (by the lemma in the Appendix) a curve $\tilde{w}$ of F lying over a curve w of $F' - (k_1 \cup \cdots \cup k_{r-2})$ is closed if and only if for each i the linking number of w and k_i is divisible by α_i, and this property E characterizes F uniquely as covering of F'. Thus the covering of S^3 determined by property E is the Poincaré space with r exceptional fibers of multiplicities $\alpha_1, \ldots, \alpha_r$. By Theorem 12, F is uniquely determined by the numbers $\alpha_1, \ldots, \alpha_r$. Therefore F is independent of the choice of the numbers m, n out of $\alpha_1, \ldots, \alpha_r$.

Theorem 17 is interesting because it deals with the homeomorphism type of certain covering spaces, which can be characterized independently of any fibration. This is so since the requirement that the knots $k_1, \ldots, k_{r-2}$ be ordinary fibers of the fibering of S^3 can be replaced by the following: $k_1, \ldots, k_{r-2}$ are pairwise disjoint simple closed curves on a torus which separates S^3 into two solid tori, and these curves are not null homotopic in either solid torus. Then it can be shown that there is a fibering of S^3 that contains these $r - 2$ curves as ordinary fibers.

The special case of Theorem 17 for $r = 3$ deserves special attention.

The *g-fold cyclic covering* of a knot k in S^3 is the branched covering with the following property: A Curve $\tilde{w}$ of the covering space which lies over a curve w of $S^3 \backslash k$ is closed if and only if the linking number of w and k is a multiple of g.[27] The special case can now be formulated as follows:

ADDENDUM TO THEOREM 17. *Let* $\alpha_1, \alpha_2, \alpha_3$ *be three pairwise coprime numbers* $\geqslant 2$. *Then the* α_3*-fold cyclic covering of the torus knot of type* $m = \alpha_1$, $n = \alpha_2$ *is a Poincaré space. The same space is obtained if* $\alpha_1, \alpha_2, \alpha_3$ *are arbitrarily interchanged.*

For this Poincaré space is the fibered Poincaré space with three exceptional fibers of multiplicities $\alpha_1, \alpha_2, \alpha_3$. Thus the Dehn trefoil space, which was obtained by drilling out and sewing back a trefoil of S^3, can be obtained as 5-fold cyclic branched covering of a trefoil or as 3-fold cyclic covering of the torus knot $m = 2$, $n = 5$ or as 2-fold cyclic covering of the torus knot $m = 3$, $n = 5$.

Finally, each fibered Poincaré space $(Oo; 0 \mid b; \alpha_1, \beta_1; \ldots; \alpha_r, \beta_r)$ can be obtained as $\alpha_1 \alpha_2 \cdots \alpha_r$-fold branched covering of S^3. For, letting a translation group of order $g = \alpha_1 \alpha_2 \cdots \alpha_r$ act on F, we get a fibered space without exceptional fibers which is S^3 by Theorems 16 and 12. This fibering of S^3 is by unknotted curves any two of which are simply linked. The branch curves in S^3 are the images of the r exceptional fibers, i.e., r unknotted and pairwise linked curves in S^3, of index $\alpha_1, \alpha_2, \ldots, \alpha_r$, respectively.

15. Spaces Which Cannot Be Fibered

Let F be a fibered space (open or closed). Let H be an ordinary fiber, O a point of H and W a closed curve starting and ending at O. Translating the fiber H along W, H comes back as $H' = H^{\pm 1}$. Thus as elements of the fundamental group, $W^{-1}HW = H^{\pm 1}$. Therefore if a manifold M can be fibered, then $\pi_1(M)$ must contain an element H such that for each element W of $\pi_1(M)$, $W^{-1}HW = H^{\varepsilon(W)}$, where $\varepsilon(W) = \pm 1$. This condition turns out to be nontrivial since we shall show that an ordinary fiber H represents the trivial element of the fundamental group only if the fibered space is S^3 or a lens space with a fibration that can be explicitly described.[28] In particular, if the fundamental group is infinite, then H is not trivial.

---

[27] Another characterization of the cyclic covering is as follows: Cut S^3 along a spanning surface of k to get a "sheet" and glue g of those sheets together cyclically. H. Kneser communicated to me that there are in general besides this cyclic covering other g-fold coverings of a knot which also have the property that for a small loop linking the knot once the g-fold multiple is the first to lift to a closed curve in the covering space. The cyclic coverings play some rôle in knot theory. See K. Reidemeister, *Abh. Math. Sem. Univ. Hamburg* **5** (1927), 7, "Knotentheorie." Berlin (1932).

[28] About lens spaces, see. DB II, §1.

First we prove a preliminary theorem.

THEOREM 18. *An open simply connected space cannot be fibered.*

Proof. Suppose F is an open simply connected fibered space with orbit surface f. Then $f \approx$ open disk. We distinguish two cases:

(a) Suppose F is without exceptional fibers. Since $\pi_1(F) = 1$, H bounds a singular disk E in F. The image on f is a singular disk e which can be covered by an orbit neighborhood ω since f is open and simply connected. E lies in a neighborhood Ω corresponding to ϑ, i.e., $H \simeq 0$ in the solid torus Ω, a contradiction.

(b) F has at least one exceptional fiber C of order α. Drilling out C we obtain a space $\bar{F}$ with orbit surface $\bar{f}$, a punctured open disk. $H_1(\bar{F})$ is free of rank 1, generated by a meridian M of the drilled-out solid torus which maps α times onto the boundary curve l of $\bar{f}$, $\alpha \geqslant 2$. The map $\bar{F} \to \bar{f}$ induces a homomorphism[29] of $H_1(\bar{F}) \to H_1(\bar{f})$ (onto). Since $H_1(\bar{f})$ is infinite cyclic, M has to map onto a generator $\pm l$ of $H_1(\bar{f})$, but $M \to \alpha l$, $\alpha > 1$, a contradiction.

Theorem 18 implies that R^3 can not be fibered. If we project as in §3 a fibering of S^3 stereographically in Euclidean space, the latter will be filled with curves which resemble closely a fibration. Only one curve, the z-axis is not closed.

Using Theorem 18 we can prove

THEOREM 19. *If in a fibered space F a fiber H or a finite multiple of H is homotopic to 0, then F is closed and $\pi_1(F)$ is finite.*

Proof. The universal covering $\tilde{F}$ of F is a fibered space [by §9, (6)] which is closed by Theorem 18 (therefore $\tilde{f} \approx S^3$) and therefore the covering $\tilde{F} \to F$ is finite sheeted.

THEOREM 20. *If F is a (closed or open) fibered space in which an ordinary fiber is homotopic to 0, then F is a Lens space. Any Lens space admits such a fibering.*

Proof. By Theorem 19, $\pi_1(F)$ is finite. We apply Theorem 9. If $f \approx S^2$ and F has three exceptional fibers, then

$$\pi_1(F) = \big\{ Q_0, Q_1, Q_2, Q_3, H : Q_0 H^b = 1 = Q_i^{\alpha_i} H^{\beta_i} \ (i = 1, 2, 3),$$

$$Q_0 Q_1 Q_2 Q_3 = 1, \ Q_j H Q_j^{-1} = H \ (j = 0, 1, 2, 3) \big\}. \tag{1}$$

$\alpha_1, \alpha_2, \alpha_3$ is one of the Platonian triples. Eliminating Q_0 and adding the relation $H^2 = 1$, we obtain a quotient group with defining relations

$$\check{Q}_i^{\alpha_i} \check{H}^{\delta_i} = \check{Q}_1 \check{Q}_2 \check{Q}_3 \check{H}^{\delta_4} = \check{H}^2 = 1, \qquad \check{Q}_i \check{H} \check{Q}_i^{-1} = \check{H} \qquad (i = 1, 2, 3). \tag{2}$$

---

[29] See Footnote 19.

Here $\delta_1, \delta_2, \delta_3, \delta_4 = 0$ or $= 1$ depending on whether $\beta_1, \beta_2, \beta_3, b$ are even or odd, respectively. Taking new generators, we can always assume that $\delta_1 = \delta_2 = \delta_3 = 1$, $\delta_4 = 0$. For in the Platonian triples $\alpha_1, \alpha_2, \alpha_3$ one exponent, say $\alpha_2 = 2$. Then $\beta_2 = 1$ ($0 < \beta_i < \alpha_i$); hence $\delta_2 = 1$. But if α_1 is odd, β_1 may be even and $\delta_1 = 0$. In this case take as new generator Q_1' defined by $\check{Q}_1 = Q_1' \check{H}$. The relation $\check{Q}_1^{\alpha_1} \check{H}^{\delta_1} = 1$ becomes $Q_1'^{\alpha_1} \check{H}^{\delta_1 + \alpha_1} = 1$ and $\alpha_1 + \delta_1 = \alpha_1$ is odd, hence $\check{H}^{\delta_1 + \alpha_1} = \check{H}$. Thus assume $\delta_1 = \delta_2 = \delta_3 = 1$. Now if $\delta_4 = 1$, we define Q_2' by $\check{Q}_2 = Q_2' \check{H}$. Then $\delta_4 = 0$ and since $\alpha_2 = 2$ the other relations are not changed. Therefore

$$Q_1'^{\alpha_1} = Q_2'^{\alpha_2} = Q_3'^{\alpha_3} = \check{H}, \qquad Q_1' Q_2' Q_3' = 1, \check{H}^2 = 1. \tag{3}$$

The groups defined by these relations are (for the Platonian triples) the binary platonian groups. In *Math. Ann.* **104**, 26, it is shown that $\check{H}$ has order 2. Therefore H does not have order 1 in $\pi_1(F)$ and $H \neq 0$ in F.

Now suppose $f \approx P^2$, hence $r = 1$ or 0. For $r = 1$, $\pi_1(F)$ has relations

$$AHA^{-1}H = 1, \qquad Q_0 Q_1 = A^2, \qquad Q_j H Q_j^{-1} = H \qquad (j = 0, 1)$$
$$Q_0 H^b = 1 = Q_1^{\alpha_1} H^{\beta_1}. \tag{4}$$

Eliminating Q_0 and adding the relation $H^2 = 1$, we obtain a quotient group with relations

$$\check{A}^2 \check{Q}_1^{-1} \check{H}^{\delta_1} = \check{Q}_1^{\alpha_1} \check{H}^{\delta_2} = 1, \qquad \check{H}^2 = 1,$$
$$\check{A} \check{H} \check{A}^{-1} = \check{H}, \qquad \check{Q}_1 \check{H} \check{Q}_1^{-1} = \check{H}.$$

Eliminating $\check{Q}_1$ we obtain the Abelian group

$$\check{H}^2 = 1, \qquad \check{A}^{2\alpha_1} \check{H}^{\delta_3} = 1.$$

$\delta_1, \delta_2, \delta_3$ are 0 or 1. In this Abelian group $\check{H}$ does not have order 1, regardless whether $\delta_3 = 0$ or $= 1$; hence $H \neq 0$ in F. If $r = 0$, we have $\alpha_1 = 1$ and obtain the same result.

The remaining case is that $f = S^2$ and F has at most two exceptional fibers. We decompose f into two disks each having at most one exceptional point. This corresponds to a decomposition of F into two solid tori V_1, V_2. Hence F is a lens space or $S^2 \times S^1$. In $S^2 \times S^1$ the fiber is not $\simeq 0$ (Theorem 19). For each lens space there are infinitely many distinct fiberings in which each ordinary $H \simeq 0$. For a lens space is determined by a simple closed curve on $\partial V_1 = \Pi_1$ which is identified with a meridian M_2 of V_2. Thus if M_1, B_1 are meridian and longitude on Π_1, the lens space is determined by the homology

$$M_2 \sim p B_1 + q M_1 \qquad \text{(on } \Pi_1\text{)}, \tag{5}$$

hence by p, q. Here $p \neq 0$; otherwise $M_2 \sim \pm M_1$ and $F \approx S^2 \times S^1$. Fiber V_1 such that

$$H \sim p B_1 + x M_1 \tag{6}$$

where $x \neq q$, $(x, p) = 1$. By Lemma 6 the fibering of the resulting lens space is uniquely determined by the fibering of V_1. Now $H \simeq 0$ since $H \sim M_2 - qM_1 + xM_1$; but M_1 and M_2 are $\simeq 0$ in the lens space. This completes the proof of Theorem 20.

By Theorems 11 and 18, S^3 is the only simply connected 3-manifold that admits a fibration. If, however, the fundamental group is not trivial, we can now state a fibration condition:

THEOREM 21. *If a (open or closed) nonsimply connected manifold M can be fibered, then $\Pi_1(M)$ has an element $H \neq 1$ such that $W^{-1}HW = H^{\varepsilon(W)}$, $\varepsilon(W) = \pm 1$ [for each $W \in \pi_1(M)$].*

For either a fiber $H \simeq 1$ in $\pi_1(M)$, then M is a lens space and $\pi_1(M)$ is cyclic, or $H \neq 1$ in $\pi_1(M)$ and the result follows from the first paragraph of this section.

Using this theorem we can exhibit infinitely many (open or closed) manifolds that cannot be fibered, namely, the connected sum of two manifolds. The connected sum of two manifolds R_A and R_B is obtained by removing from each a 3-ball and gluing together the two resulting boundary 2-spheres, which can be done in two different ways. If A and B are the fundamental groups of R_A and R_B, then the fundamental group of the connected sum is the free product $A * B$ of A and B.[30] The free product $A * B$ is defined as follows[31]: An element is an arbitrary product of finitely many elements of A and B which are called terms. Each such element which is not the identity element can be reduced to a normal form, in which terms of A and B different from the identity alternate. Two elememts of the free product are equal if and only if their normal forms agree term by term. For example,

$$A_{i_1}B_{j_1}A_{i_2}B_{j_2} \cdots A_{i_r}B_{j_r} = A'_{i_1}B'_{j_1}A'_{i_2}B'_{j_2} \cdots A'_{i_r}B'_{j_r}$$

if and only if

$$A_{i_1} = A'_{i_1}, \qquad B_{j_1} = B'_{j_1}, \qquad \cdots, \qquad B_{j_r} = B'_{j_r}.$$

Two elements are multiplied by composing the terms of the two products. We now use

LEMMA 8. *If A and B are nontrivial groups, then the free product $A * B$ has an element H as in Theorem 21 if and only if both A and B have order 2.*

Proof. It follows from the normal form of the elements of $A * B$ that $H \notin A$ and $H \notin B$, since, e.g., composing an element of A with an element $\neq 1$ of B cannot give an element of A. But since $H \notin A$, H does not commute with any nontrivial element of A, since aHa^{-1} does not have the same normal form as H. Therefore, for $a \neq 1 \in A$, $aHa^{-1} = H^{-1}$. For $a' \neq 1 \in A$,

---

[30] The proof of this claim is on p. 36 of the paper cited in Footnote 18.

[31] See C. Schreider, Die Untergruppen der freien Gruppen. *Abh. Math. Sem. Univ. Hamburg* **5** (1927), 161.

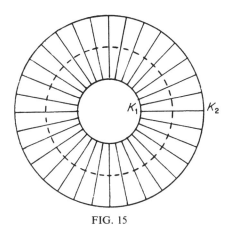

FIG. 15

$a'Ha'^{-1} = H^{-1}$, hence $a'^{-1}aHa^{-1}a' = H$, hence $a^{-1}a' = 1$, Therefore each element $a' \neq 1$ of A is $= a^{-1}$; in particular, $a^{-1} = a$, i.e., $A = \mathbb{Z}_2(a)$. The same holds for B.

Theorem 21 now implies

THEOREM 22. *The connected sum of two nonsimply connected 3-manifolds can be fibered only if both manifolds have a fundamental group of order 2.*

In the exceptional case the connected sum can be fibered, for example the sum of two projective spaces. $P^3 \# P^3$ is obtained by identifying diametrical points on the boundary spheres K_1 and K_2 of $S^2 \times I$ (see Fig. 15) since the dotted 2-sphere separates this manifold into two punctured projective spaces. The fibers are the radii of $S^2 \times I$; any two diametrical radii form one fiber. The invariants of the fibering are $(\text{On}; 1 \mid 0)$; $b = 0$ since $P^3 \# P^3$ admits a fiber preserving orientation reversing homeomorphism (reflection on the dotted S^2). Therefore by Theorem 6, $(\text{On}; 1 \mid b) = (\text{On}; 1 \mid -b)$, hence $b = -b$.

The simplest example of a space that cannot be fibered is $(S^2 \times S^1) \# (S^2 \times S^1)$. We have encountered three possible cases:

(1) F cannot be fibered.
(2) F can be fibered in only one way (Poincaré spaces).
(3) F has infinitely many fiberings (S^3). In this example all fibrations have the same orbit surface, namely S^2.

We conclude with an example of a space having two fiberings with different orbit surfaces. It is the quaternion space, with fundamental group the quaternion group. It is obtained from a cube by identifying any two opposite faces under a rotation of $\pi/2$. Since the quaternion group, which is generated by $\pm 1, \pm i, \pm j, \pm k$, has an element, namely, -1, that commutes with all others, and also another element, e.g., i, that commutes with ± 1, and

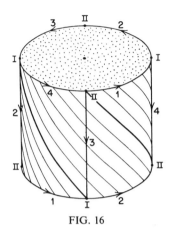

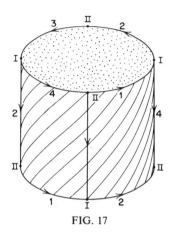

FIG. 16   FIG. 17

$\pm i$ and whose conjugate with $\pm j, \pm k$ is $-i$, one could conjecture that the space can be fibered in two different ways. This is indeed the case. We deform the cube to a cylinder where bottom and top disks are identified under a (say right-handed) rotation of $\pi/2$, and the lateral surface of the cylinder is divided by four vertical lines into four faces, where each two opposite faces are identified under a right-handed rotation of $\pi/2$ (see Fig. 16). Under the identification the lateral faces are deformed so that a vertical line becomes a quarter circle of the bottom (resp. top) disk.

If we deform the bottom disk of the cylinder under a continuous left rotation of total angle $\pi/2$ into the top disk, then each point of the bottom disk describes a screw line, in particular the center point of the bottom disk. These screw lines form the first fibering of the quaternion space. There are three 2-fold exceptional fibers: the axis and the diagonals of the pairwise corresponding faces.

The second fibering is obtained from the first by reflection on a plane through the axis, i.e., consists of right hand screw lines (see Fig. 17). There are no exceptional fibers.

The two orbit surfaces are distinct, since in the first fibering the fibers can be simultaneously oriented, in the second this is not possible. By Theorem 9 the orbit surface of the first fibering is S^2, that of the second is P^2. In the first case we can take as orbit surface a semidisk of the bottom disk, where the radii and quarter circles on the boundary have to be identified. In the second case it is the whole bottom disk with diametrical points on the boundary identified.

Appendix. Branched Coverings

1. Definition of Branched Covering

For a Euclidean 3-ball E of radius 1 let φ (geographical length), ϑ (angular height), and ρ (radius) be polar coordinates,

$$0 \leqslant \varphi < 2\pi, \qquad -\pi/2 \leqslant \vartheta \leqslant +\pi/2. \qquad 0 \leqslant \rho \leqslant 1.$$

Denote polar coordinates for a Euclidean ball $\tilde{E}$ by tildes. $\tilde{E}$ is called a *p-fold branched covering* of E if the map of $\tilde{E}$ to E is given by

$$\rho = \tilde{\rho}, \qquad \vartheta = \tilde{\vartheta}, \qquad \varphi \equiv p\tilde{\varphi} \qquad (\text{mod } 2\pi) \qquad (p > 1).$$

In both E and $\tilde{E}$, the diameter from south pole to north pole is called the branch curve. If K and $\tilde{K}$ are homeomorphic images of E and $\tilde{E}$, then $\tilde{K}$ is mapped to K via $\tilde{E}$ and E. Then $\tilde{K}$ is also called a p-fold branched covering of K, and the curves in K, $\tilde{K}$ which correspond to the branch curves of E, $\tilde{E}$, respectively, under the homeomorphisms are called the branch curves of K, $\tilde{K}$, respectively. If $\tilde{K}$ maps homeomorphically to K, we say that $\tilde{K}$ is an unbranched covering of K.

Let $k_1, \ldots, k_r$ be a finite number of simple closed curves, called knots, in a 3-manifold M with the following properties: For each point P of the knot k_i there is a neighborhood $U(P)$ in M, disjoint to k_j for $j \neq i$, which can be mapped homeomorphically to the interior of a Euclidean 3-ball so that the image of $k_j \cap U(P)$ is a diameter. $U(P)$ is called a *normal neighborhood* of P and $k_i \cap U(P)$ the *diameter* of $U(P)$. If P does not lie on a knot, we call normal any neighborhood which is homeomorphic to the interior of a 3-ball and which is disjoint from all the knots. An admissible path in M is the image under a continuous map of an oriented line segment such that it is disjoint from the knots except possibly for the endpoint.

Let $\tilde{M}$ be a 3-manifold and $\mathfrak{A}: \tilde{M} \to M$ be a continuous map. We say that the point $\tilde{P}$ of $\tilde{M}$ lies over the point P of M and that P is the projection of $\tilde{P}$ if $\mathfrak{A}(\tilde{P}) = P$. An admissible path in M is a path whose image under $\mathfrak{A}$ is admissible in M. Now $\tilde{M}$ is called a branched covering of M with branch curves $k_1, \ldots, k_r$ if the following holds (see also §9):

I.   Over each point P of M lies at least one point $\tilde{P}$ of $\tilde{M}$.

II.   If $\tilde{P}_1, \tilde{P}_2, \ldots$ are all the points which lie over P, there is a normal neighborhood $U(P)$ in M and there are normal neighborhoods $U(P_1)$ $U(P_2), \ldots$ in $\tilde{M}$ which together consist of all points lying over points of $U(P)$ and which have the following properties: (a) If P is a point on a knot k_j, then $U(\tilde{P}_i)$ is a branched or unbranched covering of $U(P)$ with $k_j \cap U(P)$ as branch curve; (b) If P does not lie on a knot, then $\mathfrak{A} \mid U(\tilde{P}_i): U(\tilde{P}_i) \to U(P)$ is a homeomorphism.

Let N be the open submanifold of M obtained from M by removing all points on the knots; let $\tilde{N}$ be the submanifold $\mathfrak{A}^{-1}(N)$ of $\tilde{M}$. Then we have the following theorems, which we state without proof:

(1) $\tilde{N}$ *is an unbranched covering of* N(§9).

(2)   *If $P = P(t)$, $0 \leqslant t \leqslant 1$, is an admissible path in M from a point $P(0)$ to a point $P(1)$, and if $\tilde{P}(0)$ is a point over $P(0)$, then there exists a unique lift $\tilde{P}(t)$ in $\tilde{M}$ which starts at $\tilde{P}(0)$ and such that $\tilde{P}(t)$ lies over $P(t)$.*

(3)   *If $\tilde{w}$ is a closed curve of $\tilde{N}$ which lies over a contractible curve in N, then $\tilde{w}$ is contractible in $\tilde{N}$.*

(4)   *If exactly n points lie over some point of N, then exactly n points lie over each point of N (n-fold covering).*

2. The Subgroup $\mathfrak{H}$ of the Fundamental Group

Let $\mathfrak{F}, \tilde{\mathfrak{F}}$ be the fundamental group of N, $\tilde{N}$, respectively. Choosing the base point $\tilde{O}$ for $\tilde{\mathfrak{F}}$ over the base point O for $\mathfrak{F}$, a homotopy class of (based) loops of $\tilde{N}$ is mapped to such a class of N. This induces an isomorphism of $\tilde{\mathfrak{F}}$ onto a subgroup $\mathfrak{H}$ of $\mathfrak{F}$. We call $\mathfrak{H}$ the subgroup of $\mathfrak{F}$ corresponding to the given covering. Note however that $\mathfrak{H}$ depends on the choice of the base point $\tilde{O}$ over O; we choose once and for all a fixed $\tilde{O}$ over O. (If we would choose another base point over O, we would get a subgroup conjugate to $\mathfrak{H}$ in $\mathfrak{F}$.) A based loop in N belongs to $\mathfrak{H}$ if and only if its lift from $\tilde{O}$ is closed in $\tilde{N}$. Decomposing $\mathfrak{F}$ into its cosets of $\mathfrak{H}$,

$$\mathfrak{F} = \mathfrak{H} + \mathfrak{H}F_2 + \mathfrak{H}F_3 + \ldots ,$$

we get a one-to-one correspondence between these cosets and the points over O as follows: Choose a path w from the coset $\mathfrak{H}F_i$ and lift it from $\tilde{O}$ to $\tilde{w}$. The endpoint of $\tilde{w}$ corresponds to the coset $\mathfrak{H}F_i$. This correspondence is apparently independent of the choice of the path w from $\mathfrak{H}F_i$. In particular, if the covering of N by $\tilde{N}$ is finite sheeted, then the number of sheets equals the index of $\mathfrak{H}$ in $\mathfrak{F}$.

3. Unique Determination of $\tilde{M}$ by $\mathfrak{H}$

For the following it is convenient to consider only a particular system of neighborhoods of the covering space. As neighborhoods of a point P of the covering space we consider only those 3-balls which lie concentrically in a normal 3-ball and which cover (branched or unbranched) a normal neighborhood of the image point P. This system of neighborhoods (for all points P of $\tilde{M}$) is equivalent to the system of all open sets of $\tilde{M}$.

If $\tilde{M}_1$ and $\tilde{M}_2$ are two branched covers of M which induce the same subgroup $\mathfrak{H}$ of $\mathfrak{F}$, then they are homeomorphic so that corresponding points have the same image in M. In order to define the homeomorphism $f : \tilde{M}_1 \to \tilde{M}_2$, join a point $\tilde{P}_1 \in \tilde{M}_1$ to $\tilde{O}_1$ by an admissible path $\tilde{a}_1$ and lift the image path a of $\tilde{a}_1$ to a path $\tilde{a}_2$ in $\tilde{M}_2$ from $\tilde{O}_2$. Let $f(\tilde{P}_1)$ be the endpoint of this lift. $f(\tilde{P}_1)$ is uniquely determined by $\tilde{P}_1$ and does not depend on the path $\tilde{a}_1$. For if $\tilde{P}_1$ does not lie over a point on a branch curve, and if $\tilde{b}_1$ is another path joining $\tilde{P}_1$ to $\tilde{O}_1$, then the path $\tilde{a}_1 \tilde{b}_1^{-1}$ is a closed curve in $\tilde{M}_1$ and therefore its image in M is contained in the subgroup $\mathfrak{H}$ of $\mathfrak{F}$; since $\tilde{M}_2$ corresponds to the same subgroup $\mathfrak{H}$ it follows that the lift $\tilde{a}_2 \tilde{b}_2^{-1}$ is a closed curve in $\tilde{M}_2$ and therefore the endpoint of $\tilde{b}_2$ is the same as that of $\tilde{a}_2$. If $\tilde{P}_1$ lies over a point on a branch curve, we deform the path $\tilde{a}_1 \tilde{b}_1^{-1}$ inside an arbitrarily small ball neighborhood $\tilde{U}_1$ of $\tilde{P}_1$ into an admissible path as follows: Choose a point $\tilde{A}_1$ on $\tilde{a}_1$ close to $\tilde{P}_1$ such that the subpath $\tilde{A}_1 \tilde{P}_1$ of $\tilde{a}_1$ lies in $\tilde{U}_1$; similarly, choose a point $\tilde{B}_1$ on $\tilde{b}_1$ shortly before $\tilde{P}_1$ and join $\tilde{A}_1$ and $\tilde{B}_1$ by a path $\tilde{v}$ inside $\tilde{U}_1$

which misses the branch curve. The corresponding detachment is done in the ground space M. The ball neighborhood $\tilde{U}_1$ is mapped to a normal neighborhood, the points $\tilde{A}_1, \tilde{B}_1$ into two points a, b close to P, and the detached ground path belongs to $\mathfrak{H}$ since it is the image of an admissible closed curve in $\tilde{M}_1$. Since we can choose $\tilde{U}_1$ arbitrarily small, we can detach the path ab^{-1} into a curve of $\mathfrak{H}$ in an arbitrarily small normal neighborhood of P. Now supposing that $\tilde{a}_2$ and $\tilde{b}_2$ lead from $\tilde{O}_2$ to different endpoints $\tilde{P}_2$ and $\tilde{Q}_2$, we could find disjoint ball neighborhoods $\tilde{U}_2$ and $\tilde{V}_2$ of $\tilde{P}_2$ and $\tilde{Q}_2$. The corresponding normal image neighborhoods U and V of P in M have a neighborhood W in common, inside which we detach the path ab^{-1}. Lifting the path a (from O to A) to $\tilde{M}_2$, we obtain a path from $\tilde{O}_2$ to a point $\tilde{A}_2$. Running from A along v to B, the lift in $\tilde{M}_2$ leads to a point $\tilde{B}_2$ which lies in $\tilde{U}_2$. On the other hand, running from O to B along b, the lift in $\tilde{M}_2$ is a path from $\tilde{O}_2$ to a point in $\tilde{V}_2$. But since the detached path ab^{-1} belongs to $\mathfrak{H}$, the latter point has to be $\tilde{B}_2$. Therefore $\tilde{U}_2$ and $\tilde{V}_2$ cannot be disjoint and $\tilde{Q}_2 = \tilde{P}_2$.

This shows that the map $f: \tilde{M}_1 \to \tilde{M}_2$ is well defined and one-to-one. To show that f is a homeomorphism, we have to find for any given neighborhood $\tilde{U}_1$ of $\tilde{P}_1$ a neighborhood $\tilde{U}_2$ of $\tilde{P}_2 = f(\tilde{P}_1)$ such that $f(\tilde{U}_2) \subset \tilde{U}_1$. If U_1 is the normal neighborhood of P in M which is (branched or unbranched) covered by $\tilde{U}_1$ and if a is a path from O to P which lifts in $\tilde{M}_1$ to a path from $\tilde{O}_1$ to $\tilde{P}_1$, then each path from O to a point P' of U_1, which agrees with a up to a point A shortly before P and from there remains inside U_1 lifts in $\tilde{M}_1$ from $\tilde{O}_1$ to a point in $\tilde{U}_1$. Now let $\tilde{U}_2$ be a ball neighborhood which is mapped into a normal subneighborhood U_2 of U_1. In $\tilde{M}_2$, a lifts to a path from $\tilde{O}_2$ to $\tilde{P}_2$, and we can get to any point $\tilde{P}_2'$ of $\tilde{U}_2$ along a path which agrees with $\tilde{a}_2$ up to a point shortly before $\tilde{P}_2$ and which from there on remains in $\tilde{U}_2$. In the ground space M, this path maps to the type of paths from O to a point P', discussed above. This lifts in $\tilde{M}_1$ to a path from $\tilde{O}_1$ to a point in $\tilde{U}_1$. Hence $f: \tilde{M}_1 \to \tilde{M}_2$ is continuous and, since the same arguments apply to the inverse map, f is a homeomorphism.

This shows that the covering $\tilde{M} \to M$ is uniquely determined by the subgroup $\mathfrak{H}$. In the same way one can show that to a given subgroup $\mathfrak{H}$ of finite index there exists a corresponding covering $\tilde{M}$.

4. Regular Coverings*

LEMMA ABOUT BRANCHED COVERINGS OF S^3 WITH ABELIAN GROUP OF COVERING TRANSLATIONS. *Let* $\tilde{M} \to M = S^3$ *be a regular finite sheeted covering branched over the knots* $k_1, \ldots, k_x$, *with group of covering translations*

---

* *Translators note:* In this section regular coverings and covering translations are discussed and it is shown that for a regular covering corresponding to the normal subgroup $\mathfrak{H}$ of $\mathfrak{F}$ the group of covering transformations is isomorphic to $\mathfrak{F}/\mathfrak{H}$. A more detailed exposition can be found in Chapter VIII, §57 of "Seifert and Threlfall: A Textbook of Topology."

Abelian and of order $g = \alpha_1 \cdots \alpha_x$. Assume: For a small loop C_i that links k_i exactly once, the lifts of $C_i^{\alpha_i}$ in $\tilde{M}$ are closed curves.[32] *Then it follows that a path $\tilde{w}$ of $\tilde{M}$ that covers a path w which misses the knots is closed if and only if for each i the linking number χ_i of w with k_i is divisible by α_i. Since this determines the subgroup $\mathfrak{H}$ of $\mathfrak{F}$ corresponding to $\tilde{M}$ there is by §3 only one covering $\tilde{M}$ with the above property.*

Proof. Every loop of $\mathfrak{F}$ lies in a certain coset of $\mathfrak{H}$ in $\mathfrak{F}$. A null homologous loop w of $\mathfrak{F}$ belongs always to $\mathfrak{H}$, since w is a product of commutators which all lie in $\mathfrak{H}$ since $\mathfrak{F}/\mathfrak{H}$ is Abelian. Hence two homologous loops of $\mathfrak{F}$ lie in the same coset. But the homology group of N is the free Abelian group generated by $C_1, \ldots, C_x$. Thus each loop w of N is homologous to a linear combination $\sum_{i=1}^{x} \chi_i C_i$, where χ_i denotes the uniquely determined linking number of w with k_i (with a suitable orientation of k_i). In particular $w \sim O$ in N if and only if all its linking numbers vanish. Therefore loops of $\mathfrak{F}$ with the same linking numbers χ_i lie in the same coset $\mathfrak{H}$ of $\mathfrak{F}$. The loop C_i need not be based at O and may thus not belong to $\mathfrak{F}$, but joining O to a point of C_i by an admissible path v_i we get a path $c_i = v_i C_i v_i^{-1}$ that belongs to $\mathfrak{F}$, is homologous to C_i in N, and whose α_ith power belongs to $\mathfrak{H}$. But $C_i^{\alpha_i}$ has linking number α_i with k_i and linking number 0 with the other knots. Therefore those loops of $\mathfrak{F}$ whose linking number χ_i is divisible by α_i (for each i) belong to $\mathfrak{H}$. Two loops w and w' with all x linking numbers congruent, i.e.,

$$\chi_i \equiv \chi_i' \pmod{\alpha_i} \qquad (i = 1, \ldots, x),$$

belong to the same coset of $\mathfrak{H}$ in $\mathfrak{F}$. Since there are only $\alpha_1 \cdots \alpha_x$ incongruent systems of linking numbers, and just as many cosets, all loops of $\mathfrak{F}$ whose linking numbers with the knots $k_1, \ldots, k_x$ are piece by piece congruent make up a coset of $\mathfrak{H}$ in $\mathfrak{F}$. In particular $\mathfrak{H}$ itself consists of all loops whose linking numbers $\chi_1, \ldots, \chi_x$ are divisible by α_1 (resp. $\alpha_2, \ldots, \alpha_x$). The theorem therefore is true for all loops based at O. But then the theorem holds also for the other loops, since each loop w in N can be deformed without crossing the knots into a loop based at O, and this neither changes its linking number with k_i nor its property of being covered by a loop of the covering space.

---

[32] It suffices to require that at least one lift of $C_i^{\alpha_i}$ is a closed curve; since the covering is regular, it then follows that all other lifts are closed curves.

INDEX TO "A TEXTBOOK OF TOPOLOGY"

Pure and Applied Mathematics

A Series of Monographs and Textbooks

Editors **Samuel Eilenberg and Hyman Bass**

Columbia University, New York

RECENT TITLES

IN PREPARATION